AF750892

DEPT. OF BOTANY
THE UNIVERSITY
MANCHESTER, 13.

PHYTOFLAGELLATES

DEVELOPMENTS IN MARINE BIOLOGY

Volume 1—Toxic Dinoflagellate Blooms, edited by Dennis L. Taylor and Howard H. Seliger, 1979

Volume 2—Phytoflagellates, edited by Elenor R. Cox, 1980

PHYTOFLAGELLATES

Editor:

ELENOR R. COX
Department of Biology, Texas A&M University, College Station, Texas

ELSEVIER/NORTH-HOLLAND
NEW YORK • AMSTERDAM • OXFORD

UML WITHDRAWN

BD

© 1980 by Elsevier North Holland, Inc.

Published by:

Elsevier North Holland, Inc.
52 Vanderbilt Avenue, New York, New York 10017

Sole distributors outside of the United States and Canada:

Elsevier/North-Holland Biomedical Press
335 Jan van Galenstraat, P.O. Box 211
Amsterdam, The Netherlands

Library of Congress Cataloging in Publication Data

Main entry under title:

Phytoflagellates.
(Developments in marine biology; v. 2 ISSN 0163-6995)

Bibliography: p.
Includes index.
1. Flagellata. I. Cox, Elenor R. II. Series.
QL368.A12P45 589'.4 79-22435
ISBN 0-444-00363-0

Manufactured in the United States of America

JOHN RYLANDS
UNIVERSITY
LIBRARY OF
MANCHESTER

Contents

PREFACE vii

PHYTOFLAGELLATES: THE ORGANISMS AND THE TAXONOMIC CATEGORIES . . .
. . . . Elenor R. Cox 1

UNICELLULAR CHLOROPHYTES . . . . Carole A. Lembi 5

UNICELLULAR CHLOROPHYTES (Snow Algae) . . . . Ronald W. Hoham 61

PRASINOPHYTES . . . . Richard E. Norris 85

COLONIAL CHLOROPHYTES . . . . Richard C. Starr 147

EUGLENOID FLAGELLATES . . . . Patricia L. Walne 165

CHRYSOPHYTES . . . . Richard N. Pienaar 213

XANTHOPHYTES . . . . David J. Hibberd 243

PRYMNESIOPHYTES (=Haptophytes) . . . . David J. Hibberd 273

EUSTIGMATOPHYTES . . . . David J. Hibberd 319

SILICOFLAGELLATES . . . . Shirley D. Van Valkenburg 335

CHLOROMONADS . . . . Peter Heywood 351

PHOTOSYNTHETIC CRYPTOPHYTES . . . . Elisabeth Gantt 381

FREE-LIVING DINOFLAGELLATES . . . Karen Steidinger & Elenor R. Cox 407

PHYLOGENY OF PHYTOFLAGELLATES . . Kenneth D. Stewart & Karl Mattox 433

GENUS INDEX 463

SUBJECT INDEX 471

Preface

You might say this book exploded from a symposium sponsored by the Phycological Society of America. Clinton Dawes, University of South Florida, then Program Chairman of the Phycological Society of America, invited Karen Steidinger, Marine Research Laboratory, Department of Natural Resources, St. Petersburg, Florida, and me to organize a symposium on the marine and fresh-water photosynthetic flagellate groups. We called the symposium, *Phytoflagellates: Form and Function*. The symposium was held at the Phycological Society of America's annual meeting with the American Institute of Biological Sciences in August, 1978, on the campus of the University of Georgia, Athens, Georgia. The invited speakers for the symposium and the titles of their presentations were: Richard E. Norris, University of Washington, Friday Harbor Marine Laboratories, Friday Harbor, Washington, *The Prasinophyte Flagellates*; Peter Heywood, Brown University, Providence, Rhode Island, *Chloromonads: Form and Function*; Patricia L. Walne, University of Tennessee, Knoxville, Tennessee, *Euglenoid Flagellates: Selected Features of Form and Function*; Stefan Gartner, Texas A&M University, College Station, Texas, *Coccolithophores: Their Evolution Through Time*; Shirley Van Valkenburg, University of Maryland, College Park, Maryland, *Silicoflagellate Ultrastructure*; Karen Steidinger, Marine Research Laboratory, Department of Natural Resources, St. Petersburg, Florida, *Dinoflagellates: Organelles and Features*; and Kenneth D. Stewart and Karl R. Mattox, Miami University, Miami, Ohio, *Phylogenetic Criteria and Relationship of Phytoflagellates*.

The audience was enthusiastic about the symposium. At the social which followed the symposium and in conversations during the subsequent days of the meeting the idea of this book began to take form. Many people said it was unfortunate that all phytoflagellate groups could not have been included and that the speakers were limited to relatively short presentations because of time constraints. Many people emphasized the importance of the phytoflagellates to both zoologists and phycologists and the fact that information about these organisms is scattered in journals of many different disciplines. With all these thoughts in mind and because no similar volume exists, the decision was made to expand the symposium into a book which would synthesize the current research and literature on the photosynthetic flagellates.

Each of the participants in the 1978 symposium was invited to revise the oral presentation into written form, to expand the information included and to reorganize the material to conform as much as possible to a uniform format.

The positive response of the participants to this task was very encouraging. In addition, other well-known researchers were invited to contribute chapters on major groups of phytoflagellates not covered at the symposium. Again, the response of those requested to help with additional chapters was enthusiastic and all contributors added another duty to their full schedules.

The contributors to this book are scientists with impressive credentials. Each author is a recognized authority on the organisms discussed and each is able to present current knowledge succinctly to others.

The contributors to this volume are: Ronald Hoham, Colgate University, Hamilton, New York; Patricia Walne, University of Tennessee, Knoxville, Tennessee; Elisabeth Gantt, Smithsonian Institution, Rockville, Maryland; Shirley Van Valkenburg, University of Maryland, College Park, Maryland; Richard Norris, University of Washington, Friday Harbor Marine Laboratories, Friday Harbor, Washington; Richard Starr, University of Texas at Austin, Austin, Texas; David J. Hibberd, The Culture Centre of Algae and Protozoa, Cambridge University, Cambridge, England; Carole Lembi, Purdue University, West Lafayette, Indiana; Richard N. Pienaar, University of Natal, Natal, South Africa; Karen Steidinger, Marine Research Laboratory, Department of Natural Resources, St. Petersburg, Florida, Elenor R. Cox, Texas A&M University, College Station, Texas; Peter Heywood, Brown University, Providence, Rhode Island; and Kenneth Stewart and Karl Mattox, Miami University, Miami, Ohio. Drs. Stewart and Mattox have written the concluding chapter which summarizes the evolutionary relationships of photosynthetic flagellates.

Obviously this book extends the scope of the 1978 Phycological Society of America symposium. It is a compilation of current information about the major groups of phytoflagellates, including recent taxonomic, ultrastructural and physiological information from the current literature. The authors have presented the known and have stressed areas that are unknown where additional research is needed. The writers of these chapters are phycologists. However, they have drawn extensively from the zoological and microbiological, as well as botanical literature for the summaries of these protists.

I must express gratitude to many people for their efforts in bringing this project to fruition. Special thanks are expressed to: Karen Steidinger, co-convener of the 1978 Phycological Society of America symposium, for her ideas and enthusiasm and for making initial contacts with the publisher; to the authors who contributed their time and scientific expertise to this project and who were patient when production bogged down at my desk; to my colleagues and friends at Texas A&M University who heard more about the "book" than they wished to hear for longer than they wished to hear it; to Caroline France who conscientiously read copy and sorted index cards; and, most especially, to Nancy Scott who typed the manuscript and kept up with a thousand details, including keeping me and all the papers organized. There would have been no book without Nancy's help.

Any royalties from the sale of this book will be paid to the Endowment Fund of the Phycological Society of America.

Elenor R. Cox
Texas A&M University
College Station, Texas 77343

Copyright 1980 by Elsevier North Holland, Inc.

PHYTOFLAGELLATES: THE ORGANISMS AND THE TAXONOMIC CATEGORIES

ELENOR R. COX
Department of Biology, Texas A&M University, College Station, Texas 77843
U.S.A.

ORGANISMS

The phytoflagellates are a diverse and probably unrelated group of eukaryotic microorganisms which occupy various niches in the freshwater or marine environment. They are usually free-living, unicellular and uninucleate, although binucleate forms (several dinoflagellates) and colonial forms do occur. The phytoflagellates have as common characteristics their motility by means of one or more flagella and their ability to photosynthesize and evolve molecular oxygen as a by-product. Several of the groups of photosynthetic flagellates, notably the dinoflagellates, euglenoids and cryptophytes have autotrophic, facultative autotrophic and heterotrophic genera. We deal here with the obligate autotrophs. All phytoflagellates have chlorophyll *a*. In addition, various groups possess different combinations of accessory photosynthetic pigments, including chlorophyll *b*, chlorophyll c_1, chlorophyll c_2, biliproteins (phycocyanin, allophycocyanin and phycoerythrin) and a variety of carotenoids (carotenes and xanthophylls). These photosynthetic pigments are located on, in or between the thylakoids of the chloroplast or chloroplasts. The basic number of flagella in the phytoflagellates is two. In some groups however, one flagellum is emergent and one is non-emergent or even reduced to a basal body; in other groups there are four or more flagella, which however, always occur in multiples of two. The flagella may be variously inserted on the cell (e.g., apically, laterally, etc.); they may be isokont or heterokont and either smooth or covered by hairs or scales of different kinds. The phytoflagellate cell body may be bounded by a plasma membrane or the plasma membrane may be covered by scales, spines, cell walls or complex loricas, thecae or pellicles. The phytoflagellates are a beautifully diverse group of organisms.

A diagram of the cellular organization of a typical phytoflagellate, *Chlamydomonas eugametos* Moewus is pictured in Figure 1. These organisms have all the organelles of eukaryotic cells: membrane bounded nucleus, Golgi apparatuses, mitochondria, endoplasmic reticulum, ribosomes, microtubules, microbodies. Flagella and membrane bounded chloroplasts containing thylakoids are present. In addition, granules, vacuoles, various kinds of ejectile organelles, eyespots (stigmata), pyrenoids, photosynthate and contractile vacuoles may be present.

TAXONOMIC CATEGORIES

As a graduate student, I often heard this statement, "nature mocks at human categories." This thought is most aptly applied to the organisms treated in this book. Phytoflagellates are variously regarded as plants, animals or protists. They are studied by phycologists, protozoologists, microbiologists, ecologists, and used as experimental organisms by cell biologists, biochemists, physiologists and geneticists. With this fact in mind, the titles of the chapters in this book—which reflect the "categories" into which we have placed the organisms—have been kept informal, reflecting neither correct botanical nor zoological nomenclature. We have chosen to present the organisms in this way to facilitate use by workers in different disciplines. Our use, however, of such chapter headings as Prasinophytes or Cryptophytes betrays our phycological orientation.

To the protozoologist the phytoflagellates comprise several different orders (about 9 or 10) of the class Flagellata and the phylum Protozoa (Grell, 1973). Botanical classification does not treat these same organisms in such a straight-forward manner. To the phycologist the phytoflagellates are grouped in several different divisions (=phyla of zoological classification): the Chlorophyta, Euglenophyta, Chrysophyta, Cryptophyta and Pyrrhophyta. Some classes of these divisions are entirely comprised of phytoflagellates, but others (like the division Chlorophyta, class Chlorophyceae), only partly so. We have also chosen to include two chapters, Xanthophytes and Eustigmatophytes, treating organisms which are not strictly phytoflagellates as we have defined the word. The chapter titles and the generally accepted botanical classes (Bold and Wynne, 1978) and zoological orders (Grell, 1973) are compared so far as is possible in Figure 2.

The reader's attention is directed to the recent review by Dodge (1979) for additional information which will supplement that presented here.

Fig. 1. Diagram of the cellular organization of *Chlamydomonas eugametos* Moewus. ce, chloroplast envelope; cv, contractile vacuole; cw, cell wall; dg, dense granule; er, endoplasmic reticulum; f, flagellum; ga, Golgi apparatus; i, inclusion; l, lipid body; m, mitochondrion; n, nucleus; ne nuclear envelope; nu, nucleolus; p, plastid; pa, papilla; pm, plasma membrane; py, pyrenoid; r, ribosomes; s, stigma; st, starch; v, vesicle. (Courtesy of Professor Patricia L. Walne.)

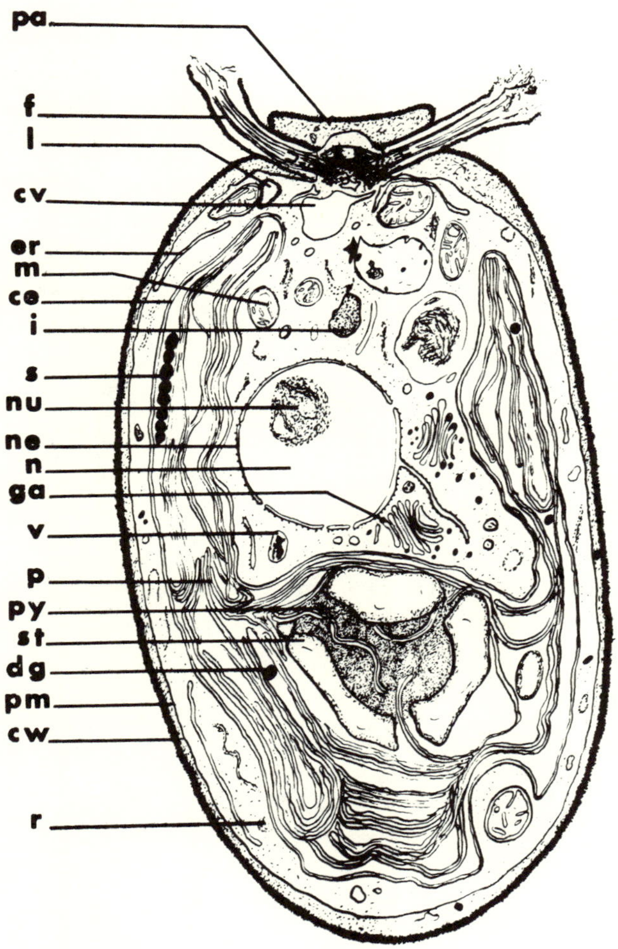
pa
f
l
cv
er
m
ce
i
s
nu
ne
n
ga
v
p
py
st
dg
pm
cw
r

Informal names used in Chapter Headings	Botanical Classification (Class)	Zoological Classification (Order)
Unicellular Chlorophytes	Chlorophyceae	Phytomonadina
Unicellular Chlorophytes (Snow Algae)	Chlorophyceae	Phytomonadina
Prasinophytes	Prasinophyceae	Phytomonadina
Colonial Chlorophytes	Chlorophyceae	Phytomonadina
Euglenoid Flagellates	Euglenophyceae	Euglenoidina
Chrysophytes	Chrysophyceae	Chrysomonadina
Xanthophytes*	Xanthophyceae	——
Prymnesiophytes (= Haptophytes)	Prymnesiophyceae	Chrysomonadina
Eustigmatophytes*	Eustigmatophyceae	——
Silicoflagellates	Chrysophyceae	Chrysomonadina
Chloromonads	Chloromonadophyceae	Chrysomonadina
Photosynthetic Cryptophytes	Cryptophyceae	Cryptomonadina
Photosynthetic Dinoflagellates	Dinophyceae	Dinoflagellata

*The Xanthophytes and Eustigmatophytes are not strictly phytoflagellates as defined in this book.

Fig. 2. Informal names used in chapter headings, botanical classification (class) and zoological classification (order).

REFERENCES

Bold, H. C. and M. J. Wynne. 1978. *Introduction to the Algae. Structure and Reproduction.* Prentice-Hall, Inc., Englewood Cliffs, New Jersey. 706 pp.

Dodge, J. D. 1979. The Phytoflagellate Fine Structure and Phylogeny. In Levandowsky, N., S. H. Hutner and L. Provasoli [Eds.] *Biochemistry and Physiology of Protozoa*, 2nd Edition. Academic Press, New York.

Grell, K. G. 1973. *Protozoology.* Springer-Verlag, New York. 554 pp.

Copyright 1980 by Elsevier North Holland, Inc.

UNICELLULAR CHLOROPHYTES

CAROLE A. LEMBI

Department of Botany and Plant Pathology, Purdue University, W. Lafayette, IN 47907, U.S.A.

INTRODUCTION

Electron microscope studies of the 1950's and early 1960's provided researchers with their first views of the fine structure of motile chlorophycean cells. Studies of *Chlamydomonas* (Sager and Palade, 1957; Gibbs *et al.*, 1958; Gibbs, 1962a, 1962b), *Polytoma* (Lang, 1963a, 1963b), *Carteria* (Lembi and Lang, 1965), the colonial Volvocales (Lang, 1963c), and the motile cells of filamentous green algae (Manton *et al.*, 1955; Manton, 1964, 1965) including the multiflagellated zoospores and spermatozoids of *Oedogonium* (Hoffman and Manton, 1962, 1963) revealed a pattern of organelle ultrastructure and organization now known to be typical of the motile green cell.

In 1967, this image was greatly enhanced with the publication of Ringo's study (1967a) of the flagellar apparatus of *Chlamydomonas*. Using glutaraldehyde as a fixative, Ringo not only provided the finest picture of the cellular organization of the chlamydomonad cell up to that time, but a detailed interpretation of the structure and organization of the flagellar axoneme, basal body, and associated cytoplasmic microtubules. Indeed, his descriptions continue to provide the factual and conceptual background for research in many areas of current interest including flagellar movement and assembly, intracellular communication and coordination, and functional and evolutionary relationships based on comparative studies of flagellar apparatus ultrastructure.

The emphasis of this review is on the developments in green phytoflagellate form and function that occurred after 1967. The organisms comprising the review (Table 1) are those that are flagellated for the majority of their life cycle. This excludes flagellated cells of green, filamentous algae and, by definition, genera once considered to be chlorophytes but which are now placed in a separate class, the Prasinophyceae.

Much of the research has been conducted on *Chlamydomonas*, specifically *C. reinhardtii*. The ease of obtaining synchronized, mass cultures of this organism on defined mineral media plus the large number of mutants available for the study of form-function relationships and genetic regulation of biochemical processes has led to a body of research surpassed by no other alga with the possible exceptions of *Chlorella* and *Euglena*. *Chlamydomonas* has also received considerable attention because it was once considered the progenitor of all other

Table 1. CLASSIFICATION AND GENERAL CHARACTERISTICS OF SELECTED UNICELLULAR CHLOROPHYCEAN FLAGELLATES

Family	Genus	No. Flagella	Presence of Chlorophyll	Presence of Cell Wall
Polyblepharidaceae	*Dunaliella*	2	+	-
	Polytomella	4	-	-
Chlamydomonadaceae	*Chlamydomonas*	2	+	+
	Carteria	4	+	+
	Chlorogonium	2	+	+
	Polytoma	2	-	+
Haematococcaceae	*Haematococcus*	2	+	+
Phacotaceae	*Dysmorphococcus*	2	+	?(loricate)

green algal cells (e.g., Scagel *et al.*, 1965; Klein and Cronquist, 1967); however, this view is no longer held by most phycologists since recent studies of mitotic and cytokinetic processes and motile cell organization suggest that the ancestry of only a limited number of algal groups can be traced through *Chlamydomonas*; namely, volvocalean, tetrasporalean and certain ulotrichalean algae (Pickett-Heaps, 1974, 1975; Stewart and Mattox, 1975).

General characteristics. The information in this section is drawn primarily from the general literature on algal morphology (Smith, 1933; Fritsch, 1935; Bold and Wynne, 1978). For a recent review of the variations in cell shape and morphological organization in the green phytoflagellates, as exemplified by the genus *Chlamydomonas*, see Ettl (1976).

The unicellular chlorophytes are mostly spherical or oblong in shape although species that are ovoid, pear-shaped, fusiform (e.g., all species of *Chlorogonium*), or lobed have also been described. One of the distinguishing characteristics of the group is the presence of 2 or 4 flagella of equal length which are anteriorly inserted in the cell, often emerging from the sides of an anterior papilla or thickening of the cell wall. The genus *Polyblepharides* is reported to have 6 or 8 flagella, but this alga is extremely rare, and detailed observations of its flagellar organization are not available.

The chloroplast is massive occupying about 40% of the cell volume in *Chlamydomonas reinhardtii* (Schötz *et al.*, 1972). In most species it is cup-shaped with a single median pyrenoid embedded in the basal portion. In other species, the chloroplast is anteriorly, laterally, or centrally placed in the cell and may appear H-shaped, asteroidal, lamellate, or be divided into discoidal units. The major pigments of the photosynthetic genera are chlorophyll *a* and *b*, β-carotene and the xanthophylls lutein, violaxanthin, neoxanthin, and zeaxanthin (Goodwin, 1974). Two genera, *Polytoma* and *Polytomella* are colorless heterotrophs although *Polytoma uvella* is capable of synthesizing a keto carotenoid of unknown chemical structure (Links *et al.*, 1960). Ketonic carotenoids are produced in large quantities in cells undergoing metabolic alterations stimulated by nitrogen deficiency or desiccation, the best example being the production of the red pigment, haematochrome or astaxanthin (3:3' diketo 4:4' dihydroxy-β-carotene) during akinete development in *Haematococcus* (Czygan, 1968, 1970; Donkin, 1976).

The heterotrophic genera utilize acetate as a carbon source in preference to glucose (Lwoff and Dusi, 1937; Pringsheim, 1937); other short chain acids and alcohols have also been found to serve as substrates for the growth of *Polytomella* (Wise, 1955, 1959; Lloyd *et al.*, 1968). Species of *Chlamydomonas*, *Chlo-*

rogonium, and *Haematococcus* exhibiting either dark growth or photoassimilation also utilize acetate preferentially (Droop, 1974). Although there are reports of a few species of *Chlamydomonas* assimilating glucose, the rate of uptake is extremely low in comparison to photosynthetic rates, and the process is probably of little significance in natural systems (Bennett and Hobbie, 1972).

The pyrenoid and stigma or red eyespot are contained completely within the confines of the chloroplast. The eyespot, a consistent feature of the motile green cell, is located anteriorly at the periphery of the chloroplast. The organelle is thought to be the site of light perception, and the classic experiments of Hartshorne (1953) showed that mutants of *Chlamydomonas* lacking eyespots were positively phototactic but reacted more slowly to light stimuli than wild type cells with eyespots. Unfortunately, nothing is known of the biochemical composition of chlorophyte eyespots or of the mechanism by which the light stimulus is transmitted to the motile organs of the cell. Eyespots have been reported to occur in *Polytoma* and *Polytomella* but may be lacking in cultured material.

Several chlorophyte species contain more than one pyrenoid per cell. The exact position of the organelle varies according to the shape and location of the chloroplast; in some species of *Chlamydomonas*, for example, it is located directly below the flagellar bases in an inverted cup-shaped chloroplast. The pyrenoid matrix is surrounded by a starch sheath which is readily visible with the light microscope after IKI staining. Starch grains are also present in the chloroplast stroma, often in large quantities in older cells. Pyrenoids are not present in heterotrophic genera.

The major storage product of the unicellular chlorophytes, including the obligate heterotrophs, is the α -(1, 4)-linked glucan, starch (Craigie, 1974). *Polytoma uvella* produces a relatively insoluble starch consisting of 16% amylose and an amylopectin similar to that of higher plants (Manners *et al.*, 1965). Low molecular weight photosynthates such as sucrose and glycerol have been detected (Craigie *et al.*, 1966, 1967; Craigie and McLachlan, 1964) and appear to play an osmoregulatory role in marine species. Osmoregulation in the marine alga, *Dunaliella*, for example, is based on the ability of the cell to regulate the synthesis and degradation of glycerol in response to varying external saline conditions (Ben-Amotz, 1975).

Cytoplasmic organelles such as the nucleus, mitochondria, Golgi apparatus, and endoplasmic reticulum are similar in structure and function to those of other eukaryotic cells. Mitochondria are always located near the flagellar bases as are two large contractile vacuoles in the freshwater species. When filled, these organelles occupy approximately 8% of the cell volume in *Chlamydomonas* (Schotz *et al.*, 1972) and presumably serve as the mechanism for osmoregulation by alternately filling with and discharging excess cellular water into the medium.

The cell wall of *Chlamydomonas reinhardtii* is non-cellulosic and composed of a crystalline glycoprotein matrix of which 30% is protein (Roberts *et al.*, 1972). The most abundant amino acid residue is hydroxyproline which accounts for 11-13% of the total protein and 3.6% of the cell wall. Sugar residues include arabinose and galactose with trace amounts of mannose. The biochemical structure of the *Chlamydomonas* cell wall differs from that of all other hydroxyproline-containing algal groups and flowering plants in that the amino acid is linked to heterooligosaccharides (primarily 4-membered chains of arabinose, galactose, and glucose) rather than to homooligosaccharides (arabinose only) (Miller *et al.*, 1972). The cell wall is fragile and much more easily disrupted than that of other algae, a condition which may be due to differences in hydroxyproline linkages. Miller *et al.* (1972) suggest that algae with fragile cell walls (such as *Chlamydomonas*) must, of necessity, rely on contractile vacuoles rather than wall pressure to maintain their osmotic balance. This view is supported by the fact that organisms with "rigid" cell walls (such as *Chlorella*) do not possess contractile vacuoles.

The biochemical composition of the cell wall of the other chlorophyte genera is similar to that of *Chlamydomonas* (Roberts, 1974). *Haematococcus* differs in general morphology from *Chlamydomonas* in that the protoplast is pulled away from the cell wall and connected to it by protoplasmic strands. For this reason, the genus is usually placed in its own family rather than in the Chlamydomonadaceae. Members of the family Phacotaceae are loricate, surrounded by nonliving layers of material of unknown chemical composition. The loricas are often impregnated with iron or manganese salts causing them to appear brown or reddish-brown in color. The protoplast in these organisms is also drawn away from the lorica. The protoplasts of *Dunaliella* and *Polytomella* are naked and enclosed by a plasma membrane that appears to be, at most, only slightly thickened.

Ultrastructure of distinctive organelles. Micrographs of the whole cell of *Chlamydomonas reinhardtii* (Ringo, 1967a; Johnson and Porter, 1968) reveal the prominant features of the "typical" green unicellular flagellate: the anterior insertion of the flagella, the nucleus positioned directly below the flagellar bases, and the cup-shaped chloroplast containing the single, large, electron-dense pyrenoid (Fig. 1). The placement of organelles in *Carteria* (Lembi,

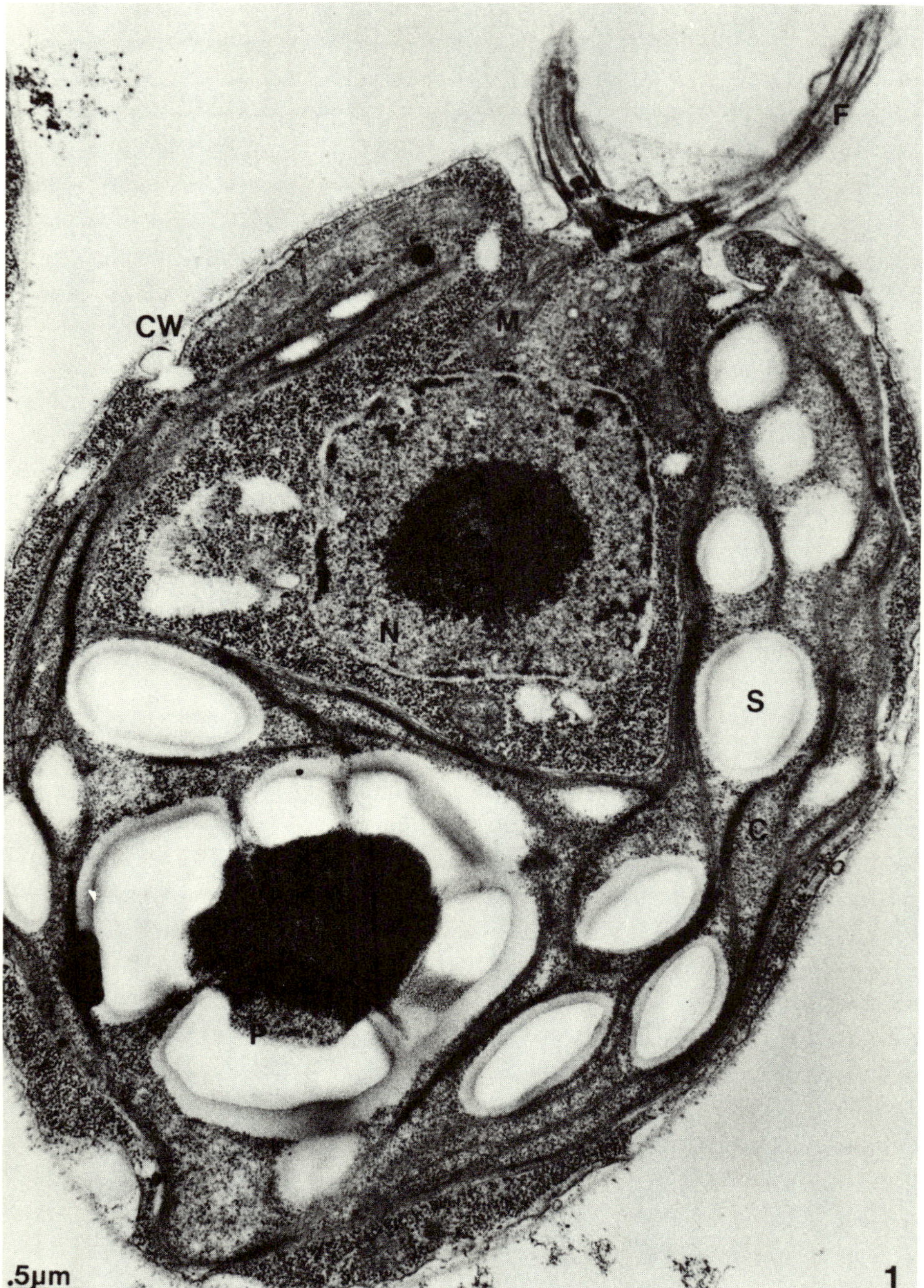
F
CW
M
N
S
C
P
.5μm
1

Fig. 1. Longitudinal section of *Chlamydomonas reinhardtii* cell showing the typical distribution of organelles. F, flagellum; M, mitochondrion; N, nucleus; C, chloroplast; P, pyrenoid with starch sheath; S, starch grain; CW, cell wall. (Reproduced with permission of Dr. D. L. Ringo, J. Cell Biol. 33: 543.)

1975), *Dunaliella* (Peterfi and Manton, 1968; Werz and Kellner, 1970; Eyden, 1975), and presumably *Haematococcus* (Lang, 1968) and *Dysmorphococcus* (Porcella and Walne, 1974) is similar to that of *Chlamydomonas* although the latter two organisms have been only partially detailed.

The obligate heterotrophic genera have been well characterized ultrastructurally: *Polytoma* by Siu *et al.* (1976a) and *Polytomella* by Moore *et al.* (1970) and Brown *et al.* (1976c). The nucleus is centrally located in the cell, and the mitochondria are found in the anterior and peripheral cytoplasm (Fig. 2). The starch-filled leucoplast profiles, which in *Polytoma* contain tubular elements suggestive of an origin from chlorophyll-containing plastids (Lang, 1963b), appear to be scattered about in the posterior half of the cell. Although thin sections indicate the presence of many separate leucoplasts, reconstruction of serial sections reveal that *Polytoma* contains a single, cup-shaped, highly perforated leucoplast (Siu *et al.*, 1976a) occupying approximately 15% of the cell volume (Gaffal, 1978). The number of mitochondria in these cells ranges from 1 to 12 (Burton and Moore, 1974; Siu *et al.*, 1976a) and probably changes with the stage of growth. The presence of higher numbers of mitochondria correlates well with the 10-15 that have been shown in model reconstructions of cells of *Chlamydomonas reinhardtii* (Arnold *et al.*, 1972; Schötz *et al.*, 1972).

Eyespot. The eyespot of the green flagellates is composed of electron-dense granules approximately 75-100 mμ in diameter arranged in layers or plates (Walne and Arnott, 1967; Makamura *et al.*, 1973). Each plate is subtended by a chloroplast lamella and the number of layers appears to be a rather consistent, species-specific character: one layer in *Chlamydomonas eugametos* (Walne and Arnott, 1967), two to three in *C. reinhardtii* (Sager and Palade, 1957), and as many as 9 rows in *C.* sp. (Lembi and Lang, 1965). The number of granules per layer can be as high as 800 (Nakamura *et al.*, 1973), and face views of the eyespot reveal the close hexagonal packing of the granules (Fig. 3). The granules are not membrane-bounded nor are they in close association with cytoplasmic microtubules as they are in the euglenoid algae. A spatial relationship between the microtubules of the flagellar rootlets and the periphery of the eyespot has been noted in the motile cells of some green algae (reviewed in Moestrup, 1978), but this has yet to be shown in *Chlamydomonas*. Functional relationships have been extremely difficult to demonstrate even in the euglenoid

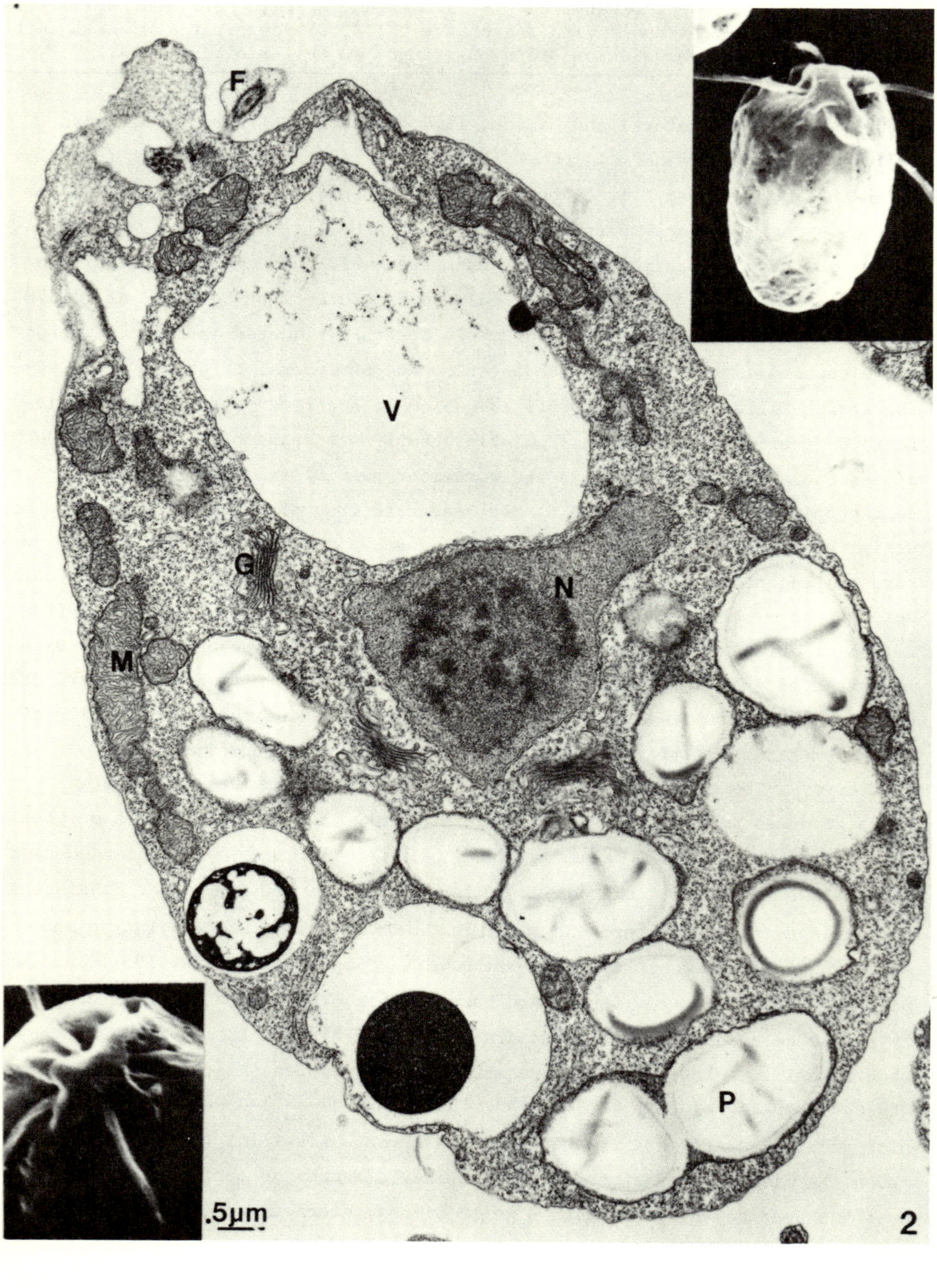

2

Fig. 2. Longitudinal section of *Polytomella agilis* cell showing the distribution of organelles. F, flagellum; M, mitochondrion; N, nucleus; G, Golgi body; V, vacuole; P, starch-filled plastid. Inserts show scanning electron micrographs of the external morphology of the cell and the insertion of the four flagella. (Reproduced with permission of Dr. D. L. Brown, J. Cell Biol. 69: 106.)

algae, and the mechanism by which eyespots capture, translate and communicate light energy as meaningful impulses to the flagellar apparatus remains one of the most elusive in flagellate research today. Hopefully, investigations of

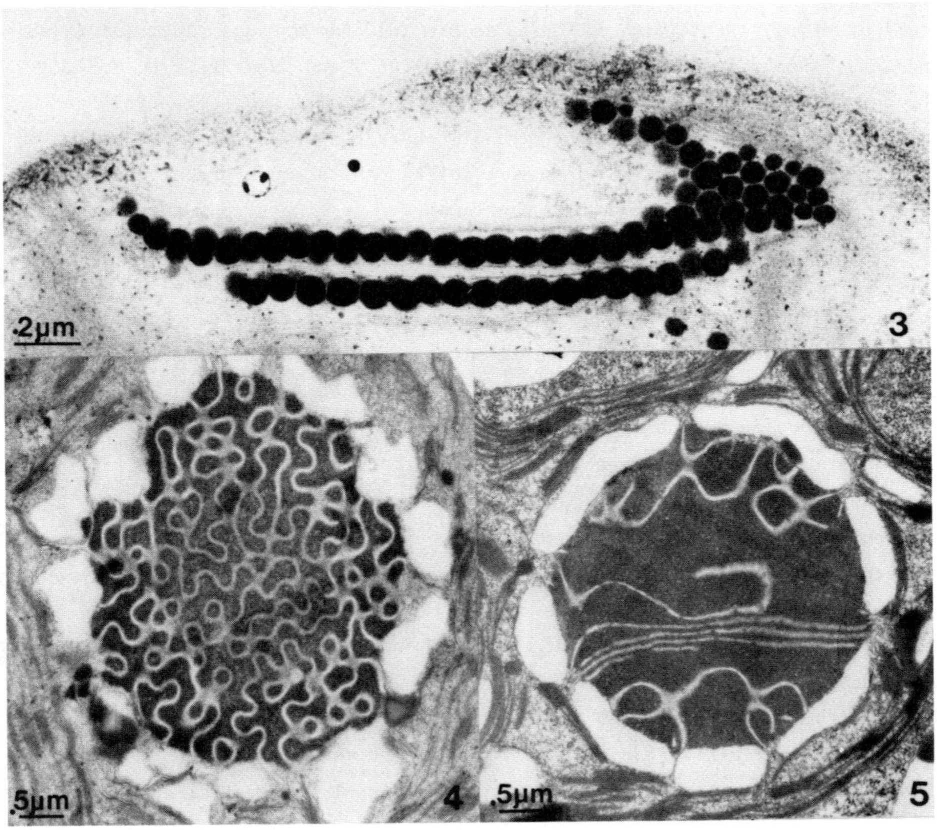

Figs. 3-5. Fig. 3. Two-layered eyespot of *Carteria radiosa* with densely staining granules shown in longitudinal and face views. The outer chloroplast membrane and subtending lamellae are faintly visible (Dr. N. L. Pearlmutter, unpublished micrograph). Figs. 4 and 5. Pyrenoid of *Carteria radiosa* (Fig. 4) and *Carteria olivieri* (Fig. 5). Note the variations in number of lamellar profiles and the shape of the sheathing starch grains.

mutant cells with abnormal eyespot or microtubule morphology will shed more light on this problem although ultrastructural studies of such mutants (e.g., Gruber and Rosario, 1974) have been few and not specifically oriented toward functional relationships.

Pyrenoid. In addition to eyespot structure, pyrenoid characteristics appear to be consistent among individuals of a species and include the number and shape of ensheathing starch grains, and the number, configuration and extent of branching of the lamellae traversing the pyrenoid matrix (Figs. 4, 5). Pyrenoid laellae are continuous with those of the chloroplast, and their configuration appears to be of several basic types: single in at least 5 species of *Carteria* Lembi and Lang, 1965; Lembi, 1975a), double in *Chlamydomonas eugametos* (Lembi and Lang, 1965), *C. moewusii* (Brown *et al.*, 1968) and *Dunaliella primolecta* (Eyden, 1975) and tubular in *Chlamydomonas* sp. (Lembi and Lang, 1965) and *C. reinhardtii* (Goodenough, 1970). Pyrenoid characteristics may prove useful in elucidating taxonomic and phylogenetic relationships among green phytoflagellates. In *Carteria*, for example, the lamellae are single, but the species can be separated into 2 groups on the basis of the other pyrenoid characteristics noted above, confirming what appear to be natural groupings based on cell shape and flagellar orientation. Caution must, however, be exercised in using these characteristics as taxonomic criteria since, at the present time, there is no information on the extent of natural or environmentally induced plasticity in either eyespot or pyrenoid morphology.

From studies on other green and prasinophyte algae it is presumed that the pyrenoid is proteinaceous and capable of carbon fixation (Holdsworth, 1971; Salisbury and Floyd, 1978b). Whether the pyrenoid is a region for the temporary storage of starch or the active, primary site of glucose polymerization has yet to be determined.

Chloroplast. The overall fine structure and physiology of the chloroplast and mitochondria is similar to that of other green plants. Mutants of *Chlamydomonas reinhardtii* have been extensively used to describe the components of the photosynthetic electron transport chain (reviewed in Levine, 1974), to eludicate functional relationships between chloroplast lamellar structure and photosynthetic activity (Goodenough and Levine, 1970; Goodenough and Staehelin, 1971; Jupin *et al.*, 1976), and to study the composition, function, and ribosomal synthesis of lamellar polypeptides (Hoober, 1972; Levine and Duram, 1973; Klein, 1974; Margulies and Michaels, 1974; Bourguignon and Palade, 1976; Givan, 1979). The discovery of non-Mendelian inheritance in *C. reinhardtii* by Sager in 1954 also insures the use of this organism in genetic investigations of mitochondrial and chloroplast inheritance (reviewed in Sager, 1972; Adams *et al.*, 1976). Ris and Plaut (1962) confirmed the presence of chloroplast DNA in three species of *Chlamydomonas* when they demonstrated that Feulgen-positive regions within the chloroplast stained as intensely as the nucleus and consisted of DNase-sensitive fibrils when viewed with the electron microscope.

DNA fibrils have been observed at the fine structural level in the leucoplasts of *Polytoma*, and DNA and ribosomal RNA have been isolated from these organelles (Siu *et al.*, 1976b). Along with the morphological evidence, the demonstration that leucoplasts, like chloroplasts, possess their own genetic complement greatly strengthens the argument that achlorophyllous genera like *Polytoma* were derived from pigmented, chlamydomonad-like ancestors. A close relationship between *Polytoma* and *Chlamydomonas* is supported by hybridization studies in which the cytoplasmic rRNAs from the two genera showed 80% homology (Siu *et al.*, 1976c). Only 50% homology between the chloroplast and leucoplast rRNAs was demonstrated indicating that the leucoplast and/or chloroplast RNA cistrons diverged during their evolution to a much greater extent than their

Cell Wall. In addition to chloroplast mutants, many other genotypes and phenotypes of *Chlamydomonas* are available for form-function investigations. Exposure of cells to mutagens such as N-methyl-N'-nitro-N-nitrosoguanidine (Gulham, 1965), UV at 235.7 nm (Ebersold, 1956; Eversole, 1956) or ethylmethane sulphonate (loppes, 1968) has resulted in mutants deficient in phosphatases, requiring metabolites such as amino acids, vitamins and acetate (reviewed in Hudock and Rosen, 1976) and altered in mitochondrial structure and function (Wiseman *et al.*, 1977). Two classes of mutants which have been most helpful in elucidating form-function relationships are cells which have an altered ability to form functional flagella and those forming abnormal cell walls. Over 70 mutants of *Chlamydomonas reinhardtii* defective in wall synthesis have been isolated (Davies and Plaskitt, 1971; Hyams and Davies, 1972) and a procedure for the rapid detection of cell wall mutants based on the color sensitive detection of phosphatase release into the medium has been developed by Loppes and Deltour (1977). Genetic analyses indicate most of the cell wall mutants are distinguished from the wild type by a single gene, and there is evidence to suggest that control of cell wall synthesis is extranuclear (Davies and Lyall, 1973; Davies and Roberts, 1976). Another benefit of having cell wall mutants for research is that organelles and subcellular components can be isolated without necessitating the prior removal of the cell wall by destructive mechanical or chemical techniques. The use of cell wall mutant CW 15 in the isolation of chromatin (Robreau and LeGal, 1974) and basal bodies (Gould, 1975) and, most recently, in protoplast fusion studies with other organisms (Fowke *et al.*, 1979) attests to the research potential of these variants.

An ordered, periodic structure to the cell wall was suggested in ultrastructural studies of *Chlamydomonas* (Lewin *et al.*, 1951; Bray *et al.*, 1974) and *Carteria* (Lembi and Lang, 1965); however, the most extensive and detailed descriptions of the substructure and biochemical composition of the chlamydomonad cell are those by Roberts and coworkers. The cell wall of *Chlamydomonas reinhardtii* is composed of 7 layers (Fig. 6) which include an amorphous inner layer (W1) of variable thickness, the central triplet (W2-W6) of constant thickness, and an outer, thin layer (W7) also of variable consistency (Roberts *et al.*, 1972). Optical

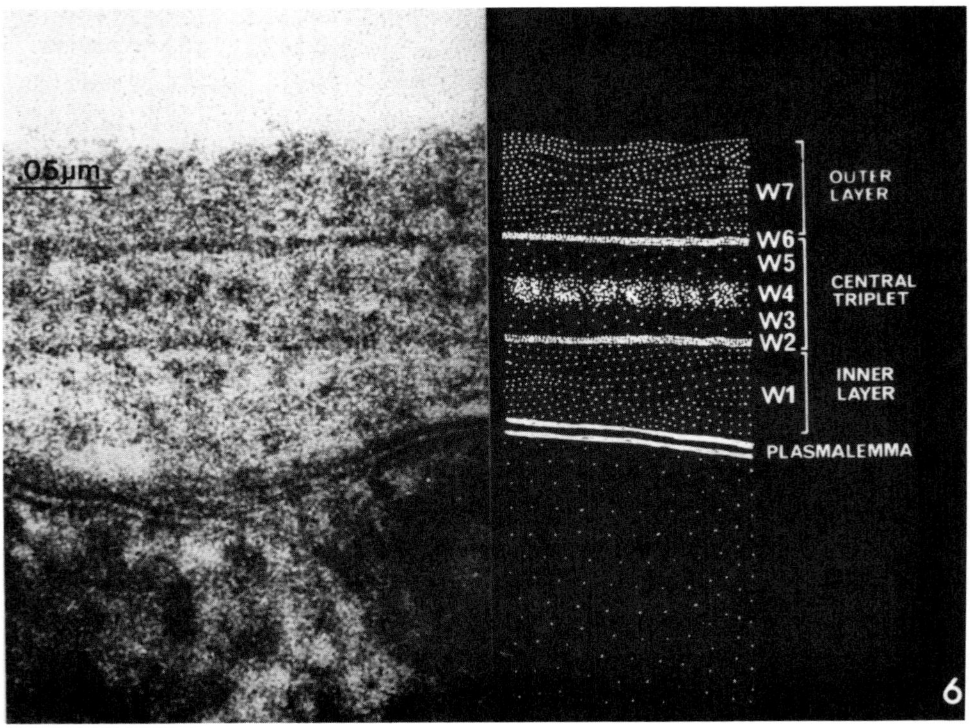

Fig. 6. Electron micrograph and corresponding diagram of cell wall layers of *Chlamydomonas reinhardtii*. (Reproduced with permission of Dr. K. Roberts, J. Ultrastr. Res. 40: 599.)

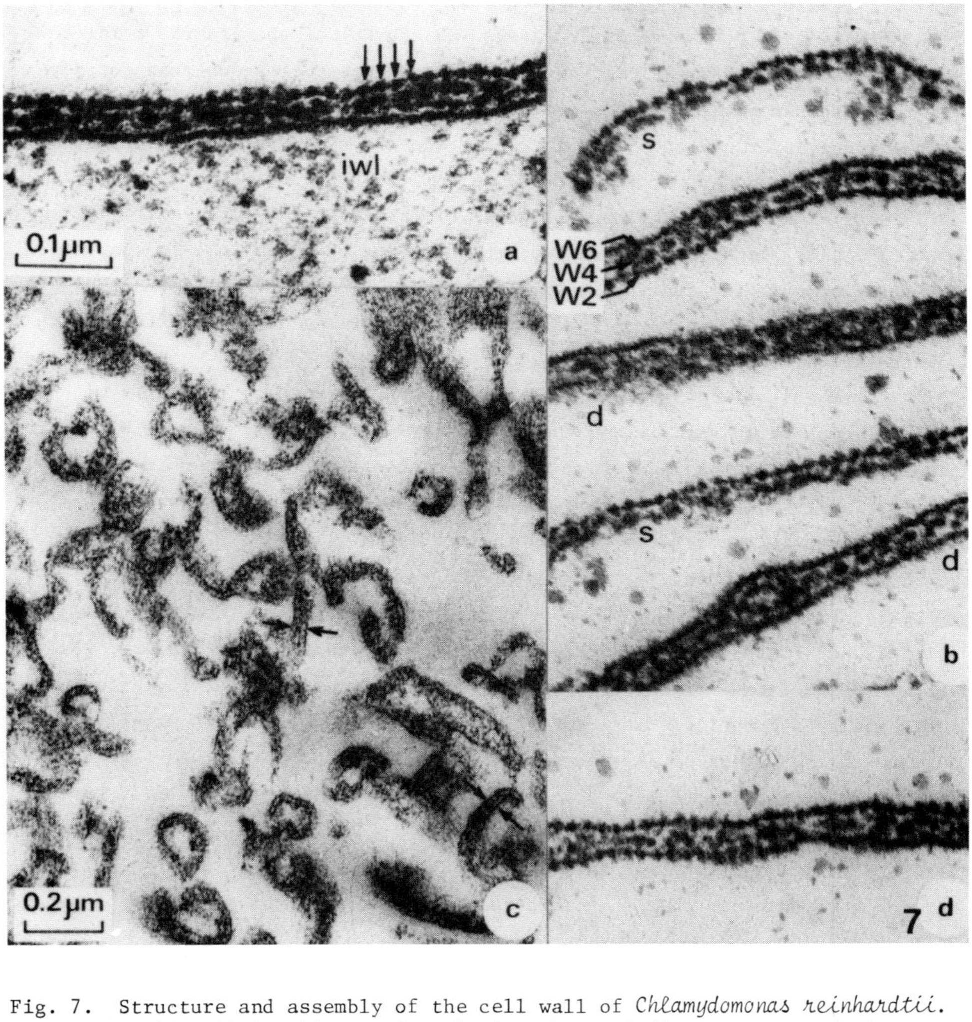

Fig. 7. Structure and assembly of the cell wall of *Chlamydomonas reinhardtii*. a. Intact cell wall in thin section showing the periodic structure of the outer component of the triplet layer. iwl, inner cell wall layer (W1). b. Layers of cast off cell walls showing W2, W4, W6 layers of the central triplet. d, "complete" triplet layer; s, triplet with one layer missing. c. Pellet of reassembled fragments of glycoprotein fraction 2BII. Segments of complete triplet layers are visible at the arrows. d. Fragment reassembled from 2Bl and 2BII showing increased periodic structure in comparison to 2BII fragments alone. (Reproduced with permission of Dr. K. Roberts, Planta 138: 91.)

diffraction techniques show the central triplet to be composed of a highly ordered crystalline lattice, the repeating morphological unit being a parallelogram with sides of 28.5 nm and 23.6 nm and an angle of 80° (Hills *et al.*, 1973; Roberts, 1974). Using a tannic acid containing fixative, Catt *et al.* (1978) demonstrated that it is the outer layers (W2 and W6) of the central triplet which are specifically periodic and crystalline in structure whereas the core (W4) is somewhat more amorphous (Fig. 7a, b). The major chemical components of the inner layer and the central triplet are the hydroxyproline-rich glycoproteins described earlier (Roberts *et al.*, 1972; Roberts, 1974; Hills *et al.*, 1975; Catt *et al.*, 1976). The cell walls of *Chlorogonium, Polytoma, Carteria, Haematococcus, Brachiomonas* and two other species of *Chlamydomonas* consist of a similar crystalline configuration although the dimensions of the morphological repeating unit are slightly different than in *C. reinhardtii* (Roberts, 1974; Roberts and Hills, 1976).

In vitro cell wall assembly has been followed by treating the walls with a salt solution which solubilizes the central triplet and outer layer (W1 remains insoluble). The solubilized glycoprotein components then reassemble upon dialysis (Hills, 1973; Roberts, 1974) and the fragments that form show the same periodic crystalline substructure as the original wall. The glycoprotein can be further separated into a major fraction (2BII) which is inherently capable of reassembly (Fig. 7c) and a higher molecular weight minor fraction (2BI) which is not (Catt *et al.*, 1978). However, if fraction 2BI is added to 2BII, the minor fraction is incorporated into the fragment structure providing increased size and periodic structure (Fig. 7d). *In vivo* assembly may first require the formation of the inner layer (W1) to act as a template for the attachment of the crystalline glycoprotein layers (Hills *et al.*, 1975; Davies and Roberts, 1976).

Synthesis of cell wall protein occurs primarily at the end of the light period in the growth cycle as the cells reach their maximum size and in the second half of the dark period which corresponds to the end of cell division and separation and the production of new walls by the daughter cells (Lang and Chrispeels, 1976). At this time, the cells devote approximately 15% of their protein synthesizing capacity to making cell wall proteins.

Flagellar apparatus. The apparatus in *Chlamydomonas reinhardtii* has been described in detail by Ringo (1967a). His schematic drawing of a longitudinal section through the basal bodies with corresponding cross-sections (Fig. 8) shows the typical 9 + 2 arrangement of axonemal tubules, the star pattern at the junction of the basal body and flagellum (originally described in *Poly-*

toma by Lang, 1963a), and the cartwheel configuration of the basal body proper consisting of triplet tubules. The surface of the flagellar membrane is covered with hairlike projections or mastigonemes. The basal bodies lie in a V configuration and are joined at the top by the striated distal fiber and at the bottom by 2 smaller proximal fibers. Particularly significant was Ringo's description of the 4 microtubule bands or "roots" that radiate from the basal bodies in a cruciate pattern, diverge, and pass along the peripheral cytoplasm of the cell. The microtubules of this system account for the total complement of cytoplasmic microtubules in the non-dividing cells. Ringo, and later other investigators using his work as a model, described the roots as consisting of 4 microtubules each (in a 3 over 1 pattern) in the region between and adjacent to the basal bodies. However, reexamination of these works by Melkonian (1977) and Moestrup (1978) as well as direct observations of the *Chlamydomonas* flagellar apparatus (Birbeck, 1976; Goodenough and Weiss, 1978) indicate that the root system is bilaterally symmetrical consisting of two root types, one with two and the other with four microtubules in a 4-2-4-2 alternating pattern (Fig. 9). The resolution of this feature brings the flagellar root system of *Chlamydomonas* and related genera such as *Volvox* and *Tetraspora* into line with the so-called primitive cruciate X-2-X-2 system which is found in the motile cells of at least 40 other species of green and prasinophyte algae (Moestrup, 1978).

As demonstrated in *Chlamydomonas* (Goodenough and Weiss, 1978) and *Polytomella* (Brown *et al.*, 1976c), the 2-membered microtubule root is associated with a finely striated fiber which is preferentially associated with one of the two microtubules (its position is diagrammatically illustrated in Fig. 9). This fiber was originally termed the dense plate by Ringo (Fig. 8) who did not envision it as extending beyond the inter-basal body region. It now seems likely that the finely striated fiber is analogous to the striated root component that extends into the cytoplasm as described in the zoospores of *Stigeoclonium* (Manton, 1964), *Fritschiella* (Melkonian, 1975) and members of the *Oedogoniales* (Hoffman, 1970; Markowitz, 1978).

The morphology of the flagellar apparatus has proven to be a valuable criterion in evaluating phylogenetic relationships among algae. The many-membered unilateral roots of the motile cells of pseudoparenchymatous and filamentous genera such as *Coleochaete* and *Klebsormidium* serve to place these organisms in an evolutionary line with the more advanced green algae, bryophytes and higher land plants (reviewed in Moestrup, 1978). Among members of the chlorophycean phytoflagellates, information currently available suggests that certain genera are closely allied to *Chlamydomonas*, whereas others, once thought to be closely

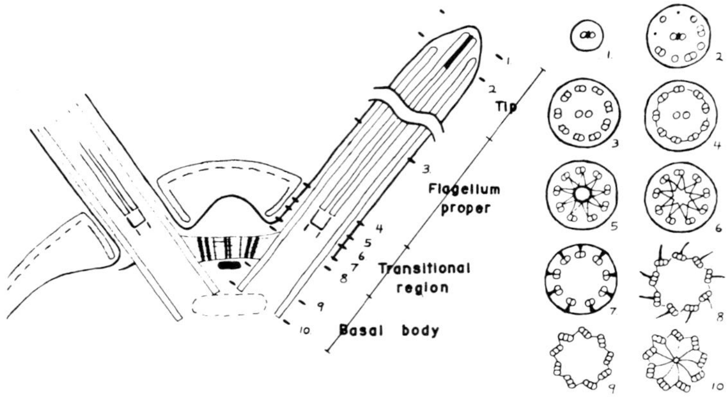

Fig. 8. Diagram of longitudinal section of *Chlamydomonas reinhardtii* flagella and basal bodies with corresponding cross-sections. The striated distal fiber connects the two basal bodies and the dashed line indicates the position of the proximal fibers which are out of the plane of the drawing. The "dense plate" located between the distal and proximal fibers is actually a cross-section of the finely striated fiber described by Goodenough and Weiss (1978). (Reproduced with permission of Dr. D. L. Ringo, J. Cell Biol. 39: 543.)

related, may actually be farther removed. Biflagellate genera in which basal body, distal fiber and microtubule root orientation appear to be similar to that of *Chlamydomonas* are the marine *Dunaliella* (Hyams and Chasey, 1974), the heterotrophic *Polytoma* (Lang, 1963a; Siu *et al.*, 1976a), the non-motile *Tetraspora* (Lembi and Walne, 1971) and probably the individual vegetative cells of motile colonial volvocalean genera such as *Eudorina* (Hobbs, 1971) and *Volvox* (Olson and Kochert, 1970). These examples indicate that alterations in the ancestral Chlamydomonad cells enabled its generic descendants (as well as a host of more closely-related species) to take advantage of a variety of habitats and life forms, yet the conservative features of the flagellar apparatus were retained, probably to insure the continuation of essential processes such as gametic union and migration to regions of optimal light and/or nutrient con-

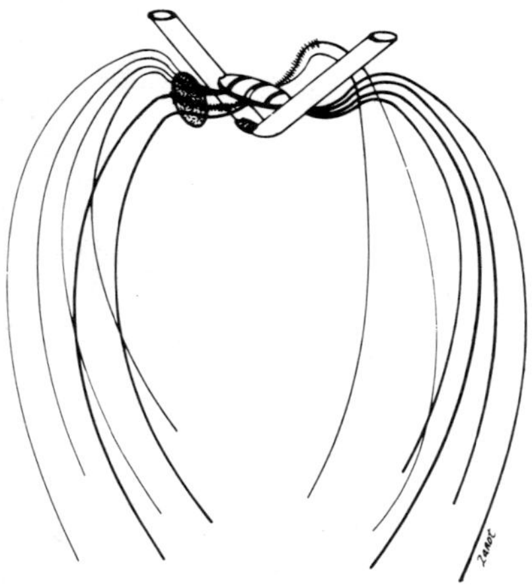

9

Fig. 9. Diagram based on Ringo's (1967a) original schematic of the cytoplasmic microtubule system of *Chlamydomonas reinhardtii* demonstrating the 4-2-4-2 alternating microtubule pattern. The finely striated fiber associated with one of the two microtubules of two-membered roots is indicated with cross lines. The stippled body is the gametic mating structure which is also in contact with the finely striated fiber. (Reproduced with permission of Dr. U. W. Goodenough, J. Cell Biol. 76: 430.)

centrations.

Thus, closely-related quadriflagellate species might be expected to show a duplication of the basal body apparatus, i.e., basal bodies joined in pairs with the same striated distal and proximal fiber connections observed in *Chlamydomonas*. The closest example of this appears to be the heterotrophic genus *Polytomella* in which opposite basal bodies are linked together (Fig. 10) and the microtubule root number is doubled (Brown *et al.*, 1976c). The basal bodies of the A pair of flagella are linked by a striated distal fiber and striated proximal fibers whereas the basal bodies of the B pair, situated slightly deeper in the cell, are connected by a single nonstriated fiber at

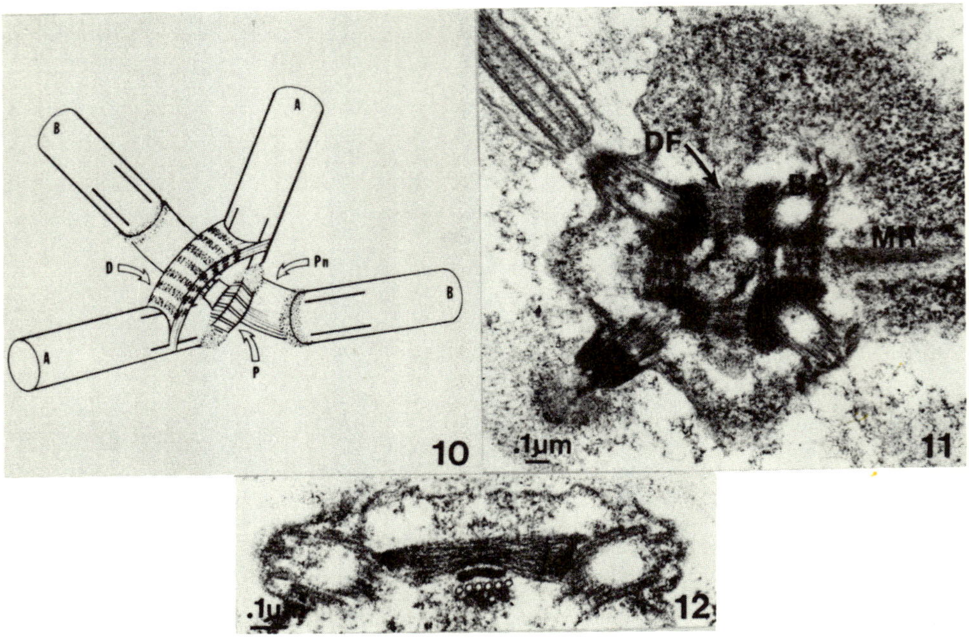

Figs. 10-12. Fig. 10. Diagram of the arrangement of the four basal bodies and their interconnecting fibers in *Polytomella agilis*. Note similarities to the interconnecting structures in *Clamydomonas*. A, basal body of the A pair of flagella; B, basal body of the B pair of flagella; D, distal striated fiber of the A pair; P, proximal striated fiber of the A pair; Pn, proximal nonstriated fiber of the B pair. (Reproduced with permission of Dr. D. L. Brown, J. Cell Biol. 69: 106.) Fig. 11. Section through four basal bodies and interconnecting distal fibers of *Carteria radiosa* (Group I). Compare with the interconnecting fiber arrangement in *Polytomella*. The distal fibers are electron-dense in the region of contact with the basal body. Two, parallel, electron-dense bars extend between opposite flagella, but it is not known whether they connect basal bodies or adjacent microtubule roots. DF, distal fiber; BB, basal body; MR, microtubule root. (Dr. N. L. Pearlmutter, unpublished micrograph.) Fig. 12. Longitudinal section of a distal fiber and microtubule root in *Carteria radiosa*. (Dr. N. L. Pearlmutter, unpublished micrograph.)

their proximal ends. The A and B pair are not connected to one another although they are linked to adjacent microtubule roots by electron-dense material.

The quadriflagellate alga expected to most resemble *Chlamydomonas* in flagellar apparatus ultrastructure, as it does in other morphological features, is *Carteria*. However, this turns out not to be the case; indeed, a diversity of flagellar apparatus structure was found among the 5 species studied by Lembi (1975a). The two species that make up Group I *Carteria* (*C.* sp. and *C. radiosa*) are spherical in shape and lack prominent anterior papillae. The basal bodies are oriented in a double-V with proximal ends close together, but rather than linking opposite basal bodies as in *Polytomella*, the distal fibers connect adjacent basal bodies to form a ring (Fig. 11) with the microtubule bands passing under the centers of the distal fibers (Fig. 12). Two other features separate these species from the typical chlamydomonad: the distal fiber is nonstriated (although the proximal fiber is), and the microtubule bands do not show the X-2-X-2 pattern but consist of as many as 6 over 1 microtubular components. A more unique basal body orientation is that of Group II *Carteria* (*C. olivieri*, *C. eugametos*, *C. crucifera*) in which the cells are elongate and display prominent anterior papillae. The basal bodies are inserted at the corners of a square so that the proximal ends are not in close proximity. The basal bodies do, however, appear to be linked in a ring by nonstriated distal fibers. The microtubule root organization is unlike that described for any other alga; it is extremely complex and requires further elaboration before the spatial relationship of its components to the basal bodies and fibers is fully understood.

Although not excluding the possibility that *Carteria* may be more distantly related to *Chlamydomonas* than other genera, these findings suggest that there is a certain amount of plasticity in the physical arrangement of flagellar components in the unicellular chlorophytes which can result in functional flagellar activity. In addition, such variations may account for the type of flagellar motion exhibited by these organisms. *Chlamydomonas reinhardtii* and *Polytomella agilis*, with oppositely linked basal bodies, show a flagellar motion similar to that of a breast stroke and swim with little or no cell rotation whereas *Carteria* species, with distal fibers linking adjacent basal bodies in a ring, often swim in a tight helical path with the flagellar appearing to display a rotatory rather than breast stroke motion. Observations on other chlorophyte species would certainly be most helpful in establishing the relationship between the mechanics of flagellar movement and the specific orientation of fibers and basal bodies.

An unusual feature of cells of *Carteria radiosa* is the presence of a rhizoplast extending toward the nucleus from the proximal fiber of each basal body (Lembi, 1975b). Considered to be a prasinophyte characteristic, the rhizoplast in organisms such as *Platymonas* may serve a contractile function in flagellar motion of the interphase cell (Salisbury and Floyd, 1978a) and provides material for spindle formation during nuclear division (Stewart *et al.*, 1974). The role of the rhizoplast in *C. radiosa* is not known although it does not appear to be present during nuclear division. Rhizoplasts have not been reported in ultrastructural studies of any other unicellular chlorophycean flagellate, but examination of Eyden's study of *Dunaliella* (1975) shows an "extra" root leading into the cytoplasm which is similar in appearance to the rhizoplast of *C. radiosa*. Although rhizoplasts have not been detected in the many studies of *Chlamydomonas*, there is a remote possibility that they are present but extremely short-lived as functioning, discrete organelles. On the other hand, the rhizoplast may be associated only with organisms of a certain cell shape in order to facilitate the flagellar stroke as suggested by Salisbury and Floyd. They may also be extremely reduced in organisms such as *Chlamydomonas* and their function as centers of spindle tubule formation replaced by the cytoplasmic region just proximal to the basal bodies (microtubule organizing center, *sensu* Pickett-Heaps, 1974).

Flagellar axoneme. Comparative studies of wild type and nonmotile mutant cells of *C. reinhardtii* have been particularly useful in identifying and characterizing the structural components essential to flagellar movement. Since Randall *et al.* (1964) described a nonmotile mutant missing the central axonemal tubules, over 50 strains showing loss of motility, abnormal swimming patterns, and stubbiness or loss of flagella have been isolated and described (Warr *et al.*, 1966; Randall *et al.*, 1967; McVittie, 1969, 1972; Goodenough and St. Clair, 1975; Randall and Starling, 1976; Gruber and Rosario, 1977). Inheritance of all flagellar genes appears to be Mendelian and phenotypic changes are due to single gene mutations (Randall and Sterling, 1976).

The structural components of the flagellar axoneme include the 2 central tubules and central sheath, the 9 outer doublets each consisting of A and B tubules, and the dynein arms, radial spokes, and nexin links (Fig. 13) (Hopkins, 1970; Allen and Borisy, 1974; Chasey, 1974; Witman *et al.*, 1978). Each of the central and outer doublet tubules are composed of 13 protofilaments (Ringo, 1967b) which in turn are made up of two tubulin proteins with molecular weights of 56,000 and 53,000 (Witman *et al.*, 1972a). Both tubulins are present in the A and B tubules of the outer doublets but only the 56,000 molecular weight protein is present in the 3 protofilaments that make up the common wall between the two tubules (Witman *et al.*, 1972b). The proteins of the radial spokes and central tubule-central sheath complex have been partially characterized (Witman *et al.*, 1978).

It is now well known that ATPase activity is associated with the dynein protein arms which extend from the A tubule toward the B tubule of the adjacent doublet, and that flagellar bending is initiated by the ATP-generated sliding of adjacent doublets relative to one another. Radial spokes and central tubule-central sheath complex also appear to be essential for the conversion of inter-

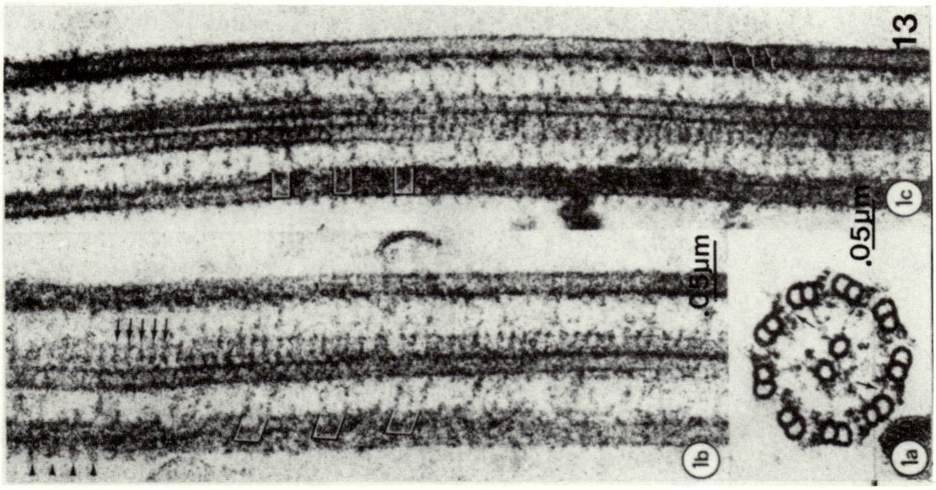

Fig. 13. Isolated axonemes of *Chlamydomonas reinhardtii*. a. cross section showing dynein arms, radial spokes, and nexin links (arrows). b. Longitudinal section. Projections forming the central sheath are seen extending from a central tubule (arrows). Outer dynein arms (arrowheads) are visible along the outside edge of the left doublet microtubule, and the radial spokes (brackets) appear as paired projections extending from the outer doublets toward the central tubules. This axoneme bent to the right below the region shown, and the tilted spokes may represent an initial stage in bend formation. c. Longitudinal section. Radial spokes are grouped in pairs (brackets), and the inner dynein arms are visible along the inner edge of the right doublet microtubule. (Reproduced with permission of Dr. G. B. Witman, J. Cell Biol. 76: 729.)

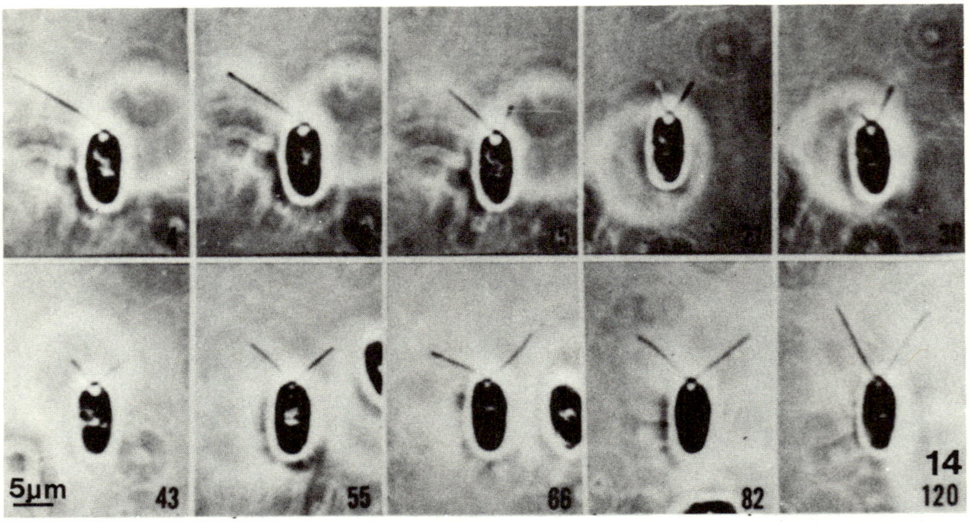

Fig. 14. Shortening and elongation of flagella in a single *Chlamydomonas reinhardtii* cell. The time (min) after amputation of the right flagellum is shown in lower right hand corner of each photomicrograph. (Reproduced with permission of Dr. J. L. Rosenbaum, J. Cell Biol. 41: 600.)

doublet sliding into axonemal bending (Witman *et al.*, 1978). Dynein ATPase is activated by Ca^{+2} and Mg^{+2} and other flagellar enzymes involved in nucleotide transformations have been isolated and characterized (Warner, 1972; Watanabe and Flavin, 1973, 1976). The flagellar surface membrane is capable of generating saltatory, bidirectional movements (Bloodgood, 1977).

A good deal of information pertinent to the understanding of the synthesis and assembly of flagellar proteins has also been obtained using *Chlamydomonas reinhardtii*. Rosenbaum *et al.* (1969) found that when one flagellum is amputated, the intact flagellum shortens while the amputated one regenerates (Fig. 14). Both flagella attain an equal intermediate length and then grow to their normal length at the same rate. These observations suggest that protein from the shortening flagellum can be readily mobilized for the regeneration of the amputated flagellum. In fact, flagella that are resorbed into the cytoplasm intact (as they are when *Polytomella* cells are exposed to hydrostatic pressure) have been shown ultrastructurally to disassemble at the same time as new flagella are being formed (Brown and Rogers, 1978). Even in the presence of cycloheximide, flagella can regenerate as long as this internalized source of structured axonemal material is present.

Cycloheximide treatment of *Chlamydomonas* cells in which both flagella have been amputated permits a small amount of flagellar regrowth (about 1/3 of normal length) indicating that a cytoplasmic pool of unstructured flagellar proteins is also present in flagellate cells to sustain the early phases of flagellar growth. Additional amounts are synthesized *de novo* (Gorovsky *et al.*, 1970). Deflagellation-induced tubulin synthesis has been detected within 15 min of amputation, and maximal rates of tubulin synthesis occur between 45-90 min after deflagellation when approximately 14% of the protein being synthesized by the cell is flagellar tubulin (Weeks and Collis, 1976). An interesting and significant sidelight to studies of flagellar elongation is the report by Solter and Gibor (1978) that flagella respond to changes in osmotic pressure: the higher the osmotic pressure, the shorter the flagella. They speculate that some flagellar length mutants may be osmoregulatory mutants, and even more importantly, they advise attention be paid to the salinity of the environment when flagellar length is used as a taxonomic character.

During assembly, tubulin is added primarily to the distal ends of flagellar tubules (Fig. 15), specifically the A tubules of the outer doublets (Allen and Borisy, 1974; Binder *et al.*, 1975). A smaller portion of the tubulin assembles at the proximal end of the central doublet; distal assembly on the central doublet appears to be mechanically prevented by the insertion of these tubules into a distal cap attached to the tip of the growing flagellum (Dentler and Rosenbaum, 1977).

Basal bodies of *Chlamydomonas* have been isolated (Snell *et al.*, 1974; Hyams and Borisy, 1975a; Gould, 1975) and the assembly of tubulin subunits onto the distal portions of isolated basal bodies has been reported (Snell *et al.*, 1974). Of even greater interest is the demonstration that the basal body roots of *Polytomella* function as organizing centers for cytoplasmic microtubules (Stearns *et al.*, 1976). *In vitro* assembly of tubulin occurs directly onto the ends of existing basal body and root microtubules and amorphous material associated with the roots.

<u>Movement patterns</u>. All members of this group move with the flagella-bearing end toward the direction of movement. The flagellar stroke of *Chlamydomonas reinhardtii* is described by Ringo (1967a) as being analogous to that of a swimmer's breast stroke; i.e., the flagella, which remain straight except at the

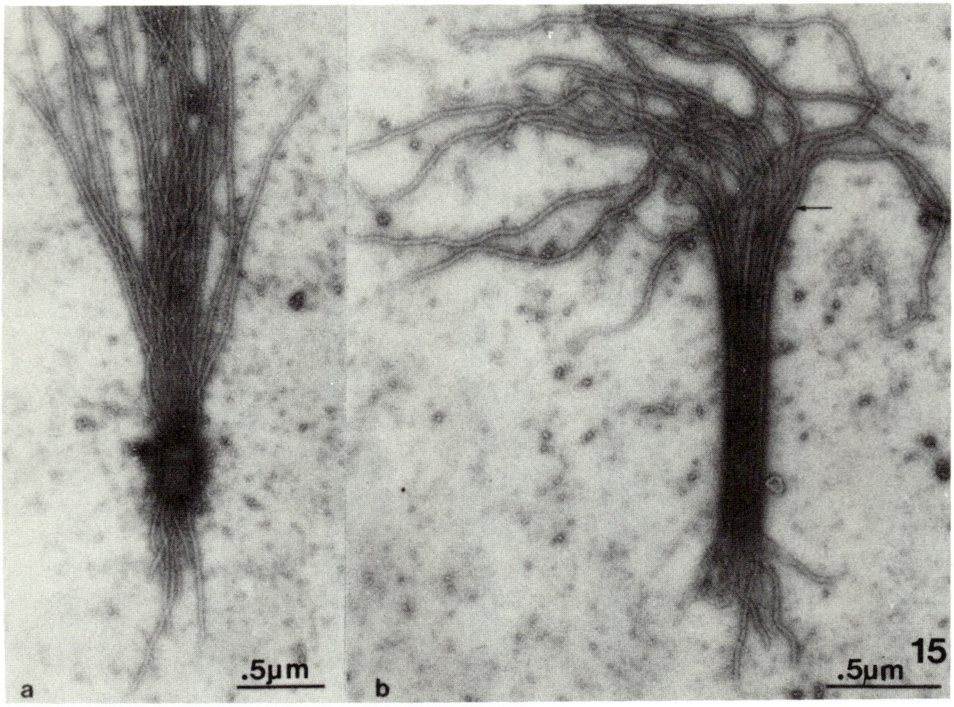

Fig. 15. Electron micrographs of a basal body(a) and axoneme(b) of *Chlamydomonas reinhardtii* negatively stained with phosphotungstic acid. Assembly of tubulin protein (arrow) occurs primarily at the distal end of these structures. (Reproduced with permission of Dr. J. L. Rosenbaum, J. Cell. Biol. 79: 500).

base, sweep backward toward the cell and then return to the original position by a bending motion which passes from the flagellar base to the tip. The cells are also capable of a slower reverse movement in which the waves of bending of the two outstretched flagella push the cell backward.

Ringo attributed the mechanical coordination of the biflagellar beat to the distal fiber connecting the two basal bodies. Support for this has been obtained in studies of the quadriflagellate *Polytomella* in which the power strokes of fiber-linked, opposite flagella are synchronized whereas the flagellar pairs, which are not fiber-linked, do not show coordinated movements in relation to one another (Brown *et al.*, 1976c). Other cytoplasmic components must be excluded as synchronizing agents since isolated flagellar apparatus of *C. reinhardtii* show repeated, coordinated swimming beats when placed in the presence of adenosine triphosphatase (Hyams and Borisy, 1975a).

Like *Chlamydomonas*, cells of *Polytomella agilis* display a breast stroke pattern of flagellar motion and swim primarily in a straight path (Brown *et al.*, 1976c). Having four flagella to accomplish the task of swimming, however, seems to provide a distinct advantage over having only two. The average speed of movement in *Polytomella* has been measured at 233 μm/sec., a rate which is slightly more than double that of the biflagellate *Polytoma* (Gittleson and Noble, 1973) and *Chlamydomonas* (Nultsch, 1974).

Flagellar activity appears to be extremely sensitive to and partly regulated by internal fluctuations in cation concentration. Hyams and Borisy (1975b) demonstrated that form and direction of flagellar motion in isolated flagellar apparatus from *Chlamydomonas* is dependent on calcium concentration; on the other hand, suffusing the system with calcium through microinjection into live cells causes an inhibition of flagellar activity (Nichols and Rikmenspoel, 1978a, 1978b). Interestingly, microinjection of either magnesium or manganese appears to initiate and enhance flagellar movement.

The flagellate greens are positively phototatic. This reaction, which is most intense during the exponential phase of growth (Stavis and Hirschberg, 1973), is usually observed at lower light intensities whereas a negative reaction occurs at higher light intensities (Nultsch, 1974). Although the chemical nature of the photoreceptor is unknown, measurements of the maximum phototactic activity at 503 nm in *Chlamydomonas* suggests a possible involvement of carotenoid pigments (Nultsch *et al.*, 1971). Mutants with altered phototactic responses have been isolated and characterized (Hudock and Hudock, 1973; Hirschberg and Stavis, 1977) while the metabolic and environmental factors which exert an influence on the phototactic response have been described (Sachs and Mayer, 1961; Marbach and Mayer, 1970; Nultsch and Throm, 1975; Nultsch, 1977). *Chlamydomonas* also exhibits negative geotaxis (Bean, 1977).

Life histories. Asexual vegetative cells are haploid. The exact number of chromosomes has been difficult to ascertain in most genera; for example, cells of *Chlamydomonas reinhardtii* are reported to have a chromosome number of either eight (Buffaloe, 1958; Levine and Folsome, 1959; Maguire, 1976) or sixteen (McVittie and Davies, 1971; Loppes *et al.*, 1972). Hastings *et al.* (1965) and Levine and Goodenough (1970) show the partial mapping of sixteen linkage groups in *C. reinhardtii*.

Asexual reproduction is by mitotic division and cytokinesis into 2, 4, or 8 daughter protoplasts which form new cell walls while still enclosed by the parent cell wall. Some species remain motile during division, but the majority lose their flagella prior to cell division. Escape of the daughter cells is by rupture or, as reported by Schlösser (1976), the enzymatic breakdown (caused by a "sporangium-autolysine") of the parent cell wall.

Synchronous cell divisions in *Chlamydomonas* are initiated by changes in temperature (Rooney *et al.*, 1971) or light quality (Carroll *et al.*, 1970) and maintained under conditions of constant temperature and alternating light-dark periods (Bernstein, 1960; Kates and Jones, 1964b; Surzycki, 1971). Cell division occurs during the dark period, and the maximum size of the vegetative cell is reached near the end of the light period (Mihara and Hase, 1971). Synchronized life cycles have greatly facilitated research on the initiation and temporal sequence of a diversity of biochemical processes including DNA, RNA, and protein synthesis (Iwanij *et al.*, 1975; Lien and Knutsen, 1976), microtubular activity (Gould and Jones, 1968), incorporation of polypeptides into chloroplast lamellae (Bourguignon and Palade, 1976), lamellar formation (Schor *et al.*, 1970), and the synthesis of chlorophyll and chlorophyll DNA (Kates and Jones, 1967) and carotenoid pigments (Francis *et al.*, 1975).

Sexual. Sexual reproduction has been reported in all genera including confirmations for *Dysmorphococcus* (Dawson and Harris, 1975) and *Polytomella* (Lewis *et al.*, 1974). Most species are isogamous although anisogamy and oogamy have been described in *Chlorogonium* (Wawrik, 1972). Although appearing morphologically similar, isogametes differ biochemically, and only complementary mating types fuse. Strains or mating types are present in both heterothallic and homothallic clones. Sexual activity in *Chlamydomonas reinhardtii* is induced by the withdrawal of nitrogen from the medium (Sager and Granick, 1954), and two gametogenic divisions ensue resulting in the formation of four gametes from each vegetative cell (Kates *et al.*, 1968). The differentiation period for gamete development is approximately 10-12 hr in synchronized cultures (Kates and Jones, 1964a) and is accompanied by the breakdown and subsequent resynthesis of nucleic acids (Siersma and Chiang, 1971) and proteins (Kates, 1966). Standardized procedures for the production of gametes in liquid and agar media are described by Martin and Goodenough (1975). These authors suggest that the processes resulting in gamete development are actually mechanisms which insure the survival of the cell during periods of nitrogen starvation.

The clumping of opposite mating types and their eventual pairing are initiated by the presence of mating type substances (gamones) on the flagellar tips (Wiese, 1969). The importance of flagella in the mating process is readily illustrated; if flagella are amputated, the induction of mating is delayed for 45 min until the flagella can regenerate to 1/4-1/2 of their original length

(Solter and Gibor, 1977; Ray *et al.*, 1978). Mating type substances have been isolated from the supernatant of gamete suspensions and are extremely active as isoagglutinins causing gametes of the complementary mating type to clump or agglutinate with one another (although they cannot fuse) (Wiese, 1965). Isoagglutinins isolated from the - strain of *Chlamydomonas eugametos* can induce the clumping of 8 mm^3/ml packed + gametes at a concentration of only 0.001 ng/ml (Wiese, 1974). These compounds have been partially characterized as high molecular weight complexes of sulfated glycoproteins with a high electronegative charge (Wiese, 1961, 1967, 1969), and the constituent amino acids and neutral sugars have also been identified (Wiese, 1974). Glutaraldehyde-fixed cells can cause agglutination (Bergman *et al.*, 1975) as can isolated gamete flagella (Claes, 1971).

Wiese has suggested that initial contact is established between a carbohydrate structure on the flagella of the + gametes of *Chlamydomonas moewusii* and a protease sensitive structure on the flagella of the - gametes (Wiese and Hayward, 1972; Wiese, 1974). Biochemical analyses of the gamone "pellet" (vesicles derived from the isolation of flagellar membranes and mastigonemes which exhibit activity as isoagglutinins) reveal no differences in glycoprotein composition between mating types of *C. reinhardtii* (Bergman *et al.*, 1975) although McLean *et al.* (1974) did find differences in buoyant densities among vesicle preparations of *C. moewusii*. Bergman *et al.* (1975) speculate that mating type specificity is due to very slight, and as yet undetectable, changes in the chemical composition of the flagellar surface.

After agglutination of the flagellar tips, the gametes come together and fuse. Early in the fusion process, the gametes of *C. reinhardtii* release a lytic factor which causes the shedding of the gamete cell wall (Claes, 1971; Goodenough and Weiss, 1975). The initiation site for cell wall lysis also appears to be located on the flagellar surface although the exact site is probably different than the agglutination site (Ray *et al.*, 1978). The fusion product is termed the zygote or zygospore, and development has been described by Cavalier-Smith (1976) in *Chlamydomonas*. Within 24 hr the newly formed zygotes expand in diameter and manufacture a thickened cell wall. Minami and Goodenough (1978) report that within 1 hr of formation, the zygote ceases synthesizing polypeptides characteristic of the vegetative and gametic cell and initiates the synthesis of zygote-specific polypeptides, particularly cell wall components. After a suitable period of dormancy (5-7 days in dark culture), the zygotes germinate upon exposure to light via a meiotic division to produce 4 or 8 haploid motile cells (Sager, 1955; Buffaloe, 1958; Maguire, 1976).

Zygotes have been reported to occasionally divide mitotically to give rise to stable diploid vegetative strains (Ebersold, 1967). An exception to the development of thick-walled, encysted zygotes is *Polytomella* in which the zygotes germinate within 3-3 1/2 hr after formation (Lewis *et al.*, 1974).

Palmelloid formation. Some species of *Chlamydomonas* and *Carteria* show a tendency toward colony formation in which vegetative cells lose their flagella and proliferate in a gelatinous matrix. These forms are often the result of subaerial exposure (Smith, 1933), and under the proper conditions (transfer to new medium or flooding), cells can revert to the motile phase. Palmellas have been artificially induced by calcium deficiency, exposure to certain organic acids (Iwasa and Murakami, 1968, 1969) and chloroplatinic acid (Nakamura *et al.*, 1973). Nakamura *et al.* (1975) believe that palmella formation is due to an abnormality in cell wall formation; however, much remains to be learned about the induction process at the biochemical level. It is likely that the tendency toward palmelloid formation in *Chlamydomonas* led to the genesis of the nonmotile, palmella-like habit of the pseudociliate tetrasporalean genera.

Resting spores. Most of the green phytoflagellates are reported to form resting spores resistant to cold, rapid drying, nutrient depletion etc. *Polytomella* and *Dunaliella* form cysts resistant to desiccation, releasing single cells when they germinate. *Chlamydomonas* forms aplanospores and akinetes; *Polytoma*, aplanospores; and *Haematococcus*, akinetes which germinate to give rise to 4, 8, or 16 zoospores or aplanospores (Smith, 1933). Again, very little is known about the environmental and physiological factors which control the initiation and germination of resting spores; this area would seem to present a fruitful avenue for new research.

Mitosis and meiosis. Mitosis. The ultrastructure of cell division in *Chlamydomonas reinhardtii* is described by Johnson and Porter (1968) and Goodenough (1970) and summarized by Pickett-Heaps (1975). A brief summary will be given here along with added details and a summary diagram of mitosis (Fig. 16) from Triemer and Brown's work (1974) on *Chlamydomonas moewusii*. Basal body replication is one of the first signs of incipient division and probably occurs at late interphase or preprophase. The exact timing of basal body replication has been difficult to establish with thin sectioning techniques because a third basal body is often noted in interphase cells. Gould (1975), applying a negative stain to whole mounts of isolated basal bodies, states that probasal bodies in the form of very thin rings of microtubules appear shortly after mitosis and remain attached by filaments to the mature basal bodies during interphase. Dur-

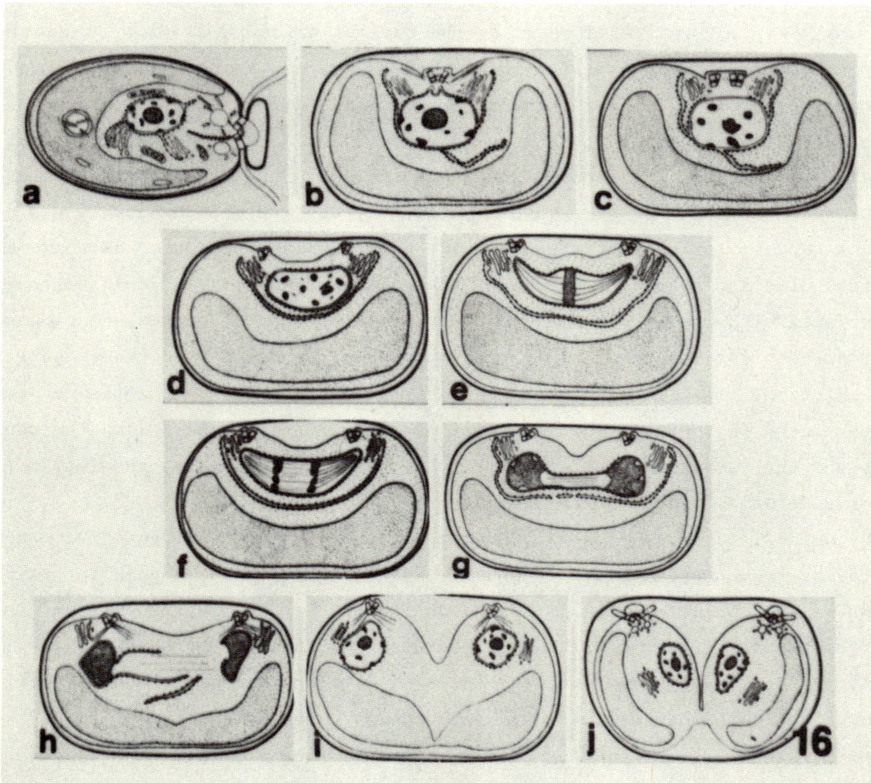

Fig. 16. Summary diagram of mitosis in *Chlamydomonas moewusii*. a. Interphase. b. Preprophase. The nucleus migrates toward the basal bodies. c. Preprophase. Basal bodies replicate and the nucleolus fragments. d. Prophase. Basal bodies migrate to the nuclear poles, and a band of endoplasmic reticulum envelops the nucleus. Spindle tubules appear in the nucleus. In C. *reinhardtii*, spindle tubules originate at the basal bodies (centrioles). e. Metaphase. Nucleus becomes crescent-shaped as chromosomes line up at the metaphase plate. The nuclear poles actually extend beyond the centrioles (not shown). f. Early anaphase. Kinetochores appear; chromosomes migrate from the metaphase plate. g. Telophase. Chromosomes move into the polar regions of the nucleus. h. Late telophase. New nuclear membrane is formed. i and j. Cytokinesis. The phycoplast develops and furrowing proceeds to divide the cytoplasm into two cells. (Reproduced with permission of Dr. R. E. Triemer, J. Phycol. 10: 419.)

ing the next cell division, the probasal structures elongate to form mature, recognizable basal bodies.

As preprophase commences, the flagella detach from the basal bodies at the transition zone. The nucleus migrates to a position directly below the basal bodies, and the nucleolus begins to disperse (Fig. 16b). Although Johnson and Porter reported the absence of basal bodies (centrioles) at the poles of the elongating nucleus, Coss (1974) and Triemer and Brown (1974) confirmed the presence of centriole pairs adjacent to the nuclear poles in *C. reinhardtii* and *C. moewusii*, respectively (Fig. 16d). The nuclear envelope remains intact (closed spindle) during karyokinesis except for the formation of polar fenestrae, or pores through which the spindle microtubules extend, presumably originating from a region at or near the centrioles. Fenestrae were not noted in *C. moewusii*, but the authors suggest that they could have been overlooked due to the rapidity of this particular stage of karyokinesis. Once the spindle microtubules are formed in prophase, the centrioles no longer remain closely associated with the poles as the nucleus becomes crescent-shaped and elongates beyond them. As described by Triemer and Brown, three-layered kinetochores and the interzonal spindle are apparent at early anaphase as the chromosomes begin to separate (Fig. 16f). The remainder of karyokinesis is marked by the separation of the two daughter nuclei, the degeneration of the interzonal spindle and the reformation of the nucleolus.

Cytokinesis occurs by the development of a phycoplast; cleavage microtubules emanating from the basal bodies precede and delineate the cleavage furrow which grows inward circumferentially around the cell. Furrowing proceeds most rapidly from the anterior end of the cell between the two sets of basal bodies (Fig. 16i). As the furrow develops, it passes through and divides the constricting chloroplast and elongated pyrenoid into the daughter halves (Fig. 17). It is assumed that only one daughter cell retains the eyespot, the other daughter cell forming a new eyespot *de novo* (Buffaloe, 1958).

The second mitosis proceeds as the first, cell walls and flagella are produced, and the daughter cells are released from the parent wall.

Cell division has been described in two wall-less genera, *Dunaliella* (Marano, 1976) and *Asteromonas* (Floyd, 1978), a phytoflagellate with probable chlorophyte affinities. In both cases, the features of cell division, including cytokinesis via a phycoplast, are similar to those in *Chlamydomonas*.

<u>Fertilization</u>. The fertilization process in *Chlamydomonas* has been well documented: in *C. reinhardtii* by Friedmann *et al.* (1968), Cavalier-Smith (1975), and Goodenough and Weiss (1975); and in *C. moewusii* by Brown *et al.* (1968) and Triemer and Brown (1975). Mesland (1976) also studied the mating process in *Chlamydomonas eugametos* with the scanning electron microscope. As

---

Fig. 17. Cytokinesis in *Chlamydomonas reinhardtii*. As the anterior furrow develops, the chloroplast also constricts dividing the elongating pyrenoid. Oblique and cross sections of the cleavage microtubules of the phycoplast are visible (arrows). C, chloroplast; P, pyrenoid; S, starch grain; F, furrow. (Micrograph courtesy of Dr. U. W. Goodenough.)

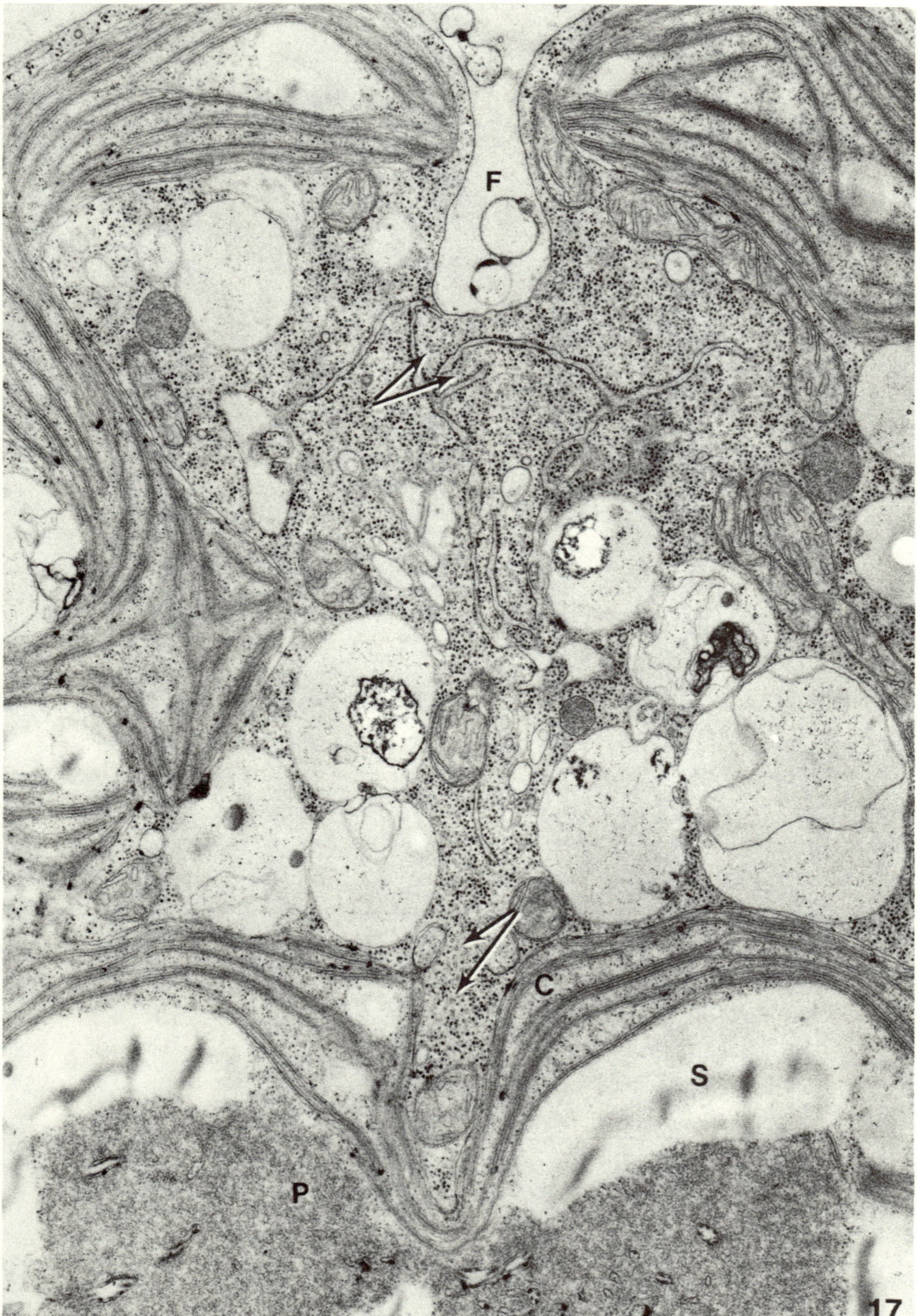
F
C
S
P
17

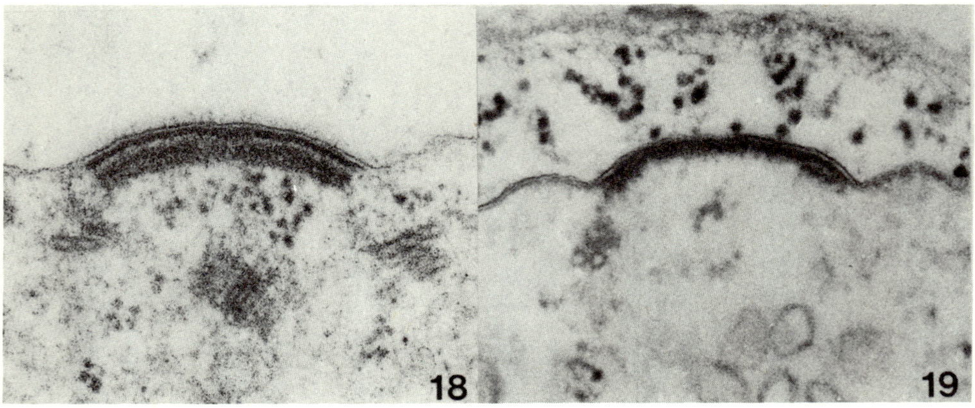

Figs. 18 and 19. Mating structure of an mt^+ (Fig. 18) and an mt^- (Fig. 19) gamete. Cytoplasm associated with the mt^+ mating structure will produce a bulge and eventually the cytoplasmic bridge linking the two gametes. (Reproduced with permission of Dr. U. W. Goodenough, J. Cell Biol. 67: 623.)

described by Goodenough and Weiss for *C. reinhardtii*, the mating process is characterized by the lysis of the gamete cell wall, the formation of specialized mating structures (Figs. 18, 19), the extension of a fertilization tube from the mt^+ mating structure (choanoid body, *sensu* Friedmann *et al.*, 1968) toward the mt^- mating structure, the formation of a cytoplasmic bridge (Fig. 20) and final union of the two protoplasts. Flagella are not detached during fusion, and quadriflagellate cells appear within 5 min of gamete mixing. Although the biochemical basis for the specialization of the mating structures is not known, the morphology of the plasma membrane overlying the mating structures differs in the two mating types when examined by freeze-fraction techniques (Weiss *et al.*, 1977).

The fertilization process in *C. moewusii* differs somewhat. The gametes re-

---

Fig. 20. Gametes of *Chlamydomonas reinhardtii* fixed 3 min after mating. A portion of the mating structure (arrow) (activated doublet zone) of the mt^+ gamete (upper cell) is visible in the cytoplasmic bridge. The mt^+ cytoplasm passes through and microfilaments emanate from the activated doublet zone to form the major portion of the bridge. (Reproduced with permission of Dr. U. W. Goodenough, J. Cell Biol. 67: 623.)

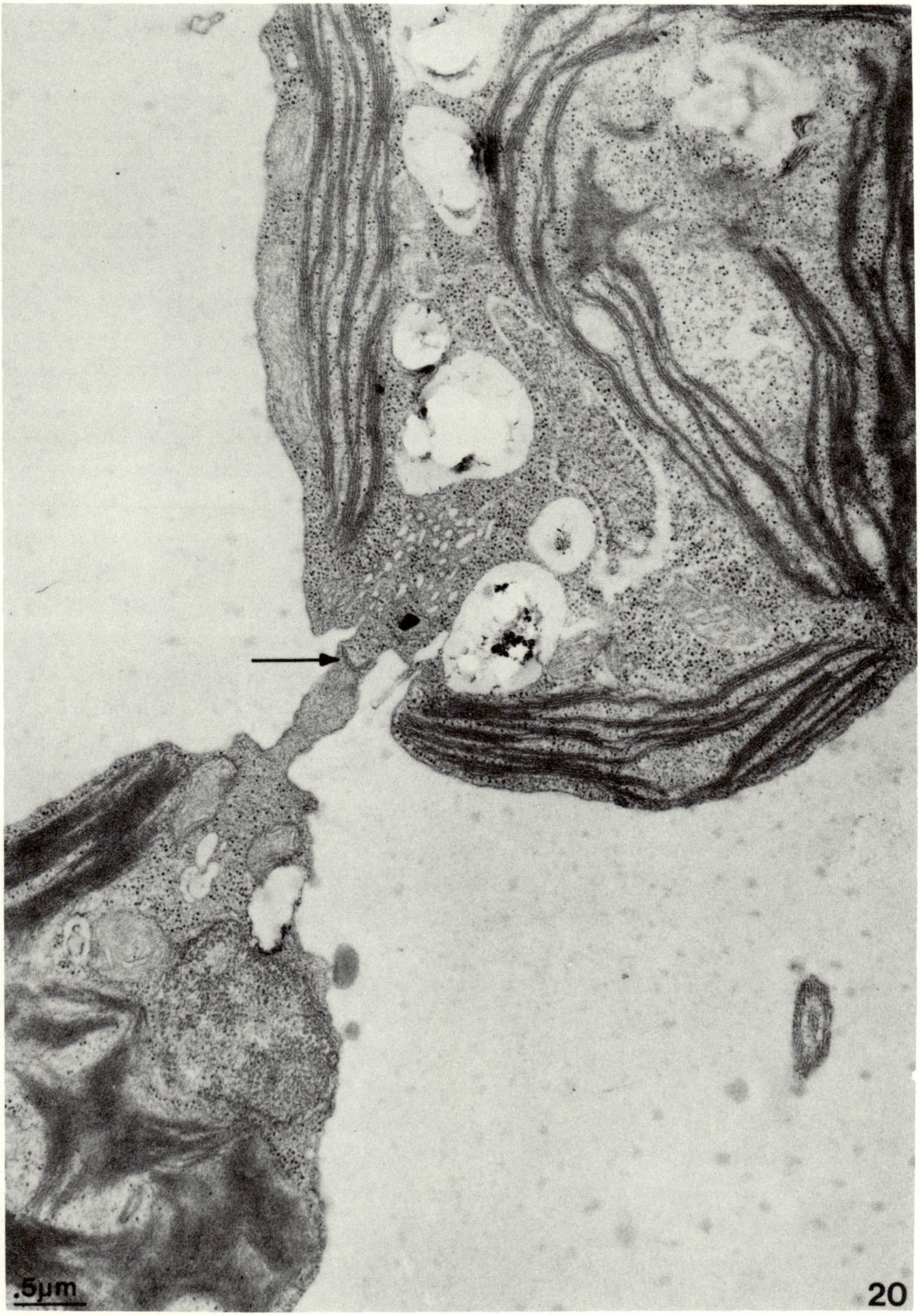
.5µm
20

tain their cell wall except at the region where the anterior papillae fuse. Mating structures were not reported by Triemer and Brown although electron-dense material does form in the cytoplasmic bridge prior to protoplast fusion. Nuclear fusion is delayed by as much as 8-10 hr after the formation of the pairs. The + gamete remains motile during fusion whereas the flagella of the - gamete trail along the sides of the cell.

Cavalier-Smith (1970) reported the fusion of chloroplasts in zygotes of *C. reinhardtii*, and the leucoplasts of *Polytoma* zygotes are also known to fuse forming a single structure (Gaffal, 1978). The zygospore wall thickens into 4-8 layers and cytoplasmic dedifferentiation occurs so that the Golgi apparatus, contractile vacuoles and endoplasmic reticulum are no longer visible. The plastid loses its chlorophyll and fills with starch grains, and the cytoplasm becomes packed with specialized granules (Cavalier-Smith, 1976). The colorless plastids of *Polytoma* contain visible ribosomes, and DNA has been detected biochemically in *Chlamydomonas* zygospores (Sueoka *et al.*, 1967).

Meiosis. The ultrastructure of meiosis in the *Chlamydomonas reinhardtii* zygospore was described by Triemer and Brown (1976). Basal bodies reappear during the resting stage or at early leptotene, and the meiotic stages are ultrastructurally similar to those of other organisms. The synaptonemal complex also resembles that of other organisms; it disappears by diplotene-diakinesis and does not appear to participate in crossing over. The division process from metaphase to cleavage is similar to that in mitosis with the formation of fenestrae at the poles of the intact nuclear envelope and development of a phycoplast resulting in cell cleavage.

Sager and coworkers (Sager, 1972) have clearly shown that in most instances the chloroplast DNA of only the maternal (mt^+) parent is expressed in the progeny, resulting in the observed non-Mendelian segregation of chloroplastidic genes. There is also evidence that the chloroplast DNA of both parents is transmitted to the zygote, that chloroplast fusion and segregation are similar to nuclear fusion and gene segregation, and that recombination occurs between the two parental chloroplast DNAs during zygote development (Chiang, 1968, 1971). This combination of events should result in typical Mendelian segregation patterns. Light and electron microscopic observations that chloroplasts do fuse after mating (Cavalier-Smith, 1970) tend to support these conclusions. Uniparental inheritance of chloroplast genes, however, is the rule rather than the exception, and some as yet undiscovered mechanism other than cytoplasmic exclusion which permits the observed preferential transmission of maternal genes, must be in effect in *Chlamydomonas*. Models to explain the physical mechanism of uniparental inheritance have been proposed by Sager and Ramanis (1973) and Gillham *et al.* (1974).

Ultrastructural alterations due to environmental changes. The onset of adverse environmental conditions results in the formation of either palmelloid colonies surrounded by a copious gelatinous matrix or encysted cells with thickened walls. Unfortunately, ultrastructural studies of these rather important phases of the life cycle have been conducted on only a few species of green flagellates.

Chloroplatinic acid-induced palmelloid cells of *Chlamydomonas* fixed with potassium permanganate appear similar to normal vegetative cells in general morphology (Nakamura *et al.*, 1975). Eyespots are present and the flagella, although motionless, show a normal basal body and 9 + 2 axonemal structure (flagella are usually lost in naturally occurring palmellas). Flagellar activity is postulated as being physically inhibited by encasement in the gelatinous layers although a physiological inhibition cannot be ruled out. This situation seems to be quite different from that of the palmelloid organism *Tetraspora* in which the pseudocilia show a typical chlamydomonad basal body organization and structure but the axoneme lacks the normal complement of tubules (Lembi and Walne, 1971). The structure of the axoneme may be reflective of the inherent ability of individual cells to revert to the motile condition; for example, cells of palmelloid *Chlamydomonas*, whether artificially or naturally induced, can swim once the palmella is dissociated (Nakamura *et al.*, 1974) whereas cell division, and presumably the biochemical reinduction of functional flagella, is required before motile cells of *Tetraspora* are released from the colony.

Encystment in *Polytomella* is characterized by the loss of flagella and the production of a thickened, four-layered cell wall of unknown chemical composition (Sheeler *et al.*, 1970; Brown *et al.*, 1976a). Rough endoplasmic reticulum and Golgi bodies proliferate during the early stages of encystment, but later become reduced in size and number as do the leucoplasts. Microtubules and basal bodies also dedifferentiate. Encystment has been beautifully characterized by Brown *et al.* (1976b). The cysts increase in size from water uptake, and a cytoplasmic polarity is clearly established as droplets of low electron density occupy that end of the cell where the cytoplasm will expand and eventually break through the weakened cell wall.

Encystment or spore formation in many species is accompanied by the formation of carotenoid pigments. In the akinetes of *Haematococcus*, ketonic carotenoids accumulate in the perinuclear cytoplasm but are not associated with any cytoplasmic organelle or vesicle (Lang, 1968). The loss of green color appears to be due to the masking of the chlorophyll by the carotenoids rather than a breakdown of photosynthetic lamellae.

Physiological and biochemical processes specific to the group. Acetate is assimilated in the green and colorless flagellates via the glyoxylate pathway, the marker enzymes malate synthetase and isocitrate lyase having been demonstrated in particulate fractions of *Polytomella* (Haigh and Beevers, 1964b; Gerhardt, 1971), *Chlamydomonas* (Haigh and Beevers, 1964a; Harrop and Kornberg, 1966; Giraud and Czaninsky, 1971), and *Chlorogonium* (Haigh and Beevers, 1964a).

Glycolate production and metabolism is also demonstrable in the autotrophic species although specific features of the pathway differ from those in higher plants. The glycolate oxidizing enzyme is a dehydrogenase rather than an oxidase (Frederick *et al.*, 1973; Merrett and Lord, 1973; Tolbert, 1974), and fractionation studies suggest that glycolate metabolism takes place in mitochondria rather than in microbodies (Gruber *et al.*, 1974; Stabenau, 1974a, 1974b). Mitochondrial localization has been confirmed cytologically in *Chlamydomonas reinhardtii* by Beezley *et al.* (1976) and it is likely that oxidation is linked to oxygen through the mitochondrial electron transport system (Stabenau, 1974a).

Microbodies have been characterized ultrastructurally in *Chlamydomonas*, *Chlorogonium*, *Polytoma*, and *Polytomella* (Gerhardt and Berger, 1971; Silverberg, 1975). Microbody-associated catalase activity (a marker for glycolate oxidase activity in higher plants) is detectable after subcellular fractionation (Stabenau, 1974a) and by a DAB-mediated cytochemical reaction (Silverberg and Sawa, 1974; Silverberg, 1975), but the enzyme in these algae appears to be nonspecialized in function and not associated with either glycolate or glyoxylate metabolism (Stabenau and Beevers, 1974).

Although glycolate metabolism is a characteristic feature of photorespiration, light/dark ratios of CO_2 evolution are much lower in the green algae than in other C_3 plants (Brown and Tregunna, 1967; Fock *et al.*, 1971; Cheng and Colman, 1974). It is now clear that green algae do not exhibit a conventional photorespiration system and indeed may not photorespire at all (Lloyd *et al.*, 1977). According to Bidwell (1977), light respiration in *Chlamydomonas* appears to be a continuation of dark respiration in the light. The export of recently fixed photosynthate from the chloroplast to the mitochondrion takes place in the dark rather than in the light, effectively reducing the amount of photosynthate available for photorespiration. It should be noted that low photorespiration rates do not indicate the presence of a C_4 pathway. Higher aquatic plants such as *Najas flexilis* and *Myriophyllum spicatum* photorespire at low rates and are still considered to be C_3 plants (Hough and Wetzel, 1972, 1978; Stanley and Naylor, 1972).

Habitat. The green phytoflagellates are found in a variety of environments: freshwater, brackish water, marine and terrestrial. The great majority are free-living, but some strains are endosymbiotic; for example, *Chlamydomonas hedleyi* in the foraminifer *Archaias angulatus* (Lee *et al.*, 1974). According to Round (1969), some forms live epibiotically in the empty thecae of *Dinobryon*, the mucilage of other algae and higher aquatic plants and attached to certain zooplankters. Within the group are members that can tolerate any of a wide range of salinities. *Dunaliella* is a frequent and prolific member of marine and estuarine sites and inland saline lakes and pools, while an unidentified species of *Chlamydomonas* has been reported as occurring each year in the plankton of Lake Onondaga (New York State, USA), a highly polluted saline lake fed by natural salt springs (Sze and Kingsbury, 1972). At the other extreme, *Chlamydomonas* has been reported in the plankton of nutrient-poor, low-saline lakes (Whitford, 1964; Round, 1969).

Although flagellates are not considered typical of the flora of rivers and streams, species of *Chlamydomonas* have been reported in the potamoplankton (Lackey, 1964), and the genus has been listed as one of the seven most important planktonic groups in the major rivers of the United States (Williams, 1964). During the years 1961-1962, the three most important components of the late summer-early fall phytoplankton maxima were *Chlamydomonas*, *Scenedesmus* and *Nitzschia*. *Chlamydomonas* sp. and *C. globosa* have also been reported in swimming pools in the Phoenix area (Adamson and Sommerfeld, 1978).

Chlamydomonads are often dominants in acid bogs (Baas Becking and Nicolai, 1934; Fott, 1956; Ettl, 1965) and Palmer (1962) lists *Chlamydomonas* as an indicator of high acidity in water supplies. Acidophilic chlamydomonads have been brought into culture (Fott and McCarthy, 1964; Cassin, 1974) and Joyon and Fott (1964) described the ultrastructure of *Carteria acidicola* noting its unusual pyrenoid which lacks a starch sheath. Cassin (1974) reported that of the two acidophilic species studied, only *Chlamydomonas acidophila* could be considered acid-loving since it grows at pH 2.0 but not at pH 7.0 or 8.0 as did *C. sphagnophila*. She postulated that acid tolerance is dependent on the exclusion of H^+ ions from the cell interior, the barrier possibly being located at the chloroplast envelope since acidophiles remain green at pHs below which phaeophytinization of chlorophyll would normally occur. H^+ exclusion by the chloroplast envelope has been advanced as the reason for the successful colonization of acid waters such as strip mine pits by green algae (Brock, 1973). Conversely, blue-green algae may be restricted to environments of pH 5.0 or higher because their photosynthetic lamellae lie in the peripheral cytoplasm unprotected from the effects of high acidity. Cassin also found *C. acidophila* to display dark growth on glucose. Acetate was assimilated but only in the presence of glucose, suggesting that the glyoxylate pathway may be modified in this alga.

Green flagellates are common in extremely polluted waters with a high organic content such as standing water in feedlots and pastures and in sewage lagoons (Fitzgerald, 1964). After compiling and tabulating the survey reports of 165 investigators, Palmer (1969) ranked *Chlamydomonas* third among the top eight genera in tolerance to organic pollution. Although the genus is near the top of the list, the highest ranked species, (*C. reinhardtii*), is only number 45 on the species list indicating that most investigators monitoring water samples do not identify *Chlamydomonas*-like organisms to species. Other green flagellates on the species list were *Chlorogonium euchlorum* and *Chlorogonium elongatum*. Obviously, organisms that are heterotrophic or photoheterotrophic would appear to be at a competitive advantage over those that are strict autotrophs, particularly when light is restricted under conditions of high turbidity and dense algal growth. Efficiency in organic carbon assimilation is high; Eppley and MaciasR (1963) noted that chlamydomonad algae assimilated most of the acetate generated by anaerobic bacteria in sewage oxidation ponds.

The appearance of motile green unicells in the phytoplankton of productive freshwater lakes and ponds has often been noted (Hutchinson, 1967), but the ecology of this group has been little studied in comparison to that of the blue-green and diatom algae. There are numerous reports of *Chlamydomonas* and *Carteria* as dominants among the small, often motile winter algae that bloom under ice (Wright, 1964; Hutchinson, 1967; Bozniak and Kennedy, 1968), but generalizations concerning the conditions under which green flagellates occur as components of the phytoplankton the rest of the year are difficult to make. Part of the problem resides in the lack of identification. Hutchinson (1967) defines the planktonic green algae as including members of the Volvocales, Chlorococcales and desmids. The limnological literature is replete with references to "planktonic green algae", "chlorophycean species" or "flagellates", and many of these collections probably do include some volvocalean unicellular flagellates. Since they may be only a minor component of the flora, however, their identification to genus is not stressed. Thus, information on the relative importance of this group in the dynamics of phytoplankton ecology and succession as well as documentation of occurrence in freshwater lakes are generally unavailable. To compound the problem, flagellates do not preserve well in most standard fixatives (with the exception of Lugol's iodine solution), and gross underestimates of population density can result by sampling for these motile organisms at fixed depths.

If chlorophycean flagellates comprise even a small portion of the "planktonic greens", their blooms in eutrophic lakes occur in the late spring or early summer following the diatom crash and prior to the blue-green blooms of midsummer (Wetzel, 1975; Bartell *et al.*, 1978). Essential nutrient concentrations are usually low although it is possible for the bloom itself to be so intense that nutrient concentrations are reduced below detectable levels. In fact, the depletion of nitrate-N and ammonium-N by these algae is believed to be one of the primary causes for the development of nitrogen-fixing blue-green algae which later dominate the mid-summer plankton. Schelske and Stoermer (1971) and Stoermer (1975) predict that high phosphorus loadings in the Great Lakes will eventually lead to the dominance of green and blue-green algae at the expense of diatom growth.

Chlamydomonad algae do occur as dominants in ponds and shallow lakes in midsummer. Happey-Wood (1976) found *Chlamydomonas* to predominate under low nutrient concentrations and the stable conditions of summer stratification in Abbot's Pool. The alga was apparently able to maintain a competitive edge over non-motile algae such as *Chlorococcum* because of its ability to migrate vertically, i.e., downward toward nutrients diffusing up from the hypolimnion. *Chlorococcum* became the dominant after the fall turnover when the increased water turbulence prevented the *Chlamydomonas* cells from orienting themselves along a light gradient and remaining in the photic zone.

It is unlikely that the heterotrophic or photoheterotropic condition is of substantial benefit to the planktonic forms of motile greens. Natural levels of organic compounds in most lakes are extremely low, and algae are not able to compete effectively with bacteria for available substrates (Wright and Hobbie, 1966). Although glucose is not a preferred substrate for most chlamydomonad species, it is still interesting to note that at natural glucose levels, Bennett and Hobbie (1973) calculated that it would take 71,300 years for a *Chlamydomonas* cell totally dependent on glucose utilization to reproduce itself.

The unicellular green flagellates utilize both nitrate and ammonium ions as nitrogen sources (Morris, 1974). Ammonium-N is preferentially used by most algae, but the growth rates on the two sources are approximately the same. Cain (1965) showed that planktonic *Chlamydomonas* species appear to utilize nitrate preferentially and Stross (1963) showed the same for *Haematococcus*. In lakes, the prevalent form of nitrogen is nitrate-N, and this is the form considered to be most limiting to green algal growth. Other sources of nitrogen such as urea (Birdsey and Lynch, 1962) and argenine (Kirk and Kirk, 1978) can be utilized by certain species when available. The importance of organic forms of nitrogen to *Haematococcus*, which is found primarily in nitrogen-rich sites such as bird baths and rock pools, has been described by Droop (1961).

Although little is known regarding the physical and chemical conditions that regulate the appearance of green flagellates in lake plankton, considerable research has been conducted on the biological interactions of these algae with other planktonic organisms. *Chlamydomonas* is adversely affected by extracellular components secreted by blue-green (Vance, 1965) and other green algae (Kroes, 1971) and may exert a profound influence on other organisms by its own secretion of extracellular inhibitors (Proctor, 1957) and potential growth stimulators such as polysaccharides, glycolate and amino acids and sugars (reviewed in Hellebust, 1974; Vogel *et al.*, 1978). In the marine environment, thiamine secreted by *Dunaliella* was shown to be utilized by *Coccolithus huxleyi* and biotin by *Amphidinium carterae* (Carlucci and Bowes, 1970). Such studies, however, have been limited to the laboratory, and the relative importance of these interactions in nature awaits further elucidation. Evidence to date indicates that the green flagellates, in contrast to blue-green algae, are highly favored as food sources by herbivorous zooplankton (Arnold, 1971; Porter, 1973).

A final note should be made of the research conducted on the response of *Dunaliella* to chlorinated hydrocarbons such as DDT and PCB (Menzel *et al.*, 1970; Bowes, 1972; Mosser *et al.*, 1972a, 1972b) and petroleum hydrocarbons (Prouse *et al.*, 1976). In most cases, *Dunaliella* exhibits tolerance to these compounds in comparison to marine diatoms indicating that substantial loading of such pollutants in estuarine environments could have significant effects on phytoplankton species composition and eventually on other organisms in the food web.

Culture media references. Most of the freshwater green flagellate species grow well in Bold's Basal Medium, and for nonsterile culture, Pringsheim's Soil-Water Medium is often used (Bold and Wynne, 1978; Starr, 1978). The mineral medium of Sueoka (1960) is preferred for the growth of *Chlamydomonas reinhardtii* for genetic studies; similarly, Sager and Granick's Medium I (1953) seems to be preferred for flagellar isolation studies. These media can be supplemented with sodium acetate to permit cultures to reach higher cell densities. Media for the culture of the marine *Dunaliella* are described by

Guillard and Ryther (f/2 medium) (1962) and Carlucci and Bowes (1970) while *Polytoma* and *Polytomella* are grown on media consisting of sodium acetate, yeast extract and tryptone (Starr, 1978). The isolation and culture of *Chlamydomonas* protoplasts is described by Gresshoff (1976) and Schlosser *et al.* (1976).

Needed research. It is not surprising that much of the research on chloroplast, flagellar apparatus, and cell wall structure, function, and assembly has been conducted on *Chlamydomonas* and its relatives. These algae provide ideal model systems in which cellular and molecular processes basic to other, more complex organisms can be characterized and defined. Thus, for researchers interested in the fundamental nature of cellular organization and metabolism, a thorough examination of the literature cited in the preceeding sections will yield many ideas for fruitful research.

It is interesting to note that even with the advances in knowledge of the past 10-15 years, many of the original questions posed by Ringo (1967a) concerning the function of the various components of the flagellar apparatus of *Chlamydomonas* have not been resolved. We still do not know which structures are responsible for the coordination of flagellar motion nor do we know how variations in flagellar apparatus organization (such as those shown in *Carteria* and *Polytomella*) relate to the type of motion expressed by these organisms. The role of cytoplasmic microtubule roots in flagellar movement remains to be determined, e.g., whether they function in the dissipation of stress from flagellar beating or in the initiation and coordination of flagellar motion. Other questions surrounding the cytoplasmic microtubules concern their possible function in the maintenance of cell shape and cell wall formation. Perhaps most intriguing of all, biochemical and structural relationships between the flagellar system and the eyespot must be ascertained. Certainly, the composition and function of both the eyespot and the pyrenoid must be elucidated with perhaps the first step being the development of methods for the isolation of these organelles into purified fractions.

With so much emphasis placed on genetic mutants in form-function research, it seems that very little effort has been expended in describing alterations in ultrastructural features of the motile cell due to environmental changes. This avenue of research would have tremendous implications in the ecology, physiology and taxonomy of these algae, and would seem to be prerequisite to any attempt to place species in natural taxonomic groups and determine their phylogenetic relationship with other chlorophyte and prasinophyte algae.

When conducting a literature search on green phytoflagellates, the paucity of information on species and genera other than *Chlamydomonas reinhardtii* is immediately apparent. Certainly not all of the research conducted on *C. reinhardtii* should be repeated on these organisms, but it would be helpful to have a complete ultrastructural "history", as it were, for at least one representative species in each genus. This compendium should include a general outline of cellular organization, and detailed descriptions of flagellar apparatus, pyrenoid, and eyespot morphology; mitosis, cytokinesis, the mating process (including biochemical phenomena such as agglutination and cell wall shedding), meiosis and resting spore formation. Much of this information is lacking in genera such as *Dysmorphococcus*, *Haematococcus* and *Chlorogonium*. In addition, processes specific to these genera require further elaboration, e.g., the mechanism of lorica formation in members of the Phacotaceae.

Fossil record. Because most of the members of this group do not form mineralized cell walls, the fossil record is scant. Schopf (1970) lists the Haematococcaceae as dating from the Jurassic period whereas members of the Chlamydomonadaceae and Phacotaceae have been recorded in the more recent fossil deposits of the tertiary period (approximately 50 million years ago). The origin of these groups is probably much earlier since algae supposedly derived from a monadal, possibly chlamydomonad-like ancestor date from the Devonian period and earlier (e.g., species of Oedogoniaceae and Ulotrichaceae).

ACKNOWLEDGEMENT

The author gratefully acknowledges the assistance of Dr. Ronald W. Hoham for providing access to his list of references on green phytoflagellates.

REFERENCES

Adams, G. M. W., K. P. Vanwinkle-Swift, N. W. Gillham and J. E. Boynton. 1976. Plastid inheritance in *Chlamydomonas reinhardtii*. In R. A. Lewin [Ed.] *The Genetics of Algae*. Univ. Calif. Press, Berkeley.

Adamson, R. P. and M. R. Sommerfeld. 1978. Survey of swimming pool algae of the Phoenix, Arizona, metropolitan area. J. Phycol. 14: 519-521.

Allen, C. and G. G. Borisy. 1974. Structural polarity and directional growth of microtubules of *Chlamydomonas* flagella. J. Mol. Biol. 90: 381-402.

Arnold, C. G., O. Schimmer, F. Schotz and H. Bathelt. 1972. Die mitochondrien von *Chlamydomonas reinhardtii*. Arch. Mikrobiol. 81: 50-67.

Arnold, D. E. 1971. Ingestion, assimilation, survival and reproduction by *Daphnia pulex* fed seven species of blue-green algae. Limnol. Oceanog. 16: 906-920.

Baas Becking, L. G. M. and E. Nicolai. 1934. On the ecology of a sphagnum bog. Blumea 1: 10-45.

Bartell, S. M., T. F. H. Allen and J. F. Koonce. 1978. An assessment of principal component analysis for description of phytoplankton periodicity in Lake Wingra. Phycologia 17: 1-11.

Bean, B. 1977. Geotactic behavior of *Chlamydomonas*. J. Protozool. 24: 394-401.

Beezley, B. B., P. J. Gruber and S. E. Frederick. 1976. Cytochemical localization of glycolate dehydrogenase in mitochondria of *Chlamydomonas*. Pl. Physiol. 58: 315-319.

Ben-Amotz, A. 1975. Adaptation of the unicellular alga *Dunaliella parva* to a saline environment. J. Phycol. 11: 50-54.

Bennett, M. E. and J. E. Hobbie. 1972. The uptake of glucose by *Chlamydomonas* sp. J. Phycol. 8: 392-398.

Bernstein, E. 1960. Synchronous division in *Chlamydomonas moewusii*. Science 131: 1528-1529.

Bergman, K., V. W. Goodenough, D. A. Goodenough, J. Jawitz and H. Martin. 1975. Gametic differentiation in *Chlamydomonas reinhardtii*. II. Flagellar membranes and the agglutination reaction. J. Cell Biol. 67: 606-622.

Bidwell, R. G. S. 1977. Photosynthesis and light and dark respiration in freshwater algae. Can. J. Bot. 55: 809-818.

Binder, L. I., W. L. Dentler and J. L. Rosenbaum. 1975. Assembly of chick brain tubulin onto flagellar microtubules from *Chlamydomonas* and sea urchin sperm. Proc. Nat. Acad. Sci. U.S.A. 72: 1122-1126.

Birbeck, T. E. 1976. Variation of the flagellar apparatuses within the genus *Chlamydomonas*. J. Phycol. Suppl. 12: 72.

Birdsey, E. C. and V. H. Lynch. 1962. Utilization of nitrogen compounds by unicellular algae. Science 137: 763-764.

Bloodgood, R. A. 1977. Motility occurring in association with the surface of the *Chlamydomonas* flagellum. J. Cell Biol. 75: 983-989.

Bold, H. C. and M. J. Wynne. 1978. *Introduction to the Algae. Structure and Reproduction*. Prentice-Hall, Inc., Englewood Cliffs, N.J.

Bourguignon, L. Y. W. and G. E. Palade. 1976. Incorporation of polypeptides into thylakoid membranes of *Chlamydomonas reinhardtii*. J. Cell Biol. 69: 327-344.

Bowes, G. W. 1972. Uptake and metabolism of 2,2-bis-(p-chlorophyenyl)-1,1,1-trichloroethane (DDT) by marine phytoplankton and its effect on growth and chloroplast electron transport. Pl. Physiol. 49: 172-176.

Bozniak, E. G. and L. L. Kennedy. 1968. Periodicity and ecology of the phytoplankton in an oligotrophic and eutrophic lake. Can. J. Bot. 46: 1259-1271.

Bray, D. F., K. Nakamura, J. W. Costerton and E. B. Wagenaar. 1974. Ultrastructure of *Chlamydomonas eugametos* as revealed by freeze-etching: cell wall, plasmalemma and chloroplast membrane. J. Ultrastr. Res. 47: 125-141.

Brock, T. D. 1973. Lower pH limit for the existence of blue-green algae: evolutionary and ecological implications. Science 179: 480-483.

Brown, D. L. and K. A. Rogers. 1978. Hydrostatic pressure-induced internalization of flagellar axonemes, disassembly, and reutilization during flagellar regeneration in *Polytomella*. Exp. Cell Res. 117: 313-324.

Brown, D. L. and E. B. Tregunna. 1967. Inhibition of respiration during photosynthesis by some algae. Can. J. Bot. 45: 1135-1143.

Brown, D. L., G. G. Leppard and A. Massalski. 1976a. Fine structure of excystment of the quadriflagellate alga, *Polytomella agilis*. Protoplasma 90: 139-154.

Brown, D. L., A. Massalski and G. G. Leppard. 1976b. Fine structure of excystment of the quadriflagellate alga, *Polytomella agilis*. Protoplasma 90: 155-171.

Brown, D. L., A. Massalski and R. Patenaude. 1976c. Organization of the flagellar apparatus and associated cytoplasmic microtubules in the quadriflagellate alga, *Polytomella agilis*. J. Cell Biol. 69: 106-125.

Brown, R. M., Jr., S. C. Johnson and H. C. Bold. 1968. Electron and phase-contrast microscopy of sexual reproduction in *Chlamydomonas moewusii*. J. Phycol. 4: 100-120.

Buffaloe, N. D. 1958. A comparative cytological study of four species of *Chlamydomonas*. Bull. Torrey Bot. Club 85: 157-178.
Burton, M. D. and J. Moore. 1974. The mitochondrion of the flagellate, *Polytomella agilis*. J. Ultrastr. Res. 48: 414-419.
Cain, J. 1965. Nitrogen utilization in 38 freshwater chlamydomonad algae. Can. J. Bot. 43: 1367-1378.
Carlucci, A. F. and P. M. Bowes. 1970. Vitamin production and utilization by phytoplankton in mixed culture. J. Phycol. 6: 393-400.
Carroll, J. W., J. Thomas, C. Dunaway and J. C. O'Kelley. 1970. Light-induced synchronization of algal species that divide preferentially in darkness. Photochem. Photobiol. 12: 91-98.
Cassin, P. E. 1974. Isolation, growth and physiology of acidophilic chlamydomonads. J. Phycol. 10: 439-447.
Catt, J. W., G. J. Hills and K. Roberts. 1976. A structural glycoprotein, containing hydroxyproline, isolated from the cell wall of *Chlamydomonas reinhardtii*. Planta 131: 165-171.
Catt, J. W., G. J. Hills and K. Roberts. 1978. Cell wall glycoproteins from *Chlamydomonas reinhardtii*, and their self-assembly. Planta 138: 91-98.
Cavalier-Smith, T. 1970. Electron microscopic evidence for chloroplast fusion in zygotes of *Chlamydomonas reinhardtii*. Nature 228: 333-335.
Cavalier-Smith, T. 1975. Electron and light microscopy of gametogenesis and gamete fusion in *Chlamydomonas reinhardtii*. Protoplasma 86: 1-18.
Cavalier-Smith, T. 1976. Electron microscopy of zygospore formation in *Chlamydomonas reinhardtii*. Protoplasma 87: 297-315.
Chasey, D. 1974. The three-dimensional arrangement of radial spokes in the flagella of *Chlamydomonas reinhardtii*. Exp. Cell Res. 84: 374-380.
Cheng, K. H. and B. Colman. 1974. Measurements of photorespiration in some microscopic algae. Planta 115: 207-212.
Chiang, K.-S. 1968. Physical conservation of parental cytoplasmic DNA through meiosis in *Chlamydomonas reinhardtii*. Proc. Nat. Acad. Sci. U.S.A. 194-200.
Chiang, K.-S. 1971. Replication, transmission and recombination of cytoplasmic DNAs in *Chlamydomonas reinhardtii*. In Boardman, N. K., A. W. Linnane and R. M. Smillie [Eds.] *Autonomy and Biogenesis of Mitochondria and Chloroplasts*. North-Holland, Amsterdam.
Claes, H. 1971. Autolysis of the cell wall from gametes of *Chlamydomonas reinhardtii*. Arch. Mikrobiol. 78: 180-188.
Coss, R. A. 1974. Mitosis in *Chlamydomonas reinhardtii* basal bodies and the mitotic apparatus. J. Cell Biol. 63: 325-329.
Craigie, J. S. 1974. Storage products. In Stewart, W. D. P. [Ed.] *Algal Physiology and Biochemistry*. Univ. Calif. Press, Berkeley.
Craigie, J. S. and J. McLachlan. 1964. Glycerol as a photosynthetic product in *Dunaliella tertiolecta* Butcher. Can. J. Bot. 42: 777-778.
Craigie, J. S., J. McLachlan, R. G. Ackman and C. S. Tocher. 1967. Photosynthesis in algae. III. Distribution of soluble carbohydrates and dimethyl-β-propiothetin in marine unicellular Chlorophyceae and Prasinophyceae. Can. J. Bot. 45: 1327-1334.
Craigie, J. S., J. McLachlan, W. Majak, R. G. Ackman and C. S. Tocher. 1966. Photosynthesis in algae. II. Green algae with special reference to *Dunaliella* ssp. and *Tetraselmis* ssp. Can. J. Bot. 44: 1247-1254.
Czygan, F. 1968. Sekundär-Carotinoide in Grunalgen. I. Chemie, Vorkommen und Faktoren, welche die Bildung dieser Polyene beeinflussen. Arch. Mikrogiol. 61: 81-102.
Czygan, F. 1970. Blutregen und Blutschnee: Stickstoffmangel-Zellen von *Haematococcus pluvialis* und *Chlamydomonas nivalis*. Arch. Mikrobiol. 74: 69-76.

Davies, D. R. and V. Lyall. 1973. The assembly of a highly ordered component of the cell wall: the role of heritable factors and of physical structure. Molec. Gen. Genet. 125: 21-34.

Davies, D. R. and A. Plaskitt. 1971. Genetical and structural analysis of cell-wall formation in *Chlamydomonas reinhardtii*. Genet. Res. 17: 33-43.

Davies, D. R. and K. Roberts. 1976. Genetics of cell wall synthesis in *Chlamydomonas reinhardtii*. In Lewin, R. A. [Ed.] *The Genetics of Algae*. Univ. Calif. Press, Berkeley.

Dawson, J. and D. O. Harris. 1975. Preliminary studies of the genus *Dysmorphococcus* Takeda (Phacotaceae). J. Phycol. Suppl. 11: 3.

Dentler, W. L. and J. L. Rosenbaum. 1977. Flagellar elongation and shortening in *Chlamydomonas*. III. Structures attached to the tips of flagellar microtubules and their relationship to the directionality of flagellar microtubules assembly. J. Cell Biol. 74: 747-759.

Donkin, P. 1976. Ketocarotenoid biosynthesis by *Haematococcus lacustris*. Phytochem. 15: 711-715.

Droop, M. R. 1961. *Haematococcus pluvialis* and its allies. III. Organic nutrition. Rev. Algol., N.S. 4: 247-258.

Droop, M. R. 1974. Heterotrophy of carbon. In Stewart, W. D. P. [Ed.] *Algal Physiology and Biochemistry*. Univ. Calif. Press, Berkeley.

Ebersold, W. T. 1956. Crossing over in *Chlamydomonas reinhardtii*. Am. J. Bot. 43: 408-410.

Ebersold, W. T. 1967. *Chlamydomonas reinhardtii*: heterozygous diploid strains. Science 157: 447-449.

Eppley, R. W. and F. M. MaciasR. 1963. Role of the alga *Chlamydomonas mundana* in anaerobic waste stabilization lagoons. Limnol. Oceanogr. 8: 411-416.

Ettl, H. 1965. Beitrag zur Kenntnis der Morphologie der Gattung *Chlamydomonas* Ehrenberg. Arch. Protistenk. 108: 271-430.

Ettl, H. 1976. Die Gattung *Chlamydomonas*. Beihefte Nova Hedw. 49: 1-1122.

Eversole, R. A. 1956. Biochemical mutants of *Chlamydomonas reinhardtii*. Am. J. Bot. 43: 404-407.

Eyden, B. P. 1975. Light and electron microscope study of *Dunaliella primolecta* Butcher (Volvocida). J. Protozool. 22:

Fitzgerald, G. P. 1964. The biotic relationships within water blooms. In Jackson, D. F. [Ed.] *Algae and Man*. Plenum Press, N.Y.

Floyd, G. L. 1978. Mitosis and cytokinesis in *Asteromonas gracilis*, a wall-less, green monad. J. Phycol. 14: 440-445.

Fock, H., D. T. Canvin and B. R. Grant. 1971. Effects of oxygen and carbon dioxide on photosynthetic O_2 evolution and CO_2 uptake in sunflower and *Chlorella*. Photosynthetica 5: 389-394.

Fott, B. 1956. Flagellaten aus extrem sauren Gewässern. Preslia 28: 145-150.

Fott, B. and A. J. McCarthy, S. J. 1964. Three acidophilic volvocine flagellates in pure cultures. J. Protozool. 11: 116-120.

Fowke, L. C., P. M. Gresshoff and H. J. Marchant. 1979. Transfer of organelles of the alga *Chlamydomonas reinhardtii* into carrot cells by protoplast fusion. Planta 144: 341-347.

Francis, G. W., L. P. Strand, T. Lien and G. Knutsen. 1975. Variations in carotenoid content of *Chlamydomonas reinhardtii* throughout the cell cycle. Arch. Microbiol. 104: 249-254.

Frederick, S. E., P. J. Gruber and N. E. Tolbert. 1973. The occurrence of glycolate dehydrogenase and glycolate oxidase in green plants: an evolutionary survey. Pl. Physiol. 52: 318-323.

Friedmann, I., A. L. Colwin and L. Colwin. 1968. Fine-structural aspects of fertilization in *Chlamydomonas reinhardtii*. J. Cell Sci. 3: 115-128.

Fritsch, F. E. 1935. *The Structure and Reproduction of the Algae. Vol. I.* Cambridge University Press, London.

Gaffal, K. P. 1978. Configural changes in the plastidome of *Polytoma papillatum* after completion of cytokinesis and during fusion of the gametes. Protoplasma 94: 175-191.

Gerhardt, B. 1971. Zur lokalisation von enzymen der microbodies in *Polytomella caeca*. Arch. Mikrobiol. 80: 205-218.

Gerhardt, B. and C. Berger. 1971. Microbodies und Diaminobenzidin-Reaktion in den Acetat-Flagellaten *Polytomella caeca* und *Chlorogonium elongatum*. Planta 100: 155-166.

Gibbs, S. P. 1962a. The ultrastructure of the pyrenoids of green algae. J. Ultrastr. Res. 7: 262-272.

Gibbs, S. P. 1962b. The ultrastructure of the chloroplasts of algae. J. Ultrastr. Res. 7: 418-435.

Gibbs, S. P., R. A. Lewin and D. E. Philpott. 1958. The fine structure of the flagellar apparatus of *Chlamydomonas moewusii*. Exp. Cell Res. 115: 619-622.

Gillham, N. W. 1965. Induction of chromosomal and non-chromosomal mutations in *Chlamydomonas reinhardtii* with N-methyl-N'-nitro-N'nitrosoguanidine. Genetics 52: 529-537.

Gillham, N. W., J. E. Boynton and R. W. Lee. 1974. Segregation and recombination of non-Mendelian genes in *Chlamydomonas*. Symposium on Non-Chromosomal Inheritance: XIII Int. Congr. Genetics 78: 439-457.

Giraud, G. and Y. Czaninski. 1971. Localisation ultrastructurale d'activites oxydasiques chez le *Chlamydomonas reinhardtii*. C. R. Hebd. Seanc. Acad. Sci. Paris 273: 2500-2503.

Gittleson, S. M. and R. M. Noble. 1973. Locomotion in *Polytomella agilis* and *Polytoma uvella*. Trans. Am. Micros. Sco. 92: 122-128.

Givan, A. L. 1979. Ribulose bisphosphate carboxylase from a mutant strain of *Chlamydomonas reinhardtii* deficient in chloroplast ribosomes. Planta 144: 271-276.

Goodenough, U. W. 1970. Chloroplast division and pyrenoid formation in *Chlamydomonas reinhardtii*. J. Phycol. 6: 1-6.

Goodenough, U. W. and R. P. Levine. 1970. Chloroplast structure and function in *ac-20*, a mutant strain of *Chlamydomonas reinhardtii*. III. Chloroplast ribosomes and membrane organization. J. Cell Biol. 44: 547-562.

Goodenough, U. W. and H. S. St. Clair. 1975. *Bald-2*: a mutation affecting the formation of doublet and triplet sets of microtubules in *Chlamydomonas reinhardtii*. J. Cell Biol. 66: 480-491.

Goodenough, U. W. and L. A. Staehelin. 1971. Structural differentiation of stacked and unstacked chloroplast membranes. J. Cell Biol. 48: 594-619.

Goodenough, U. W. and R. L. Weiss. 1975. Gametic differentiation in *Chlamydomonas reinhardtii*. III. Cell wall lysis and microfilament-associated mating structure activation in wild-type and mutant strains. J. Cell Biol. 67: 623-637.

Goodenough, U. W. and R. L. Weiss. 1978. Interrelationships between microtubules, a striated fiber, and the gametic mating structure of *Chlamydomonas reinhardtii*. J. Cell Biol. 76: 430-438.

Goodwin, T. W. 1974. Carotenoids and biliproteins. In Stewart, W. D. P. [Ed.] *Algal Physiology and Biochemistry*. Univ. Calif. Press, Berkeley.

Gorovsky, M. A., K. Carlson and J. L. Rosenbaum. 1970. Simple method for quantitative densitometry of polyacrylamide gels using fast green. Anal. Biochem. 35: 359-370.

Gould, R. B. and R. F. Jones. 1968. Electron-microscopic observations of microtubules during the synchronized vegetative cell cycle of *Chlamydomonas reinhardtii*. Anat. Rec. 160: 355.

Gould, R. R. 1975. The basal bodies of *Chlamydomonas reinhardtii*. J. Cell Biol. 65: 65-74.

Gresshoff, P. M. 1976. Culture of *Chlamydomonas reinhardtii* protoplasts in defined media. Aust. J. Pl. Physiol. 3: 457-464.

Gruber, H. E. and B. Rosario. 1974. Variation in eyespot ultrastructure in *Chlamydomonas reinhardtii* (*ac-31*). J. Cell Sci. 15: 481-494.

Gruber, H. E. and B. Rosario. 1977. Ultrastructural variations in the flagellar apparatus of *Chlamydomonas reinhardtii* (*ac-31*). J. Submicro. Cytol. 9: 103-113.

Gruber, P. J., S. E. Frederick and N. E. Tolbert. 1974. Enzymes related to lactate metabolism in green algae and lower land plants. Pl. Physiol. 53: 167-170.

Guillard, R. R. L. and J. H. Ryther. 1962. Studies of marine planktonic diatoms. I. *Cyclotella nana* Hustedt, and *Detonula confervacea* (Cleve) Gran. Can. J. Microbiol. 8: 229-239.

Haigh, W. G. and H. Beevers. 1964a. The occurrence and assay of isocitrate lyase in algae. Arch. Biochem. Biophys. 107: 147-151.

Haigh, W. G. and H. Beevers. 1964b. The glyoxylate cycle in *Polytomella caeca*. Arch. Biochem. Biophys. 107: 152-157.

Happey-Wood, C. M. 1976. Influence of stratification on the growth of planktonic Chlorophyceae in a small body of water. Br. Phycol. J. 12: 371-381.

Harrop, L. C. and H. L. Kornberg. 1966. The role of isocitrate lyase in the metabolism of algae. Proc. Roy. Soc. B 166: 11-29.

Hartshorne, J. N. 1953. The function of the eyespot in *Chlamydomonas*. New Phytol. 52: 292-297.

Hastings, P. J., E. E. Levine, E. Cosbey, M. O. Hudock, N. W. Gillham, S. J. Suzycki, R. Loppes and R. P. Levine. 1965. The linkage groups of *Chlamydomonas reinhardtii*. Microbiol. Gen. Bull. 23: 17-19.

Hellebust, J. A. 1974. Extracellular products. In *Algal Physiology and Biochemistry*. Univ. Calif. Press, Berkeley.

Hills, G. J., M. Gurney-Smith, and K. Roberts. 1973. Structure, composition and morphogenesis of the cell wall of *Chlamydomonas reinhardtii*. II. Electron microscopy and optical diffraction analysis. J. Ultrastr. Res. 43: 179-192.

Hills, G. J., J. M. Phillips, M. R. Gay and K. Roberts. 1975. Self-assembly of a plant cell wall *in vitro*. J. Mol. Biol. 96: 431-441.

Hirschberg, R. and R. Stavis. 1977. Phototaxis mutants of *Chlamydomonas reinhardtii*. J. Bacteriol. 129: 803-808.

Hobbs, M. J. 1971. The fine structure of *Eudorina illinoiensis* (Kofoid) Paschen. Br. Phycol. J. 6: 81-103.

Hoffman, L. R. 1970. Observations on the fine structure of *Oedogonium*. IV. The striated component of the compound flagellar "roots" of *O. cardiacum*. Can. J. Bot. 48: 189-196.

Hoffman, L. R. and I. Manton. 1962. Observations on the fine structure of the zoospore of *Oedogonium cardiacum* with special reference to the flagellar apparatus. J. Exp. Bot. 13: 443-449.

Hoffman, L. R. and I. Manton. 1963. Observations on the fine structure of *Oedogonium*. II. The spermatozoids of *O. cardiacum*. Am. J. Bot. 50: 455-463.

Holdsworth, R. H. 1971. The isolation and partial characterization of the pyrenoid protein of *Eremosphaera viridis*. J. Cell Biol. 51: 499-512.

Hoober, J. K. 1972. A major polypeptide of chloroplast membranes of *Chlamydomonas reinhardtii*. Evidence for synthesis in the cytoplasm as a soluble component. J. Cell Biol. 52: 84-96.

Hopkins, J. M. 1970. Subsidiary components of the flagella of *Chlamydomonas reinhardtii*. J. Cell Sci. 7: 823-839.

Hough, R. A. and R. G. Wetzel. 1972. A ^{14}C-assay for photorespiration in aquatic plants. Plant Physiol. 49: 987-990.

Hough, R. A. and R. G. Wetzel. 1978. Photorespiration and CO_2 compensation point in *Najas flexilis*. Limnol. Oceanogr. 23: 719-724.

Hudock, G. A. and M. O. Hudock. 1973. Phototaxis: isolation of mutant strains of *Chlamydomonas reinhardtii* with reversed sign of response. J. Protozool. 20: 139-140.

Hudock, G. A. and H. Rosen. 1976. Formal genetics of *Chlamydomonas reinhardtii* In Lewin, R. A. [Ed.] *The Genetics of Algae*. Univ. Calif. Press, Berkeley.

Hutchinson, G. E. 1967. A Treatise on Limnology. Vol. II. Introduction to Lake Biology and the Limnoplankton. John Wiley and Sons, Inc., N.Y.

Hyams, J. S. and G. G. Borisy. 1975a. Flagellar coordination in *Chlamydomonas reinhardtii*: isolation and reactivation of the flagellar apparatus. Science 189: 891-893.

Hyams, J. S. and G. G. Borisy. 1975b. The dependence of the waveform and direction of beat of *Chlamydomonas* flagella on calcium ions. J. Cell Biol. 67: 186a.

Hyams, J. S. and D. Chasey. 1974. Aspects of the flagellar apparatus and associated microtubules in a marine alga. Exp. Cell Res. 84: 381-387.

Hyams, J. S. and D. R. Davies. 1972. The induction and characterisation of cell wall mutants of *Chlamydomonas reinhardtii*. Mutation Res. 14: 381-389.

Iwanij, V., N. H. Chua and P. Siekevitz. 1975. Synthesis and turnover of ribulose bisphosphate carboxylase and of its subunits during the cell cycle of *Chlamydomonas reinhardtii*. J. Cell Biol. 64: 572-585.

Iwasa, K. and S. Murakami. 1968. Palmelloid formation of *Chlamydomonas*. I. Palmelloid induction by organic acids. Physiol. Plant. 21: 1224-1233.

Iwasa, K. and S. Murakami. 1969. Palmelloid formation of *Chlamydomonas*. II. Mechanism of palmelloid formation by organic acids. Physiol. Plant. 22: 43-50.

Johnson, U. G. and K. R. Porter. 1968. Fine structure of cell division in *Chlamydomonas reinhardtii*. J. Cell Biol. 38: 403-425.

Joyon, L. and B. Fott. 1964. Quelques particularites infrastructurales du plaste des *Carteria* (Volvocales). J. Microsc. 3: 159-166.

Jupin, H., A. Picaud, J. Garnier and N. Hauswirth. 1976. Relations entre l'ultrastructure plastidale et l'etat du photosysteme I chez *Chlamydomonas reinhardtii* (souche sauvage et mutant *F1* 5). C. R. Hebd. Seanc. Acad. Sci. Paris 283: 627-630.

Kates, J. R. 1966. Biochemical aspects of synchronized growth and differentiation in *Chlamydomonas reinhardtii*. Ph.D. Thesis. Princeton Univ. Princeton, N.J.

Kates, J. R. and R. F. Jones. 1964a. The control of gametic differentiation in liquid cultures of *Chlamydomonas reinhardtii*. J. Cell. Comp. Physiol. 63: 157-164.

Kates, J. R. and R. F. Jones. 1964b. Variation in alanine dehydrogenase and glutamate dehydrogenase during synchronous development of *Chlamydomonas*. Biochim. Biophys. Acta 86: 438-447.

Kates, J. R. and R. F. Jones. 1967. Periodic increase in enzyme activity in synchronized cultures of *Chlamydomonas reinhardtii*. Biochim. Biophys. Acta 145: 153-158.

Kates, J. R., K. S. Chiang and R. F. Jones. 1968. Studies on DNA replication during synchronized vegetative growth in gametic differentiation in *Chlamydomonas reinhardtii*. Exp. Cell Res. 49: 121-135.

Kirk, D. L. and M. M. Kirk. 1978. Amino acid and urea uptake in ten species of Chlorophyta. J. Phycol. 14: 198-203.

Klein, R. M. and A. Cronquist. 1967. A consideration of the evolutionary and taxonomic significance of some biochemical, micromorphological, and physiological characters in the Thallophytes. Quart. Rev. Biol. 42: 105-296.

Klein, S. M. 1974. Polypeptide composition of photosynthetic membranes from *Chlamydomonas reinhardtii* and *Anabaena variabilis*. Pl. Physiol. 53: 777-778.

Kroes, H. W. 1971. Growth interactions between *Chlamydomonas globosa* Snow and *Chlorococcum ellipsoideum* Deason and Bold under different experimental conditions, with special attention to the role of pH. Limnol. Oceanogr. 16: 869-879.

Lackey, J. B. 1964. The ecology of plankton algae. In Jackson, D. F. [Ed.] *Algae and Man*. Plenum Press, N.Y.

Lang, N. J. 1963a. An additional ultrastructural component of flagella. J. Cell Biol. 19: 631-634.

Lang, N. J. 1963b. Electron microscopic demonstration of plastids in *Polytoma*. J. Protozool. 10: 333-339.

Lang, N. J. 1963c. Electron microscopy of the Volvocaceae and Astrephomenaceae. Am. J. Bot. 50: 280-300.

Lang, N. J. 1968. Electron microscopic studies of extraplastidic astaxanthin in *Haematococcus*. J. Phycol. 4: 12-19.

Lang, W. C. and M. J. Chrispeels. 1976. Biosynthesis and release of cell wall-like glycoproteins during the vegetative cell cycle of *Chlamydomonas reinhardtii*. Planta 129: 183-189.

Lee, J. J., L. J. Crockett, J. Hagen and R. J. Stone. 1974. The taxonomic identity and physiological ecology of *Chlamydomonas hedleyi* sp. nov., algal flagellate symbiont from the foraminifer *Archaias angulatus*. Br. Phycol. J. 9: 407-422.

Lembi, C. A. 1975a. The fine structure of the flagellar apparatus of *Carteria*. J. Phycol. 11: 1-9.

Lembi, C. A. 1975b. A rhizoplast in *Carteria radiosa* (Chlorophyceae). J. Phycol. 11: 219-221.

Lembi, C. A. and N. J. Lang. 1965. Electron microscopy of *Carteria* and *Chlamydomonas*. Am. J. Bot. 52: 464-477.

Lembi, C. A. and P. L. Walne. 1971. Ultrastructure of pseudocilia in *Tetraspora lubrica* (Roth) Ag. J. Cell Sci. 9: 569-579.

Levine, R. P. 1974. Mutant studies on photosynthetic electron transport. In Stewart, W. D. P. [Ed.] *Algal Physiology and Biochemistry*. Univ. Calif. Press, Berkeley.

Levine, R. P. and H. A. Duram. 1973. The polypeptides of stacked and unstacked *Chlamydomonas reinhardtii* chloroplast membranes and their relation to photosystem II activity. Biochem. Biophys. Acta 325: 565-572.

Levine, R. P. and C. E. Folsome. 1959. The nuclear cycle of *Chlamydomonas reinhardtii*. Zeit. Verebungsl. 98: 192-202.

Levine, R. P. and U. W. Goodenough. 1970. The genetics of photosynthesis of the chloroplast in *Chlamydomonas reinhardtii*. Ann. Rev. Genet. 4: 397-408.

Lewin, R. A., M. J. Owen and J. L. Melnick. 1951. Cell-wall structure in *Chlamydomonas*. Exp. Cell Res. 2: 708-710.

Lewis, E., G. Munger, R. Watson and D. Wise. 1974. Life cycle of *Polytomella caeca* (Phytomonadida, Polyblepharidae). J. Protozool. 21: 647-649.

Lien, T. and G. Knutsen. 1976. Synchronized cultures of a cell wall-less mutant of *Chlamydomonas reinhardtii*. Arch. Microbiol. 108: 189-194.

Links, J., A. Verloop and E. Havinga. 1960. The carotenoids of *Polytoma uvella*. Arch. Mikrobiol. 36: 306-324.

Lloyd, N. D. H., D. T. Canvin and D. A. Culver. 1977. Photosynthesis and photorespiration in algae. Pl. Physiol. 59: 936-940.

Lloyd, D., D. A. Evans and S. E. Venables. 1968. Propionate assimilation in the flagellate *Polytomella caeca*. Biochem. J. 109: 897-907.

Loppes, R. 1968. Ethyl methanesulfonate; an effective mutagen in *Chlamydomonas reinhardtii*. Molec. Gen. Genet. 102: 229-231.

Loppes, R. and R. Deltour. 1977. A biochemical assay for detecting cell-wall mutants of *Chlamydomonas reinhardtii*. Pl. Sci. Lett. 8: 261-265.

Loppes, R., R. Matagne and P. J. Strukert. 1972. Complementation at the *Arg.*-7 locus in *Chlamydomonas reinhardtii*. Heredity 28: 239-251.

Lwoff, A. and H. Dusi. 1937. La pyrimidine et le thiazol, facteurs de croissance pour le flagellé *Polytomella caeca*. Compt. Rend. 205: 756.

McLean, R. J., C. J. Laurendi and R. M. Brown, Jr. 1974. The relationship of gamone to the mating reaction in *Chlamydomonas moewusii*. Proc. Nat. Acad. Sci. U.S.A. 71: 2610-2613.

Maguire, N. 1976. Mitotic and meiotic behavior of the chromosomes of the octet strain of *Chlamydomonas reinhardtii*. Genetica 46: 479-502.

Manners, D. J., G. A. Mercer, J. R. Stark and J. F. Ryley. 1965. Studies on the metabolism of the protozoa. The molecular structure of a starch-type polysaccharide from *Polytoma uvella*. Biochem. J. 97: 530-532.

Manton, I. 1964. Observations on the fine structure of the zoospore and young germling of *Stigeoclonium*. J. Exp. Bot. 15: 399-411.

Manton, I. 1965. Some phyletic implications of flagellar structure in plants. Adv. Bot. Res. 2: 1-34.

Manton, I., B. Clarke and A. D. Greenwood. 1955. Observations with the electron microscope on biciliate and quadriciliate zoospores in green algae. J. Exp. Bot. 6: 126-128.

Marano, F. 1976. Étude ultrastructurale de la division chez *Dunaliella*. J. Microsc. Biol. Cell 25: 279-282.

Marbach, I. and A. M. Mayer. 1970. Direction of phototaxis in *Chlamydomonas reinhardtii* and its relation to cell metabolism. Phycologia 9: 255-261.

Margulies, M. M. and A. Michaels. 1974. Ribosomes bound to chloroplast membranes in *Chlamydomonas reinhardtii*. J. Cell Biol. 60: 65-77.

Markowitz, M. M. 1978. Fine structure of the zoospore of *Oedocladium carolinianum* (Chlorophyta) with special reference to the flagellar apparatus. J. Phycol. 14: 289-302.

Martin, N. C. and U. W. Goodenough. 1975. Gametic differentiation in *Chlamydomonas reinhardtii*. I. Production of gametes and their fine structure. J. Cell Biol. 67: 587-605.

McVittie, A. C. 1969. Studies on flagella-less, stumpy and short flagellum mutants of *Chlamydomonas reinhardtii*. Proc. Roy. Soc. B. 173: 59-60.

McVittie, A. C. 1972. Flagellum mutants of *Chlamydomonas reinhardtii*. J. Gen. Microbiol. 71: 525-540.

McVittie, A. C. and D. R. Davies. 1971. The location of the Mendelian linkage groups in *Chlamydomonas reinhardtii*. Molec. Gen. Genet. 112: 225-228.

Melkonian, M. 1975. The fine structure of the zoospores of *Fritschiella tuberosa* Iyeng. (Chaetophorineae, Chlorophyceae) with special reference to the flagellar apparatus. Protoplasma 86: 391-404.

Melkonian, M. 1977. The flagellar root system of zoospores of the green alga *Chlorosarcinopsis* (Chlorosarcinales) as compared with *Chlamydomonas* (Volvocales). Plant Syst. Evol. 128: 79-88.

Menzel, D. W., J. Anderson and A. Randtke. 1970. Marine phytoplankton vary in their response to chlorinated hydrocarbons. Science 167: 1724-1726.

Merrett, M. J. and J. M. Lord. 1973. Glycolate formation and metabolism by algae. New Phytol. 72: 751-768.

Mesland, D. A. M. 1976. Mating in *Chlamydomonas eugametos*. A scanning electron microscopical study. Arch. Microbiol. 109: 31-35.

Mihara, S. and E. Hase. 1971. Studies on the vegetative life cycle of *Chlamydomonas reinhardtii* Dangeard in synchronous culture. I. Some characteristics of the cell cycle. Pl. Cell Physiol. 12: 225-236.

Miller, D. H., D. T. A. Lamport and M. Miller. 1972. Hydroxyproline heterooligosaccharides in *Chlamydomonas*. Science 176: 918-920.

Minami, S. A. and U. W. Goodenough. 1978. Novel glycopeptide synthesis induced by gametic cell fusion in *Chlamydomonas reinhardtii*. J. Cell Biol. 77: 165-181.

Moestrup, Ø. 1978. On the phylogenetic validity of the flagellar apparatus in green algae and other chlorophyll *a* and *b* containing plants. Bio Systems 10: 117-144.
Moore, J., M. H. Cantor, P. Sheeler and W. Kahn. 1970. The ultrastructure of *Polytomella agilis*. J. Protozool. 17: 671-676.
Morris, I. 1974. Nitrogen assimilation and protein synthesis. In Stewart, W. D. P. [Ed.] *Algal Physiology and Biochemistry*. Univ. Calif. Press, Berkeley.
Mosser, J. L., N. S. Fisher, T. C. Teng and C. F. Wurster. 1972a. Polychlorinated biphenyls: toxicity to certain phytoplankters. Science 175: 191-192.
Mosser, J. L., N. S. Fisher and C. F. Wurster. 1972b. Polychlorinated biphenyls and DDT alter species composition in mixed cultures of algae. Science 176: 533-535.
Nakamura, K., D. F. Bray, J. W. Costerton and E. B. Wagenaar. 1973. The eyespot of *Chlamydomonas eugametos*: a freeze-etch study. Can. J. Bot. 51: 817-819.
Nakamura, K., W. L. Hepler, E. J. Shaskin and D. W. Brooks. 1974. Antialgal activity of chloroplatinic acid on *Chlamydomonas eugametos*. Can. J. Bot. 52: 715-717.
Nakamura, K., D. F. Bray and E. B. Wagenaar. 1975. Ultrastructure of *Chlamydomonas eugametos* palmelloids induced by chloroplatinic acid treatment. J. Bacteriol. 121: 338-343.
Nichols, K. M. and R. Rikmenspoel. 1978a. Control of flagellar motion in *Chlamydomonas* and *Euglena* by mechanical microinjection of Mg^{2+} and Ca^{2+} and by electric current injection. J. Cell Sci. 29: 233-247.
Nichols, K. M. and R. Rikmenspoel. 1978b. Control of flagellar motility in *Euglena* and *Chlamydomonas*. Exp. Cell Res. 116: 333-340.
Nultsch, W. 1974. Movements. In Stewart, W. D. P. [Ed.] *Algal Physiology and Biochemistry*. Univ. Calif. Press, Berkeley.
Nultsch, W. 1977. Effect of external factors on phototaxis of *Chlamydomonas reinhardtii*. II. Carbon dioxide, oxygen and pH. Arch. Microgiol. 112: 179-185.
Nultsch, W. and G. Throm. 1975. Effect of external factors on phototaxis of *Chlamydomonas reinhardtii*. I. Light. Arch. Microbiol. 103: 175-179.
Nultsch, W., G. Throm and I. Von Rinscha. 1971. Phototaktische Untersuchen an *Chlamydomonas reinhardtii* Dangeard in homokontinuierlicher Kultur. Arch. Mikrobiol. 80: 351-369.
Olson, L. W. and G. Kochert. 1970. Ultrastructure of *Volvox carteri*. II. The kinetosome. Arch. Mikrobiol. 74: 31-40.
Palmer, C. M. 1962. Algae in Water Supplies. Public Health Survice Publ. No. 657., Robert A. Taft Sanitary Engineering Center, Cincinnati, 58 p
Palmer, C. M. 1969. A composite rating of algae tolerating organic pollution. J. Phycol. 5: 78-92.
Peterfi, L. S. and I. Manton. 1968. Observations with the electron microscope on *Asteromonas gracilis* Artari emend. (*Stephoptera gracilis* (Artari) Wisl.), with some comparative observations on *Dunaliella* sp. Br. Phycol. Bull. 3: 423-440.
Pickett-Heaps, J. D. 1974. The evolution of mitosis and the eukaryotic condition. BioSystems 6: 37-48.
Pickett-Heaps, J. D. 1975. *Green Algae. Structure, Reproduction and Evolution in Selected Genera*. Sinauer Assoc., Inc., Sunderland, Mass.
Porcella, R. A. and P. L. Walne. 1974. Ultrastructure of lorica and vegetative cells of *Dymorphococcus globosus* Bold and Starr. J. Phycol. Suppl. 10: 18.
Porter, K. G. 1973. Selective grazing and differential digestion of algae by zooplankton. Nature 244: 179-180.

Pringsheim, E. G. 1937. Beitrage zur Physiologie saprophytischen Algen und Flagellaten. II. *Polytoma* and *Polytomella*. Planta 26: 665-691.
Proctor, V. W. 1957. Studies of algal antibiosis using *Haematococcus* and *Chlamydomonas*. Limnol. Oceanogr. 2: 125-139.
Prouse, N. J., D. C. Gordon and P. D. Keizer. 1976. Effects of low concentrations of oil accommodated in sea water on the growth of unialgal marine phytoplankton cultures. J. Fish. Res. Bd. Can. 33: 810-818.
Randall, J. and D. Starling. 1976. Genetic determinants of flagellum phenotype in *Chlamydomonas reinhardtii*. In Lewin, R. A. [Ed.] *The Genetics of Algae*. Univ. Calif. Press, Berkeley.
Randall, J., T. Cavalier-Smith, A. McVittie, J. R. Warr and J. M. Hopkins. 1967. Developmental and control processes in the basal bodies and flagella of *Chlamydomonas reinhardtii*. Dev. Biol. Suppl. 1: 43-83.
Randall, J., J. R. Warr, J. M. Hopkins and A. VcVittie. 1964. A single gene mutation of *Chlamydomonas reinhardtii* affecting motility: a genetic and electron microscopic study. Nature 203: 912-914.
Ray, D. A., K. M. Solter and A. Gibor. 1978. Flagellar surface differentiation. Evidence for multiple sites involved in mating of *Chlamydomonas reinhardtii*. Exper. Cell Res. 114: 185-189.
Ringo, D. L. 1967a. Flagellar motion and fine structure of the flagellar apparatus in *Chlamydomonas*. J. Cell Biol. 33: 543-571.
Ringo, D. L. 1967b. The arrangement of subunits in flagellar fibers. J. Ultrastr. Res. 17: 266-277.
Ris, H. and W. Plaut. 1962. Ultrastructure of DNA-containing areas in the chloroplast of *Chlamydomonas*. J. Cell Biol. 13: 383-391.
Roberts, K. 1974. Crystalline glycoprotein cell walls of algae: their structure, composition and assembly. Phil. Trans. Roy. Soc. Lond. B 268: 129-146.
Roberts, K. and G. J. Hills. 1976. The crystalline glycoprotein cell wall of the green alga *Chlorogonium elongatum*: a structural analysis. J. Cell Sci. 21: 59-71.
Roberts, K., M. Gurney-Smith and G. J. Hills. 1972. Structure, composition and morphogenesis of the cell wall of *Chlamydomonas reinhardtii*. I. Ultrastructure and preliminary chemical analysis. J. Ultrastruct. Res. 40: 599-613.
Robreau, G. and Y. Le Gal. 1974. Conditions for isolating chromatin from *Chlamydomonas reinhardtii* CW 15. C.R. Hebd. Sean. Acad. Sci. Paris. 279: 1923-1926.
Rooney, D. W., B. C. Yen and D. J. Mikita. 1971. Synchronization of *Chlamydomonas* division with intermittent hypothermia. Exp. Cell Res. 65: 94-98.
Rosenbaum, J. L., J. E. Moulder and D. L. Ringo. 1969. Flagellar elongation and shortening in *Chlamydomonas*. J. Cell Biol. 41: 600-619.
Round, F. E. 1969. *Introduction to the Lower Plants*. Plenum Press, N.Y.
Sachs, T. and A. M. Mayer. 1961. Studies on the relation between metabolism and phototaxis of *Chlamydomonas snowiae*. Phycologia 1: 149-159.
Sager, R. 1954. Mendelian and non-Mendelian inheritance of streptomycin resistance in *Chlamydomonas*. Proc. Nat. Acad. Sci. U.S.A. 40: 356-362.
Sager, R. 1955. Inheritance in the green alga *Chlamydomonas reinhardtii*. Genetics 40: 476-489.
Sager, R. 1972. *Cytoplasmic Genes and Organelles*. Academic Press, N.Y.
Sager, R. and S. Granick. 1953. Nutritional studies with *Chlamydomonas reinhardtii*. Ann. N.Y. Acad. Sci. 56: 831-838.
Sager, R. and S. Granick. 1954. Nutritional control of sexuality in *Chlamydomonas reinhardtii*. J. Gen. Physiol. 37: 729-742.
Sager, R. and G. E. Palade. 1957. Structure and development of the chloroplast in *Chlamydomonas*. I. The normal green cell. J. Biophys. Biochem. Cytol. 3: 463-488.

Sager, R. and Z. Ramanis. 1973. The mechanism of maternal inheritance in *Chlamydomonas*: biochemical and genetic studies. Theoret. Applied Genet. 43: 101-108.

Salisbury, J. L. and G. L. Floyd. 1978a. Calcium-induced contraction of the rhizoplast of a quadriflagellate green alga. Science 202: 975-977.

Salisbury, J. L. and G. L. Floyd. 1978b. Molecular, enzymatic and ultrastructure characterization of the pyrenoid of the scaly green monad *Micromonas squamata*. J. Phycol. 14: 362-368.

Scagel, R. F., R. J. Bandoni, G. E. Rouse, W. B. Schofield, J. R. Stein and T. M. C. Taylor. 1965. *An Evolutionary Survey of the Plant Kingdom*. Wadsworth Publ. Co., Inc., Belmont, Calif.

Schelske, C. L. and E. F. Stoermer. 1971. Eutrophication, silica depletion, and predicted changes in algal quality in Lake Michigan. Science 173: 423-424.

Schlösser, U. 1976. Entwicklungstadien- und sippenspezifische Zellwand-Autolysine bei der Friesetzung von Fortpflanzungs zellen un der Gattung *Chlamydomonas*. Ber. Deutsch. Bot. Ges. 89: 1-56.

Schlösser, U., H. Sachs and D. G. Robinson. 1976. Isolation of protoplasts by means of a "species-specific" autolysine in *Chlamydomonas*. Protoplasma 88: 51-64.

Schopf, J. W. 1970. Pre-Cambrian micro-organisms and evolutionary events prior to the origin of vascular plants. Biol. Rev. 45: 319-352.

Schor, S. L., P. Siekewitz and G. E. Palade. 1970. Cyclic changes in thylakoid membranes of synchronized *Chlamydomonas reinhardtii*. Proc. Nat. Acad. Sci. U.S.A. 66: 174-180.

Schötz, F., H. Bathelt, C. Arnold and O. Schimmer. 1972. Die Architekur und Organisation der *Chlamydomonas*: Zelle Ergebnisse der Elektronenmikroskopie von Serienschnitten und der daraus resultierenden driedimensionalen Rekonstruktion. Protoplasma 75: 229-254.

Sheeler, P., M. Canton and J. Moore. 1970. Studies on the growth and encystment of *Polytomella agilis*. Protoplasma 69: 171-185.

Siersma, P. W. and K.-S. Chiang. 1971. Conservation and degradation of cytoplasmic and chloroplast ribosomes in *Chalmydomonas reinhardtii*. J. Mol. Biol. 58: 167-185.

Silverberg, B. A. 1975. An ultrastructural and cytochemical characterization of microbodies in the green algae. Protoplasma 83: 269-295.

Silverberg, B. A. and T. Sawa. 1974. Cytochemical localization of oxidase activities with diaminobenzidine in the green alga *Chlamydomonas dysosmos*. Protoplasma 81: 177-184.

Siu, C.-H., H. Swift and K.-S. Chiang. 1976a. Characterization of cytoplasmic and nuclear genomes in the colorless alga *Polytoma*. I. Ultrastructural analysis of organelles. J. Cell Biol. 69: 352-370.

Siu, C.-H., H. Swift and K.-S. Chiang. 1976b. II. General characterization of organelle nucleic acids. J. Cell Biol. 69: 371-382.

Siu, C.-H., H. Swift and K.-S. Chiang. 1976c. III. Ribosomal RNA cistrons of the nucleus and leucoplast. J. Cell Biol. 69: 383-392.

Smith, G. M. 1933. *The Fresh Water Algae of the United States*. McGraw-Hill Book Co., Inc., New York, N.Y.

Snell, W. J., W. L. Dentler, L. T. Haimo, L. I. Binder and J. L. Rosenbaum. 1974. Assembly of chick brain tubulin onto isolated basal bodies of *Chlamydomonas reinhardtii*. Science 185: 357-360.

Solter, K. M. and A. Gibor. 1977. Evidence for role of flagella as sensory transducers in mating of *Chlamydomonas reinhardtii*. Nature 265: 444-445.

Solter, K. M. and A. Gibor. 1978. The relationship between tonicity and flagellar length. Nature 275: 651-652.

Stabenau, H. 1974a. Localization of enzymes of glycolate metabolism in the alga *Chlorogonium elongatum*. Pl. Physiol. 54: 921-924.

Stabenau, H. 1974b. Verteilung von Microbody-Enzymen aus *Chlamydomonas* in Dichtegradienten. Planta. 118: 35-42.

Stabenau, H. and H. Beevers. 1974. Isolation and characterization of microbodies from the alga *Chlorogonium elongatum*. Pl. Physiol. 53: 866-869.

Stanley, R. A. and A. W. Naylor. 1972. Photosynthesis in Eurasian watermilfoil (*Myriophyllum spicatum* L.). Pl. Physiol. 50: 149-151.

Starr, R. C. 1978. The Culture Collection of Algae at the University of Texas at Austin. J. Phycol. Suppl. 14: 47-100.

Stavis, R. and R. Hirschberg. 1973. Phototaxis in *Chlamydomonas reinhardtii*. J. Cell Biol. 59: 367-377.

Stearns, M. E., J. A. Connolly and D. L. Brown. 1976. Cytoplasmic microtubule organizing centers isolated from *Polytomella agilis*. Science 191: 188-191.

Stewart, K. D. and K. R. Mattox. 1975. Comparative cytology, evolution and classification of the green algae with some consideration of the origin of other organisms with chlorophylls *a* and *b*. Bot. Rev. 41: 104-135.

Stewart, K. D., K. R. Mattox and C. D. Chandler. 1974. Mitosis and cytokinesis in *Platymonas subcordiformis*, a scaly green flagellate. J. Phycol. 10: 65-80.

Stoermer, E. F. 1975. Phytoplankton as indicators of water quality in the Laurentian Great Lakes. J. Phycol. Suppl. 11: 23.

Stross, R. G. 1963. Nitrate preference in *Haematococcus* as controlled by strain, age of inoculum and pH of the medium. Can. J. Microbiol. 9: 33-40.

Sueoka, N. 1960. Mitotic replication of deoxyribonucleic acid in *Chlamydomonas reinhardtii*. Proc. Nat. Acad. Sci. U.S.A 46: 83-91.

Sueoka, N., K. S. Chiang and J. R. Kates. 1967. Deoxyribonucleic acid replication in meiosis of *Chlamydomonas reinhardtii*. I. Isotopic transfer experiments with a strain producing eight zoospores. J. Molec. Biol. 25: 47-66.

Surzycki, S. 1971. Synchronously grown cultures of *Chlamydomonas reinhardtii*. In San Pietro, A. [Ed.] *Methods in Enzymology. Vol. 23, Part A*. Academic Press, N.Y.

Sze, P. and J. M. Kingsbury. 1972. Distribution of phytoplankton in a polluted saline lake, Onondaga Lake, New York. J. Phycol. 8: 25-37.

Tolbert, N. E. 1974. Photorespiration. In Stewart, W. D. P. [Ed.] *Algal Physiology and Biochemistry*. Blackwell, Oxford, England.

Triemer, R. E. and R. M. Brown, Jr. 1974. Cell division in *Chlamydomonas moewusii*. J. Phycol. 10: 419-433.

Triemer, R. E. and R. M. Brown, Jr. 1975. The ultrastructure of fertilization in *Chlamydomonas moewusii*. Protoplasma 84: 315-325.

Triemer, R. E. and R. M. Brown, Jr. 1976. Ultrastructure of meiosis in *Chlamydomonas reinhardtii*. Br. Phycol. J. 12: 23-44.

Vance, B. D. 1965. Composition and succession of cyanophycean water blooms. J. Phycol. 1: 81-86.

Vogel, S. L., H. L. Frisch and I. J. Gotham. 1978. Qualitative assay of dissolved amino acids and sugars excreted by *Chlamydomonas reinhardtii* (Chlorophyceae) and *Euglena gracilis* (Euglenophyceae). J. Phycol. 14: 403-406.

Walne, P. L. and H. J. Arnott. 1967. The comparative ultrastructure and possible function of eyespots: *Euglena granulata* and *Chlamydomonas eugametos*. Planta 77: 325-353.

Warner, F. D. 1972. Macromolecular organization of eukaryotic cilia and flagella. Adv. Cell Mol. Biol. 2: 193-235.

Warr, J. R., A. McVittie, J. Randall and J. M. Hopkins. 1966. Genetic control of flagellar structure in *Chlamydomonas reinhardtii*.Genet. Res. Camb. 7: 335-351.

Watanabe, T. and M. Flavin. 1973. Two types of ATPase from flagella of *Chlamydomonas reinhardtii*. Biochem. Biophys. Res. Commun. 52: 195-201.

Watanabe, T. and M. Flavin. 1976. Nucleotide-metabolizing enzymes in *Chlamydomonas* flagella. J. Biol. Chem. 251: 182-192.

Wawrik, F. 1972. Oogamie bei *Chlorogonium maximum* Skuja. Arch. Protistenk. 114: 489-494.

Weeks, D. P. and P. S. Collis. 1976. Induction of microtubule protein synthesis in *Chlamydomonas reinhardtii* during flagellar regeneration. Cell 9: 15-27.

Weiss, R. L., D. A. Goodenough and U. W. Goodenough. 1977. Membrane differentiations at sites specialized for cell fusion. J. Cell Biol. 72: 144-160.

Werz, G. and G. Kellner. 1970. Die Struktur des Golgi-Apparates bei gefriergeatzten *Dunaliella*-Zellen. Protoplasma 69: 351-364.

Wetzel, R. G. 1975. *Limnology*. W. B. Saunders Co., Philadelphia.

Whitford, L. A. 1964. The phytoplankton of an extremely oligotrophic lake. In Chandra, L. [Ed.] *Advancing Frontiers of Plant Sciences*. Instit. Adv. of Sci. and Culture, New Delhi, India.

Wiese, L. 1961. Gamone. Fortschr. Zool. 13: 119-145.

Wiese, L. 1965. On sexual agglutination and mating type substances (gamones) in isogamous heterothallic Chlamydomonads. I. Evidence of the identity of the gamones with the surface components responsible for sexual flagella contact. J. Phycol. 1: 46-74.

Wiese, L. 1967. The mating type substance of the male sex in *Chlamydomonas eugametos*. J. Phycol. Suppl. 3: 9.

Wiese, L. 1969. Algae. In Metz, C. B. and A. Monroy [Eds.] *Fertilization. Comparative Morphology, Biochemistry and Immunology. Vol. 2*. Academic Press, N.Y.

Wiese, L. 1974. Nature of sex specific glycoprotein agglutins in *Chlamydomonas*. Ann. N.Y. Acad. Sci. 234: 383-395.

Wiese, L. and P. C. Hayward. 1972. On sexual agglutination and mating type substances in isogamous dioecious Chlamydomonads. III. The sensitivity of sex cell contact to various enzymes. Amer. J. Bot. 59: 530-536.

Williams, L. G. 1964. Possible relationships between plankton-diatom species numbers and water-quality estimates. Ecology 45: 809-823.

Wise, D. L. 1955. Carbon sources for *Polytomella caeca*. J. Protozool. 2: 156-158.

Wise, D. L. 1959. Carbon nutrition and metabolism of *Polytomella caeca*. J. Protozool. 6: 19-23.

Wiseman, A., N. W. Gillham and J. E. Boynton. 1977. Nuclear mutations affecting mitochondrial structure and function in *Chlamydomonas*. J. Cell Biol. 73: 56-77.

Witman, G. B., K. Carlson, J. Berliner and J. L. Rosenbaum. 1972a. *Chlamydomonas* flagella. I. Isolation and electrophoretic analysis of microtubules, matrix, membranes and mastigonemes. J. Cell Biol. 54: 507-539.

Witman, G. B., K. Carlson and J. L. Rosenbaum. 1972b. *Chlamydomonas* flagella. II. The distribution of tubulins 1 and 2 in the outer doublet microtubules. J. Cell Biol. 54: 540-555.

Witman, G. B., J. Plummer and G. Sander. 1978. *Chlamydomonas* flagellar mutants lacking radial spokes and central tubules. J. Cell Biol. 76: 729-747.

Wright, R. T. 1964. Dynamics of a phytoplankton community in an ice-covered lake. Limnol. Oceanogr. 9: 163-178.

Wright, R. T. and J. E. Hobbie. 1966. Use of glucose and acetate by bacteria and algae in aquatic ecosystems. Ecology 47: 447-464.

Copyright 1980 by Elsevier North Holland, Inc.
Cox, ed. Phytoflagellates

UNICELLULAR CHLOROPHYTES - SNOW ALGAE

RONALD W. HOHAM
Department of Biology, Colgate University, Hamilton, New York 13346, U.S.A.

INTRODUCTION

The dominant algal vegetation found in snow belongs to the Chlamydomonadaceae in the Chlorophyta. Thus the dominant phytoflagellates found in snow are unicellular greens. However, other phytoflagellates are represented in the cryophilic environment including euglenoids, chrysophytes, dinoflagellates and cryptomonads, but these will not be dealt with in this chapter. In addition, the snow microflora may include blue-greens, xanthophytes, diatoms and non-flagellated greens. Snow algae were first recorded by Aristotle, but were later brought to the attention of the scientific world in 1819 (Bauer, 1819). Between 1819 and 1963, a number of publications covered the floristics of snow algae (Kol, 1968). Fukushima's (1963) comprehensive report on the ecology, life histories, and distribution of snow algae in Japan was a major contribution to the study of these organisms. Since 1963, snow algae have been studied from the standpoint of their ecology, life histories, a better knowledge of their taxonomy and distribution, ultrastructure and physiology. Even though much information remains to be learned about these organisms which occupy an extreme environment, considerable data has accumulated since the publication of Fukushima (1963).

General characteristics. A discussion of the general characteristics of snow algae belonging to the Chlamydomonadaceae includes the number and insertion of flagella, pigments, cell walls, and food reserves. The number of flagella is 2 or 4 per cell (2 in *Chloromonas*, *Chlamydomonas* and *Smithsonimonas* and 4 in *Carteria* and *Chlainomonas*). All five genera are placed in the Chlamydomonadaceae of the Chlorophyta (Bourrelly, 1966; Kol, 1968). The flagella are inserted anteriorly, one on each side of an apical dome or papilla in the biflagellate genera. In *Chlainomonas*, flagella are inserted at 90° angles with respect to the one on either side (Fig. 1; Hoham, 1974a, b). Also, the quadriflagellate planozygotes of *Chloromonas brevispina* (Fritsch) Hoh., Roem. and Mull. have a flagellar insertion similar to that described for *Chlainomonas* or *Tetratoma* (Hoham *et al.*, 1979).

The red pigment in *Chlamydomonas nivalis* (Bau.) Wille was identified by Viala (1966) as the xanthophyll, astaxanthin. Czygan (1970) reported that the

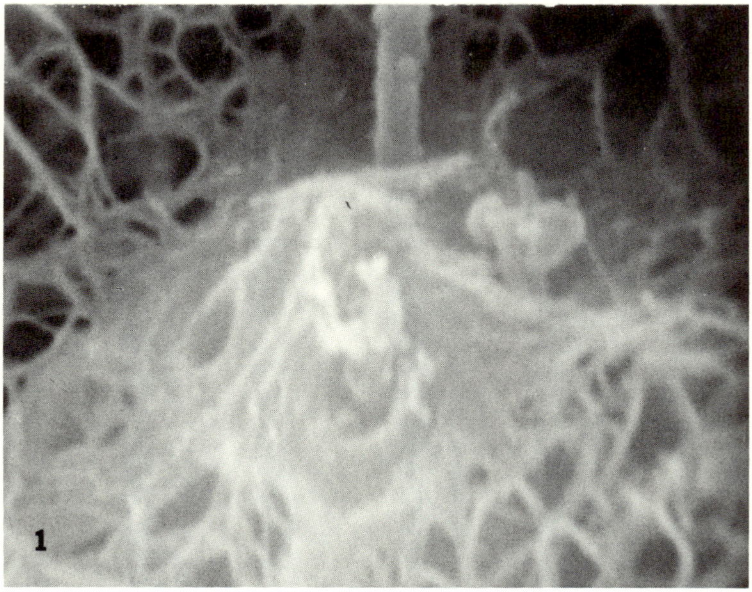

Fig. 1. Flagellar apparatus in *Chlainomonas kolii*. Three of four flagella shown (two severed above apertures) (X 14,200) (After Hoham and Rosowski).

red pigments in *Chlamydomonas nivalis* belong to the keto-carotenoids. Czygan (1970) concluded that polyene biosynthesis paralleled chlorophyll decomposition and both were related to nitrogen deficiency. In the laboratory, Czygan (1970) showed a pigment shift in resting spores of *Chlamydomonas nivalis* which were placed in nutrient solution for four weeks. A general disappearance of red carotenoids was noted (echinenon, canthaxanthin, astaxanthin, astacin) and were replaced by increasing concentrations of β- and α- carotene, lutein, zeaxanthin, violaxanthin, neoxanthin, as well as chlorophylls *a* and *b*. However, Czygan's experiments were conducted under a maximum of 6000 lux in the laboratory. Mosser *et al.* (1977) recorded a concentration of 86,000 lux in the field for *Chlamydomonas nivalis*. This raises the question whether or not the pigment shift was due to the added nitrogen in the nutrient solution or a lower light intensity in the laboratory compared to that found in nature. Also, the red pigment, astaxanthin, probably exists in *Chlainomonas rubra* (Stein and Brooke) Hoh. and *C. kolii* (Hardy and Curl) Hoh., but this needs verification. Other secondary carotenoids (yellows, oranges, pinks) characteristic of zygotes of *Chloromonas nivalis* (Chod.) Hoh. and Mull. (Hoham and Mullet, 1977, 1978) and *Chloromonas brevispina* (Hoham *et al.*, 1979) have not been identified. These carotenoids are dissolved in lipids at the polar regions of the zygote, and their synthesis appears to be related to increasing light intensity (Hoham and Mullet, 1977; Hoham *et al.*, 1979). In another species of snow algae, *Cryocystis granulosa* Kol cell type (a stage in the life cycle of a volvocalean alga, Hoham, unpubl.), cells when first-formed are green, then become yellow-green and eventually orange. The orange carotenoids are dissolved in lipid granules throughout the cell. In this species, the development of the carotenoids appears to be less related to light intensity and more to the aging of the cells.

The wall structure has not been examined for its chemical composition in any snow alga, but a distinct wall is present in *Chloromonas*, *Chlamydomonas* and *Carteria*. The "wall structure" in *Chlainomonas kolii* is most peculiar because the protoplast is surrounded by a gelatinous sheath and two external envelopes. The inner envelope appears somewhat rough and the outer envelope is comprised of a mosaic pattern (Fig. 2). When the outer envelope is cast off, it collapses.

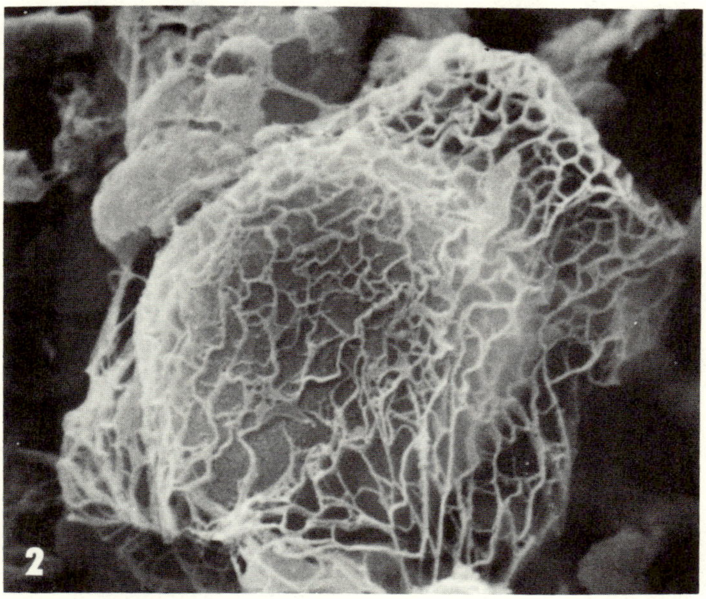

Fig. 2. Two-envelope structure in *Chlainomonas kolii*. Outer envelope a mosaic and inner envelope is somewhat rough (X 5,750) (After Hoham and Rosowski).

Perhaps it is proteinaceous, but this has not been verified (Hoham, 1974b). In *Chlainomonas rubra*, there is only a single envelope surrounding the gelatinous material (Stein and Brooke, 1964; Hoham, 1974a). The wall structure of snow algal zygotes in the Chlamydomonadaceae is ornamented. The primary wall may be comprised of flanges, protuberances or spines, and the inner wall appears to be smooth (Kol, 1968; Hoham, 1975b; Hoham and Mullet, 1977; Hoham *et al.*, 1979). The flanges in *Chloromonas nivalis* may be made of parallel plates (Fig. 3; Hoham and Mullet, 1977), and the spines in *Chloromonas brevispina* usually have undulating ridges (Fig. 4; Hoham *et al.*, 1979).

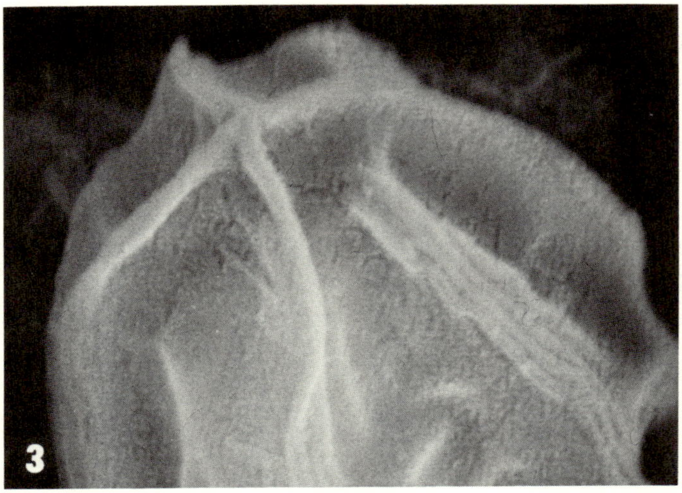

Fig. 3. Zygote of *Chloromonas nivalis*. Flanges of wall may be comprised of several plates (X 7,750) (After Hoham and Mullet; permission of *Phycologia*).

The common food reserves in snow algae belonging to the Chlamydomonadaceae are starch and lipid (Hoham, 1975b; Hoham and Mullet, 1977; Hoham *et al.*, 1979). Starch is commonly found in rapidly dividing cells in nature or in newly formed zygotes. Lipids are often found in mature zygotes and zygospores or in vegetative cells that have been exposed to subfreezing conditions in nature (Hoham, 1975a) or in the laboratory (Hoham, 1975b). As mentioned above, there may be secondary pigments (carotenoids) that develop in the lipid in maturing zygotes.

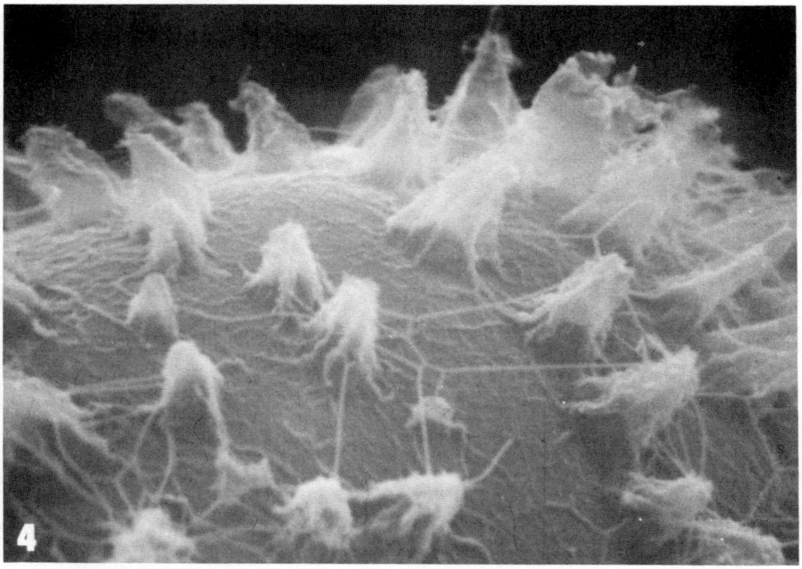

Fig. 4. Zygote of *Chloromonas brevispina*. Spines of wall with undulating ridges (X 14,000) (After Hoham, Roemer and Mullet; permission of *Phycologia*).

Ultrastructure. Only two papers have been published pertaining to the ultrastructure of snow algae in the Chlamydomonadaceae. Fjerdingstad *et al.* (1974) published electron micrographs of *Chlamydomonas nivalis* (East Greenland material). One micrograph was of vegetative cells and the other showed cell division. These micrographs were quite small and distinctive organelles were not clear. However, an undulating membrane was indicated surrounding the vegetative cells of *Chlamydomonas nivalis*. Stein and Bisalputra (1969) reported on crystalline bodies in the chloroplast of an unidentified species of *Chloromonas* from snow (the organism was published as a *Chlamydomonas* sp., but the cells lacked pyrenoids). The composition of these crystalline inclusions was suggested to be proteinaceous. The inclusions were hexagonal and varied in size from 0.2 - 0.9 μm (Fig. 5). Larger inclusions were observed in older cultures and in those grown in mineral medium at 0° or 5°C. The inclusions were generally smaller in cultures grown at 10°C. This was the first report of such an inclusion in an algal chloroplast. There appeared to be no structural relationship between the chloroplast thylakoids and the inclusion. There also ap-

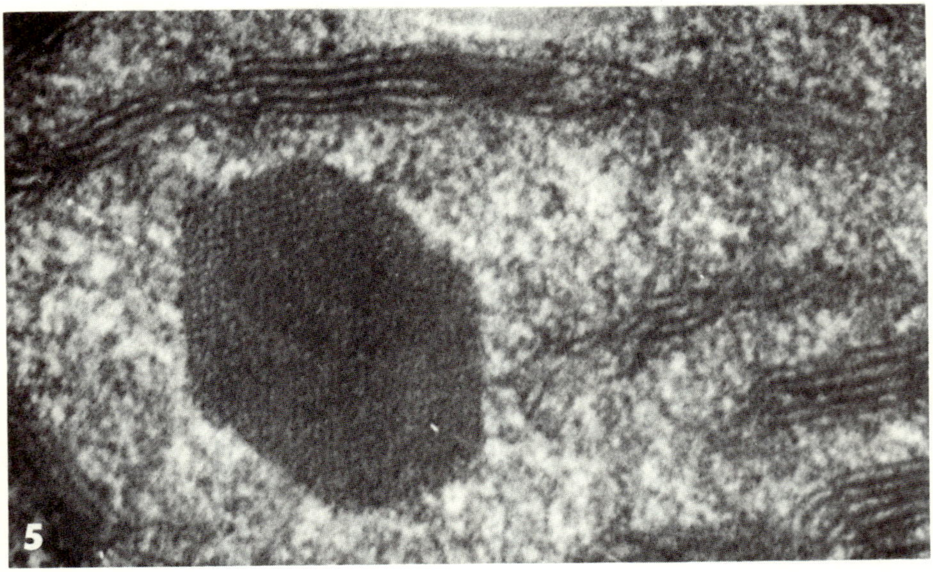

Fig. 5. Crystalline inclusion in chloroplast of *Chloromonas* sp., (published as *Chlamydomonas* sp.) (X 113,900) (After Stein and Bisalputra; permission of NRC of Canada, *Canadian Journal of Botany*).

peared to be a less dense chloroplast matrix surrounding the inclusion, and the inclusion was not membrane bound. It was suggested that the crystalline bodies were involved in the storage of photosynthetates.

Movement patterns. Phototactic response is well documented in *Chloromonas pichinchae* (Lagerh.) Wille, a species lacking eyespots (Hoham, 1975b). Vegetative cells move through liquid meltwater surrounding the snow crystals migrating towards the surface of snowbanks, usually at dawn and dusk when light intensities are minimum. Due to higher light intensities recorded at midday, cells become negatively phototactic, migrating to 10 to 15 cm below the snowbank surface. Lateral migrations of populations of *C. pichinchae* were recorded as much as 20 cm in 24 h. The particular wavelength and intensity of light responsible for these behavioral patterns has not been determined. However, negative phototaxis was recorded in the laboratory at 400 ft. c. under fluorescent lighting. Phototactic responses were also recorded for *Chlamydomonas nivalis* (Hardy and Curl, 1972), *Chlainomonas rubra* (Hoham, 1974a), *Chloromonas nivalis* (Hoham and Mullet, 1977) and *Chloromonas brevispina* (Hoham *et al.*, 1979).

Vegetative cells of *C. pichinchae* move with their anterior ends directed forward except in aging cultures in the laboratory where cells move with their posterior ends directed forward (Hoham, 1975b). Planozygotes of *Chloromonas nivalis* and *C. brevispina* move with their posterior ends directed forward (Hoham and Mullet, 1977; Hoham *et al.*, 1979). In the two species of the quadriflagellate *Chlainomonas* recorded from snow, *C. rubra* and *C. kolii*, cells display jerk-type motions with little distance gained through such movements.

Life histories. The life histories of the volvocalean algae studied thus far are quite complex. It is now known that a number of snow algae that were once assigned to the Chlorococcales are zygotic stages of different species of *Chloromonas*. Included in this list are *Scotiella* (Hoham, 1975b; Hoham and Mullet, 1977, 1978), *Cryocystis*, *Cryodactylon*, *Oocystis* and *Trochiscia* (Hoham *et al.*, 1979). There is also evidence that one species of *Carteria* described from snow is the planozygote of a *Chloromonas* (Hoham, unpubl.). In one life history study (Hoham *et al.*, 1979), the zygotes of *Chloromonas brevispina* were identical to seven taxa previously described from snow as belonging to the Chlorococcales (Fig. 6). This is interesting because the zygotes of *Chloromonas* and *Chlamydomonas* have not been studied in most species (Huber-Pestalozzi, 1961; Ettl, 1970, 1976). It is still not clear as to how much variation occurs in the life cycle of *Chlamydomonas nivalis*. A number of life histories have been suggested (Wille, 1903; Fukushima, 1963; Garric, 1965; Kobayashi, 1967; Stein and Amundsen, 1967; Kol, 1968; Hoham, 1971; Kawecka and Drake, 1978). Historically, this organism was placed in ten different genera until it was recognized as a species of *Chlamydomonas* by Wille (1903). Kol (1968) suggested that one must consider *C. nivalis* as a collective species covering several different kinds of *Chlamydomonas*. This would appear to be the case since the papers mentioned above include at least 15 different morphological cell types for this one species.

In those species of snow algae where the life cycles have been studied critically, the active or vegetative phase may last less than one week in the field during any one year. Snow algal blooms are initiated when air temperatures remain above freezing for several consecutive days, a meltwater develops in the snowbank, nutrients and dissolved gases are available to the algae, and light penetrates through the snow (Pollock, 1970; Curl *et al.*, 1972; Thomas, 1972; Hoham, 1974a, b, 1975b; Hoham and Mullet, 1977; Hoham *et al.*, 1979). With a combination of the above factors, the resting zygospores germinate at

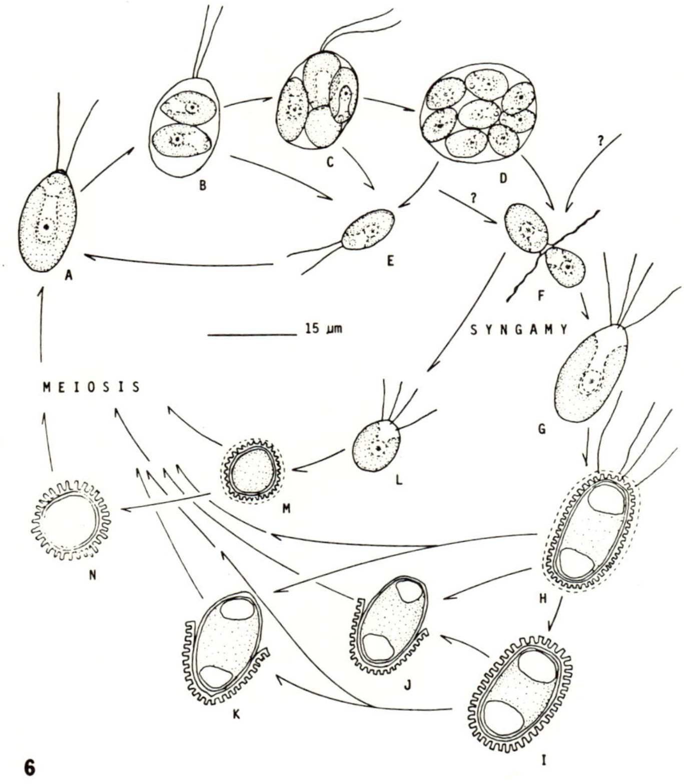

Fig. 6. Life cycle of *Chloromonas brevispina*. Vegetative cell (A). Cleavage of vegetative cells (B-D). Zoospore (E). Anisogamy (F). Larger planozygote formed first (G). Smaller planozygote formed later (L). Zygotes identical to taxa previously placed in Chlorococcales, *Cryocystis brevispina*, H; *Cryocystis japonica*, I; *Cryodactylon glaciale*, J; *Oocystis lacustris* f. *nivalis*, K; *Trochiscia nivalis*, M; *Trochiscia rubra* (if pink or red) or *Trochiscia cryophila* (if green) N. (After Hoham, Roemer and Mullet; permission of *Phycologia*.)

the snow-soil interface (old snow-new snow interface in persistent snowfields) releasing zoospores that swim in the liquid meltwater towards the surface of the snowbank causing visible blooms only a few days after germination (Fig. 7). The zoospores become the vegetative phase of the life cycle dividing to form additional daughter cells. Some daughter cells are asexual (zoospores) and others are sexual (gametes). The stages leading to syngamy have been discussed

Fig. 7. Vertical cut through snowbank at 1387 m, south of Cayuse Pass, Washington, showing diffuse areas and horizontal bands of snow algae. (After Hoham, Roemer and Mullet; permission of *Phycologia*).

in a number of snow algae, *Chlamydomonas nivalis* (Fukushima, 1963; Kawecka and Drake, 1978), *Chloromonas pichinchae* (Hoham, 1975b), *Chloromonas nivalis* (Hoham and Mullet, 1977, 1978) and *Chloromonas brevispina* (Hoham *et al.*, 1979). It is not understood if nutrient depletion in the liquid meltwater initiates syngamy, but the formation of new zygotes usually occurs within a few days after germination of the previous year's zygotes (Hoham, 1975b). Unless the investigator visits the snowbank during the germination process, the only stages of the life cycle likely to be found will be the new zygotes. This explains in part why so many zygotes were designated in the past to species status and placed in the Chlorococcales. The newly formed zygotes persist in the snow until the banks melt away, and the zygotes eventually come to rest on the soil. In persistent snowfields, the zygotes remain dormant on the old snow. In either situation, the snow algae are covered over with new snow during autumn, winter, and spring where they lay dormant until germination takes place again the following spring.

Mitosis and meiosis. Neither mitosis nor meiosis have been examined at the ultrastructural level in snow algae in the Chlamydomonadaceae. The processes and time of day that cell division occurs have been documented in a number of species. Formation of gametes occurs at the beginning of the night cycle in *Chloromonas pichinchae* (Hoham, 1975b), during the middle to end of the night cycle in *Chloromonas brevispina* (Hoham *et al.*, 1979) and during the end of the night cycle in *Chlamydomonas nivalis* var. *kobayashi* Fuk. (Fukushima, 1963). In *Chlainomonas rubra*, cell division occurs at the beginning of the night cycle. However, cell division in this species was reported to be external rather than internal (Hoham, 1974a). It was suggested that some of these external cell divisions might lead to autogamy. In *Chlamydomonas nivalis*, several types of sexual reproduction were observed (Kawecka and Drake, 1978). One type was somewhat similar to the autogamy suggested in *Chlainomonas rubra*.

Meiosis occurs in several species of snow algae 2 to 4 months after zygotes are produced. In those species studied, it was observed that the freezing process after a prolonged dormancy expedites meiosis in a large percentage of zygotes (*Chlainomonas rubra*, Hoham, 1974a; *Chloromonas pichinchae*, Hoham, 1975b; *Chloromonas nivalis*, Hoham and Mullet, 1977; *Chloromonas brevispina*, Hoham *et al.*, 1979).

Physiological ecology and habitat. One question frequently raised pertains to whether or not snow algae are optimally suited for growing in temperatures at or near 0°C. Most studies attempting to answer this question have been done in the laboratory growing the algae in defined media, using a light regime and photoperiod, and exposing the organisms to different temperatures. Hoham (1975a) suggested that true snow algae have optimum growth below 10°C and do not grow at temperatures above 10°C. He selected 10°C as the maximum temperature because some snow algae (*Raphidonema nivale* Lagerh., Hoham, 1975a) may continue dividing during summer months on the soil after the snow melted. The surface temperature of the soil at midday under tree canopies was recorded at 13-15°C. In the laboratory, *R. nivale* grew well at 1-15°C, but died at 20°C, and is probably a facultative cryophile (able to grow in snow and on soil). However, *Chloromonas pichinchae* would not grow at temperatures above 10°C in the laboratory, and the vegetative condition was recognized only at temperatures between 1-5°C. Therefore, *C. pichinchae* is probably an obligative cryophile (growing only in snow), not being able to survive in the vegetative condition on the soil during summer months at 13-15°C (North Fork Teanaway Valley, Washington, 1387 M). Morita (1975) defined a strict or true psychrophile (cryophilic organism) with an optimum temperature for growth of 15°C or lower, and a minimal temperature for growth at 0°C or lower. Morita's work pertains primarily to bacteria. A number of isolates in the Chlamydomonadaceae have optimum temperatures as determined in the laboratory that would fall into either definition listed above. *Chloromonas pichinchae* grew at temperatures of 1 and 5°C, but cultures became palmelloid at temperatures approaching 10°C (Hoham, 1975a, b). *Chloromonas* sp. (published as *Chlamydomonas* sp.) was reported to have an optimum temperature of 5°C (Stein and Bisalputra, 1969). Under laboratory conditions using a cooling stage, cells of *Chlainomonas rubra* and *C. kolii* shed their flagella at temperatures above 4°C, and cells of *C. rubra* were damaged when frozen at -1°C showing a punctured plasmalemma and outer envelope (Hoham, 1975a). Under these conditions, cells of *C. rubra* often released large lipid bodies external to the protoplast. Using laboratory cultures, Czygan (1970) reported optimum temperatures of 4°C for several strains of *Chlamydomonas nivalis*.

Mosser *et al.* (1977) using ^{14}C ($^{14}C\text{-}HCO_3^-$ and $^{14}CO_2$) in the field, reported that *Chlamydomonas nivalis* photosynthesized optimally at 10 or 20°C but retained substantial activity at temperatures as low as 0 or -3°C. Using ^{14}C, primary productivity has been determined in snow algal populations. Similar amounts of μgC fixed per mm^3 cell volume per h were obtained using *Chlamydomonas nivalis* (0.05-0.97, Mosser *et al.*, 1977) or mixed populations of snow algae containing *Chlamydomonas* sp. (0.04-1.85, Komárek *et al.*, 1973) or *Chlamydomonas nivalis* (0.002-0.86, Fogg, 1967). Thomas (1972) reported higher amounts of fixed carbon (5.7-34.2) for *C. nivalis*, but he questioned these values which were based on what he thought was a high assay of CO_2 concentration in the snow meltwater.

Using proton-induced x-ray emission spectroscopy, the accumulation of trace metals in snow has been reported from Greenland, Washington, U.S.A. and Spitzbergen (Erik Fjerdingstad, 1973; Erik Fjerdingstad *et al.*, 1974; Hoham *et al.*, 1977; Einer Fjerdingstad *et al.*, 1978). High levels of trace metals were recorded from snow near Kulusuk, Greenland: Si (69 ppm), P (22), S (30), Cl (56), K (30), Ca (20), Ti (5), V (0.6), Mn (0.4), Fe (38), Ni (0.07), Cu (0.4), Zn (0.9), Se (0.04), Y (0.04), Mo (0.04) and Ba (2) (Erik Fjerdingstad, 1973; Erik Fjerdingstad *et al.*, 1974). In a similar study, lesser amounts of the same trace metals were reported from snow in East Greenland (Hudson Land) and Spitzbergen (Sveagruva coal mine region) (Einer Fjerdingstad *et al.*, 1978). The alga, *Chlamydomonas nivalis*, and particulate matter were dried and the concentrations of the trace metals within them were compared to those recorded in the snow meltwater (Table 1).

Table 1. Concentration ratio of some heavy metals in the alga, *Chlamydomonas nivalis* and particulate matter, compared to values recorded from snow meltwater (Einer Fjerdingstad *et al.*, 1978).

Element	Spitzbergen	East Greenland
S	317	840
Cl	27	114
Cr	> 5000	> 3000
Ca	3800	190
Ti	6800	11200
Mn	5400	4600
Ni	> 4200	140
Cu	2300	2200
Zn	> 900	> 500
Br	92	
Rb	2800	6500
Sr	5700	1000
Y	> 4600	> 3500
Zr	18000	> 18000
Nb	> 2300	
Pb	1600	> 4400

In another study, high concentrations of several trace metals were recorded from each of three areas, Washington, U.S.A., East Greenland and Spitzbergen (Hoham *et al.*, 1977). This study raised the question of the probability that different species of snow algae or different strains of the same species may have different nutritional requirements.

Dust and coniferous litter covering snowbanks are potentially important sources of nutrition for algae living in snow (Hoham, 1976). Metabolites from conifers are leached from trees when precipitation falls through the overlying branches (Tamm, 1950, 1951; Tukey and Morgan, 1964). In addition, dust nuclei in the atmosphere around which snowflakes form provide nutrients for algae at the time when a meltwater forms throughout the snowbank (Pollock, 1970). Some nutrients may be supplied to the lower 15 cm of the snowbank by capillarity from the soil, but the major nutrient source is the debris at the surface of the snowbank since water flow in the snowbank is either gravitational or lateral from the surface to ground level (Hoham, 1975b; Hoham and Mullet, 1977; Hoham *et al.*, 1979).

One snow alga in the Chlamydomonadaceae, *Chloromonas pichinchae*, was investigated for the effects that extracts from coniferous litter and different snow meltwaters would have on its growth and development (Hoham, 1976). The alga was collected from snow under coniferous trees near Mt. Stuart, Washington, and isolated into axenic culture. Laboratory experiments indicated positive growth

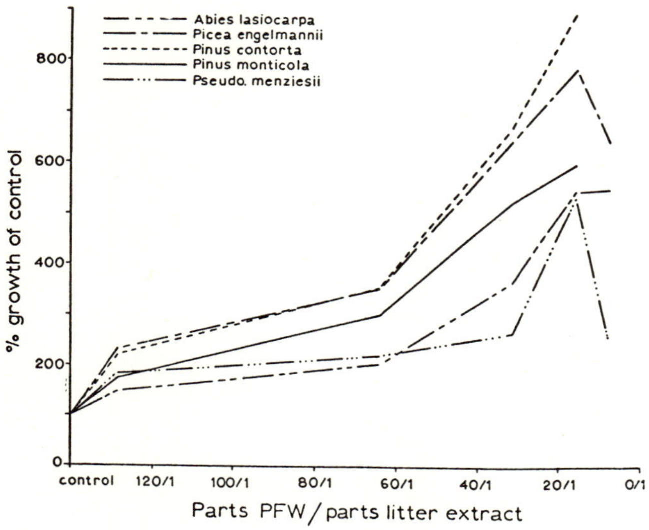

Fig. 8. Effect of leaf litter extract upon growth of *Chloromonas pichinchae* after 13 days. Control = PFW medium only. (After Hoham; permission of *Artic and Alpine Research*, University of Colorado).

responses for *C. pichinchae* grown in extracts of coniferous leaves (Fig. 8), bark (Fig. 9) and pollen, and the species of conifers used were the same as those found adjacent to the snowbanks studied. *Chloromonas pichinchae* responded differently to snow meltwater collected at three elevations near Mt. Stuart. The alga grew best in snow meltwater collected under a coniferous tree canopy (Fig. 10), and this sample contained the highest concentration of inorganic PO_4-P. Similar coniferous extract experiments were conducted on another snow alga collected from the same snowbanks near Mt. Stuart, Washington (Hoham, 1976). *Raphidonema nivale* Lagerh., a ulotrichalean green, was inhibited by some of the coniferous extracts, and the results of the laboratory experiments showed best growth for *C. pichinchae*. These results corresponded to observations made in the field where *C. pichinchae* was much more abundant than *R.*

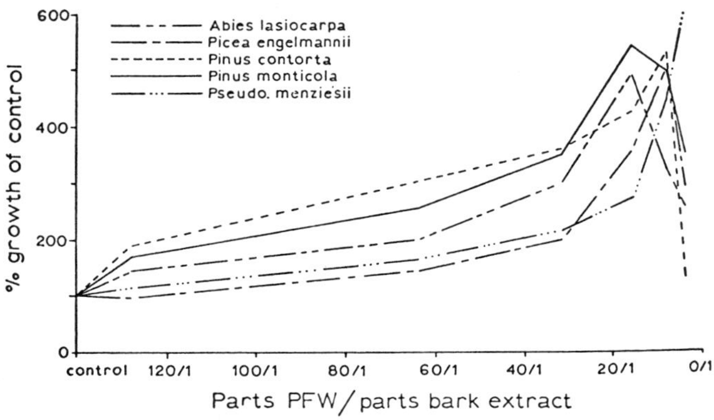

Fig. 9. Effect of bark extract upon growth of *Chloromonas pichinchae* after 13 days. Control = PFW medium only. (After Hoham; permission of *Arctic and Alpine Research*, University of Colorado).

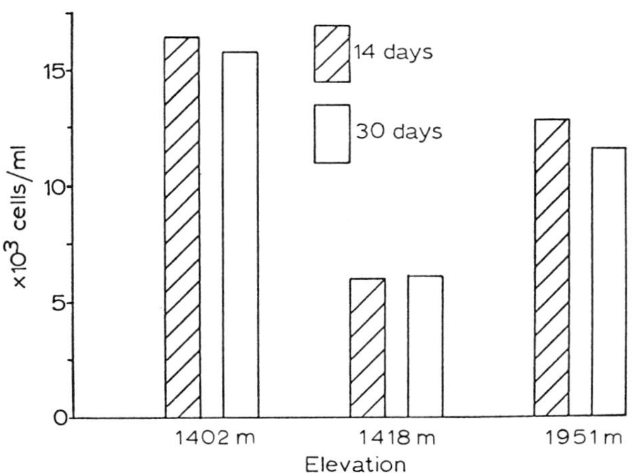

Fig. 10. Effect of snow meltwater from different localities near Mt. Stuart, Washington, upon growth of *Chloromonas pichinchae*. (After Hoham; permission of *Arctic and Alpine Research*, University of Colorado).

nivale in snowbanks located under coniferous tree canopies.

A variety of inorganic and organic metabolites, including forms of nitrogen and phosphorus, are leached from pine, spruce and Douglas fir during falling precipitation (Tamm, 1950, 1951; Tukey and Morgan, 1964). These metabolites could accumulate in underlying snowbanks. The snow alga, *Chloromonas pichinchae*, has the capacity to utilize inorganic and organic forms of phosphorus and nitrogen for growth (Hoham, 1971). Under axenic conditions using two-week old starved cultures in the laboratory, *C. pichinchae* assimilated inorganic NH_4-N (NH_4Cl; Fig. 11), organic L-glutamic acid-N ($C_5H_9NO_4$; Fig. 12), inorganic PO_4-P (KH_2PO_4:K_2HPO_4; 1:1; Fig. 13) and organic PO_4-P ($Na_2C_3H_5[OH]_2PO_4$; Fig. 14).

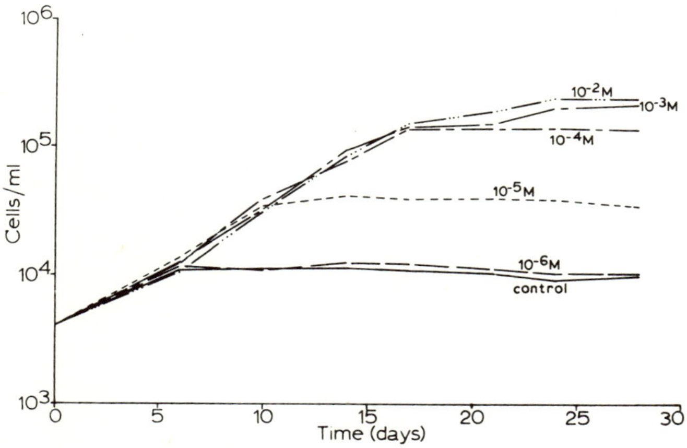

Fig. 11. Effect of NH_4-N (NH_4Cl) upon growth of *Chloromonas pichinchae*. Control = PFW - inorganic-N. (After Hoham, 1971).

The distribution of snow algal species may be related to exposure, shading, elevation, snow duration and nutrients. In the American Southwest, green snow was usually found several centimeters below the surface of snowbanks in the forests, orange snow at or near the surface of snowbanks adjacent to trees, and red snow was characteristic of the open exposures, often above timberline (Hoham and Blinn, 1979). Green and orange snow were also dominant in the forested areas and red snow in the open exposures of the American Pacific Northwest (Hoham, 1971), Japan (Fukushima, 1963) and Europe (Kol, 1968). Pollock (1970) concluded that green snow was characteristic of shaded snowbanks, red snow received sunlight for at least one-half the day, and orange snow received

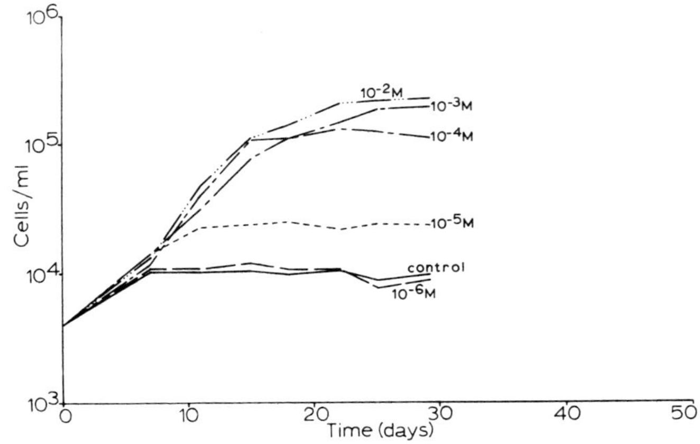

Fig. 12. Effect of glutamic acid-N ($C_5H_9NO_4$) upon growth of *Chloromonas pichinchae*. Control = PFW - inorganic-N. (After Hoham, 1971).

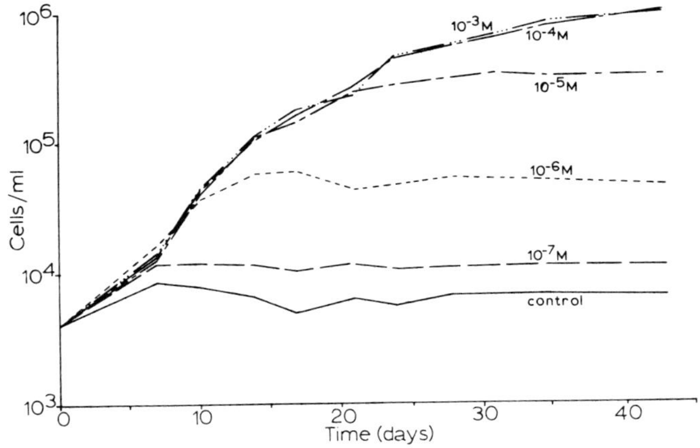

Fig. 13. Effect of inorganic PO_4-P (KH_2PO_4:K_2 HPO_4; 1:1) upon growth of *Chloromonas pichinchae*. Control = PFW - PO_4. (After Hoham, 1971).

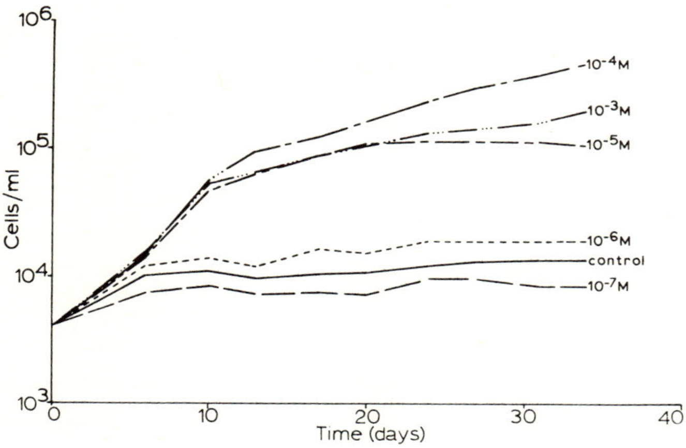

Fig. 14. Effect of organic PO_4-P ($Na_2C_3H_5[OH]_2PO_4$) upon growth of *Chloromonas pichinchae*. Control = PFW - PO_4. (After Hoham, 1971).

an intermediate amount of sunlight.

In the American Southwest, species distribution was related mostly to exposure and shading, but also to elevation, nutrients and snow duration (Hoham and Blinn, 1979). *Chloromonas nivalis* was the dominant species in the Southwest causing green to orange coloration in mostly shaded regions of the forests. *Cryocystis granulosa* Kol cell type (a stage in the life cycle of a volvocalean alga, Hoham, unpubl.) caused orange to orange-red coloration of snow near tree canopies, but usually received more irradiation than populations of *Chloromonas nivalis*. *Cryocystis granulosa* cell type causing orange snow and *Carteria nivale* Kol - *Scotiella polyptera* Fritsch cell types (stages in the life cycle of a biflagellate volvocalean alga, Hoham, unpubl.) causing green snow were commonly found in shallow, residual snowbanks (mostly under 30 cm deep) near tree canopies. The tree canopies were more open above C. *granulosa* when compared to *Carteria nivale-Scotiella polyptera*, thus C. *granulosa* appeared to tolerate more irradiation. *Chloromonas nivalis* and *Cryosystis granulosa* frequently grew together in the same snowbanks, but often one dominated at a particular collecting site. *Cryocystis granulosa* dominated at lower elevations and *Chloromonas nivalis* at higher elevations on several mountains in Arizona (Mt. Union,

2377-2407 m; Bill Williams Mtn., 2469-2515 m; San Francisco Mtns., 2499-2682 m). *Chlamydomonas nivalis* causing red snow dominated in the open exposures, but was usually replaced by *Chloromonas nivalis* causing orange to green snow in more shaded areas adjacent to the open exposures.

Kol (1934a, b, 1935a, b, 1939, 1941) and Sieminska (1951a, b) proposed that species of snow algae were restricted to certain snows by pH. They proposed that snowfields should be divided into silicotrophic (pH 4.5-5.8) and calcitrophic (pH 6.0-6.5). Later, Kol (1968) proposed an intermediate type of snowfield (pH 5.7-6.2). These proposals suggested that the underlying rocks influenced the pH in the overlying snow and the kinds of species of snow algae that could grow there. Kol (1968) suggested that several species of *Chloromonas* and *Chlamydomonas* were restricted to one or two types of the snow. Since the movement of snow meltwater is gravitational or lateral (Hoham, 1975b; Hoham and Mullet, 1977; Hoham *et al.*, 1979), it is difficult to see how underlying rock could influence the pH of overlying snow. Fukushima (1963) tested Kol's silicotrophic-calcitrophic theory in Japan and found that species of snow algae were distributed by exposure to sunlight or shading and not to underlying rocks. Fukushima (1963) reported green snow and red snow over neutral rocks (probably Kol's intermediate type, pH 5.7-6.2), and found the green snow in shaded areas and the red snow in open exposures. Garric (1965) also concluded that the silicotrophic-calcitrophic theory of cryoflora types was not valid for distribution of snow algae in the Pacific Northwest (U.S.A.). He suggested that the degree of illumination was probably the most critical factor.

It has not been determined if snow algal species are restricted to a certain pH range. Hoham (1971) reported pH readings of 4.0-6.0 in Pacific Northwest (U.S.A.) snow where readings tended to be more acid under coniferous trees (pH often 4.0-4.5). A species of snow algae isolated from snow under these trees was tested in the laboratory for optimum pH with other environmental conditions kept constant. *Chloromonas pichinchae* grew best at pH 6.0 in a range from 4.0-7.0 (Fig. 15). The pH range used included that found in nature and in the defined growth medium (PFW). The pH points were readjusted every 48 h over the 23 day period and pH fluctuations were never higher than 0.5 (usually 0.1). The results indicated that *C. pichinchae* grew optimally in an acid pH (6.0) in the laboratory, and it grew in an acid pH (usually 4.9-5.2) in nature (Hoham, 1975b).

Snow algae in the Chlamydomonadaceae are worldwide in distribution. Kol (1968) reported them from western North America, Greenland, western and eastern Europe, Siberia, Japan, western South America and the Antarctic peninsula.

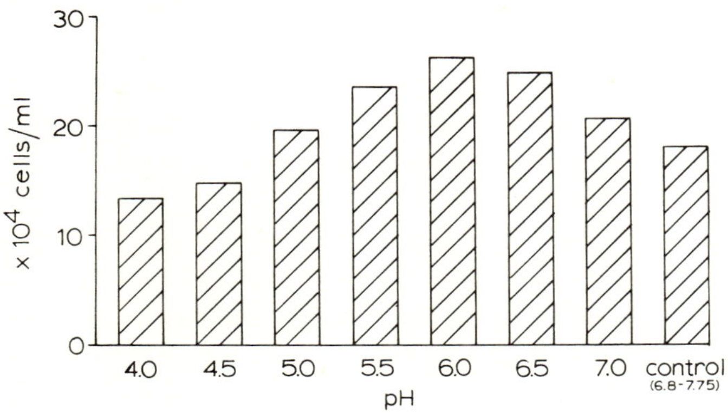

Fig. 15. Effect of pH upon growth of *Chloromonas pichinchae* after 23 days. (After Hoham, 1971).

Marchant (pers. comm., 1977) reports their presence in eastern Australia. Other records near Antarctica include the Balleny Islands (Kol and Flint, 1968) and South Orkney Islands (Fritsch, 1912; Fogg, 1967). In the northeastern United States, Hoham (unpubl.) found representatives in the central New York highlands, Adirondack Mtns., N.Y., and the White Mtns., New Hampshire. Kol and Peterson (1976) reported findings from glaciers in equatorial New Guinea. *Chlamydomonas nivalis* is found worldwide near sea level in polar regions to over 5000 m in mountainous areas (Kol, 1968). Fukushima (1963) reported snow algae as low as 200 m above sea level (Shiretoko Peninsula) to 3400 m (Mt. Fuji) in Japan.

For cryophilic members of the Chlamydomonadaceae, several factors are necessary for their appearance in snow (Fukushima, 1963; Pollock, 1970; Hoham, 1975b; Hoham and Mullet, 1977; Hoham *et al.*, 1979). Snow algae do not become visible until April-May at lower elevations and June-July at higher elevations in north temperate latitudes when air temperatures remain above freezing causing a melt-water in the snowbanks. At this time, resting spores germinate at the snow-soil or new snow-old snow interface (other factors such as light, photoperiod, available nutrients, and dissolved gases are also important). The amount of time required for zygote germination at the snow-soil interface till the production of new zygotes in the overlying snow may be less than one week (Hoham, 1975b; Hoham and Mullet, 1977; Hoham *et al.*, 1979). If species of snow algae are to persist in the same locality, the snowbanks must be deep enough each year to allow for the completion of the life cycles. In areas where snowfall is sporadic, snow algae probably will not be found, or only on rare occasions.

Several species of snow algae in the Chlamydomonadaceae are primary producers in the overall ecosystem. Populations of rotifers and protozoa (primary consumers) graze upon the algae (Pollock, 1970; Hardy and Curl, 1972; Hoham *et al.*, 1979). Other organisms in the ecosystem may include bacteria, fungi, other algae and other invertebrates.

Culture media references. A number of conventional media may be used for growing snow algae (see Rosowski and Parker, 1971), but some of these may be only partially successful or not at all. Stein (1963) published two acid media for snow algae, BMT (pH 4) and Tris (pH 4.5) pertaining to a study on a *Chromulina* (Chrysophyta) from snow. Attempts to secure cultures of *Chlainomonas rubra* (published as *Sphaerellopsis rubra*) in the acid media mentioned above were not successful (Stein and Brooke, 1964). Stein and Amundsen (1967) published a culture medium for snow algae, but the main results of the cultures studied were not reported. Czygan (1970) reported growing *Chlamydomonas nivalis* in a laboratory nutrient solution after Kessler and Czygan (1970). PFW medium for snow algae was successfully used to grow *Chloromonas pichinchae* (Hoham, 1973, 1975b), but was less successful for *Chloromonas brevispina* (Hoham *et al*, 1979). M-1 medium, based upon actual concentrations of nutrients found in snow, was used to establish *Chloromonas brevispina* in culture (Hoham *et al.*, 1979), and under certain conditions, *Chlamydomonas nivalis* (Hoham, unpubl.).

Needed research. Since Fukushima's publication in 1963, considerable information has accumulated on the systematics, life histories, and physiological ecology of snow algae. Areas of future research should include more on their biochemistry and physiology, ultrastructure, culture studies, and a better understanding of distribution. In the areas of biochemistry and physiology, studies should include cell wall chemistry, identifications of secondary pigments and the significance of their development, lipid metabolism as related to starch synthesis and temperature, general cellular metabolism and enzymology. A study of enzyme systems could be revealing as we currently do not know if snow algae in the Chlamydomonadaceae possess temperature sensitive enzymes, but some species probably do (*Chloromonas pichinchae*, Hoham, 1975a, b). Hardy and Curl (1972) stated that the snow alga, *Chromulina chionophilia* (a golden alga), possessed a photoreactivation enzyme that could repair damage done by ultraviolet radiation at higher elevations to chlorophyll and other pigments. More ultrastructural studies are needed for comparative purposes with other species in the Chlamydomonadaceae. The presence of unique crystalline bodies in the chloroplast of a species of *Chloromonas* from snow is of interest (Stein and Bisalputra, 1969). Studies in laboratory cultures have often been frustrating. Creating a snowbank in the laboratory (a modified growth chamber) may alleviate some of the problems of understanding growth requirements. The development of new media, such as M-1 (Hoham *et al.*, 1979), based upon known chemical composition of snow may be necessary for certain species. It appears that the chemical composition of snow varies considerably as it pertains to snow algae in different geographical regions (Einer Fjerdingstad *et al.*, 1978). Snow algae, as a group of organisms, are probably more widely distributed than is realized. Only recently have they been found in New York and New Hampshire (this report) and New Mexico (Hoham and Blinn, 1979). In the United States, it is possible that populations will be located in the near future from Maine to the southern Appalachians, and in the wooded areas of the Northern Great Lakes States and adjacent provinces in Canada. We know nothing about the distribution of snow algae in central Asia, mainland China or the Himalayas. Mt. Kilimanjaro may represent a suitable habitat in Africa. Perhaps snow algae may be located on some of the high peaks in central Mexico, and this seems possible since Kol and Peterson (1976) found them in glaciers in equatorial New Guinea.

Fossil record. There are no known fossils belonging to this group of algae. Kol (1968) suggested that present day populations of snow algae may be remnants of much larger populations that existed through the ice ages. Perhaps snow algae were derived from species of soil algae through natural selection as glaciers and semi-permanent snowfields were more abundant worldwide during the ice ages.

ACKNOWLEDGEMENTS

The Author wishes to thank Mr. Warren Wheeler, Colgate University, for preparing the photographic work and Dr. Carole Lembi, Purdue University, for reading the manuscript.

REFERENCES

Bauer, F. 1819. Microscopical observations on the red snow. Quart. J. Lit. Sci. Arts, 7: 222-229.

Bourrelly, P. 1966. *Les Algues d'Eau Douce.* I. Les algues vertes. Boubée & Cie, Paris. 511 pp.

Curl, H., Jr., J. T. Hardy, and R. Ellermeier. 1972. Spectral absorption of solar radiation in alpine snowfields. Ecology, 53: 1189-1194.

Czygan, F.-C. 1970. Blutregen und Blutschnee: Stickstoffmangelzellen von *Haematococcus pluvialis* und *Chlamydomonas nivalis*. Arch. Mikrobiol., 74:

69-76.
Ettl, H. 1970. Die Gattung *Chloromonas* Gobi *emend.* Wille (*Chlamydomonas* und die nachstverwandten Gattungen I). Nova Hedwigia, No. 34: 1-283.
Ettl, H. 1976. Die Gattung *Chlamydomonas* Ehrenberg (*Chlamydomonas* und die nachstverwandten Gattungen II). Nova Hedwigia, No. 49: 1-1122.
Fjerdingstad, Einer, L. Vanggaard, K. Kemp, and Erik Fjerdingstad. 1978. Trace elements of red snow from Spitsbergen with a comparison with red snow from East-Greenland (Hudson land). Arch. Hydrobiol., 84: 120-134.
Fjerdingstad, Erik. 1973. Accumulated concentrations of heavy metals in red snow algae in Greenland. Hydrologie, 35: 247-251.
Fjerdingstad, Erik, K. Kemp, Einer Fjerdingstad and L. Vanggaard. 1974. Chemical analyses of red 'snow' from East-Greenland with remarks on *Chlamydomonas nivalis* (Bau.) Wille. Arch. Hydrobiol., 73: 70-83.
Fogg, G. E. 1967. Observations on the snow algae of the South Orkney Islands. Phil. Trans. Roy. Soc. London, Ser. B, 252: 279-287.
Fritsch, F. E. 1912. Freshwater algae collected in the South Orkneys by Mr R. N. Rudmose Brown, B.Sc., of the Scottish National Antarctic Expedition, 1902-1904. J. Linn. Soc., Bot., 40: 293-338.
Fukushima, H. 1963. Studies on cryophytes in Japan. J. Yokohama Munic. Univ., Ser. C, Nat. Sci., 43: 1-146.
Garric, R. K. 1965. The cryoflora of the Pacific Northwest. Am. J. Bot., 52: 1-8.
Hardy, J. T. and H. Curl, Jr. 1972. The candy-colored, snowflaked alpine biome. Nat. Hist., 81 (9): 74-78.
Hoham, R. W. 1971. Laboratory and field studies on snow algae of the Pacific Northwest. Ph.D. thesis, University of Washington, Seattle. (Libr. Congr. Card No. Mic., 71-28,423), 207 pp. (Univ. Microfilms, Ann Arbor, Mich. [Diss. Abstr., 32B, 2557B]).
Hoham, R. W. 1973. Pleiomorphism in the snow alga, *Raphidonema nivale* Lagerh. (Chlorophyta), and a revision of the genus *Raphidonema* Lagerh. Syesis, 6: 243-253.
Hoham, R. W. 1974a. New findings in the life history of the snow alga, *Chlainomonas rubra* (Stein & Brooke) comb. nov. (Chlorophyta, Volvocales). Syesis, 7: 239-247.
Hoham, R. W. 1974b. *Chlainomonas kolii* (Hardy et Curl) comb. nov. (Chlorophyta, Volvocales), a revision of the snow alga, *Trachelomonas kolii* Hardy et Curl (Euglenophyta, Euglenales). J. Phycol., 10: 392-396.
Hoham, R. W. 1975a. Optimum temperatures and temperature ranges for growth of snow algae. Arct. Alp. Res., 7: 13-24.
Hoham, R. W. 1975b. The life history and ecology of the snow alga *Chloromonas pichinchae* (Chlorophyta, Volvocales). Phycologia, 14: 213-226.
Hoham, R. W. 1976. The effect of coniferous litter and different snow meltwaters upon the growth of two species of snow algae in axenic culture. Arct. Alp. Res., 8: 377-386.
Hoham, R. W. and D. W. Blinn. 1979. Distribution of cryophilic algae in an arid region, the American Southwest. Phycologia, 18: 133-145.
Hoham, R. W., K. Kemp, Erik Fjerdingstad and Einer Fjerdingstad. 1977. Comparative study of trace elements and nutrients in green snow (Washington, U.S.A.) and red snow (Greenland and Spitzbergen). J. Phycol., 13 (Suppl.), 30 (Abstr.).
Hoham, R. W. and J. E. Mullet. 1977. The life history and ecology of the snow alga *Chloromonas cryophila* sp. nov. (Chlorophyta, Volvocales). Phycologia, 16: 53-68.
Hoham, R. W. and J. E. Mullet. 1978. *Chloromonas nivalis* (Chod.) Hoh. & Mull. comb. nov., and additional comments on the snow alga, *Scotiella*. Phycologia, 17: 106-107.

Hoham, R. W., S. C. Roemer and J. E. Mullet. 1979. The life history and ecology of the snow alga *Chloromonas brevispina* comb. nov. (Chlorophyta, Volvocales). Phycologia, 18: 55-70.

Huber-Pestalozzi, G. 1961. Das Phytoplankton des Susswassers Systematik und Biologie. In Thienemann, A. [Ed.] *Die Binnengewässer*, Vol. 16 (Part 5). E. Schweizerbart'sche Verlagsbuchhandlung, Stuttgart. 744 pp.

Kawecka, B. and B. G. Drake. 1978. Biology and ecology of snow algae. 1. The sexual reproduction of *Chlamydomonas nivalis* (Bauer) Wille (Chlorophyta, Volvocales). Acta Hydrobiol., 20: 111-116.

Kessler, E. and F.-C. Czygan. 1970. Physiologische und biochemische Beitrage zur Taxonomie der Gattung *Chlorella*. IV. Verwertung organischer Stickstoffverbindungen. Arch. Mikrobiol., 70: 211-216.

Kobayashi, Y. 1967. Coloured snow with *Chlamydomonas nivalis* in the Alaskan arctic and Spitsbergen. Bull. Nat. Sci. Mus., 10: 207-210.

Kol, E. 1934a. Biologie de la cryovégétation des Alpes valaisannes et du massif du Mont-Blanc. Bull. Soc. Bot. Genève, 25: 287-292.

Kol, E. 1934b. Kryobiologische Studien I. Verhandl. Internat. Ver. Theor. Angew. Limnol., 6: 275-282.

Kol, E. 1935a. Kryobiologische Studien am Jungfraujoch (3470 m) und in dessen Umgebung. Beih. Bot. Centralbl., 53: 34-47.

Kol, E. 1935b. Ueber die Kryovegetation des Retyezát und der umliegenden Gebirge in Transsylvanien. Verhandl. Internat. Ver. Theor. Angew. Limnol., 7: 475-486.

Kol, E. 1939. Zur Schneevegetation Patagoniens. Ark. Bot. (Stockholm), 29: 1-4.

Kol, E. 1941. The green snow algae of Yellowstone National Park. Am. J. Bot., 28: 185-191.

Kol, E. 1968. Kryobiologie. Biologie und Limnologie des Schnees und Eises I. Kryovegetation. In Elster, H.-J. and W. Ohle [Eds.] *Die Binnengewässer*, Vol. 24, E. Schweizerbart'sche Verlagsbuchhandlung, Stuttgart. 216 pp.

Kol, E. and E. A. Flint. 1968. Algae in green ice from the Balleny Islands, Antarctica. New Zeal. J. Bot., 6: 249-261.

Kol, E. and J. A. Peterson. 1976. Cryobiology. In Hope, G. S., J. A. Peterson, U. Radok and I. Allison [Eds.] *The Equatorial Glaciers of New Guinea, Results of the 1971-1973 Australian Universities' Expeditions to Irian Jaya: Survey, Glaciology, Meteorology, Biology and Palaeoenvironments*. A. A. Balkema, Rotterdam. pp. 81-91.

Komárek, J., F. Hindák and P. Javornický. 1973. Ecology of the green kryophilic algae from Belanské Tatry Mountains (Czechoslovakia). Arch. Hydrobiol., Suppl. 41, Algol. Stud., 9: 427-449.

Marchant, H. J. 1977. Personal communication, The IXth International Seaweed Symposium, Santa Barbara, California (August, 1977).

Morita, R. Y. 1975. Psychrophilic bacteria. Bact. Rev., 39: 144-167.

Mosser, J. L., A. G. Mosser and T. D. Brock. 1977. Photosynthesis in the snow: The alga *Chlamydomonas nivalis* (Chlorophyceae). J. Phycol., 13: 22-27.

Pollock, R. 1970. What colors the mountain snow? Sierra Club Bull., 55: 18-20.

Rosowski, J. R. and B. C. Parker [Eds.]. 1971. *Selected Papers in Phycology*, Dept. Bot., Univ. Nebraska, Lincoln. 876 pp.

Sieminska, J. 1951a. The red snow in Tatra (*Chlamydomonas nivalis* Wille). Acta Soc. Bot. Polon., 21: 1-4.

Sieminska, J. 1951b. Red snow below the Szpiglasowa pass in the High Tatra. Acta Soc. Bot. Polon., 21: 231-233.

Stein, J. R. 1963. A *Chromulina* (Chrysophyceae) from snow. Can. J. Bot., 41: 1367-1370.

Stein, J. R. and C. C. Amundsen. 1967. Studies on snow algae and fungi from the Front Range of Colorado. Can. J. Bot., 45: 2033-2045.

Stein, J. R. and T. Bisalputra. 1969. Crystalline bodies in an algal chloroplast. Can. J. Bot., 47: 233-236.

Stein, J. R. and R. C. Brooke. 1964. Red snow from Mt. Seymour, British Columbia. Can. J. Bot., 42: 1183-1188.

Tamm, C. O. 1950. Growth and plant nutrient concentration in *Hylocomium proliferum* (L.) Lindb., in relation to tree canopy. Oikos, 2: 60-64.

Tamm, C. O. 1951. Removal of plant nutrients from tree crowns by rain. Physiol. Plant., 4: 184-188.

Thomas, W. H. 1972. Observations on snow algae in California. J. Phycol., 8: 1-9.

Tukey, H. B., Jr. and J. V. Morgan. 1964. The occurrence of leaching from above-ground plant parts and the nature of the material leached. Proc. XVI Internat. Hort. Congr., 4: 153-160.

Viala, G. 1966. L'astaxanthine chez le *Chlamydomonas nivalis* Wille. Compt. Rend. Hebd. Séances Acad. Sci., 263: 1383-1386.

Wille, N. 1903. Algologische Notizen XI. Nyt. Mag. Naturvidensk, 41: 109-162.

Copyright 1980 by Elsevier North Holland, Inc.

PRASINOPHYTES

RICHARD E. NORRIS

Friday Harbor Laboratories, University of Washington, Friday Harbor, Washington 98250, U.S.A.

ABSTRACT

Structure of green flagellates that have been assigned to, or associated with, Prasinophyceae is reviewed from the point of view of light and electron microscopy. Possible phylogenetic linkages of various characteristics are emphasized, including their biochemical products. It is concluded that possible evolutionary connections among presently known species indicate that prasinophyte flagellates are the most primitive of the green algal flagellates, and that they show close relationships to one another. Recognition of this group of green flagellates as a class, Prasinophyceae, within the Chlorophyta is supported.

INTRODUCTION

Prasinophyceae are mostly unicellular, flagellate algae that contain both chlorophylls *a* and *b* as well as starch granules in their chloroplasts. They occur in both hemispheres, mostly in temperate regions, and are found in freshwater, brackish and marine habitats. They were first recognized as different from most other green algae in a series of publications by Chadefaud (for a review see Chadefaud, 1977). Chadefaud recognized the group (calling it various names including "Chlorophytes Prasinates") before electron microscopy had been used with algae, and he characterized the group as differing from most green algal flagellates by the following characteristics:

1. cells having a grooved region at the place of flagellar attachment to the cell ("dépression infundibuliform" or "fosse vestibulaire"),
2. presence of four parabasal bodies which probably represent dictyosomes of the Golgi apparatus,
3. presence of muciferous bodies or trichocysts.

At that time (1941) Chadefaud assigned the genera *Pyramimonas* Schmarda (as *Pyramidomonas*), *Platymonas* G. S. West[1], *Chlorodendron* Senn and *Prasinocladus* Kuckuck to the "Prasinates".

---

1. The epithet *Platymonas* is used in this chapter instead of *Tetraselmis* Stein 1878 because there is inconclusive evidence that the two genera should be united (Parke and Green, 1976).

In 1960 Chadefaud (p. 331 and 336) retained "Prasinophycinées" as a group separate from the rest of the green algae, but it is difficult to explain why, in the same publication (p. 390 and 391), he places the prasinophyte genera in the Prasinovolvocales of the Euchlorophyceae (containing genera having pyramidal cells: *Pyramimonas*, *Platymonas*, *Prasinocladus*, *Chlorodendron* and *Polyblepharides*; globular cells: *Prasinochlamydomonas* and *Prasinocarteria*; and dorsiventral cells: *Trichloris* and *Dangeardinella*).

Christensen (1962) assigned the pyramidal and globular cell types to the class Prasinophyceae of the Chlorophyta, including in it the genera *Pyramimonas*, *Tetraselmis* (*Platymonas*), *Prasinocladus* and *Halosphaera*, but he separated the dorsiventral types into a separate class, the Loxophyceae which contained the following genera: *Bipedinomonas*, *Pedinomonas*, *Micromonas*, *Nephroselmis* and *Mesostigma*. Christensen's analysis of the two classes brought to the forefront the evidence gained from electron microscopy that supported separation of these groups from the main body of green algae. Subsequent publications usually do not recognize Loxophyceae as a class separate from Prasinophyceae. Bourrelly (1970) included the following genera in the Prasinophyceae in his revision of the freshwater green algae: *Pocillomonas* Steinecke, *Pyramimonas* Schmarda, *Trichloris* Scherffel et Pascher, *Gyromitus* Skuja[2], *Streptomonas* Klebs, *Tetraselmis* Stein, *Prasinochlamydomonas* Chadefaud, *Pedinomonas* Korshikov, *Nephroselmis* Stein, *Scourfieldia* G. S. West, *Myochloris* Belcher et Swale, *Mesostigma* Lauterborn, *Monomastix* Scherffel, *Chloronephris* Pacher et Jahoda, *Prasinochloris* Belcher, *Hyalocardium* Ettl, *Pseudocardiomonas* Bourrelly, and *Protoaceromonas* Skvortzov. Bourrelly (1970) lists three other genera that may have affinities with Prasinophyceae: *Kuzininia* Skvortzov, *Meyeriella* Skvortzov, and *Protochroomonas* Skvortzov. Bourrelly (1966) had previously indicated that another genus, *Swarchewskiella* Jasnitsky, may have affinities with Prasinophyceae.

Parke and Green (1976) list species in the following prasinophyte genera that occur in the sea around Great Britain: *Pedinomonas* Korshikov, *Bipedinomonas* N. Carter, *Mantoniella* Desikachary, *Micromonas* Manton et Parke, *Nephroselmis* Stein, *Pseudoscourfieldia* Manton, *Pachysphaera* Ostenfeld, *Pterosperma* Pouchet, *Sphaeropsis* Meunier, *Asteromonas* Artari, *Chloraster* Ehrenberg, *Stephanoptera* P. A. Dangeard, *Halosphaera* Schmitz, *Platymonas* G. S. West, *Prasinocladus* Kuckuck and *Tetraselmis* Stein.

---

2. *Gyromitus* Skuja has been studied with the electron microscope by Swale and Belcher (1974, 1975) and found to have no characteristics linking it with prasinophyte flagellates.

Other genera that may be allied with Prasinophyceae are *Spermatozopsis* Korshikov (Manton, 1965, 1970) and *Chaetosphaeridium* Klebahn (Moestrup, 1973, 1974).

It is doubtful whether or not all the genera listed above are related to the Prasinophyceae, but it is recognized that a prominant group of green and colorless flagellates exists that is not closely related to Volvocalean types in the Chlorophyta. Recent classification schemes recognize this fact in various ways. Stewart and Mattox (1978) express the attitude that these are primitive green flagellate types, but they do not recognize them as a separate group because various species of flagellates have characteristics from which Chlorophyceae, Charophyceae and Ulvaphyceae each may have been derived. Round (1971, 1973) elevates prasinoid algae to phylum status as one of four chlorophyll *b* containing phyla, Chlorophyta, Charophyta, Prasinophyta and Euglenophyta. One of the most recent analyses of algal phyla, van den Hoek (1978) retains Prasinophyceae as a class in the Chlorophyta. Most of these conclusions have been decided without complete overall analysis of the characteristics of Prasinophyceae.

The highly important place that prasinophyte flagellates have in phylogeny of green plants is becoming more evident (Pickett-Heaps and Marchant, 1972; Taylor, 1976, 1978; Chadefaud, 1977; Stewart and Mattox, 1978). It is the intent of the following review to amalgamate former characterizations of various members of the Prasinophyceae providing a basis on which further analyses of the organisms may be judged.

CHARACTERIZATION OF LIVING CELLS

Flagellate cells of Prasinophyceae are omnifarious from the point of view of the characteristics that are visible with the light microscope. Cells of different species may have from one to eight flagella. Flagella may be equal or unequal in length and they may be borne posteriorally, laterally or anteriorally on the cells. Usually the flagella are attached to the cell at a depressed place in the cell that may be a deep flagellar pit in species having four or more anterior flagella. Along with the lateral to terminal position of flagella in different species, the symmetry of cells is also quite variable ranging from those with bilateral symmetry to those having radial symmetry when viewed from the polar regions. Flattening or compression of cells is common in species having bilateral symmetry (*Bipedinomonas*) and in some types having terminal (polar) insertion of flagella (*Pseudoscourfieldia*).

Flagellate cells may be very small (1.0 µm in *Micromonas*) or quite large (up to 40 μm in *Pyramimonas*). Phycoma stages of *Halosphaera viridis* may be as large as 800 μm in diameter.

Contents of flagellate cells as seen with the light microscope usually consist of a single chloroplast, often lobed in concert with cell structure, a single nucleus located near the flagellar base, an eyespot and pyrenoid present within the chloroplast, and starch granules in various positions inside the chloroplast. Chloroplasts in most species are parietal in position and contain a prominent pyrenoid.

Anisokonty, heterodynamic activity and posteriorally directed flagella are considered to be characteristics indicative of prasinophyte flagellates having one or two flagella (and *Trichloris*), the species having four or more flagella usually having them attached anteriorally, and having homodynamic activity. Cells with fewer than four flagella often are bilaterally symmetrical in Prasinophyceae, and may have strongly flattened cells in a plane parallel to flagellar movement. Cells usually rotate on their axis as they move through the water, but such action may change with different attitudes and speed of swimming.

It is also characteristic of many prasinophyte flagellates to have periods of inactivity alternating with active swimming. During these resting periods the cells usually hold their flagella in positions characteristic for each species. Smaller cells may be able to attach to the water-air interface and thereby hang suspended in a favorable lighted environment without expending energy by swimming. Such cells are part of the hypo-neustonic population. Similarly, some pelagic species of *Pyramimonas* have been observed to swim toward the surface and then stop all flagellar movement with the recurved flagella acting as a parachute as the cell slowly sinks. Such behavior undoubtedly is characteristic of highly evolved behavioral patterns in reaction to environmental variations and stresses.

Behavior of motile cells in the algae is a subject that has not been given the attention that it deserves. Documentation of behavior in prasinophyte cells has been carefully provided by Parke and Adams (1961), Parke and Rayns (1964), Belcher (1964, 1965, 1968b, 1969), Manton *et al.* (1965), Parke (1966) and Manton (1975). In these publications attention is given to descriptions of motile cell behavior in various prasinophytes, but it will not be possible to estimate the value of this character in analysis of flagellate relationships until many more observations are documented.

FLAGELLA

Prasinophyte flagella can be classified into two major categories, those having flagellar scales and those without such appendages. Species without flagellar scales may have two rows of extremely thin hairs along the flagellar length (*Pedinomonas minor*, Ettl and Manton, 1964; Pickett-Heaps and Ott, 1974; *P. tuberculata*, Manton and Parke, 1960; *P. major*, Belcher, 1968b). These flagellar hairs have been described as more slender than those of bacterial flagella and Pickett-Heaps and Ott (1974) found them on approximately 20% of the cells that they examined. Flagellar hairs have not been found on the flagellum of *Micromonas pusilla*. Fibrillar to granular material is deposited on the surface of flagella belonging to *Asteromonas gracilis* (Peterfi and Manton, 1968).

Flagellar shape has been lucidly discussed by Manton (1965) in which the conclusion is made that the term 'acronematic' should not be used because its past usage for ill-defined structures renders it meaningless. Most prasinophyte flagella do not have terminal hair-points or what has been termed an acronematic condition. Their flagella terminate abruptly and are broadly rounded at the end. A small group of prasinophyte flagellates, however, has flagella with a bipartite condition, the distal one-third to approximately one-eighth of the flagellar length being thinner than the proximal part. In the species in which such flagella have been critically examined, the hair-point is covered by the plasmalemma, the hair-point being caused by the central pair of microtubules in the flagella being prolonged beyond the peripheral ring of microtubules. Such a membrane-bound extremity is vastly different from the 'hair-points' of some brown algal and chrysophycean flagellates in which the terminal sector consists of extracellular hairs. In the past, both of these types have been termed acronematic, hence the confusion in usage of that term. Manton (1965), in referring to small green flagellates, predicted that bipartite flagella "may prove to be useful phyletic indicators within this group when more is known." Unfortunately, at this time little more is known about the presence or absence of bipartite flagella in green monads, but the information that is accumulating is important. It is interesting to note that as far back as 1949 Owen, using light microscopy, reported on flagellates with and without bipartite structure. He showed that *Scourfieldia complanata* and *Pyramimonas montana* have bipartite flagella along with various species of volvocalean flagellates including species of *Chlamydomonas*, *Dunaliella*, *Carteria* and *Polytoma*. Furthermore, Owen used this flagellar characteristic to propose a phylogenetic scheme, based on one proposed by G. S. West in 1916, in which the *Pyramimonas*-type is basic in evolution of green flagellates, and the

Chlamydomonas-type is a more highly derived form.

Using electron microscopy, the following prasinophyte flagellate cells are known to have bipartite flagella: *Micromonas pusilla* (Manton, 1959), *Pedinomonas minor* (Ettl and Manton, 1964; Pickett-Heaps and Ott, 1974), *Pedinomonas major* (Belcher, 1968b), *Monomastix* aff. *minuta* (Manton, 1970), *Spermatozopsis exsultans* (Manton, 1965), and *Scourfieldia caeca* (figured but not discussed by Manton, 1975). Belcher (1966) showed that the flagellar tips of *Prasinochloris sessilis* motile cells are thinner than the proximal part, but electron microscopy was not used in this study.

Prasinophyte flagella usually contain only the axoneme inside the plasmalemma, but the winged flagellum of *Pedinomonas major* is an exception. It is reported to contain two darkly staining fibrils adjacent to the wings and outside the ring of microtubules (Belcher, 1968b).

Flagellar Appendages. Thick tubular hairs are regularly found on prasinophyte flagella on which there are scales (except *Mesostigma viride* Manton and Ettl, 1965). These hairs have been compared to similar hairs found on certain cryptomonads (Manton and Parke, 1960; Parke and Manton, 1965; Manton, 1965). The hairs often are curved, having tapering ends and usually show internal transverse striations (Manton and Parke, 1960; Manton, 1965). They seem to be more or less irregularly placed along the length of the flagella probably as a result of their described caducous nature. Parke *et al.* (1978) describe two types of flagellar hairs in some *Pterosperma* flagellate cells. Short fragile caducous hairs are present on all species, but some species have additional longer, thicker hairs that have characteristic cross segmental patterning. These larger hairs occur only near flagellar bases. Because flagellar hairs are not a local specialization of the flagellar surface, but, instead, are formed in Golgi vesicles in a way similar to flagellar and cell body scales and later deposited on the flagellar surface, it seems appropriate to call them 'hairscales' (Moestrup and Thomsen, 1974). Function of hairs and hairscales in Prasinophyceae is obscure.

Flagella having a covering of organic scales are common in Prasinophyceae, and this characteristic is one of the most distinctive of the class. Stewart and Mattox (1975, 1978) expressed the opinion that scaly green monads, however, might not form a natural group because such scaly flagellates were also discovered in other more highly advanced green algae which they place in the Ulvaphyceae and Charophyceae. Additional discussion of this matter will be given in the conclusions to this chapter.

Species having a single layer of scales, in addition to hairscales, on the flagellar membrane are probably more primitive than species having two or more layers of scales (Fig. 1). In the green flagellates there are seven species known at the present time with only one layer of scales on the flagella: *Mantoniella squamata*, *Nephroselmis gilva*, *Nephroselmis* sp. (Leadbeater, 1974), *Dolichomastix nummulifera*, *D. lepidota*, *D. eurylepidea* and *Mesostigma viride*. Scales on the flagella of *Mantoniella* and *Dolichomastix* are the same structure as scales on the cell body, with slight variations in size and shape. This is an unusual feature for scale-covered cells because most species show considerable differences in the structure of flagellar and cell body scales. The scales of *Mantoniella squamata* are 0.15 to 0.4 µm in diameter, having orbital spider's-web structure of slightly mineralized fibers (Manton and Parke, 1960; Parke and Rayns, 1964; Barlow, 1977). Species of *Dolichomastix* have similarly shaped scales within the same size range for *D. nummulifera* and *D. lepidota*, but *D. eurylepidea* may have flagellar scales as large as 0.7 µm (Manton, 1977).

Nephroselmis gilva has flagellar scales that are slightly smaller than the ones on the cell body, 0.22 and 0.23 µm respectively (Parke and Rayns, 1964). The flagellar scales show a hexagonal outline with a spider's-web pattern, whereas the body scales are octagonal with the same pattern. The main distinguishing feature of the flagellar scales, however, is an adnate, horizontally directed spine on the distal side of the scale, arising centrally and terminating outside the margin at one of the corners of the hexagonal scale. Scales on an unnamed species attributed to *Nephroselmis* by Leadbeater (1974) are similar, but not identical to those of *N. gilva*. Similar scales having an adnate spine are present on flagella of *Pyramimonas*, *Halosphaera*, *Pachysphaera*, and *Pterosperma* and, as stated by Manton (1977), "a character as unusual as adnate spines of the type described is likely to be phyletically significant." Flagellar scales on these organisms are arranged in nine longitudinal rows, the scales being imbricated in a regular manner with the spines directed distally on the flagella. Such an arrangement of polarized flagellar scales seems to be a more highly evolved characteristic.

Mesostigma viride is unusual among the scaly green flagellates in being the only species known that has flagellar scales but seems to be without hairscales. Flagellar scales in *Mesostigma* are very small (approximately 0.03 µm wide) and are subrhomboid in shape. Manton and Ettl (1965) describe them as having somewhat of a maple leaf outline. There is a small knob in the center of the scale and the spine-tipped lobes are directed distally when the scales are in place on the flagellum (Manton and Ettl, 1965, Fig. 19). Although Manton and Ettl state that these same scale types also cover the plasmalemma on the upper surface of the cell and in the flagellar pit, it seems that these body scales are not as lobed as are the ones covering the flagella, and they are more square in outline rather than subrhomboid (see their Figs. 19, 20, and 31).

Other scale-bearing prasinophyte flagellates have a layer of sometimes square flagellar scales (Norris and Pearson, 1975; Pennick, 1977a), but mostly pentagonal scales that are compactly aligned next to the plasmalemma. These scales may be hexagonal (Hori and Chihara, 1974a; Norris and Pearson, 1975) in some species, and it is interesting to note that scales in this layer on the flagella seem to be more flexible in shape than are the scales in the corresponding layer on the cell body. The flagellar scales in this layer have a low rim, a high central boss and a pentagonal electron-translucent configuration on the floor of the scale surrounding the protuberance. Norris and Pienaar (1978) termed these scales as Type 1 flagellar scales in *Pyramimonas*. This scale layer completely surrounds all the flagella on the cell. The scales are produced in the Golgi cisternae and often are stored inside the scale reservoir, if it is present. It is believed that some species deposit the flagellar scales at a point adjacent to the flagellar roots by merging of scale-containing vesicles, derived from the Golgi, with the plasmalemma. In some species of *Pyramimonas*, a channel has been observed connecting the scale reservoir with the outside medium at a point near the flagellar bases, and it is believed that scales may migrate through this channel onto the flagella.

The outermost layer of flagellar scales in prasinophytes (Type 2 of Norris and Pienaar, 1978) may be of three different types, according to the species on which they occur. In *Pseudoscourfieldia marina* (Manton, 1975) and *P. longifilis*[3] (Manton *et al.*, 1965 - as *Heteromastix longifilis*, Norris - unpublished) these scales are small and fusiform. Similar fusiform rod-shaped scales are on the flagella as Type 2 scales of *Prasinocladus* (Parke and Manton, 1965; Hori and Chihara, 1974a) and *Platymonas* (Manton and Parke, 1965; Norris - unpublished).

---

[3]This taxon in recent times has been placed in *Heteromastix* (Rayns *in* Parke and Dixon, 1964), *Nephroselmis* (Norris, 1967) and *Bipedinomonas* (Parke and Green, 1976). It is now evident that it has scales and cell structure most similar to *Pseudoscourfieldia marina* Manton, 1975, and it is hereby placed in that genus: *Pseudoscourfieldia longifilis* (Butcher) comb. nov. Basionym: *Anisomonas longifilis* Butcher, 1959, p. 38, pl. 8, Fig. 3.

Fig. 1. Chart showing possible phylogenetic relationships among types of Prasinophyceae for which scale structure is known. Flagellar hairscales are not shown. B = Cell Body Scales, F = Flagellar Scales. No attempt has been made to show size relationships among the scales. Layer relationships among scales as designated by Norris and Pienaar, 1978 are from 1 to 3, reading left to right.

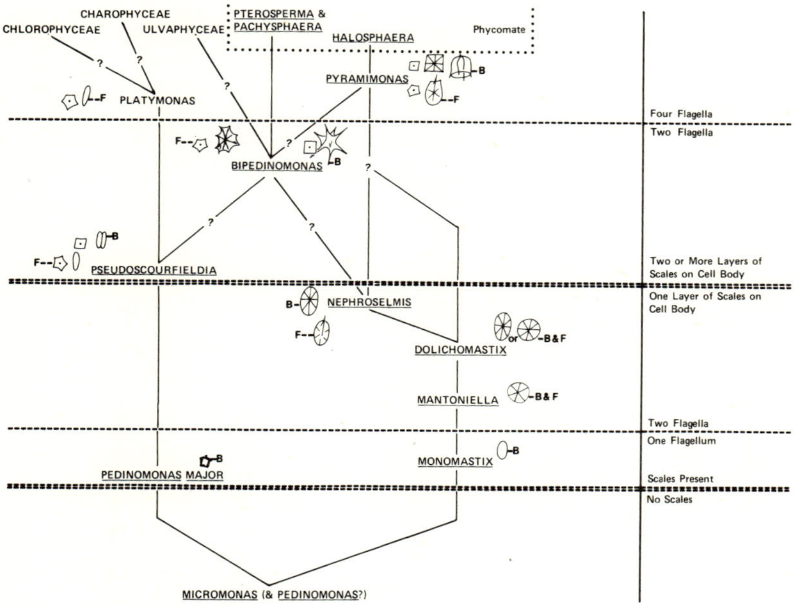
CHAROPHYCEAE
CHLOROPHYCEAE
ULVAPHYCEAE
PTEROSPERMA &
PACHYSPHAERA
HALOSPHAERA
Phycomate
PYRAMIMONAS
PLATYMONAS
Four Flagella
Two Flagella
BIPEDINOMONAS
PSEUDOSCOURFIELDIA
Two or More Layers of Scales on Cell Body
One Layer of Scales on Cell Body
NEPHROSELMIS
DOLICHOMASTIX
MANTONIELLA
Two Flagella
One Flagellum
MONOMASTIX
PEDINOMONAS MAJOR
Scales Present
No Scales
MICROMONAS (& PEDINOMONAS?)

FIGURE 1

Bipedinomonas angulata[4] and *B. rotunda* have outer flagellar scales (Type 2) somewhat similar to the scales on *Mantoniella*, but smaller and with radiating fibrils prolonged into five or six marginal protuberances (Manton *et al.*, 1965; Mattox and Stewart, 1977). Two layers of flagellar scales are described for *B. rotunda* (Manton *et al.*, 1965), but these authors mention that there are three layers of flagellar scales in *B. angulata*. It is impossible to determine the exact number of scale layers from their Fig. 42, which is a shadow-cast whole mount of a flagellum, but three layers of flagellar scales are shown in Fig. 43 for an unnamed species of *Heteromastix*. Mattox and Stewart (1977) show figures (1, 10 and 16) of *Bipedinomonas angulata* (as *Heteromastix*) in which there seem to be only two layers of scales on the flagella. This organism is from the University of Texas Culture Collection of Algae (UTEX 1026) and was isolated by Dr. R. C. Starr. The specimens described by Manton *et al.* (1965) under the name *Heteromastix angulata* were isolated and identified by Dr. H. Ettl from a pond in Czechoslovakia. It is possible that the speciners used by Manton *et al.* and the ones used by Mattox and Stewart represent entirely different species.

Spider's web-like scales having three marginal spines lying adjacent to one another on the polarized scale are described by Swale and Belcher (1968) for *Pyramimonas tetrarhynchus* flagellar scales (Type 2). These three-spined scales seem to be unique for this species of prasinophytes.

CELL COVERING

In early stages of recognition of the genera now placed in Prasinophyceae, most species were considered to be without a cell covering. Electron microscopy, however, revealed a covering of spectacular scales of varying form that cover the cell body in many species, and the scaly cell covering has become an important characteristic in the recognition of Prasinophyceae.

There are a few species of prasinophyte flagellates that electron microscopy has shown not to have a scaly covering on the cell body, but, for other reasons, they have been associated with the prasinophytes. It remains uncertain, however, that these naked species have a distinct phyletic relationship with the scaly monads. *Micromonas pusilla*, *Pedinomonas minor*, *P. noctilucae* and *Scourfieldia caeca* have been shown by electron microscopy to have no visible cover-

---

[4]*Bipedinomonas angulata* (Korshikov) comb. nov. Basionym: *Heteromastix angulata* Korshikov, 1923, p. 154, Fig. 42-58. As explained by Parke and Green (1976), *Heteromastix* Korshikov, 1923, is a later homonym of *Heteromastix* H. James-Clark, 1865, p. 146, Figs. 71 & 88.

ing on the cell bodies (Manton, 1959; Ettl and Manton, 1964; Pickett-Heaps and Ott, 1974; Manton, 1975; Sweeney, 1976). Peterfi and Manton (1968) showed that *Asteromonas gracilis* motile cells have a covering of varying amounts of fibrillar material. They believed that this deposit may be mucilage that is fibrillar because of the fixation artifact effect.

Pedinomonas tuberculata, a species having small tubercules protruding from the cell body, was shown by Manton and Parke (1960) to have the cell membrane covered by spicules and traces of amorphous material. Inside the cell membrane the tubercules contained bodies that gave positive reactions to cellulose, calcite and quartz. It may be speculated that this species represents a prototype having barely recognizable scales on the surface of the cell.

Cell Body Scales in One Layer. Scales on chlorophyll *b* containing green flagellates were first noted by Manton and Parke (1960) on *Mantoniella squamata* (as *Micromonas*). In that species, scales were found lying in an imbricated pattern over the entire cell and flagellar surfaces. The structure of the scales is consistently fibrillar and the pattern is like an orbital spider's web. The size of the scales varies considerably, but their structure has little variation, and they were shown to originate in Golgi cisternae. Although prasinophyte scales have been generally recognized as being organic, it is especially interesting that Manton and Parke (1960) described the scales on *Mantoniella* as being slightly mineralized. Only one type of body scale occurs on cells of *Mantoniella*, and it has a structure identical to those scales on the flagella.

Dolichomastix, a genus described by Manton (1977), also has a single layer of scales on the cell. The form of the scales resembles those of *Mantoniella* in being circular to elliptical in outline and in having concentric and radiating fibrils. *Dolichomastix* differs from *Mantoniella* in having two very long, almost equal, flagella. It should be emphasized that living cells of *Dolichomastix* have not been observed and that one cannot be certain that the species in this genus are photosynthetic or that they contain chlorophyll *b*. Their position within the Prasinophyceae must remain doubtful until these characteristics as well as others are determined.

Nephroselmis gilva Parke *et* Rayns (1964) and *Nephroselmis* sp. (Leadbeater, 1974) also have a single layer of scales covering the cell. In these species, however, the flagellar scales are slightly different in structure from the body scales, having an adnate spine on their distal side. *Nephroselmis* scales otherwise, as far as these species are concerned, have cell body scales similar in structure to those of *Mantoniella* and *Dolichomastix*.

The only other prasinophyte flagellates for which a single layer of cell body scales has been demonstrated are *Pedinomonas major* and *Monomastix* aff. *minuta*. Belcher (1968b) found that the plasmalemma of *Pedinomonas major* is covered by a layer of tiny hexagonal scales. Small, rimless oval scales were demonstrated to be present on the cell body of *Monomastix* aff. *minuta* by Manton (1970) forming a layer that is recognizable with light microscopy. In neither of these species are the flagella covered with a layer of scales.

The information presented above is summarized in Fig. 1.

Cell Body Scales in More Than One Layer. The next step in evolutionary advance from the putative primitive type with one layer of body scales on prasinophyte flagellates is one of major importance for, along with elaboration of more than one layer of scales, there emerges one uniform basic scale type in all the more advanced prasinophytes. This basic type of scale always occurs in a dense layer adjacent to the plasmalemma and is a small square scale designated Type 1 body scale by Norris and Pienaar (1978). Although similar to their Type 1 flagellar scale, the Type 1 body scale differs in being square in outline and having a higher rim.

It is possible to propose two ways in which the more advanced species having two scale layers may have been evolved: 1) by continued development of the *Pedinomonas major* type scale, enlarging it slightly and adding more fibrillar material thereby forming the basic Type 1 body scale, with subsequent development of a second scale type that is deposited on the outside; 2) by modification of the structure of the *Mantoniella squamata*, *Nephroselmis gilva*, or *Dolichomastix* type scale as the Type 2 body scale, and development of a new scale, the small square Type 1 body scale, many of which are inserted in a layer beneath the larger scales.

It is difficult to speculate on possible factors that may have guided development of a second layer of scales, but the protective function seems to be reasonable. The addition of a second layer not only adds protection to the cell membrane, but the second and third layers of scales also often seem to be attached to the underlying scales in such a way that they may lend stability to their position on the cell.

Two biflagellate species, *Pseudoscourfieldia marina* and *P. longifilis*, have two layers of body scales, but they have bilaterally symmetrical lobing of the scales in the outer layer. The Type 2 body scale in these species is different in the two species, but this type shares the same basic quality of a radially symmetrical network of fibrils onto which bilateral symmetry of scale lobing has been superimposed. It seems possible, therefore, that they may be derived, as a slightly more advanced type, from the orbital scale type found in *Mantoniella*.

Pyramimonas is a large genus of quadriflagellate prasinophytes, a genus that often is referred to as typical for this group. Most species of *Pyramimonas* have three layers of scales on the cell body: Type 1 body scales composed of rimmed square scales with a small central boss, Type 2 body scales which are usually larger square scales having variable fibrillar patterns, and Type 3 body scales that have coronate shapes often distinctive for certain groups of species. The Type 2 body scales seem to be the most distinctive diagnostic characteristic for species in *Pyramimonas* (Norris and Pienaar, 1978), and coronate scales seem to differ among groups of species in that genus.

Pyramimonas virginica seems to be exceptional in *Pyramimonas* in having only two layers of body scales (Pennick, (1977a), but it is possible that *P. reticulata* may also have only two body scale layers (Belcher, 1968a). In addition to his study of *P. virginica*, Pennick has also reported on scales of other *Pyramimonas* species: *P. obovata* (Pennick *et al.*, 1976), *P. grossii* (Pennick and Clarke, 1976), *P. amylifera* (Pennick, 1978), and *P. orientalis* (Pennick *et al.*, 1978). Norris and Pienaar (1978) reviewed the cell structure of five *Pyramimonas* species, with emphasis on scale types, (*P. amylifera*, *P. parkeae*, *P. orientalis*, *P. obovata* - as *P.* aff. *disomata*, and *P.* aff. *pluriloculata*) and reviewed earlier studies on the genus. Maiwald (1971) reported that a freshwater species of *Pyramimonas*, *P. montana*, is completely without a scale covering. The type species of *Pyramimonas*, also a freshwater species, *P. tetrarhynchus* Schmarda, has three layers of body scales (Swale and Belcher, 1968; Swale, 1973; Moestrup and Walne, 1979). It seems unlikely that *P. montana* belongs within this genus, but it probably belongs to a volvocalean genus, perhaps *Pyramichlamys* Ettl.

Several small-celled species of *Pyramimonas* (*P. obovata*, *P. orientalis*, *P. grossii* and *P.* aff. *pluriloculata*) have the Type 1 body scale layer restricted to the flagellar end of the cell, mostly within the flagellar pit. This arrangement of body scales seems to be a derived condition, and it is also interesting to note that these species are the ones most likely to be found in truly marine planktonic environments. The larger species having all three body scale layers completely covering the cell are either freshwater species (*P. tetrarhynchus*) or occur mostly in brackish-water situations (*P. parkeae*, *P. amylifera* and *P. virginica*).

A peculiar conical scale represents a fourth type of body scale in the *Pyramimonas*-like motile stages of *Halosphaera russellii*, but this scale occurs only as part of the outermost layer of scales within the flagellar pit (Pennick, (1977b).

The quadriflagellate motile cells of the phycomate genera *Pterosperma* and *Pachysphaera* have a cell body with bilateral symmetry. The body scales are in three layers, the innermost layer having a form similar to *Pyramimonas* Type 1 body scales, but the outer two layers of scales have a structure quite different from scales in a similar position on *Pyramimonas* cells (Leadbeater, 1974; Parke *et al.*, 1978). In *Pterosperma* and *Pachysphaera* flagellates these scales are of an orbital structure, often having eight radiating ribs and a slight boss at their center. The scales in the two layers are similar in form, varying mostly in size, the outermost layer often having slightly larger scales with a somewhat different pattern. Because of the form of the flagellate cells and the striking similarity between the Type 2 body scales, *Pterosperma* and *Pachysphaera* seem to have stronger affinities with the *Pseudoscourfieldia-Bipedinomonas* line of evolution than with the off branching *Pyramimonas-Halosphaera* lineage (Fig. 1).

The body scale layers on *Mesostigma viride* are described by Manton and Ettl (1965). They show that the innermost layer (Type 1) is identical to the Type 1 flagellar scales in this species, having a subrhomboid outline. This layer of scales does not occur on the cell's lower surface. The Type 2 body scales are naviculoid in shape having a concentric pattern and a slightly thickened margin. The spectacular Type 3 body scales of *Mesostigma viride* are basket-shaped in a conical form, having a square orifice directed outward on the cell. The base plate is closed and structureless, but the sides of the scales are elegantly ornamented by a meshwork of fibrils. The biflagellate *Mesostigma* probably represents a phylogenetic blind alley, but the presence of a flagellar pit and the elaborate Type 3 body scales suggest a possible linkage with *Pyramimonas*.

Platymonas and *Prasinocladus* are the only thecate or walled flagellate cells that have been associated with the Prasinophyceae. Chihara and Hori (1972) and Hori and Chihara (1974a) suggested that the two genera probably should be united into one. The flagellate cells of this complex have four scale-covered flagella emerging from a flagellar pit, and it is sometimes suggested, therefore, that they may be derived from *Pyramimonas*-like ancestors. Manton and Parke (1965) showed that the theca of *Platymonas tetrathele* is made from small stellate scales that are deposited by the Golgi cisternae into vesicles that then deposit them onto the cell surface. At that time the scales are fused together to form a bi-layered theca that surrounds the entire cell except for a slit at the base of the flagellar pit through which the four flagella emerge. Although the bi-layered theca of *Platymonas tetrathele* suggests possible dimorphic components in the development of the theca, Manton and Parke (1965) were explicit in their statement that there is one type of stellate particle from which the theca is formed. In studying *Platymonas convolutae*, however, Parke and Manton (1967) demonstrated that new thecae are made from two different kinds of particles independently formed within Golgi cisternae. Tufts of fibrils form the theca proper, and opaque spheres develop an extra superficial layer on the theca that appears alveolar in surface view. The theca of *Prasinocladus marinus* was shown to have a three-layered structure (Parke and Manton, 1965), but the innermost layer, a fibrillar one, probably represents a break-down of thecal material because it occurs only on discarded thecae. Evidence was found by Manton and Parke (1965) suggesting that the pyrenoid may influence fusion of stellate particles to form the theca in *Platymonas tetrathele*.

Attempts to form an alliance of the walled *Platymonas*-type prasinophyte cell with scale bearing genera may end with the *Pyramimonas*-type being considered as an ancestral form (Chadefaud, 1976; Stewart and Mattox, 1978). The elaborate array of body scales on *Pyramimonas*, as well as different Type 2 flagellar scales from those found on *Platymonas*, suggests that this association is not very close. *Pyramimonas* probably represents a highly derived type of cell within the Prasinophyceae (Fig. 1) not closely allied with *Platymonas*.

The one or two particle type of formation of thecae in *Platymonas* suggests that the phylogenetic lineage to that genus may have been directed from species having one or two body scale layers. Species having such scale layers are biflagellate, and among these types *Pseudoscourfieldia* has characteristics suggesting a possible phylogenetic linkage with *Platymonas* (Fig. 1). The presence of Type 1 flagellar scales in *Platymonas* along with the rod-shaped Type 2 flagellar scales, the same as in *Pseudoscourfieldia*, is convincing evidence of possible close relationship. An additional characteristic suggesting that *Platymonas* may have a more direct phylogenetic link with *Pseudoscourfieldia* than with other existing genera in the Prasinophyceae is the presence of peculiar, thick and curly hair-like appendages on the theca at the base of the flagellar pit in *Platymonas* (Manton and Parke, 1965; Schnepf and Maiwald, 1970; Norris - unpublished), and similar hairs being present on the scale layers within the flagellar grooves of *Pseudoscourfieldia marina* (Manton, 1975) and *P. longifilis* (Norris - unpublished). These thick hairs are present on many species of *Platymonas*, but *Pseudoscourfieldia* is the only scaly genus on which they have been observed.

Cell Covering in Aflagellate Stages. Aflagellate stages in Prasinophyceae seem to be either palmelloid or walled, a characteristic that they share with most other unicellular Chlorophyta. As mentioned above, *Platymonas* and *Prasinocladus* are the only known genera in which motile cells have a wall-like structure. Species of *Platymonas* often lose their flagella and remain non-motile for long periods of time, but during this non-motile period the cells often divide. Cyst stages in *Platymonas* are known for several species, but analysis of the cyst wall material, which often is ornamented with small protuberances, has not been done.

The theca of *Platymonas subcordiformis* was shown by Lewin (1958) to be composed of a polysaccharide which had galactose and uronic acid as major components with traces of arabinose and amino acids. Gooday (1971) examined the theca of *Platymonas tetrathele* and described its composition as pectin-like with galactose, galacturonic acid and arabinose as the major components. Gooday stated that perhaps Lewin's amino acid record referred to protein contamination in his preparations. Gooday also stated that the extracellular polysaccharide produced by *P. tetrathele* probably is the same as that found in the theca. Manton *et al.*, (1973) demonstrated that calcium is present in the theca of *Platymonas tetrathele*, probably as the calcium salt of galacturonic acid. They pointed out that accumulation of this salt, when compared with pectins of other more highly evolved plants, indicated that this biochemical activity was evolved very early in green algal evolution.

The phycoma of *Halosphaera* has a thick two-layered wall that may be without markings or variously marked with nodules, poroids or punctae on its outer surface. Parke and den Hartog-Adams (1965) reported that the inner wall is pectin-like and the outer wall is of an unknown composition, but that it may be a material similar to sporopollenin. Boalch and Mommaerts (1969) found that the inner wall of *Halosphaera parkeae* acts as a differentially permeable membrane. Jux (1969b) reported granular particles in *Halosphaera's* wall which supports earlier reports that siliceous inclusions occur in the outer wall (reviewed by Boalch and Parke, 1971).

Pachysphaera has a phycoma much like *Halosphaera's* in general appearance, and the wall may be poroid, punctate, nodulate or smooth (Parke, 1966). It is composed of two layers, the inner is pectic and the outer is composed of a complex lipoid substance (Wall, 1962). Jux (1969a) speculated that the inner wall of *Pachysphaera marshalliae* may be cellulose.

Pterosperma has a phycoma having ornamentation similar to *Pachysphaera*, but the presence of an ala or alae on the external face of the outer wall in *Pterosperma* is the main characteristic separating the genera. Parke *et al.* (1978)

showed that both the ala and the outer wall become yellow after treatment with KI_3. They also showed that *Pterosperma* phycomata also have an inner wall with the properties of a pectic-like material.

Species of Prasinophyceae that produce palmella stages have not been analyzed for the composition of the gelatin produced by the cells.

The above data do not seem to be sufficient to make any leading comments concerning the phylogeny of wall characteristics within the Prasinophyceae. The possible exception is that the apparent general absence of cellulose in this group may indicate its primitive nature among chlorophyll *b* containing plants.

INTERNAL CELL CHARACTERISTICS

Basal Bodies and Associated Flagellar Roots. The flagellar basal body apparatus (kinetosome) in prasinophyte flagellates is not well described for most species. The region is complex and difficult to describe from randomly sectioned specimens, but when detailed attention is given to its structure the information is of primary interest in studies of function and phylogeny of these flagellates. The basal body apparatus seems to form a functional unit governing motility of the cells that consists of several organelles and their associated satellite material. Prasinophyte flagellates have one of the most complex basal body systems so far as its structure is concerned, but it may be one of the most easily interpreted systems functionally because its units often are quite separate and distinct. The organelles in the basal body complex (Fig. 2) are: 1) the basal bodies themselves along with their associated transition regions to the flagella proper, and the associated satellite material which includes the synistosome (Norris and Pearson, 1975), 2) the rhizoplast, banded rootlet, or descending rootlets, 3) the microtubular rootlets or ascending rootlets, and 4) in some species, the microbody. Information on some, but not all of these organelles is available for most species that have been examined with the electron microscope, but little speculation has been forthcoming about the importance of these characteristics in the phylogeny of the organisms.

Position of basal bodies in relation to one another may have phylogenetic significance (Lembi, 1975). Unfortunately, too little is known at this time to make general conclusions, but some trends can be pointed out concerning this characteristic in prasinophytes.

Biflagellate species of prasinophytes or species having a second non-emergent basal body have basal bodies arranged at close to right angles to one another or at an angle more than 45°. Most prasinophyte flagellates having four or more flagella have the basal bodies arranged parallel to one another, but the proximal ends often may slightly converge toward one another. Lembi (1975) portrayed the orientation of basal bodies for two species of *Carteria*, a volvocalean flagellate, and showed that they have basically very different and probably not closely related systems. Lembi concluded that her group I type basal body apparatus was somewhat similar to that of *Chlamydomonas*, but only distantly related, whereas group II showed characteristics that led her to state that it

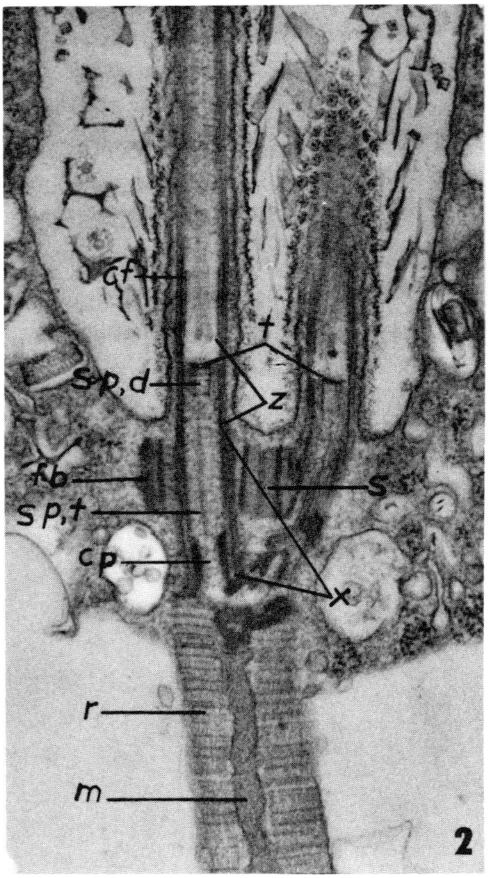

Fig. 2. Longitudinal section through base of flagellar pit of *Pyramimonas parkeae* showing transition zones and basal body apparatuses for two flagella (32,000x). cf = coiled fiber; cp = cartwheel pattern; fb = fibrillar band; m = microbody; r = rhizoplast; s = synistosome; sp,d = stellate pattern with doublet microtubules; sp,t = stellate pattern with triplet microtubules; t = transverse plate; x = region with triplet microtubules; z = transition zone.

may be derived from an asymmetric prasinophyte. Lembi's type II basal body apparatus in *Carteria* may reflect an arrangement of basal bodies that has been found in biflagellate prasinophytes, but more information is needed.

The parallel arrangement of basal bodies characteristic of quadriflagellate species of prasinophytes seems to be unique in the green monads, but much more investigation of the orientation of basal bodies is necessary before relationships can be estimated. The angled basal bodies of *Pyramimonas montana* (Maiwald, 1971) and *Pyramimonas reticulata* (Belcher, 1968a) suggests that they are not related to prasinophytes as we now understand the basal body system. This evidence supports earlier conclusions mentioned above that the former species be removed from Prasinophyceae.

It is interesting to note that the basal bodies of *Mesostigma viride*, a biflagellate species, are approximately parallel rather than angled as in other biflagellate prasinophytes, a characteristic lending support to speculation that *Mesostigma* may be more closely related to *Pyramimonas*-type prasinophytes rather than the other biflagellate species.

The Transition Zone. Manton (1964) and Dodge (1973) reviewed the structure of the transition zone in green algae, stating that most species have a basal disc at the junction between the free part of the flagellum and the basal body, and that a complex stellate structure connecting the ring of doublet microtubules is present on both sides of the basal disc. The stellate structure was first discovered in green algae by Lang (1963) for *Polytoma*. In sectional view the stellate structure of the transition zone in most green algae has an H-shape, the basal plate being the cross-bar, with the proximal part somewhat smaller than the distal part, as was illustrated by Manton (1964) for *Chlorogonium* (a relative of *Chlamydomonas*). The parallel lines of the H-shaped stellate structure, as viewed in longitudinal section, represent cylindrical fibrillar material. Several species that are sometimes assigned to the Prasinophyceae have the same type of arrangement of stellate material in the transition region: *Pyramimonas montana* (Maiwald, 1971), *Pedinomonas major* (Belcher,(1968b), *Pedinomonas noctilucae* (Sweeney, 1976) and *Micromonas pusilla* (Norris - unpublished). *Bipedinomonas rotunda* has only a very small proximal stellate section of doublets (Manton *et al.*, 1965), but most of the remaining uniflagellate, biflagellate or quadriflagellate prasinophyte species seem to have only distal stellate patterns in relation to the basal disc. It remains possible that additional research on this region in biflagellate species may show a more uniform structure in the transition region than is apparent at this time.

Species of *Pyramimonas* and *Platymonas* in which the transition region has been described seem to have no basal disc or only diffuse or incomplete septa. The central paired microtubules of the flagellum terminate in an electron transparent region inside the flagellum, and proximal to this region the stellate configuration connecting doublet microtubules characteristic for green algal flagella extends to approximately the plane of the cell surface. Below the surface of the cell the basal body becomes a ring of triplet microtubules that are connected to one another by strands of fibrillar material forming a stellate pattern similar to, but less regular than in the more distal region of doublet microtubules. Each fibril seems to connect with the central microtubule (B-tubule) in each triplet, but in some figures the fibrils seem to branch, a branch fibril connecting to each microtubule in the triplet. The stellate pattern in the triplet region of *Platymonas* extends for a much shorter distance than in *Pyramimonas*. In this region of *Pyramimonas parkeae* the triplets are also attached by fibrillar bundles to the synistosome and to the fibrillar band (Norris and Pearson, 1975). Proximally to the region of the synistosome and fibrillar band the pitch of the microtubular triplets increases and the cart-wheel configuration of basal bodies is established. The basal bodies terminate in this form. In *Pyramimonas parkeae* the open ended basal bodies are penetrated to a point just inside the triplets by a thin layer of electron dense material that attaches to branches of the rhizoplast.

Moestrup and Thomsen (1974) described a coiled fiber in *Pyramimonas orientalis* extending distally for a short distance around the terminal part of the central pair of flagellar microtubules. This structure is known to occur in flagella of Chrysophyta but had not previously been recorded for the green algae. The same structure has been observed in both longitudinal and cross sections in my material of *Pyramimonas parkeae* and *P. obovata*, and is present in some species of *Platymonas* (Norris - unpublished). It seems probable that the material identified as the stellate structure in *Pedinomonas minor* by Pickett-Heaps and Ott (1974) is actually a coiled fiber. In their longitudinal section there is a region proximal to the ends of the paired central microtubules that is probably the stellate structure instead.

Rhizoplast. The presence of a rhizoplast (banded rootlet) has been an important characteristic of prasinophyte flagellates in separating them from most of the Volvocales. Lembi (1975), however, described a rhizoplast in *Carteria radiosa*, a volvocalean flagellate having a distal fiber similar to *Chlamydomonas*. A rhizoplast was also shown to be present in *Dunaliella primolecta* by Hyams and Chasey (1974) and Eyden (1975), and its presence was suggested in *D. salina* (Trezzi *et al.*, 1964). Moestrup (1978) summarized the presence or absence of rhizoplasts in other green algal flagellate cells, and it can be observed in his tables that these structures are present in diverse types of green algae including filamentous and ulvoid species. The rhizoplast has been shown to be present in all scale-bearing Prasinophyceae that have been studied ultrastructurally except *Mantoniella squamata* and *Mesostigma viride*. In uniflagellate and biflagellate species there is a single rhizoplast usually extending from the basal bodies to a position in association with the nucleus and the chloroplast. In *Pedinomonas minor* the rhizoplast extends along the nuclear membrane next to the plasmalemma (Pickett-Heaps and Ott, 1974), whereas in *Monomastix* aff. *minuta* it penetrates more deeply into the cell alongside the inner face of the nucleus (Manton, 1970). Mattox and Stewart (1977) described a narrow and delicate rhizoplast for *Bipedinomonas angulata* that penetrates the cell along the nuclear envelope and terminates by attaching to a microbody located near the chloroplast's inner face. A similar rhizoplast was shown to be present in *Scourfieldia caeca* by Manton (1975), anchoring near the middle of the inner side of the chloroplast, but this rhizoplast branches as it approaches the basal bodies, a branch lodging at the base of each of the two basal bodies. A similar branching rhizoplast is found in *Pseudoscourfieldia marina*, but it is much more massive than the one in *Scourfieldia caeca* (Manton, 1975). *Pseudoscourfieldia longifilis* has a branching rhizoplast that is associated with a microbody near the point where it attaches to the inner chloroplast surface (Norris - unpublished).

Quadriflagellate species have two rhizoplasts per cell, one associated with each pair of basal bodies. Norris and Pearson (1975) showed how the rhizoplast branches immediately adjacent to the basal body pairs, and segments attach to each of the basal bodies. In *Pyramimonas* the rhizoplasts cover a centrally located microbody and terminate, after straddling the nucleus, in a centrally located cavity of the chloroplast. The rhizoplast of *Pyramimonas amylifera* branches proximally, apparently forming several attachment points on the nuclear and chloroplast surfaces.

The rhizoplasts of *Pyramimonas* are more delicate than the massive rhizoplasts found in *Platymonas* where they are almost cylindrical distally (near the basal bodies) and taper to splayed ends that attach near or to the plasmalemma near the middle of each broad face of the cell (Salisbury and Floyd, 1978). The plasmalemma is usually indented at the place where the rhizoplasts attach (Norris - unpublished). The rhizoplasts of *Platymonas* also cause indentations on the nucleus where they pass by, and a microbody usually is associated with the proximal regions of the rhizoplast (Norris - unpublished). In *Platymonas* a major part of the distal thick end of the rhizoplasts attach to the base of the microtubular roots that are the half-desmosomes of Schnepf and Maiwald (1970).

Until more information is available about rhizoplast form in additional species, little can be said about phylogenetic trends of this characteristic in Prasinophyceae. It seems, however, that a single rhizoplast, attaching to paired basal bodies may be a primitive characteristic with proximally branching types showing a slight advancement. Paired rhizoplasts evidently reflect doubling of flagellar number, from two to four, and these types may have developed from a branching rhizoplast as is found in *Pseudoscourfieldia* (Fig. 3). Branching of the rhizoplast occurs in *Pyramimonas* and *Platymonas* at points adjacent to their associated basal body pairs.

Platymonas, with its massive rhizoplast attaching to the plasmalemma, probably is the most advanced type in Prasinophyceae, although substantiating research on this structure is needed. Function of the rhizoplast has received considerable attention recently, speculation and experimentation receiving special stimulus from the work of Salisbury and Floyd (1978). Their experiments clearly demonstrate that the rhizoplast is a contractile organelle, a cyclic contraction and extension being induced by calcium and adenosine triphosphate. Salisbury and Floyd (1978) review the speculations on rhizoplast functions, anchoring organelles for the basal bodies that may also absorb and distribute stress generated by flagellar movements and/or a structure transmitting signals from other organelles in the cell. To these speculations they add the possible functions of initiation of flagellar power and recovery strokes, coordination of the stroke cycle and directional control of planar flagellar beat. It should be added that the rhizoplast may also function in stabilizing the positions of some major organelles within the cell such as the nucleus and chloroplast lobes. Also, Stewart *et al.*, (1974) showed convincing evidence that rhizoplasts of *Platymonas subcordiformis* become transformed into a "granular mass" from which the spindle microtubules emanate during nuclear division. They also suggest that the material in these spindle poles is consumed in the formation of the spindle microtubules, a hypothesis supported by the cyclic disappearance of rhizoplasts during mitosis when many new microtubules are formed in the cell. The rhizoplast does, indeed, seem to be a most important organelle within the prasinophyte flagellate cell, assisting in the maintenance of cell polarity as well as in the formation of mitotic spindles. Also, this material of the rhizoplast which is now known to be contractile, may have an important role in not only movement of flagella, but also in movement of particles within the cell, including, perhaps, the chromosomes during mitosis.

Moestrup (1978) measured and recorded the periodicity of rhizoplast crossbanding for the species in which it could be ascertained. Although this characteristic may be important in identifying certain types of organisms, now that

---

Fig. 3. Possible evolutionary tendencies of prasinophyte flagellates from the point of view of flagellar roots.

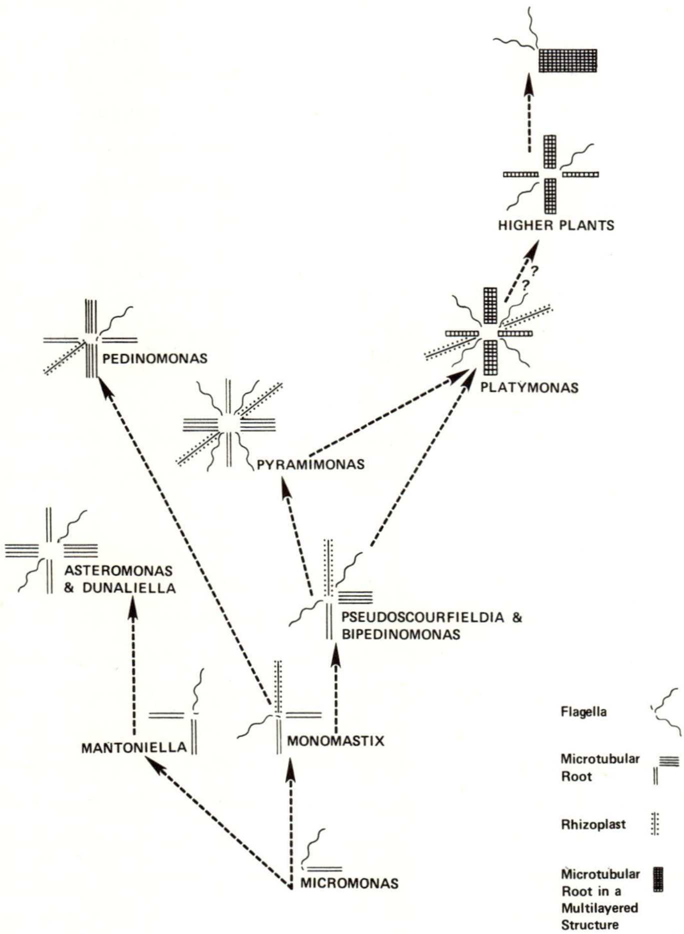
HIGHER PLANTS
?
?
PEDINOMONAS
PLATYMONAS
PYRAMIMONAS
ASTEROMONAS
& DUNALIELLA
PSEUDOSCOURFIELDIA &
BIPEDINOMONAS
MANTONIELLA
MONOMASTIX
MICROMONAS
Flagella
Microtubular
Root
Rhizoplast
Microtubular
Root in a
Multilayered
Structure

FIGURE 3

the rhizoplast is known to be contractile, the present information needs to be reassessed in that context.

<u>Microtubular Roots</u>. Groups of microtubules, usually associated with electron dense fibrillar material, are well known to be part of the basal body apparatus for chlorophyll *b* containing green flagellates. Symmetrical green flagellates have almost consistently been shown to have four microtubular rootlets, two identical sets, emanating from the basal body system in a cruciate pattern. These rootlets follow the contour of the cell body and terminate toward the posterior part of the cell. In some species numerous cytoplasmic microtubules lie immediately beneath the plasmalemma forming a supplementary structural system in addition to the rootlet microtubules which lie just inside this layer. Brown *et al.*, (1976) showed that cytoplasmic microtubules attach to the microtubular rootlets in *Polytomella agilis*, a volvocalean flagellate. Considerable attention has recently been given to microtubular rootlets in green algae, particularly in an attempt to assess their phylogenetic importance (Moestrup, 1978; Stewart and Mattox, 1978). Moestrup (1978) presents a very thorough review of the numbers of microtubules in rootlets for the species that have been analyzed, and Stewart and Mattox (1978) give their interpretation of how this characteristic may be used in showing phylogenetic relationships.

Microtubular roots are anchored to the middle or lower mid-region of the basal body apparatus in the dense fibrillar material that encloses this area. In contrast to the rhizoplast, which is the other type of rootlet attached to the basal bodies, microtubular rootlets do not penetrate deeply within the protoplast, but, instead, they lie close to the plasmalemma and follow the contour of the cell. Because of this fact, it is believed that the shapes of these wall-less cells are regulated by the microtubules of the rootlets. Some species probably have no additional structural microtubules other than those in the rootlets, but, as mentioned above, other species have additional cytoplasmic microtubules that branch from the rootlets and add a structural component. *Pyramimonas* is a good example of a large cell that has many additional structural cytoplasmic microtubules, whereas most of the small asymmetrical species such as *Nephroselmis* appear to have only the rootlet microtubules serving a structural function. Some of these species, however, do have a layer of cytoplasmic microtubules: *Monomastix* aff. *minuta* (Manton, 1970), *Mesostigma viride* (Manton and Ettl, 1965) and *Bipedinomonas angulata* (Mattox and Stewart, 1977).

Little is known about the number of microtubules in asymmetrical species of prasinophytes, but Moestrup (1978) summarized information for four species (*Mantoniella squamata*, *Monomastix* aff. *minuta*, *Bipedinomonas rotunda* and *Pedinomonas minor*). To these may be added information on *Bipedinomonas angulata* (Mattox and Stewart, 1977) who described rootlets with seven or two to three microtubules. With the information that is available it is suggested that asymmetrical prasinophyte cells may have only one set of microtubular roots, a set usually consisting of two roots, one having two microtubules and the other having a varying number of microtubules, but usually more than two. *Bipedinomonas rotunda*, for example, has one basic two-microtubular rootlet and one rootlet having seven microtubules. *Mantoniella* has the basic two microtubular rootlet and another with four microtubules. *Monomastix* aff. *minuta* has two rootlets with two microtubules in each.

Pedinomonas minor seems to be unusual among the asymmetrical green flagellates in having two sets of microtubular roots, and this information may reflect a more complicated phylogeny in which this species may be a derived type rather than basically a primitive species. Additional research on this characteristic in asymmetrical prasinophyte flagellates will probably reveal information of great importance to our understanding of prasinophyte phylogeny.

It is my opinion that symmetrical species of Prasinophyceae, in which two sets of microtubular rootlets have consistently been shown to be present, represent an advanced stage in phylogeny of these flagellates. Duplication of the sets has probably occurred repeatedly in evolution of different genera by incomplete cytokinesis and without complete duplication of cell organelles. It may be no accident, therefore, that most symmetrical cells in the Prasinophyceae have four flagella, each pair of flagella having one set of microtubular rootlets. Furthermore, it can be shown that biflagellate symmetrical cells in the green algae have two sets of microtubular rootlets, and it may be speculated that these types have been derived from quadriflagellate species, but the duplicate rootlet system was retained even though flagella were lost.

Because two microtubule rootlets are found in every species of green flagellate cell (except some highly advanced types of zoospores), this type of rootlet must be a basic type in this group. It may be hypothesized, therefore, that the most primitive green flagellate cell, probably having only a single flagellum, had a two microtubule rootlet associated with its basal body. *Micromonas pusilla* may be this type of primitive green flagellate, and sections of this organism often do show a single pair of microtubules beneath the plasmalemma near the basal body (Norris - unpublished). Additional investigations utilizing serial sections of *Micromonas pusilla* are needed before final analysis of this interesting organism in its possible role as the most primitive of green flagellate cells. Particular attention should be given to its microtubular root system.

Stewart and Mattox (1978) express the opinion that the flagellar pit is a primitive characteristic of green flagellates, and that zoospores and gametes of more highly evolved green algae have lost the flagellar pit. Microtubular rootlets in species having a flagellar pit extend anteriorly along the pit membrane, and Stewart and Mattox (1978) hypothesize that flagellate cells without a pit have descending microtubular rootlets that evolved from "half" a cell with a flagellar pit, thereby losing the pit characteristic. More importantly, Stewart and Mattox also state that they believe that the descending microtubular rootlet, which in some more highly evolved species becomes a multilayered structure, is derived by a conversion of the rhizoplast to a microtubular rootlet, and they provide evidence that such a conversion from fibrillar to microtubular structure may be possible.

I cannot deny the hypothesis that Stewart and Mattox (1978) propose, but I believe that an equally supportable speculation is possible and may bear more supporting evidence from our present knowledge of these organisms. Basic to my hypothesis is the assumption that the rhizoplast and microtubular rootlets are two separate systems that have not been converted one to another. Also, I consider it basic that the microtubular rootlets almost always show strong affinity to regions of the cell near the plasmalemma, they are not deeply embedded in the protoplast as is the rhizoplast. It is probably because of this strong affinity between the microtubular rootlets and the plasmalemma that these rootlets are thought to be structural in function, giving form to a wall-less cell. Therefore, the ascending nature of microtubular rootlets in species having a flagellar pit is more incidental than phylogenetically basic. It is structurally natural for the microtubular rootlets to be descending along the plasmalemma in species not having a flagellar pit such as *Asteromonas*, *Dunaliella* and even the walled *Chlamydomonas*. I propose, therefore, that more advanced green algal flagellate cells, whether or not they are symmetrical or asymmetrical, have retained the same type of microtubular rootlet system that is found in more primitive species, such as the prasinophytes, and that this microtubular system has not been derived through conversion of rhizoplast into microtubular structure. My hypothesis states that in more highly advanced flagellates the rhizoplast may have been lost or may have been converted into lamellate structures present in the basal body apparatus.

The very interesting problem of derivation of the "multilayered structure"

(MLS) (often called "viergruppe" by European authors) is discussed by Stewart and Mattox (1978) and by Moestrup (1978). The MLS is found in flagellate cells of bryophytes and other higher plants as well as in zoospores or gametes of *Klebsormidium*, *Coleochaete*, *Chaetosphaeridium*, *Chara*, *Nitella* and *Trentepohlia*. The multilayered structure consists of three or four layers of alternating densely and lightly staining material, but one of the layers, the innermost, is composed of a band of parallel microtubules (approximately 60 in *Chaetosphaeridium*). Stewart and Mattox (1978) state that multilayered structures are unknown in scaly green flagellates, thereby leaving a gap in the proposed phylogenetic scheme in which they are linked to the Charophyceae. Evidently they have overlooked the microtubular rootlet system of *Platymonas* which was described by Schnepf and Maiwald (1970) as a "halbdesmosome". In this system fibrils connect the wall at the base of the flagellar pit to a sub-plasmalemmal multilayered structure in which the innermost layer consists of either two or four (in some species five) microtubules. I have found that this structure is persistent within the many species of *Platymonas* that I have studied, as well as in *Prasinocladus*. Each cell has four microtubular rootlets ascending to the plasmalemma at the base of the flagellar pit where they terminate in a multilayered structure. There are two sets of microtubular rootlets in each cell, one having two microtubules and the other having four or sometimes five microtubules. In addition to the microtubules in the rootlets described above, *Platymonas* also has a system of sub-plasmalemmal microtubules somewhat similar to those found in *Pyramimonas*.

A summary of the possible evolutionary tendencies of the prasinophyte microtubular rootlet system, leading into the Charophyceae, is provided by the diagram in Fig. 3. The most primitive type possibly is exemplified by *Micromonas pusilla* in which two microtubules serve as the rootlet for a single flagellar basal body. A rhizoplast is not present in this species, but is present in *Monomastix* aff. *minuta*, a species having a second basal body as well as a second two microtubular rootlet. In biflagellate asymmetrical prasinophytes the cell size is usually greater than in the uniflagellate types. Coordinated with the increase in cell size and flagellar number is an increase in numbers of microtubules in one of the rootlets. These may be the first sign of supplementary microtubules (usually called cytoplasmic microtubules) that are much more highly manifested in the large-celled quadriflagellate species. It is to be expected that increase in numbers of microtubules should be coactive with increase in cell size, but it is particularly interesting to note that one member of the rootlet set retains the conservative number of two microtubules while the second one has an irregular number sometimes even within the same species. This conservative rootlet seems to persist throughout the evolution of the prasinophytes. It is not denied that a great deal more attention is needed for the correct analysis of this characteristic in green flagellates, but it is hoped that the ideas presented here may serve to stimulate this type of research.

Special association of a microtubular rootlet with the stigmatal region of the chloroplast has not been demonstrated for most Prasinophyceae, although such an association is suggested for *Pedinomonas minor* (Ettl and Manton, 1964; Pickett-Heaps and Ott, 1974).

Chloroplast. Most species that have been attributed to the Prasinophyceae have a photosynthetic system, usually present as a single body, the chloroplast. Flagellate species usually have a single chloroplast that often is massive, occupying most of the interior space of the cell. Chloroplasts of flagellate cells are bowl-shaped and are invariably oriented with their open end facing the basal body apparatus, even when this system is lateral to the long axis of the cell. It seems probable that this orientation is in response to phototactic sensitivities of the cell that are regulated by the basal body system and the flagella. In such a position the chloroplast may be placed in its most advantageous position for activation of its photoreactive system, including the stigma.

The part of the chloroplast that lies opposite to the basal body apparatus is usually the thickest region and encloses the pyrenoid. The margins of the bowl-shaped chloroplast follow the contours of the cell and often are shallowly to deeply lobed, the lobes often following similar lobing of the cell itself. Chloroplasts may occupy from approximately 30% to 90% of the cell periphery in flagellate cells with large celled species having a higher percentage cover than small celled species.

On the level of the light microscope the main structural characteristics of prasinophyte chloroplasts are mostly caused by arrangements of starch grains within the chloroplast membranes. The starch grains that surround the pyrenoid are usually curved or cup-shaped, fewer in number than in other green algae, and their reaction to staining with KI_3 is somewhat different from most other members of the Chlorophyta. Most prasinophytes that have been tested for starch reaction with KI_3 do not have their granules turning the blue-black color that is typical for other green plants; instead, starch grains of prasinophytes usually turn a reddish-purple or reddish-brown color with iodine. Susuki (1974), working with *Platymonas*, demonstrated the presence of starch and D-mannitol, the starch being similar to that of other green plants. Salisbury and Floyd (1978) found that the starch shell of the pyrenoid in *Mantoniella squamata* is composed primarily of amylose rich starch.

Small celled species seem to have their starch reserves restricted to the pyrenoid starch shell, additional starch grains rarely being found in other parts of the chloroplast. Larger cells, such as species of *Pyramimonas* and *Platymonas* have numerous additional starch grains distributed in other parts of the chloroplast. Starch grains that are independent of the pyrenoid are not cup-shaped, but have a form different from those that surround the pyrenoid.

Globular lipid droplets often are found in prasinophyte cells, and these globules may be colored by an accumulation of carotinoid pigments in some species. The lipid droplets may occur most anywhere in the chloroplast or in the cytoplasm. Most flagellate species of prasinophytes have other lipid droplets in definitely organized eyespots within the chloroplast, and they may be in various positions on the peripheral side. No prasinophyte species is known with certainty to have extra-chloroplastic eyespots, but some of the illustrations in Butcher (1959) indicate that this may be possible in some asymmetrical species, a condition that would be most unusual for green algae, and one tends to believe that the illustrations are in error or that the flagellates illustrated are not members of Chlorophyta. Several asymmetrical prasinophytes are reputed to be without stigmata, and a few species of *Pyramimonas* and *Platymonas* are described without stigmata. Other species of these genera, however, have been described with one or more than one eyespot. The presence of two stimata lying side by side but on different lobes of the chloroplast is commonly found in several of the small-celled species of *Pyramimonas*. On a fine structural level, the eyespot has no structural difference from other green algal eyespots. They are usually composed of a single layered or a two-layered plate of globules that lies immediately within the chloroplast membranes on the side adjacent to the plasmalemma. *Pyramimonas parkeae* has an eyespot consisting of two to four layers of lipid globules. It is in the cell lobe in which the nucleus is also present, and Norris and Pearson (1975) noticed that the fibrillar material in which the lipid globules lie is more electron dense on the interior side of each row of droplets. This material may be significant in functioning or positioning of the stigma. The many layered eyespot of *Pyramimonas montana* (Maiwald, 1971) is unusual for Prasinophyceae.

Halldal (1961) used *Platymonas subcordiformis* for many of his classic experiments on phototaxis. He concluded that a carotenoprotein is the photo-receptive pigment complex involved in photic orientation of that species.

Arrangement of thylakoids in prasinophytes is the same as in other green algae. Thylakoids occasionally occur singly, but mostly they are in lamellae of from two to five, and interconnecting thylakoids are commonly found. In *Pyramimonas reticulata* the thylakoids are stacked into grana, a characteristic more like higher plants than most green algae (Belcher 1968a). This characteristic, among others of that species, may indicate that this taxon belongs in another genus outside the Prasinophyceae.

The pyrenoid seems to be the only part of the prasinophyte chloroplast that has characteristics that may be considered as distinctive for this group. Chloroplasts having a single pyrenoid are the most common types in Prasinophyceae, but a few species of *Pyramimonas* are reported to be without pyrenoids, whereas other species in that genus are reported to have two or four pyrenoids. A second very small peripheral pyrenoid-like structure was reported for *Bipedinomonas angulata* in addition to a larger, more typical pyrenoid (Mattox and Stewart, 1977).

Pyrenoids having a solid core (matrix) are described for *Micromonas pusilla* and *Mantoniella squamata* (Manton, 1959; Manton and Parke, 1960). Salisbury and Floyd (1978) examined the chemical nature of the pyrenoids in *Mantoniella* and found them to be composed of complex polypeptides and to have ribulose diphosphate carboxylase activity.

Pyrenoids of the remaining prasinophytes have the core invaded by either thylakoids or by cytoplasmic channels that are membrane bound. These two types of pyrenoid intrusions seem to be unrelated phylogenetically, and probably reflect two independent lines of evolution as it shows in Fig. 4. Few thylakoids invade the pyrenoids of *Pedinomonas* and *Bipedinomonas*, and they penetrate from any direction. Pyrenoids having similar irregular thylakoidal invasions are also found in *Mesostigma* and in some species of *Pyramimonas*. Other species of *Pyramimonas* have thylakoids forming a regular pattern within the pyrenoid, parallel thylakoids in *P. parkeae* and undulating thylakoids in *P. amylifera*. In *Pyramimonas obovata* and *P. virginica* thylakoids invade the pyrenoid as a group, all penetrating at a point near the nucleus inside the chloroplast cup region.

Pit-like intrusions of the pyrenoid, that are membrane-bound cytoplasmic papillae, occur in *Pseudoscourfieldia marina*. It is also interesting to note that a mitochondrial lobe also seems to fill the papillum in some places (Man-

---

Fig. 4. Possible evolutionary tendencies of prasinophyte flagellates from the point of view of the pyrenoid.

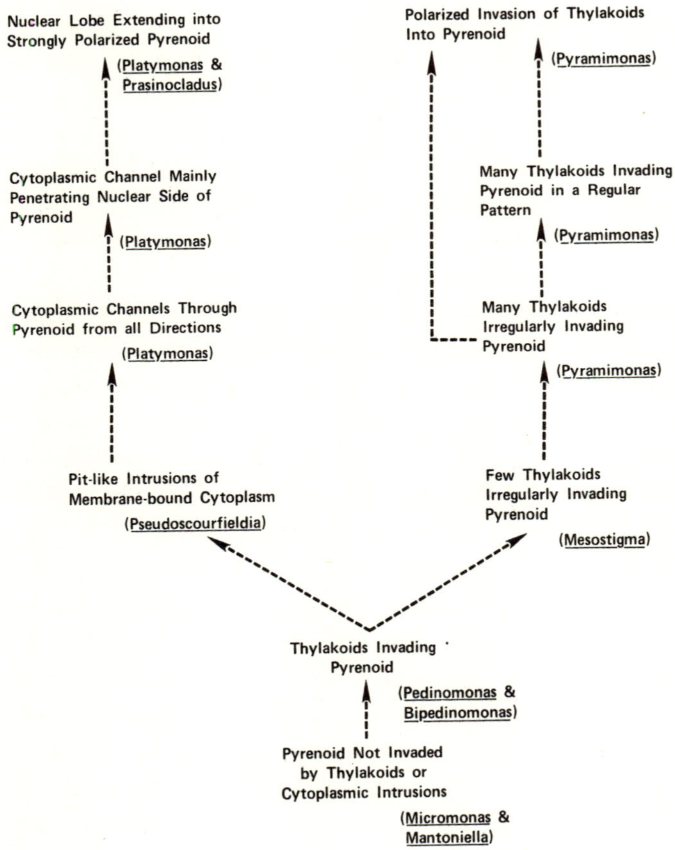
Nuclear Lobe Extending into Strongly Polarized Pyrenoid
(Platymonas & Prasinocladus)
Cytoplasmic Channel Mainly Penetrating Nuclear Side of Pyrenoid
(Platymonas)
Cytoplasmic Channels Through Pyrenoid from all Directions
(Platymonas)
Pit-like Intrusions of Membrane-bound Cytoplasm
(Pseudoscourfieldia)
Polarized Invasion of Thylakoids Into Pyrenoid
(Pyramimonas)
Many Thylakoids Invading Pyrenoid in a Regular Pattern
(Pyramimonas)
Many Thylakoids Irregularly Invading Pyrenoid
(Pyramimonas)
Few Thylakoids Irregularly Invading Pyrenoid
(Mesostigma)
Thylakoids Invading Pyrenoid
(Pedinomonas & Bipedinomonas)
Pyrenoid Not Invaded by Thylakoids or Cytoplasmic Intrusions
(Micromonas & Mantoniella)

FIGURE 4

ton, 1975). Some species of *Platymonas* and *Prasinocladus ascus* also have cytoplasmic papillae penetrating the pyrenoid from all directions, but mitochondrial lobes have not been seen within them (Chihara and Hori, 1972; Hori and Chihara, 1974b). Other species of *Platymonas* have the pyrenoid penetrated by cytoplasm concentrated in one area that is adjacent to the nucleus. A large branching lobe of cytoplasm deeply penetrates the pyrenoid in these species. A few other species of *Platymonas* and *Prasinocladus marinus* have nuclear lobes penetrating the pyrenoid along with the lobe of cytoplasm (Parke and Manton, 1965; Hori and Chihara, 1974a).

Many different kinds of green algae are known to have thylakoidal penetrations of the pyrenoid, but canaliculi in pyrenoids filled with membrane-bounded cytoplasm are relatively rare, having been reported in *Ankyra* (Swale and Belcher, 1971) and in the Oedogoniales (Hoffman, 1968).

Nucleus. Nuclei in prasinophyte cells are little different from other green algal nuclei apparently always containing a nucleolus that may or may not be visible with the light microscope. Only uninucleate flagellate prasinophyte cells are known. Unlike *Chlamydomonas* where species are known to have nuclei in most any position within the cell (Ettl, 1976), prasinophyte nuclei are mostly associated with the flagellar end of the cell, often having distinct attachments and orientation to parts of the basal body complex. In most uniflagellate and biflagellate species the nuclei are laterally positioned in the flagellar end of the cell, the opposite side of the cell sometimes being occupied by a lobe of the chloroplast. Nuclei in such positions in *Pedinomonas tuberculata* (Manton and Parke, 1960), *Monomastix* aff. *minuta* (Manton, 1970) and *Bipedinomonas rotunda* (Manton *et al.*, 1965) are strongly compressed against the plasmalemma and are discoid in shape. Uncompressed nuclei in other species are variously shaped, often being spheroidal, ellipsoidal or angular with small lobes.

Nuclei of most *Pyramimonas* species are anterio-lateral in position, often at least partly extending into one of the anterior lobes of the cell above the flagellar pit. The nucleus of flagellate cells of *Halosphaera viridis*, *H. minor* and *H. russellii* have a centro-lateral position (Parke and den Hartog-Adams, 1965; Pennick, 1977b). Manton *et al.* (1963) and Parke and den Hartog-Adams, (1965) describe uniflagellate as well as beflagellate cells in cultures of *H. russellii*, but Parke and den Hartog-Adams (1965) state that the uniflagellate cells "can revert to the 'normal' four-flagella type cell and its progeny can transform into cysts." Such evidence seems to strengthen my proposal that quadriflagellate genera such as *Pyramimonas* and *Halosphaera* may be derived from species with fewer flagella. The anterio-lateral position of the nucleus in such quadriflagellate species may also be related to the derivation of such cells from types with fewer flagella, probably by incomplete division of the cells as basal bodies are replicated.

Platymonas and *Prasinocladus* have centrally located nuclei that lie adjacent to the pyrenoid. The nucleus of *Prasinocladus marinus* has a large lobe that deeply penetrates the pyrenoid part of the chloroplast.

The nuclei of *Pyramimonas reticulata* (Belcher, 1968a) and *P. montana* (Maiwald, 1971) are centrally located in the cell and do not seem to have an intimate association with the basal body apparatus as is found in other species of *Pyramimonas*. The position of the nuclei in these species in relation to the Golgi apparatus is also different from other *Pyramimonas*, and these differences give additional support to my contention that the two species named above should be removed from *Pyramimonas* to another genus, probably *Pyramichlamys* Ettl.

Mitochondria. Prasinophyte mitochondria usually occur in central positions in the cell or toward the base of the flagella. They are usually not found in peripheral positions in the cell as in some species of *Chlamydomonas* (Arnold, 1972; Ettl, 1976). *Platymonas convolutae* (Parke and Manton, 1967) and zoospores of *Halosphaera* (Manton *et al.*, 1963; Pennick, (1977b) have mitochondria located in a space between the chloroplast and plasmalemma, an unusual situation for mitochondria in prasinophyte cells. In species of *Pyramimonas* where extensive cytoplasmic space may exist between chloroplast lobes, mitochondria may also be expected to be present in these positions (see *P.* aff. *disomata*, Norris and Pienaar, 1978). In most prasinophytes the mitochondria tend to be located adjacent to the inner side of the chloroplast, the organelles often being so close together that part of the chloroplast and mitochondrial membranes touch. A single mitochondrion is reported in cells of *Micromonas pusilla* (Manton, 1959), *Pedinomonas minor* (Ettl and Manton, 1964; Pickett-Heaps and Ott, 1974), *Pedinomonas major* (Belcher, 1968b). *Monomastix* aff. *minuta* (Manton, 1970), *Mesostigma viride* (Manton and Ettl, 1965), *Mantoniella squamata* (Manton and Parke, 1960; Barlow, 1977), and *Pseudoscourfieldia marina* (Manton, 1975). Sweeney (1976) reported up to five mitochondrial profiles at the flagellar end of the cell in sections of *Pedinomonas noctilucae*, but it seems possible that these could be profiles of a single convoluted mitochondrion. *Nephroselmis gilva* (Parke and Rayns, 1964), *Bipedinomonas angulata* (Mattox and Stewart, 1977) and species of *Pyramimonas*, *Platymonas* and *Prasinocladus* (Manton, 1966; Belcher, 1968a; Hori and Chihara,1974a,1974b; Norris and Pearson, 1975; Pennick, 1977a) have several mitochondria in each cell, all of which are located adjacent to the interior side of the chloroplast or near to the flagellar pit region of the cell. *Asteromonas gracilis* also has mitochondria in similar positions (Peterfi and Manton, 1968). In *Pyramimonas amylifera* Manton (1966) reported a mitochondrial layer that lies adjacent to the basal body system. The mitochondrion of *Pseudoscourfieldia marina* apparently forms a "peg-like" extension that penetrates deeply into the core of the pyrenoid (Manton, 1975), a most unusual structure that tempts one to speculate on interactions between the enzyme systems of the pyrenoid and mitochondrion.

Mitochondria in prasinophyte cells are usually elongate, sausage-shaped structures that sometimes show branching. The mitochondrion of *Pedinomonas minor* is reported to be a curved disc-shape (Ettl and Manton, 1964).

Mitochondrial cristae in prasinophytes are usually reported to be flattened, but they are never so flat as in the Euglenophyceae where they often appear to be discoidal on a constricted base. Prasinophyte mitochondrial cristae seem to be mostly slightly flattened tubules that may sometimes become more expanded in certain areas, even appearing to be inflated in some profiles. The only tubular cristae in prasinophytes are reported for *Micromonas pusilla* (Manton, 1959) and *Pyramimonas montana* (Maiwald, 1971) but it seems possible that their tubular appearance may be due to fixation problems. Cristae often are curved in one direction within mitochondria, and the matrix within the cristae is of less density in electron microscope preparations than is the matrix outside the cristae. Mitochondria having "bristly" surfaces on the cristae are sometimes seen in Prasinophyceae. Fine fibrils that probably represent mitochondrial DNA have been seen in *Prasinocladus ascus* (Hori and Chihara, 1974b).

Information on prasinophycean mitochondria is incomplete, the observations on them being incidental, usually, to studies with other objectives. There may be no obvious phylogenetic trend in mitochondrial structure or position within the cell in prasinophyte flagellates, but when compared with other green flagellates it seems entirely possible that evolutionary trends may be revealed using these characteristics. At this time prasinophyte flagellate mitochondria seem to have characteristics similar to those of the more primitive species of *Chlamydomonas* according to Ettl's scheme (1976).

Golgi Apparatus, Endoplasmic Reticulum and Scale Reservoir. The Golgi apparatus is a prominent part of prasinophyte flagellate cells, usually being situated near the basal body system adjacent to the nucleus. The number of dictyosomes seems to be a relatively constant characteristic for each species. Generally there are two to several dictyosomes in a species having a flagellar pit, whereas there is a single dictyosome in most uni-and biflagellate species.

Bipedinomonas rotunda is exceptional in being biflagellate and having two to three dictyosomes (Manton *et al.*, 1965). The number of dictyosomes in *Mesostigma viride*, a species having a flagellar pit, is uncertain but it is likely to be more than one (Manton and Ettl, 1965).

The dictyosome maturing face is invariably directed toward the flagellar pit in scale producing species having this cell depression. *Pyramimonas montana*, a species having neither a typical flagellar pit nor scales, has the dictyosome maturing face directed toward the nucleus, greatly differing in this way from other species of *Pyramimonas* (Maiwald, 1971). *Pyramimonas reticulata* also differs from other species in that genus by having the dictyosome situated posterior to the nucleus, and not in a position oriented toward the flagellar pit (Belcher, (1968a).

In uni-and biflagellate species dictyosome maturing faces usually are directed away from the nucleus or at least perpendicular to the nuclear membrane. The maturing face is directed toward the nucleus, however, in *Monomastix* aff. *minuta* (Manton, 1970) and *Nephroselmis gilva* (Parke and Rayns, 1964), a situation similar to the arrangement in *Asteromonas gracilis*, where there are several dictyosomes (Peterfi and Manton, 1968), *Dunaliella primolecta* (Eyden, 1975), *D. salina* (Trezzi *et al.*, 1964) and *D. bioculata* (Marano, 1976), although these scale-less volvocalean species should be examined more carefully for this characteristic.

Pedinomonas minor and *Pseudoscourfieldia marina* have their dictyosome forming faces very close to the nuclear membrane, a position that may be considered to be uncomplicated and, perhaps, primitive. Endoplasmic reticulum in these species seems to remain mostly close to the nucleus. In other biflagellate species the endoplasmic reticulum becomes more widely distributed, perhaps allowing the dictyosome forming face to become more remote from the nucleus and turned in a different direction in relation to the nucleus. The forming face, therefore, may become perpendicular to the nuclear surface as in *Pseudoscourfieldia longifilis* (Norris - unpublished), *Bipedinomonas rotunda* (Manton, *et al.*, 1965) and *Bipedinomonas angulata* (Manton *et al.*, 1965). The development of such peridictyosomal endoplasmic reticulum allows the dictyosome to be more remotely positioned in relation to the nucleus as is sometimes seen in *Mantoniella squamata* (Manton and Parke, 1960; Barlow, 1977), and in species having more than one dictyosome. Peridictyosomal endoplasmic reticulum also allows for flexibility in directions of dictyosome forming faces, even allowing them to be directed toward the nucleus as in *Monomastix* aff. *minuta* and *Nephroselmis gilva*.

Norris and Pearson (1975) speculated that dictyosome cisternae may be derived from a periplastidial endoplasmic reticulum system as well as, or instead of, the perinuclear endoplasmic reticulum in *Pyramimonas parkeae*. Moestrup and Walne (1979) describe the dictyosomes of *Pyramimonas tetrarhynchus* as being derived from an extensive endoplasmic reticulum system extending from the nuclear envelope and along the inner side of the chloroplast. Undoubtedly the endoplasmic reticulum system in *Pyramimonas* is continuous with the nuclear membrane system and probably not continous with the chloroplast membranes. It seems entirely possible, however, that organic materials may be contributed to the endoplasmic reticulum system by vesicles derived from the chloroplast membranes, although transfer of organic materials in solution through the membranes also is possible.

It is well recognized that prasinophyte cell scales are manufactured in dictyosome cisternae (Manton *et al.*, 1963; Manton *et al.*, 1965; Manton and Ettl, 1965; Manton, 1966; Manton, 1968; Manton, 1970; Maiwald, 1971; Moestrup and Thomsen, 1974; Norris and Pearson, 1975; Manton, 1975; Pennick *et al.*, 1976; Pennick, 1977a; Norris and Pienaar, 1978; Moestrup and Walne, 1979). Also, it has been shown that scale-like particles formed in Golgi cisternae of *Platymonas* are structural elements used by the cell in formation of its theca (Manton and Parke, 1965; Parke and Manton, 1967).

One of the most remarkable abilities of prasinophyte cells is in the internal formation of scales that are then deposited on the surface of the cell. Early stages in formation of scales in Golgi cisternae are illustrated and discussed by Norris and Pearson (1975) and Moestrup and Walne (1979). Norris and Pearson (1975) also showed that scales of more than one type may be formed in a single dictyosome cisterna, an observation confirmed by Moestrup and Walne (1979). Most remarkable in scale formation are species of *Pyramimonas* in which six different types of scales (including hair scales) are manufactured and then deposited not only in two different regions of the cell surface, flagella and cell body, but also the scales are aligned in specific layers on the cell membrane.

Manton (1966), Norris and Pearson (1975) and Moestrup and Walne (1979) demonstrated that the two larger types of body scales and the limuloid flagellar scales are always formed with their dorsal sides facing the forming face of the dictyosome. Apparently there is no such precise orientation of the small body and flagellar scales. These small scales are always formed in peripheral regions of dictyosome cisternae, whereas the large scales are formed in central regions of the cisternae. Hairscales seem to develop near peripheral regions of the cisternae, but not as marginal as the small scales.

Norris and Pearson (1975) describe a peripheral endoplasmic reticulum system in *Pyramimonas parkeae* that contains a precipitated fibrillar material when viewed in sections for the electron microscope. This material may be an adhesive that is secreted outside the plasmalemma, causing the scales to adhere to the surface of the cell.

Manton *et al.* (1963) proposed that scales manufactured by dictyosomes in *Halosphaera* zoospores are carried to the cell surface singly in small vesicles or they may accumulate in small, membrane bounded reservoirs that then fuse with the plasmalemma, thereby releasing the scales. Manton (1966) showed that *Pyramimonas amylifera* produces a complex scale reservoir having internal villi on which scales are arranged. She also showed that the scale reservoir has a duct opening into the flagellar pit near its base. Manton proposed that flagellar scales are liberated to the flagellar membrane at the base of the flagellar pit intermittantly rather than continuously, and that they assume positions on the flagella from this location. Not finding cell body scales in the scale reservoir of *Pyramimonas amylifera*, Manton suggested that body scales were released to the plasmalemma surface by "some process of pinocytosis in reverse." Because the plasmalemma is underlain by numerous microtubules in most anterior regions of the cell except the flagellar pit, Manton suggested that such liberation of body scales also occurred in the flagellar pit. Manton (1968) proposed a similar method of scale deposition for *Pyramimonas tetrarhynchus*. In this species, however, there is a simpler scale reservoir without villi, and body scales often accumulate in larger vesicles that fuse with the outer cell body plasmalemma, thereby liberating the scales. She noted that such vesicles had access to the plasmalemma only through gaps in the superficial endoplasmic reticulum and wondered if the endoplasmic reticulum may have a role in body scale liberation.

Maiwald (1971), Moestrup and Thomsen (1974), Norris and Pearson (1975) and Pennick *et al.* (1976) have shown that body scales as well as flagellar scales are to be found within the scale reservoir in several different species of *Pyramimonas* (*P. orientalis*, *P. parkeae* and *P. obovata*). Speaking collectively of all scale types produced by cells of *P. grossii*, *P. virginica* and the *Pyramimonas*-like motile stage of *Halosphaera russellii*, Pennick and Clarke (1976), Pennick (1977a) and Pennick (1977b) state that all types of scales are "transported to the scale reservoir, from whence they emerge into the flagellar pit and onto the body surface." It seems possible, therefore, that most species of *Pyramimonas*, as well as the motile cells of *Halosphaera*, have all of their scales distributed to the cell surface through a scale reservoir. This type of scale production is different from biflagellate and uniflagellate prasinophytes in which no scale reservoir has been demonstrated, the scales probably being directly released onto the plasmalemma through a process resembling pinocytosis in reverse. It is interesting to note that *Pyramimonas amylifera* may be an exception among species of that genus because body scales have not been found inside the reservoir by Manton (1966) nor by Norris and Pienaar (1978). Pennick (1978) did not clarify the problem, but inferred that all types of scales occur in the scale reservoir, by the statement that "The three layers of body scales reach the surface via the flagellar pit." In *P. amylifera* the body scales may be released directly onto the cell body surface in the region of the flagellar pit, as proposed by Manton (1966), whereas the flagellar scales are released onto the flagellar surface after first being deposited into the scale reservoir. Norris and Pienaar (1978) proposed that there may be a cyclic deposition of body scales and flagellar scales into the scale reservoir of *P. amylifera*, a cycle that may be coordinated with the division of the cell. They proposed that body scales may be deposited into the scale reservoir only during or just preceding the dark period, a time when the cells are preparing for division or dividing. Observations that may confirm or contradict this proposal are needed.

The scale reservoir is distinctly an invagination of the cell surface at the base of the flagellar pit in *Pyramimonas*. It must be interpreted, therefore, that scales are released onto the cell surface when they are deposited into the scale reservoir by vesicles derived from the dictyosome cisternae. Scale reservoirs in some species of *Pyrimimonas* seem to be simple bladder-like invaginations of the cell. *Pyramimonas amlifera* has numerous internal villi in its scale reservoir, (Manton, 1966; Norris and Pienaar, 1978), whereas the scale reservoir of *P. parkeae* has many duct-like extensions radiating throughout the posterior part of the cell (Norris and Pearson, 1975). It was proposed by Norris and Pearson (1975) and Norris and Pienaar (1978) that scale reservoirs, having their membranes extended into villi or ducts in some species, are incorporated into the formation of daughter cell membranes as the mother cell divides, the scales, therefore, being in position or positioned within the reservoir as cytokinesis commences and continues. It is also interesting to propose that flagellar scale types and body scale types are positioned within the reservoir in different cyclic patterns, a proposal that corresponds with observations by Pearson and Norris (1975) on *P. parkeae* in which generation of new flagella was one of the earliest stages in cell division, occurring before nuclear division. Significant support for this hypothesis is provided by observations on cell division with light microscopy on *Pyramimonas tetrarhynchus* by Belcher (1969). She stated that the scale reservoir, conspicuous as the chloroplast begins to divide and the new flagella begin to form, is not visible as new flagella develop, but apparently begins to reform at a stage late in cytokinesis. It is obvious that additional investigations on this problem are needed, and that they will probably provide some basic information to studies on division of plant cells.

Vacuoles, Mucilage Vesicles, Trichocysts and Other Vesicular Systems. Small-celled marine species of prasinophytes have little vesicular content that is not part of the endoplasmic reticulum system. Similar freshwater species, however, such as *Pedinomonas minor*, *P. major*, *Pyramimonas* spp. and *Monomastix* aff. *minuta* have a contractile vacuolar system that is observable in specimens sectioned for electron microscopy, as well as in living cells. Contractile vacuoles may be large or small according to their status of expansion or contraction, and are completely without solid contents. They are associated with the nucleus and Golgi system and are located at the flagellar end of the cell. In species having a flagellar pit, such as *Mesostigma viride* and *Pyramimonas tetrarhynchus*, the contractile vacuoles empty into the flagellar pit. Ettl and Ettl (1959) describe the contractile vacuoles for *Platymonas*, showing them to be discharged into the flagellar pit.

When viewed with the light microscope, prasinophyte cells often have highly refractive globular material within the cell, particularly in anterior regions. These globules may be vacuolar, lipid droplets, trichocysts (ejectosomes), or mucilage vesicles. Lipid material has been recorded in most prasinophyte species, present in the cytoplasm as well as inside the chloroplast. Vacuoles containing amorphous solid material, as it appears in sections for electron microscopy, are common in species of prasinophyte flagellates. The material may be accumulated metabolites that will be utilized by the cell, perhaps during cell division, or the contents may be the remains of degenerating organelles that are being converted to other cell products. This may particularly be the function of a large vacuole that contains membrane fragments that was found in sections of *Pyramimonas parkeae* (Norris and Pearson, 1975).

Pyramimonas reticulata (Belcher, 1968a) and *P. montana* (Maiwald, 1971), two doubtful members of that genus, have mucilage vesicles located sometimes in layers immediately underneath the plasmalemma. A mucilage layer envelopes the cells in these species, and it may be produced by these vesicles.

Using light microscopy, Parke and Adams (1961) and Parke and den Hartog-Adams (1965) described ejectile muciferous organelles in motile cells of *Halo-*

sphaera. Chadefaud (1941) had also described such ejectile organelles, ejected as rods, threads or balloons when treated with vital stains. The muciferous bodies were peripheral in the cell, often occurring in rows located between the lobes of the chloroplast. Although these structures have not been redescribed with electron microscopy for the same species in which they were found by Chadefaud or Parke and den Hartog-Adams, trichocysts have been found in three species of *Pyramimonas*, *P. grossii* (Manton, 1969), *P. parkeae* (Norris and Pearson, 1975) and *P. virginica* (Pennick, 1977a).

Pyramimonas trichocysts are all much alike in structure, and closely resemble the ejectosomes of cryptomonads. A membrane-bound vesicle surrounds coiled membrane-like material that encloses a group of small globules (see Fig. 26 of Norris and Pearson, 1975). As in cryptomonads, the ejected organelle has the vesicular membrane fused with the plasmalemma and the ribbons of coiled material unravelling from the center of the coil. Unlike the cryptomonads, prasinophyte ejectosomes do not have a narrow bent distal tip on the expired ejectosome. Ejection of these organelles is stimulated by various factors that seem to place cells under stress such as high temperatures, bright light and various chemicals. *Pyramimonas* trichocysts are located in peripheral regions of the cell near the flagellar end. It is inconceivable from the presently known evidence that a trichocyst could penetrate a chloroplast as is illustrated in a diagram by Pennick and Clarke (1976) for *P. grossii*. Manton (1969) showed that the trichocyst ribbon of *P. grossii* consists of two layers that are particularly visible in expired trichocysts, the convex layer being denser than the concave layer.

Norris and Pearson (1975) discussed the possible origin of the trichocysts from a lobe of the nuclear envelope in *Pyramimonas parkeae*, but additional stages in their development need to be discovered before the process can be understood. Function of trichocysts in Prasinophyceae is unknown.

Trichocysts were described for *Platymonas* (Ettl and Ettl, 1959), but the subsequent studies on this genus using electron microscopy have not confirmed this observation. The bodies described by Ettl and Ettl (1959), which often reflect the cell's dorsiventrality, probably are mucilage vesicles, but this speculation needs confirmation for that genus. Trichocyst-like organelles have also been found in *Monomastix* (Scherffel, 1912; Belcher, 1965; Manton, 1970). Belcher (1965) provided extensive detail from light microscopy of trichocyst discharge in *Monomastix*, and Scherffel (1912) mentioned that the discharged material from trichocysts in that genus is a pectose-like mucilage. Manton's electron microscope photographs of *Monomastix* trichocyst vesicles

(1970) show an amorphous darkly staining content rather than the coiled material as is found in *Pyramimonas*. Barlow (1977) illustrated conical vesicles occurring perpendicular to the plasmalemma in *Mantoniella*. The contents of these vesicles has a tendency to align parallel to the long axis of the vesicle in some places, but Barlow did not see the contents of these vesicles released from the cell.

MITOSIS AND CYTOKINESIS

Cytokinesis, studied with light microscopy, has been the subject of several studies on prasinophyte flagellates (Korshikov, 1923; Zimmermann, 1925; Bretschneider, 1925; Lambert, 1930; N. Carter, 1937; Hollande *et al.*, 1954; Belcher, 1964, 1969; Ettl and Manton, 1964; Norris, 1967; Ettl, 1967a; to mention a few). Species without a wall or with a covering of scales seem to remain motile throughout cytokinesis, whereas the walled cells of *Platymonas* and *Prasinocladus* lose their flagella for long periods of time, during which time the cells often divide. Walled prasinophyte cells do not divide while swimming.

Chromosomes have rarely been reported for Prasinophyceae, the number of chromosomes being difficult to estimate, usually, because they often do not separate using conventional methods of staining (Proskauer, 1950; Ettl and Manton, 1964). Chromosome numbers have been estimated a few times for various prasinophyte flagellates, two chromosomes for *Pedinomonas minor* (Ettl and Manton, 1964), ten to twelve chromosomes for *Platymonas tetrathele* (Zimmermann, 1925).

Manton (1959) reported on division stages for *Micromonas pusilla* using electron microscopy, and noted that sequential division of organelles occurs before a new flagellum is formed, and the cell begins to separate into daughter cells. Manton also noted that the mitochondrion divides after the chloroplast, and it may be an active agent in assisting chloroplast and cell divisions because it is closely associated with the pyrenoid during division stages.

Four species of scaly green flagellates have been the subject of detailed studies on mitosis and cytokinesis using the electron microscope: *Platymonas subcordiformis* (Stewart *et al.*, 1974), *Pyramimonas parkeae* (Pearson and Norris, 1975), *Bipedinomonas angulata* (Mattox and Stewart, 1977) and *Mantoniella squamata* (Barlow, 1977). In addition, two possibly related scale-less monads, *Pedinomonas minor* (Pickett-Heaps and Ott, 1974) and *Asteromonas gracilis* (Floyd, 1979) have been equally well examined for these characteristics. These publications reveal that there is considerable variation in mitosis and cytokinesis in prasinophyte flagellates, but many more species should be examined before accurate assessments on relationships between the genera can be made. At the present time it can be said that *Platymonas subcordiformis*, because of its development of a phycoplast, has the most advanced type of cell division among these scaly falgellates. Also, it may be proposed that the biflagellate scaleless genera *Dunaliella* and *Asteromonas*, which also have phycoplasts, may have been derived from a walled type of cell such as *Platymonas* or *Chlamydomonas* (Floyd, 1979).

From the information available on the prasinophyte species it may be possible to propose that trends in the evolution of the mitotic apparatus and sequence of events in cytokinesis are becoming evident. In this estimation the discovery by Barlow (1977) that *Mantoniella squamata* cytokinesis resembles the precocious furrow type demonstrated in the Ulvaphyceae (Stewart and Mattox, 1978) is highly significant. It may also be significant that Barlow did not find vesicles in *Mantoniella* cells that accumulate scales, but that the scales seem to be released singly. All other scale bearing species that have been investigated seem to have at least accumulation of small numbers of scales in single vesicles and, in some, such as in the genus *Pyramimonas*, large numbers of scales are found inside vesicles and also in the scale reservoir, an invagination of the cell surface. There is a strong coordination of these factors, production of scale containing vesicles or storage areas and furrowing that occurs at cytokinesis.

Mantoniella squamata, having no special internal vesicle system or reservoir that was prepared during interphase, requires comparatively long periods of time for the cells to separate during cytokinesis, and this separation seems to be dependent in some degree upon the activity of separated flagellar systems on the potential daughter cells. Also, the division of the cell into two parts requires early separation of organelles and subsequent development of an early division furrow. It seems possible in such a cell as *Mantoniella squamata* that an internal structure must be required to maintain the cell in a position whereby the organelles are separated after they have divided. This structure seems to be the microbody which appears to be holding the daughter chloroplasts apart by bracing between the two pyrenoids. Such maintenance of organelles in separate cell lobes along with the organized separation of the chromatin by mitosis that is directed toward or in the vicinity of the basal bodies prepares the cell for the process of stretching and separation which is assisted by the activity of the two pairs of flagella and deposition of scale vesicles as they are formed by the Golgi apparatus. The fact that the final separation of the daughter nuclei in *Mantoniella squamata* is dependent upon the furrowing of the cell is a very significant point. This condition may be a primitive type of cytokinesis in the prasinophyte flagellates. An advancement may be in cell division of *Bipedinomonas angulata* (Mattox and Stewart, 1977) in which most events are similar to and in the same sequence as in *Mantoniella*, but the daughter nuclei separate well before cell division is complete and vesicles having many scales are formed during interphase. One would expect cytokinesis to be completed in a comparatively shorter period of time in *Bipedinomonas angulata* than in *Mantoniella*, but such data are not available at this time.

My observations on *Pseudoscourfieldia longifilis* show that it may have a division sequence and system somewhat intermediate between *Mantoniella squamata* and *Bipedinomonas angulata*. This species differs from *Mantoniella* in having an open spindle during mitosis and in producing vesicles containing several scales. Otherwise it has division characteristics similar to *Mantoniella* including a precocious furrow that separates the chromatin during telophase. It should be emphasized, however, that the main resemblance between the ulvaphycean precocious furrow and the *Mantoniella* and *Pseudoscourfieldia longifilis* precocious furrows is in the assistance the furrow seems to provide in separation of chromatin in late mitosis. Synchronization of organelle divisions with cytokinesis in Ulvaphyceae seems to be much more developed than in the small prasinophyte species.

Division of *Pyramimonas* cells is more advanced than the species mentioned above because extensive interphase preparations for cytokinesis have occurred in the formation of a scale reservoir. Also, in the species that have been studied, the nucleus of *Pyramimonas* completely divides before cytokinesis is completed, but the daughter nuclei remain close to one another. Participation of the microbody in separation of the daughter chloroplasts is evident in *P. parkeae* (Pearson and Norris, 1975).

Participation of other organelles in cell division probably occurs but their role has not been analyzed. Mitochondria often are present near the microbody in dividing chloroplasts of *Pseudoscourfieldia longifilis* (Norris - unpublished) and may play an important role in separation of the daughter cells. It has been demonstrated that the rhizoplast seems to convert to a microtubular organizing center at the poles of spindles in *Platymonas subcordiformis* (Stewart *et al.*, 1974). *Pyramimonas parkeae* retains a visible rhizoplast during nuclear division stages and Pearson and Norris (1975) reported that spindle microtubules attach at points on the rhizoplast. If the rhizoplast is involved primarily with spindle microtubules, as it seems, then the basal body microtubular rootlets may be involved in separation of the daughter cells. At the present time one can only speculate that this may be true, but their position in the cell would seem to indicate that their role may be significant.

It is of particular interest that cleavage microtubules develop only in the wall-forming scaly green monad, *Platymonas subcordiformis*, and in genera possibly derived from walled prasinophytes, *Asteromonas* and possibly *Dunaliella* The phycoplast microtubules seem to play a role in directing the flow of scale-containing vesicles from the vicinity of the Golgi apparatus to the plane of cytokinesis in *Platymonas*. Phycoplasts occur in species in which the spindle collapses during late anaphase, suggesting that the spindle microtubules may be transposed to the position of the cleavage plane.

In summary one must first emphasize that there is great need for many more studies on division of cells among the prasinophyte flagellates. It is interesting to note, however, that the available information seems to indicate that trends can be traced from cells having little organization for nuclear division and cytokinesis, as in *Mantoniella*, to those having interphase activities oriented toward preparation for division as in *Pyramimonas*. In both of these types partitioning of organelles is aided by the microbody which usually divides late in the mitotic cycle, and by independent activities of the separated flagellar groups that cause twisting and pulling apart of the daughter cells. It is highly significant that these prasinophyte cells are motile throughout their division stages. (It is noteworthey that the wall-less, phycoplast forming genera, *Asteromonas* and *Dunaliella* also remain motile while dividing.) Also, it is significant that these species grow during the division cycle, a process that is probably necessary for the cells to divide, particularly the growth that is attributable to formation of vesicles that deposit new plasmalemma. The more primitive type of prasinophyte probably is in a more or less continuous state of division, some part of cytokinesis being prepared for at most any time during the cell cycle except, possibly, immediately after separation of daughter cells. Division of organelles is more sequential than synchronized in these organisms. A trend toward synchronization of organelle division with mitosis may be an indication of a phylogenetic advancement. Also, such synchronization may be a result of structural changes in the cell, especially the development of a confining wall. The walls of *Platymonas* do not grow as the protoplast grows as they do in at least some species of *Chlamydomonas*. Instead, *Platymonas* cells, as they produce new protoplasm, tend to release themselves from the wall and then build a new one that may be slightly larger. *Platymonas* has a phylogenetically early type of eleutheroschisis (Groover and Bold, 1969), and it is of particular interest that many green algae having eleutheroschisis, or at least wall growth synchronized with mitosis, also have a phycoplast. The confining wall, as is found in such species, may have been at least partly responsible for more synchronization of organelle division and mitosis than is found in wall-less species, channeling cell activities toward climactic events such as mitosis and wall formation. The most climactic event in *Mantoniella's* cell cycle may be the separation of daughter cells, all the previous changes in the cell structure, including mitosis, occurring without particular synchronization, except in a sequential manner.

Platymonas subcordiformis, on the other hand, a species that is non-motile during division stages, has basal bodies remaining at interphase positions during mitosis, which may be a condition that evolved because flagellar action was not needed to separate cells at late division stages. Additional information from electron microscopy on division of organelles is needed to determine how well synchronized division of the cell and its organelles may be. Microbodies are present in *Platymonas* cells (Norris - unpublished), but their role in cytokinesis has not been studied. There are several reports on cell division in *Prasinocladus* using light microscopy (reviewed by Lambert, 1930; and Proskauer, 1950). The description of cell division in *Prasinocladus* by Lambert (1930) is of particular interest because the process in a living cell is shown to occur rapidly, within a time span of only a few minutes, a cell constriction separating the chloroplast and its pyrenoid into two. Lambert's observations strongly indicate that organelle division is synchronized with cytokinesis, the two events occurring within a very short period of time.

Pedinomonas minor does not seem to be as primitive as *Mantoniella* in these characteristics because, even though the flagellar apparatus replicates early in the division cycle, the chloroplast and mitochondrion do not divide until the nuclear division is in anaphase. Also, in this species the chloroplast division is attended by a microbody and sometimes a mitochondrion. Daughter nuclei of *Pedinomonas minor* separate by internal forces of the cell well before the cell is pulled apart by the flagellar activity combined with furrowing on the cell membrane, as well as rotation of the cell (Pickett-Heaps and Ott, 1974). Fig. 5 provides a summary of possible phylogenetic relationships in presently known prasinophytes using characteristics of cytokinesis and mitosis, and related phenomena in development of daughter cells.

LIFE HISTORIES

There is a great need for more information on life histories of prasinophyte flagellates. Sexual reproduction in scale-bearing Prasinophyceae is unknown except for a report by Korshikov (1923) in which he described sexual fusion and encystment for a species of *Heteromastix*. Two scale-less genera, *Pedinomonas* and *Asteromonas*, that are often associated with prasinophytes, and a variety of a scale-less species of *Pyramimonas*, *P. reticulata* var. *conica*, have reported sexual reproduction (Korshikov, 1923; Gorbunova, 1961; Ettl, 1965). Ettl (1967b) reviewed reports of possible sexual reproduction for *Pedinomonas*, but did not confirm that it actually is convincingly demonstrated for that genus.

Asexual cysts or palmella stages probably occur in most, if not all, freshwater species of Prasinophyceae and in some marine species. Belcher (1970) tested resistance of dry cysts to heat and showed that *Monomastix*, *Spermatozopsis* and *Pedinomonas minor* have remarkable abilities to withstand high temperatures. Four species of *Pyramimonas* are reported by Belcher (1966) to be capable of forming cysts, *P. amylifera*, *P. grossii*, *P. reticulata* and *P. tetrarhynchus*. The latter two species occur in freshwater and have small, thick-walled cysts with ornamented walls that are different from cysts in marine species which are usually larger and smooth-walled.

Fig. 5. Diagrams showing relationships of 1) sequence of organelle division, 2) furrowing of cell, 3) increasing complexity of scale vesicles and reservoir that expand the cell membrane as cells grow, and 4) cytokinesis in five scale-bearing or thecate prasinophyte flagellates, *Mantoniella squamata* (after Barlow, 1977), *Pseudoscourfieldia longifilis* (Norris - unpublished), *Bipedinomonas angulata* (after Mattox and Stewart, 1977), *Pyramimonas parkeae* (after Pearson and Norris, 1975), and *Platymonas subcordiformis* (after Stewart *et al.*, 1974). It may be seen from the diagrams that *Mantoniella* has organelle divisions not closely synchronized with nuclear division, it has single scales in the vesicles, and the final stage of nuclear division is assisted by cytokinesis. *Pseudoscourfieldia longifilis* has similar division, but there are several scales in the vesicles *Bipedinomonas angulata* also has several scales in the vesicles, but the nucleus completes its division before cytokinesis occurs. Also, the microbody divides at the same time as the chloroplast in this species. In *Pyramimonas parkeae* the scale vesicles form a scale reservoir that becomes an invagination of the cell surface. Unfolding of the reservoir and its ducts probably occurs at cytokinesis when the new exterior membranes of daughter cells are formed by this system. Nuclear division is complete by the time cytokinesis occurs. Synchronization of events is more evident than in the former three species. Synchronization of organelle divisions with nuclear division is pronounced in *Platymonas subcordiformis*, and a phycoplast is developed. It is hypothesized that evolutionary complexity increases in the diagrams from the most primitive on the left to more complex on the right. B = basal bodies; C = chloroplast; Cy = complete cytokinesis; F = furrowing; G = Golgi apparatus; M = microbody; N = nucleus; N_c = nuclear division, closed spindle but with polar fenestrae; N_o = nuclear division with open spindle; N_v = nuclear division with vesiculate nuclear envelope; P = pyrenoid; Ph = phycoplast; R = scale reservoir; S = stigma; S_v = scale vesicle; W = cast off wall.

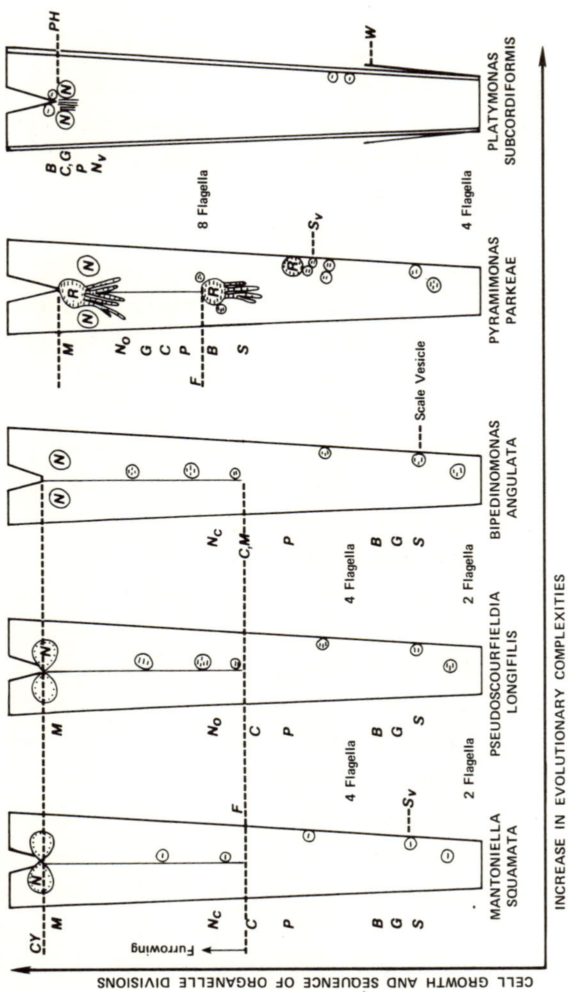
CELL GROWTH AND SEQUENCE OF ORGANELLE DIVISIONS
INCREASE IN EVOLUTIONARY COMPLEXITIES
Furrowing
MANTONIELLA SQUAMATA
PSEUDOSCOURFIELDIA LONGIFILIS
BIPEDINOMONAS ANGULATA
PYRAMIMONAS PARKEAE
PLATYMONAS SUBCORDIFORMIS
4 Flagella
2 Flagella
8 Flagella
Scale Vesicle

FIGURE 5

Belcher (1966) reported six species of *Pyramimonas* that are capable of producing palmella stages, a non-motile phase in which cells may divide. All of these species except *P. grossii* occur in freshwater. *Mantoniella squamata* seems to be exceptional among the marine scaled prasinophyte flagellates in having cells that divide in a non-motile palmelloid condition as well as in the motile state.

Belcher (1966) also described a new coccoid genus, *Prasinochloris*, that produces quadriflagellate scaled motile cells resembling *Pyramimonas*. Apparently *Prasinochloris* reproduces only by formation of the flagellate cells, whereas a species of *Ecballocystis*, also having zoospores resembling *Pyramimonas*, is capable of dividing in its non-motile condition (George, 1956). Neither of these species was observed to divide in the motile stage.

Species of *Halosphaera*, on the other hand, have motile cells with all the characteristics of *Pyramimonas* including the ability to divide (Parke and Adams, 1961; Manton *et al.*, 1963; Parke and den Hartog-Adams, 1965; Boalch and Mommaerts, 1969). These motile cells, many of them produced from large cysts, are capable of dividing and maintaining the motile stage indefinitely, and it can be readily understood that there may be difficulty in distinguishing between cyst-forming species of *Pyramimonas* and the genus *Halosphaera*.

Pachysphaera and *Pterosperma* are cyst-forming genera that also have quadriflagellate zoospores, but the form of these motile cells is very different from *Pyramimonas* (Parke, 1966; Leadbeater, 1974; Parke *et al.*, 1978). Little is known about the conditions controlling cyst formation and development of zoospores in these genera, but it is very interesting to note that cyst formation and liberation of the motile phase seem to be on a lunar cycle (Parke and Boalch, 1973; Parke *et al.*, 1978). *Pachysphaera* and *Pterosperma* motile cells divide and, similar to *Halosphaera*, may be maintained in cultures in the motile condition for long periods of time (Parke *et al.*, 1978).

Platymonas and *Prasinocladus* have walled cells and species in these genera often remain non-motile, having lost their flagella, for long periods of time. The protoplast may also periodically free itself from its wall and, appearing to contract, form a new wall within the parent theca. Instead of contracting, the cell may actually be growing, releasing its old wall which then expands, and forming a new wall that remains at least partially inside or attached to the old theca. Walls that are formed by repeatedly being cast off by a cell often accumulate and remain attached to the cell, thereby forming a short or elongated stalk as is evident in *Prasinocladus*. Daughter cells often remain within the parent theca, thereby forming a branched stalk of discarded thecae as the cells continue to grow.

Division of *Platymonas* cells occurs in the dark period for an unnamed species studied by Tanoue and Aruga (1975). Small, thick-walled cysts having ornamented walls are known to form in many species of *Platymonas* and *Prasinocladus* (Butcher, 1952, 1959; Parke and Manton, 1967; McLachlan and Parke, 1967; Tanoue and Aruga, 1975; Kobara and Hori, 1975). Such cysts form mostly in old cultures from six to eight weeks old (Tanoue and Aruga, 1975), and germination of cysts results in the formation of four quadriflagellate cells (Tanoue and Aruga, 1975; Kobara and Hori, 1975).

PIGMENTS AND OTHER BIOCHEMICAL CHARACTERISTICS

Most prasinophytes have photosynthetic pigment systems that are similar to the majority of green algae. Moreover, most of the unusual pigment systems of the green algae are known in species of prasinophytes, supporting evidence to the hypothesis that prasinophytes provide a gene pool from which all other green plants may have evolved. Ricketts (1966a, 1966b, 1967, 1970, 1971) has summarized the data on pigment composition of chloroplasts for prasinophyte algae. Additional research in this field is needed, however, particularly for the data to be applied to more recently proposed phylogenetic schemes.

With our present knowledge of pigments it is especially interesting to note that certain prasinophyte species have siphonein and siphonoxanthin in place of lutein which is present in most prasinophytes and other green algae. Another small group of prasinophyte species lacking lutein has the unusual xanthophyll, micronone. The species in both of these aberrant groups, possessing unusual xanthophyll pigments for green algae, also have a protochlorophyll-like pigment, magnesium 2,4, divinylphaeoporphyrin a_5 monomethyl ester, which is sometimes referred to as a chlorophyll *c*-like pigment (Ricketts, (1966a). This pigment is present in addition to chlorophylls *a* and *b* in these few species, but it is not present in the majority of prasinophyte species for which it has been tested. It is interesting to note that the micronone containing species, *Micromonas pusilla* and *Mantoniella squamata*, contain fewer xanthophylls than other prasinophytes and seem to be the most primitive types, from the point of view of pigments, among the prasinophyte flagellates. The siphonein and siphonoxanthin containing prasinophytes, except *Pyramimonas amylifera*, are among the more highly advanced asymmetrical flagellate types. As Ricketts has stated (1970), it is difficult to explain the intrageneric variability in pigments of *Pyramimonas*. Pigment analysis may be the first evidence revealed implying that *Pyramimonas* may be polyphyletic.

Ricketts (1966b) noted that the xanthophylls of *Micromonas pusilla* are mostly in a high state of oxidation, a characteristic similar to fucoxanthin, peridinin and echinenone, xanthophylls that are found in other algal phyla. Ricketts (1966a) also noted that all freshwater prasinophyte species that were tested for magnesium 2,4 divinylphaeoporphyrin a_5 monomethyl ester, did not have it, implying, perhaps, that species in freshwater habitats may have a more active enzyme system in the main biosynthetic pathway to chlorophyll *a*. At the same time, however, it should be noted that a considerable number of marine species of prasinophytes also do not have this protochlorophyll-like pigment, implying that the chlorophyll *a* forming system can be as active in the marine environment as in the freshwater.

Several biochemical characteristics unite the prasinophytes into a group that seems to be separate from the rest of the green algae (Stewart, 1974; Ragan and Chapman, 1978). In most green algae that have been examined, sucrose, fructose, glycerol and glucose are the major soluble photosynthetic products. Glucose is the only one of these compounds that has been found in prasinophytes, and it is present in only six out of thirteen prasinophytes that were tested. Sucrose, a common polysaccharide in many green algae, has not been demonstrated in Prasinophyceae (Ragan and Chapman, 1978). Prasinophyceae have mannitol as their principal soluble photosynthate, a compound not found in most green algae, but it has been reported in *Enteromorpha* and *Cladophora*. Strains of *Nannochloris* and *Stichococcus* have also been shown to produce mannitol as their main photosynthate, but these species have been transferred to the Eustigmatophyceae (Antia *et al*., 1975; Brown and Hellebust - personal communication).

The species of prasinophytes for which mannitol has been demonstrated are of diverse types, including several species of *Platymonas*, *Prasinocladus*, a species of *Pyramimonas*, *Micromonas pusilla* and *Mantoniella squamata* (Craigie *et al*., 1967). Characterization of prasinophytes using this characteristic should await additional investigations on more species, particularly in the genus *Pyramimonas*, among some additional biflagellate species and in the phycoma forming types.

Dimethyl-B-propiothetin (DMPT) is a volatile compound contributing to the fishy odor of *Polysiphonia lanosa*, and it has also been found in *Enteromorpha intestinalis* and in *Acrosiphonia centralis* (Challenger *et al*., 1957; Schiff, 1962). DMPT was found to be present in all prasinophytes tested (13 isolates) except in an unknown species of *Pyramimonas*, whereas it was present in only one (a species of *Chlorococcum*) out of ten species of tested Chlorophyceae

(Craigie *et al.*, 1967). Additional investigations are necessary, but it seems to be developing that presence of mannitol and DMPT in cells may indicate a prasinophyte affiliation.

In the green algae for which glycolate enzymology has been studied it appeared for a time that phycoplast producing species have a glycolate dehydrogenase system, whereas phragmoplast types have glycolate oxidase. Floyd and Salisbury (1977) examined nine species of Prasinophyceae for these enzymes and found only glycolate dehydrogenase to be present. Their conclusion is that evolutionary divergence of glycolate enzyme systems occurred after divergence of cytokinesis and spindle characteristics in primitive green algae.

Most Chlorophyceae that have been examined have a specialized enzyme system for the utilization of urease as a nitrogen source in their metabolism. Ten species of Chlorophyceae have been shown to contain ATP: urea amido-lyase (ADP) instead of urease. Prasinophyceae, however, contain urease but not ADP, thereby containing a system of urea utilizing enzymes similar to Xanthophyceae, Prymnesiophyceae and diatoms rather than the ADP system present in Chlorophyceae (Morris, 1974).

It may be significant that the hydrocarbon, n-heneicosahenaene, (C-21:6), has been found to be the most abundant hydrocarbon in *Platymonas* sp. (Lee and Loeblich, 1971). This hydrocarbon is absent in all other green algae tested, except for *Codium fragile*. It is of particular interest that this hydrocarbon is correlated in most other species with the presence of fucoxanthin and chlorophyll *c*. Lee and Loeblich (1971) suggest that the green algal ability to form this molecule may have an evolution independent of the other algal phyla in which it is found.

Most marine prasinophyte flagellates have been cultured in seawater enrichment media, whereas the freshwater species have been grown in defined media. Five freshwater species were shown to have no organic requirements in their media, their axenic growth being totally autotrophic (Ricketts, 1974). Axenic cultures of eight marine species of *Platymonas* and *Prasinocladus* also showed no requirements for vitamins in Ricketts' cultures, but nine marine species require "vitamins" in their media. It is interesting to note that most of these species could not be grown axenically. Growth of *Micromonas pusilla*, *Mantoniella squamata*, *Pseudoscourfieldia longifilis*, *Pyramimonas obovata* and "*Heteromastix* sp. (P198)" showed a requirement for a "vitamin" mixture containing vitamin B_{12}, thiamin, niacin, pantothenate, p-aminobenzoic acid, biotin, folic acid, inositol and thymine. An axenic culture of *Pyramimonas amylifera* (as *Asteromonas propulsa*) also had a requirement for this mixture in its medium

(Ricketts, 1974). Ricketts' data also indicate the "*Heteromastix* sp. (P198)", *Micromonas pusilla* and *Pyramimonas amylifera* require thymine.

Provasoli and Carlucci (1974) report *Stephanoptera gracilis* (possibly *Asteromonas*) as the only prasinophyte tested that does not require any vitamins. *Platymonas tetrathele* is listed as requiring vitamin B_{12}, whereas *Pyramimonas inconstans* requires not only vitamin B_{12} but also thiamine. Although several Chlorophyceae also require these vitamins, additional investigations on prasinophytes may reveal a trend that may be of phylogenetic significance.

CONCLUSIONS

A group of green flagellates can be identified as primitive within the green algae on the basis of several characteristics: 1) a scaly covering, probably a precursor to cell wall, 2) several internal structures, especially those rootlets associated with the basal body system, 3) pigment composition, 4) methods of mitosis and cytokinesis. These characteristics, and others, are reviewed above for these green flagellates, sometimes referred to as Prasinophyceae. Recent speculation on how these flagellates may be related to other algae (Chadefaud, 1977), and how these flagellates are involved in evolution of higher green algae (Stewart and Mattox, 1978) provide ideas that do not seem to be given complete support by the information in the review provided above.

In separating the green algae into three major lines of evolution, Chlorophyceae, Charophyceae and Ulvaphyceae, Stewart and Mattox (1978) propose that scaly green flagellates are ancestral types for all three classes, and that the radially symmetrical scaly flagellate cell (as viewed from the polar ends) is probably basic in phylogeny of these groups. Chadefaud (1977) also proposes that asymmetrical prasinophyte flagellates are derived from radially symmetrical types such as *Pyramimonas*.

Stewart and Mattox (1978) do not incorporate the scaly green flagellates into the three green algal classes that they recognize. Instead, these flagellates occupy *in limbo* positions in their evolutionary diagram. From the evidence that is reviewed above, I believe that there are convincing evolutionary linkages between scaly green flagellates so that they should be considered a class unto themselves as proposed by Christensen (1962) and reviewed by Chadefaud (1977). This class, Prasinophyceae, is the most important class in the green algae from the point of view of evolutionary lines leading to Ulvaphyceae, Charophyceae and Chlorophyceae. The members of the class are united by having flagellate, wall-less cells that are either covered by a layer or layers of scales, or show direct evolutionary linkages with walled or naked cells. A primitive and basic class, as is the Prasinophyceae, should be expected to contain a variety of specializations, so variations in cytokinesis, symmetry and number of flagella and flagellar rootlets are characteristics not to be viewed primarily as extreme diversities separating them into different major taxonomic groups. Instead, these characteristics should be viewed as relevant to an understanding of the way in which these flagellates may be important to the evolution of other green algal classes. It is already quite firmly recognized that a basic green flagellate group exists, many of them are scaly and some of them show distinct evolutionary linkages to each other and to more highly advanced green algae. This basic group, the Prasinophyceae, should be accepted. It may be easier for•some to recognize the Prasinophyceae if the walled and phycoplast forming types are removed to one of the other classes. Such a decision should be deferred, however, until cytokinesis has been examined in a more complete array of *Platymonas* and *Prasinocladus* species. At the present time, therefore, I accept them as Prasinophyceae that show phylogenetic linkage with other wall-

I find that within the Prasinophyceae there seems to be good evidence supporting a hypothesis that the most primitive of the Prasinophyceae are the biflagellate and uniflagellate species. The evidence is: 1) the evolutionary lineage from one layer of scales to two or more layers begins in biflagellate types, 2) the species having only one type of scale, a primitive characteristic, are biflagellate, 3) primitive pigment systems occur mainly in biflagellate or uniflagellate species, 4) a progressive series of cytokinesis from sequential division of organelles previous to cytokinesis, to synchronization of organelle division immediately before mitosis and cytokinesis has been demonstrated, and biflagellate types are distinctly more primitive than quadriflagellate species, 5) the precocious furrowing type of cytokinesis and late separation of chromatin by cytokinesis, a primitive process, occurs only in biflagellate species so far as we know, 6) mostly quadriflagellate species have complex cell structures that probably are not basically important to the cell and that would be expected in more highly evolved cells (trichocysts, scale reservoirs, muciferous bodies, large vacuoles).
the cells (trichocysts, scale reservoirs, muciferous bodies, large vacuoles).

Stewart and Mattox (1978) consider the flagellar pit to be a primitive characteristic in green flagellates that has been lost in subsequent evolution of zoospores in filamentous types. Also, they propose that the rhizoplast has been converted to a multilayered system in this same evolutionary process. Their proposal, as is reviewed above, is that zoospores of charophytes may have evolved from a *Pyramimonas*-like cell that has halved, thereby losing the flagellar pit and becoming biflagellate. During this evolutionary process the *Pyramimonas* microtubular rootlets and sub-plasmalemmal microtubules are lost, and the rhizoplast converts to a new microtubular system of the multilayered type in more advanced green algae.

This hypothesis, involving the elaborate transformation of one organelle type into another, is unnecessary in the phylogenetic schemes that I propose Also, conversion of a quadriflagellate to a biflagellate cell would involve division of the nucleus prior to division of other organelles, particularly delaying basal body replication, a process that usually does not occur in these organisms except, perhaps, in abnormal situations. Cells that are in continuous preparation for cytokinesis, sequential division of organelles beginning as daughter cells grow, have early initiation of a second flagellar apparatus. If cytokinesis is inhibited before mitosis occurs in a uniflagellate species such as *Micromonas pusilla*, the resulting cell would be biflagellate. *Scourfieldia caeca*, an organism studied by Manton (1975), represents a cell that may have been formed by such a process. Similar impediment of cell division in uniflagellate or biflagellate cells without a distinct flagellar groove may have produced a bilaterally symmetrical cell such as *Spermatozopsis*, having two or four flagella. (Kalina, 1965, redescribed *Spermatozopsis* as a biflagellate genus, quadriflagellate cells being division stages or the result of gamete fusion.) If, however, a biflagellate bilaterally symmetrical cell having a flagellar depression has its division arrested at a later stage, after some organelles have separated but before nuclear division, and the cell has become anteriorally lobed, a cell such as *Pyramimonas* could have evolved, the flagellar pit being formed by the united grooves of what were potentially daughter cells. Similar failure of complete cell division of such a quadriflagellate cell could have formed the species of *Pyramimonas* having eight flagella. It may be no evolutionary accident, therefore, that the most common green flagellates have mostly two or four flagella and a symmetrical cell when seen in polar view. Also, such an evolutionary process may explain the readjustment of flagellar attachment from lateral to polar and the subsequent change in motility from posteriorally directed flagella to those directed anteriorally in most biflagellate and quadriflagellate species. Triflagellate and hexaflagellate species are exceedingly rare and are probably the result of failure of basal body development at critical stages in development of ancestors of such species.

It is interesting to note that most diversity of the green monads occurs in freshwater or brackish water environments where environmental factors that may lead to an arrest of the cell cycle are much more variable than in the more gradually alterable marine habitats.

REFERENCES

Antia, N. J., T. Bisalputra, J. Y. Cheng and J. P. Kalley. 1975. Pigment and cytological evidence for reclassification of *Nannochloris oculata* and *Monallantus salina* in the Eustigmatophyceae. J. Phycol. 11: 339-343.

Arnold, C. G., O. Schimmer, F. Schötz and H. Bathelt. 1972. Die Mitochondrien von *Chlamydomonas reinhardii*. Arch. Mikrobiol. 81: 50-67.

Barlow, S. 1977. Fine structure and cell division of the scale-covered green flagellate *Mantoniella squamata* (Manton and Parke) Desikachary. Master of Science Thesis, University of Washington.

Belcher, J. H. 1964. Observations on *Spermatozopsis exsultans* Korshikov in culture. Nova Hedwigia 8: 127-133.

Belcher, J. H. 1965. An investigation of three clones of *Monomastix* Scherffel by light microscopy. Nova Hedwigia 9: 72-82.

Belcher, J. H. 1966. *Prasinochloris sessilis* gen. et sp. nov., a coccoid member of the Prasinophyceae, with some remarks upon cyst formation in *Pryamimonas*. Br. Phycol. Bull. 3: 43-51.

Belcher, J. H. 1968a. A study of *Pyramimonas reticulata* Korshikov (Prasinophyceae) in culture. Nova Hedwigia 15: 179-190.

Belcher, J. H. 1968b. A morphological study of *Pedinomonas major* Korshikov. Nova Hedwigia 16: 131-139.

Belcher, J. H. 1969. Further observations on the type species of *Pyramimonas* (*P. tetrarhynchus* Schmarda) (Prasinophyceae): an examination by light microscopy, together with notes on its taxonomy. J. Linn. Soc. Bot. 62: 241-253.

Belcher, J. H. 1970. The resistance to desiccation and heat of the asexual cysts of some freshwater Prasinophyceae. Br. Phycol. J. 5: 173-177.

Boalch, G. T. and J. P. Mommaerts. 1969. A new punctate species of *Halosphaera*. J. Mar. Biol. Assn. U. K. 49: 129-139.

Boalch, G. T. and M. Parke. 1971. The Prasinophycean genera (Chlorophyta) possibly related to fossil genera, in particular the genus *Tasmanites*. Proc. II Planktonic Conf., Roma 1970, [Ed. A. Farinacci] pp. 99-105.

Bourrelly, P. 1966. Les algues d'eau douce. Initiation à la systématique. I. Les algues vertes. N. Boubeé et Cie., Paris. 511 pp.

Bourrelly, P. 1970. Les algues d'eau douce. Initiation à la systématique. III. Les algues bleues et rouges, les Eugléniens, Peridiniens et Cryptomonadines. N. Boubeé et Cie., Paris. 512 pp.

Bretschneider, L. H. 1925. *Pyramimonas utrajectina* spec. nov., eine neue Polyblepharididae. Arch. Protistenk. 53: 124-130.

Brown, D. L., A. Massalski and B. Patenaude. 1976. Organization of the flagellar apparatus and associated cytoplasmic microtubules in the quadriflagellate alga *Polytomella agilis*. J. Cell Biol. 69: 106-125.

Butcher, R. W. 1952. Contributions to our knowledge of the smaller marine algae. J. Mar. Biol. Assn. U. K. 31: 175-191.

Butcher, R. W. 1959. An introductory account of the smaller algae of British coastal waters. Part I: Introduction and Chlorophyceae. Fishery Investigations. Ministry of Agriculture, Fisheries and Food, Series IV, 1, 74 pp.

Carter, N. 1937. New or interesting algae from brackish water. Arch. Protistenk. 90: 1-68.

Chadefaud, M. 1941. Sur l'organisation et la position systématique des flagellés du genre *Pyramidomonas*. La Rev. Scient. 79: 113-114.

Chadefaud, M. 1960. Les végétaux non vasculaires (Cryptogamie). In Chadefaud, M. and L. Emberger, *Traité de Botanique*. Systématique 1: 1-1018.

Chadefaud, M. 1976. De leur origine a celle des plantes superieures. Ann. Univ. A.R.E.R.S. Reims 14: 37-48.

Chadefaud, M. 1977. Les Prasinophycées. Remarques historiques, critiques et phylogénétiques. Bull. Soc. Phycol. France 22: 1-18.

Challenger, F., R. Bywood, P. Thomas and B. J. Hayward. 1957. The natural occurrence and chemical reactions of some thetins. Arch. Biochem. Biophys. 69: 514-523.

Chihara, M. and T. Hori. 1972. The fine structure of *Prasinocladus ascus* and *Platymonas* species found in Japan, with special reference to their taxonomy. Proc. 7th Internat. Seaweed Symp., Sapporo, 1971. Univ. Tokyo Press pp. 188-191.

Christensen, T. 1962. Systematisk Botanik, Alger. In Böcher, T. W., M. Lange and T. Sørensen, *Botanik*. Bind II, Nr. 2, Munksgaard, København. 178 pp.

Craigie, J. S., J. McLachlan, R. G. Ackman and C. S. Tocher. 1967. Photosynthesis in algae. III. Distribution of soluble carbohydrates and dimethyl-β-propiothetin in marine unicellular Chlorophyceae and Prasinophyceae. Can. J. Bot. 45: 1327-1334.

Dodge, J. D. 1973. The fine structure of algal cells. Academic Press, New York, London. 261 pp.

Ettl, H. 1965. Neue Arten der Gattung *Pyramimonas* Schmarda. Hydrobiologia 26: 484-500.

Ettl, H. 1967a. Der Verlauf der Zellteilung bei zwei marinen *Heteromastix*-Arten. Int. rev. ges. Hydrobiol. 52: 441-445.

Ettl, H. 1967b. Die Gattung *Pedinomonas* Korschikoff. Arch. Protistenk. 110: 1-11.

Ettl, H. 1976. Die Gattung *Chlamydomonas* Ehrenberg. Beih. Nova Hedwigia, Heft 49, 1122 pp.

Ettl, H. and O. Ettl. 1959. Zur Kenntnis der Klasse Volvophyceae. II. (Neue oder wenig bekannte Chlamydomonadalen). Arch. Protistenk. 104: 51-112.

Ettl, H. and I. Manton. 1964. Die feinere Struktur von *Pedinomonas minor* Korschikoff. Nova Hedwigia 8: 421-451.

Eyden, B. P. 1975. Light and electron microscope study of *Dunaliella primolecta* Butcher (Volvocida). J. Protozool. 22: 336-344.

Floyd, G. 1979. Mitosis and cytokinesis in *Asteromonas gracilis*, a wall-less green monad. J. Phycol. 14: 440-445.

Floyd, G. and J. L. Salisbury. 1977. Glycolate dehydrogenase in primitive green algae. Amer. J. Bot. 64: 1294-1296.

George, E. A. 1956. The occurrence of *Pyramimonas*-like zoospores in the green alga *Ecballocystis*. New Phytol. 55: 382-385.

Gooday, G. W. 1971. A biochemical and autoradiographic study of the role of the Golgi bodies in thecal formation in *Platymonas tetrathele*. J. Exper. Bot. 22: 959-971.

Gorbunova, N. N. 1961. Polovi protsess in *Asteromonas gracilis* Artari. Bot. Zh., Kyyiv., 46: 993-998.

Groover, R. D. and H. C. Bold. 1969. Phycological studies. VIII. The taxonomy and comparative physiology of the Chlorosarcinales and certain other edaphic algae. Univ. Texas Publ., No. 6907, Austin, 165 pp.

Halldal, P. 1961. Ultraviolet action spectra of positive and negative phototaxis in *Platymonas subcordiformis*. Physiol. Plant. 14: 133-139.

Hoffman, L. R. 1968. Observation on the fine structure of *Oedogonium*. IV. The mature pyrenoid of *Oe. cardiacum*. Trans. Am. Micro. Soc. 87: 178-185.

Hollande, A., M. Enjumet and D. LaFaurie. 1954. Contribution a l'étude des Vovocales: étude des genres *Platymonas* West et *Stephanoptera* Dangeard. Ann. Sci. Nat., Zool., 11e ser., 16: 283-292.

Hori, T. and M. Chihara. 1974a. Light and electron microscope observations on the developmental sequence of *Prasinocladus marinus* (Class Prasinophyceae). Sci. Rep. Tokyo Kyoiku Daigaku, Sec. B. 15: 265-271.

Hori, T. and M. Chihara. 1974b. Studies on the fine structure of *Prasinocladus ascus* (Prasinophyceae). Phycologia 13: 307-315.
Hyams, J. and D. Chasey. 1974. Aspects of the flagellar apparatus and associated microtubules in a marine alga. Experimental Cell Research 84: 381-387.
Jux, U. 1969a. Über den Feinbau der Zystenwandung von *Pachysphaera marshalliae*. Palaeontographica, Abt. B. 125: 104-111.
Jux, U. 1969b. Über den Feinbau der Zystenwandung von *Halosphaera* Schmitz, 1878. Palaeontographica, Abt. B. 128: 48-55.
Kalina, T. 1965. Zur Morphologie und Taxonomie der Gattung *Spermatozopsis* Korschikow (Volvocales), *Spermatozopsis acidophila* sp. n. Preslia 37: 9-12.
Kobara, T. and T. Hori. 1975. On the formation and germination of cyst in *Platymonas* and *Prasinocladus* (Prasinophyceae). Bull. Jap. Soc. Phycol. 23: 111-115.
Korshikov, A. 1923. Protochlorinae, eine neue Gruppe der grunen Flagellaten. Arch. Russ. Protistol. 2: 148-169.
Lambert, F. D. 1930. On the structure and development of Prasinocladus. Zeitschr. Bot. 23: 227-244.
Lang, N. J. 1963. An additional ultrastructural component of flagella. J. Cell Biol. 19: 631-634.
Leadbeater, B. S. C. 1974. Ultrastructural observations on nanoplankton collected from the coast of Jugoslavia and the Bay of Algiers. J. Mar. Biol. Assn. U. K. 54: 179-196.
Lee, R. F. and A. R. Loeblich III. 1971. Distribution of 21:6 hydrocarbon and its relationship to 22:6 fatty acid in algae. Phytochemistry 10: 593-602.
Lembi, C. A. 1975. A rhizoplast in *Carteria radiosa* (Chlorophyceae). J. Phycol. 11: 219-221.
Lewin, R. A. 1958. The cell walls of *Platymonas*. J. Gen. Microbiol. 19: 87-90.
Maiwald, M. 1971. A comparative ultrastructural study of *Pyramimonas montana* Geitler and a *Pyramimonas spec*. Arch. Protistenk. 113: 334-344.
Manton, I. 1959. Electron microscopical observations on a very small flagellate: the problem of *Chromulina pusilla* Butcher. J. Mar. Biol. Assn. U. K. 38: 319-333.
Manton, I. 1964. The possible significance of some details of flagellar bases in plants. J. Roy. Microscop. Soc. 82: 279-285.
Manton, I. 1965. Some phyletic implications of flagellar structure in plants. In Preston, R. D. [Ed.] *Advances in Botanical Research*, Vol. 2. Academic Press, London and New York. pp. 1-34.
Manton, I. 1966. Observations on scale production in *Pyramimonas amylifera* Conrad. J. Cell Sci. 1: 429-438.
Manton, I. 1968. Observations on the microanatomy of the type species of *Pyramimonas* (*P. tetrarhynchus* Schmarda). Proc. Linn. Soc. Lond. 179: 147-152.
Manton, I. 1969. Tubular trichocysts in a species of *Pyramimonas* (*P. grossii* Parke). Osterr. Bot. Z. 116: 378-392.
Manton, I. 1970. Electron microscopical observations on a clone of *Monomastix* Scherffel in culture. Nova Hedwigia 14: 1-11.
Manton, I. 1975. Observations on the microanatomy of *Scourfieldia marina* Throndsen and *Scourfieldia caeca* (Korsch.) Belcher et Swale. Arch. Protistenk. 117: 358-368.
Manton, I. 1977. *Dolichomastix* (Prasinophyceae) from arctic Canada, Alaska and South Africa: a new genus of flagellates with scaly falgella. Phycologia 16: 427-438.
Manton, I. and H. Ettl. 1965. Observations on the fine structure of *Mesostigma viride* Lauterborn. J. Linn. Soc., Bot. 59: 175-184.

Manton, I., K. Oates and G. Gooday. 1973. Further observations on the chemical composition of thecae of *Platymonas tetrathele* West (Prasinophyceae) by means of the x-ray microanalyser electron microscope (EMMA). J. Exp. Bot. 24: 223-229.

Manton, I., K. Oates and M. Parke. 1963. Observations on the fine structure of the *Pyramimonas* stage of *Halosphaera* and preliminary observations on three species of *Pyramimonas*. J. Mar. Biol. Assn. U. K. 43: 225-238.

Manton, I. and M. Parke. 1960. Further observations on small green flagellates with special reference to possible relatives of *Chromulina pusilla* Butcher. J. Mar. Biol. Assn. U. K. 39: 275-298.

Manton, I. and M. Parke. 1965. Observations on the fine structure of two species of *Platymonas* with special reference to flagellar scales and the mode of origin of the theca. J. Mar. Biol. Assn. U. K. 45: 743-754.

Manton, I., D. G. Rayns and H. Ettl. 1965. Further observations on green flagellates with scaly flagella: the genus *Heteromastix* Korshikov. J. Mar. Biol. Assn. U. K. 45: 241-255.

Marano, F. 1976. Étude ultrastructurale de la division chez *Dunaliella*. J. Microscop. Biol. Cell. 25: 279-282.

Mattox, K. R. and K. D. Stewart. 1977. Cell division in the scaly green flagellate *Heteromastix angulata* and its bearing on the origin of the Chlorophyceae. Am. J. Bot. 64: 931-945.

Moestrup, Ø. 1973. New observations on scales in green algae. Br. Phycol. J. 8: 214.

Moestrup, Ø. 1974. Ultrastructure of the scale-covered zoospores of the green alga *Chaetosphaeridium*, a possible ancestor of the higher plants and bryophytes. Biol. J. Linn. Soc. 6: 111-125.

Moestrup, Ø. 1978. On the phylogenetic validity of the flagellar apparatus in green algae and other chlorophyll *a* and *b* containing plants. BioSystems 10: 117-144.

Moestrup, Ø. and H. A. Thomsen. 1974. An ultrastructural study of the flagellate *Pyramimonas orientalis* with particular emphasis on Golgi apparatus activity and the flagellar apparatus. Protoplasma 81: 247-269.

Moestrup, Ø. and P. L. Walne. 1979. Studies on scale morphogenesis in the Golgi apparatus of *Pyramimonas tetrarhynchus* (Prasinophyceae). J. Cell Sci. 36: 437-459.

Morris, I. 1974. Nitrogen assimilation and protein synthesis. In Stewart, W. D. P. [Ed.] *Algal physiology and biochemistry*. Blackwell, Oxford. pp. 583-609.

McLachlan, J. and M. Parke. 1967. *Platymonas impellucida* sp. nov. from Puerto Rico. J. Mar. Biol. Assn. U. K. 47: 723-733.

Norris, R. E. 1967. Micro-algae in enrichment cultures from Puerto Penasco, Sonora, Mexico. Bull. S. Calif. Acad. Sci. 66: 233-250.

Norris, R. E. and B. R. Pearson. 1975. Fine structure of *Pyramimonas parkeae*, sp. nov. (Chlorophyta, Prasinophyceae). Arch. Protistenk. 117: 192-213.

Norris, R. E. and R. N. Pienaar. 1978. Comparative fine-structural studies on five marine species of *Pyramimonas* (Chlorophyta, Prasinophyceae). Phycologia 17: 41-51.

Owen, H. M. 1949. Flagellar structure: II. The flagellum as a taxonomic character. Trans.Am. Microscop. Soc. 68: 261-274.

Parke, M. 1966. The genus *Pachysphaera* (Prasinophyceae). In Barnes, H. [Ed.] *Some Contemporary Studies in Marine Science*. Allen and Unwin, London. pp. 555-563.

Parke, M. and I. Adams. 1961. The *Pyramimonas*-like motile stage of *Halosphaera viridis*. Bull. Res. Counc. Israel 10D: 94-100.

Parke, M. and G. T. Boalch. 1973. The life history of the Pterospermataceae. Br. Phycol. J. 8: 214.

Parke, M., G. T. Boalch, R. Jowett and D. S. Harbour. 1978. The genus *Ptero-sperma* (Prasinophyceae): species with a single equatorial ala. J. Mar. Biol. Asso. U. K. 58: 239-276.

Parke, M. and P. S. Dixon. 1964. A revised check-list of British marine algae. J. Mar. Biol. Assn. U. K. 44: 499-542.

Parke, M. and J. C. Green. 1976. Chlorophyta, Prasinophyceae. In Parke, M. and P. S. Dixon [Eds.] *Check-list of British Marine Algae* - Third Revision. J. Mar. Biol. Assn. U.K. 56: 564-566.

Parke, M. and I. den Hartog-Adams. 1965. Three species of *Halosphaera*. J. Mar. Biol. Assn. U. K. 45: 537-557.

Parke, M. and I. Manton. 1965. Preliminary observations on the fine structure of *Prasinocladus marinus*. J. Mar. Biol. Assn. U. K. 45: 525-536.

Parke, M. and I. Manton. 1967. The specific identity of the algal symbiont in *Convoluta roscoffensis*. J. Mar. Biol. Assn. U. K. 47: 445-464.

Parke, M. and D. G. Rayns. 1964. Studies on marine flagellates. VII. *Nephro-selmis gilva* sp. nov. and some allied forms. J. Mar. Biol. Assn. U. K. 44: 209-217.

Pearson, B. R. and R. E. Norris. 1975. Fine structure of cell division in *Pyramimonas parkeae* Norris and Pearson (Chlorophyta, Prasinophyceae). J. Phycol. 11: 113-124.

Pennick, N. C. 1977a. Studies of the external morphology of *Pyramimonas*. 4. *Pyramimonas virginica* sp. nov. Arch. Protistenk. 119: 239-246.

Pennick, N. C. 1977b. The external morphology of the *Pyramimonas*-like motile stage of *Halosphaera russellii* Parke. Arch. Protistenk. 119: 388-394.

Pennick, N. C. 1978. Studies of the external morphology of *Pyramimonas*. 5. *P. amylifera* Conrad. Arch. Protistenk. 120: 142-147.

Pennick, N. C. and K. J. Clarke. 1976. Studies of the external morphology of *Pyramimonas*. 3. *Pyramimonas grossii* Parke. Arch. Protistenk. 118: 285-290.

Pennick, N. C., K. J. Clarke and J. H. Belcher. 1978. Studies of the external morphology of *Pyramimonas*. 1. *P. orientalis* and its allies in culture. Arch. Protistenk. 120: 304-311.

Pennick, N. C., K. J. Clarke and J. P. Cann. 1976. Studies of the external morphology of *Pyramimonas*. 2. *Pyramimonas obovata* N. Carter. Arch. Protistenk. 118: 221-226.

Peterfi, L. S. and I. Manton. 1968. Observations with the electron microscope on *Asteromonas gracilis* Artari emend. (*Stephanoptera gracilis* (Artari) Wisl.), with some comparative observations on *Dunaliella* sp. Br. Phycol. Bull. 3: 423-440.

Pickett-Heaps, J. D. and H. J. Marchant. 1972. The phylogeny of the green algae: a new proposal. Cytobios 6: 255-264.

Pickett-Heaps, J. D. and D. W. Ott. 1974. Ultrastructural morphology and cell division in *Pedinomonas*. Cytobios 11: 41-58.

Proskauer, J. 1950. On *Prasinocladus*. Am. J. Bot. 37: 59-66.

Provasoli, L. and A. F. Carlucci. 1974. Vitamins and growth regulators. In Stewart, W. D. P. [Ed.] *Algal physiology and biochemistry*. Blackwell, Oxford. pp. 741-787.

Ragan, M. A. and D. J. Chapman. 1978. A biochemical phylogeny of the protists. Academic Press, New York, San Francisco, London. 317 pp.

Ricketts, T. R. 1966a. Magnesium 2,4-divinylphaeoporphyrin a_5 monomethyl ester, a protochlorophyll-like pigment present in some unicellular flagellates. Phytochemistry 5: 223-229.

Ricketts, T. R. 1966b. The carotenoids of the phytoflagellate, *Micromonas pusilla*. Phytochemistry 5: 571-580.

Ricketts, T. R. 1967. Further investigations into the pigment composition of green flagellates possessing scaly flagella. Phytochemistry 6: 1375-1386.

Ricketts, T. R. 1970. The pigments of the Prasinophyceae and related organisms. Phytochemistry 9: 1835-1842.

Ricketts, T. R. 1971. Identification of xanthophylls KI and KIS of the Prasinophyceae as siphonein and siphonaxanthin. Phytochemistry 10: 161-164.
Ricketts, T. R. 1974. The cultural requirements of the Prasinophyceae. Nova Hedwigia 25: 683-690.
Round, F. E. 1971. The taxonomy of the Chlorophyta. II. Br. Phycol. J. 6: 235-264.
Round, F. E. 1973. The biology of the algae. St. Martin's Press, New York. 278 pp.
Salisbury, J. L. and G. L. Floyd. 1978a. Calcium-induced contraction of the rhizoplast of a quadriflagellate green alga. Science 202: 975-977.
Salisbury, J. L. and G. L. Floyd. 1978b. Molecular, enzymatic and ultrastructure characterization of the pyrenoid of the scaly green monad *Micromonas squamata*. J. Phycol. 14: 362-368.
Scherffel, A. 1912. Zwei neue trichocystenartige Bildungen fuhrende Flagellaten. Arch. Protistenk. 27: 94-128.
Schiff, J. A. 1962. Sulfur. In Lewin R. A. [Ed.] *Physiology and Biochemistry of Algae*. Academic Press, New York and London. pp. 239-246.
Schnepf, E. and M. Maiwald. 1970. Halbdesmosomen bei Phytoflagellaten. Experientia 26: 1343.
Stewart, K. D. and K. R. Mattox. 1975. Comparative cytology, evolution and classification of the green algae with some consideration of the origin of other organisms with chlorophylls *a* and *b*. Bot. Rev. 41: 104-135.
Stewart, K. D. and K. R. Mattox. 1978. Structural evolution in the flagellated cells of green algae and land plants. BioSystems 10: 145-152.
Stewart, K. D., K. R. Mattox and C. D. Chandler. 1974. Mitosis and cytokinesis in *Platymonas subcordiformis*, a scaly green monad. J. Phycol. 10: 65-79.
Stewart, W. D. P. [Ed.]. 1974. *Algal physiology and biochemistry*. Botanical Monographs, Vol. 10, Blackwell, Oxford, London, Edinburgh, Melbourne. 989 pp.
Suzuki, H. 1974. Starch-type polysaccharide and mannitol in *Platymonas*. Phytochemistry 13: 1159-1160.
Swale, E. M. F. 1973. A third layer of body scales in *Pyramimonas tetrarhynchus* Schmarda. Br. Phycol. J. 8: 95-99.
Swale, E. M. F. and J. H. Belcher. 1968. The external morphology of the type species of *Pyramimonas* (*P. tetrarhynchus* Schmarda) by electron microscopy. Proc. Linn. Soc. Lond. 179: 77-81.
Swale, E. M. F. and J. H. Belcher. 1971. Investigation of a species of *Ankyra* Fott by light and electron microscopy. Br. Phycol. J. 6: 41-50.
Swale, E. M. F. and J. H. Belcher. 1974. *Gyromitus disomatus* Skuja - a free-living colourless flagellate. Arch. Protistenk. 116: 211-220.
Swale, E. M. F. and J. H. Belcher. 1975. *Gyromitus limax* nov. sp. - free-living colourless amoebo-flagellate. Arch Protistenk. 117: 20-26.
Sweeney, B. M. 1976. *Pedinomonas noctilucae* (Prasinophyceae), the flagellate symbiotic in *Noctiluca* (Dinophyceae) in Southeast Asia. J. Phycol. 12: 460-464.
Tanoue, E. and Y. Aruga. 1975. Studies on the life cycle and growth of *Platymonas* sp. in culture. Jap. J. Bot. 20: 439-460.
Taylor, F. J. R. 1976. Flagellate phylogeny: a study in conflicts. J. Protozool. 23: 28-40.
Taylor, F. J. R. 1978. Problems in the development of an explicit hypothetical phylogeny of the lower eukaryotes. BioSystems 10: 67-89.
Trezzi, F., M. G. Galli and E. Bellini. 1964. L'infrastruttura di *Dunaliella salina*. G. Bot. Ital. 71: 127-136.
van den Hoek, C. 1978. Algen, Einführung in die Phykologie. Georg Thieme Verlag, Stuttgart. 481 pp.

Wall, D. 1962. Evidence from recent plankton regarding the biological affinities of *Tasmanites* Newton 1875 and *Leiosphaeridia* Eisenack 1958. Geol. Mag. 99: 353-362.

West, G. S. 1916. Algae. Vol. 1 (only vol. published). Cambridge Univ. Press. 475 pp.

Zimmermann, W. 1925. Helgoländer Meeresalgen I-VI. Beiträge zur Morphologie, Physiologie und Oekologie der Algen. Wiss. Meeresunters. Abt. Helgoland, N.F. 16: Abh. 1, 1-25.

Copyright 1980 by Elsevier North Holland, Inc.
Cox, ed. Phytoflagellates

COLONIAL CHLOROPHYTES

RICHARD C. STARR
Department of Botany, The University of Texas at Austin, Austin, Texas 78712, U.S.A.

ABSTRACT

The colonial green flagellates provide a rich source of algal material for investigations in the origin and mechanism of cell differentiation in simple multicellular organisms. The variety of cell shapes, the diverse arrangements of cells in colonies, the nutrition and especially the variations in the sexual process exhibited among the different genera contribute to their usefulness in research in many disciplines. All genera can be easily grown in bacterized or axenic culture, and the majority are available from the culture collections of algae in Cambridge, Göttingen and Austin.

INTRODUCTION

Anyone who has had the occasion to examine microscopically water from ponds and lakes is familiar with one or more of the colonial green flagellates belonging to the order Volvocales of the green algae. More than likely the observed flagellate will be a member of the so-called *Gonium-Volvox* series belonging to the family Volvocaceae, other colonial green flagellates with more restrictive environmental requirements being found occasionally. The first recorded observation is given in a letter from Leeuwenhoek to the Royal Society in 1700 (Dobell, 1932) describing the form and reproduction of an "animacule" whose identity with the genus *Volvox* leaves little doubt. Linnaeus (1758) used the name *Volvox* as a genus of the algae, but it was not until 1832 that the presently accepted circumscription of the genus was made by Ehrenberg, by which time the genera *Gonium* (Müller, 1773) and *Pandorina* (Bory, 1824) had already been described. In the ensuing years additional genera have been recognized. These may be assigned to four families as is done in this chapter, or more conservatively to two (Bourrelly, 1966), depending on the significance one attributes to cell shape and/or developmental patterns.

FAMILY VOLVOCACEAE

Eleven genera are included in this family that is characterized by having the cells embedded in a mucilaginous matrix, and by having in the development of the colony a plakeal stage (curved plate) followed by inversion to form a globoid, except in *Gonium* which remains a plate. Members of this family are often considered to show a hypothetical progression of increasing complexity of colony organization and of the sexual process known as the *Gonium-Volvox* series.

Gonium Müller (Müller, 1773; Stein, 1958b) has colonies of 4 (*G. sociale* [Duj.] Warming), 8 (*G. octonarium* Pocock), 16 (*G. pectorale* Müller), or 32 cells (*G. multicoccum* Pocock) arranged in the form of a plate (Fig. 1, 2) with the flagellated poles of the cells facing the same direction. Sexual reproduction involves the formation of isogametes. As with all the colonial flagel-

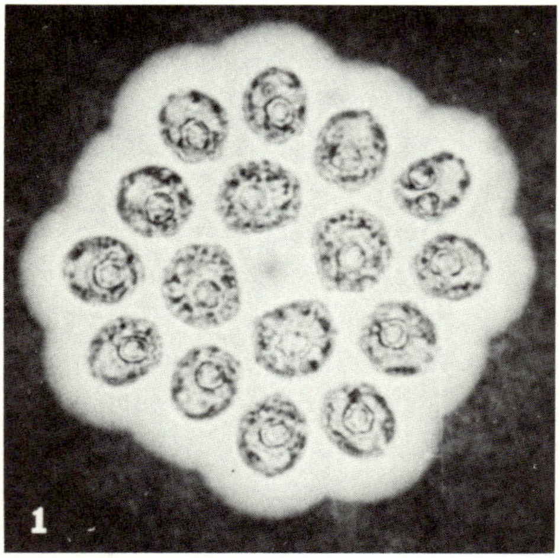

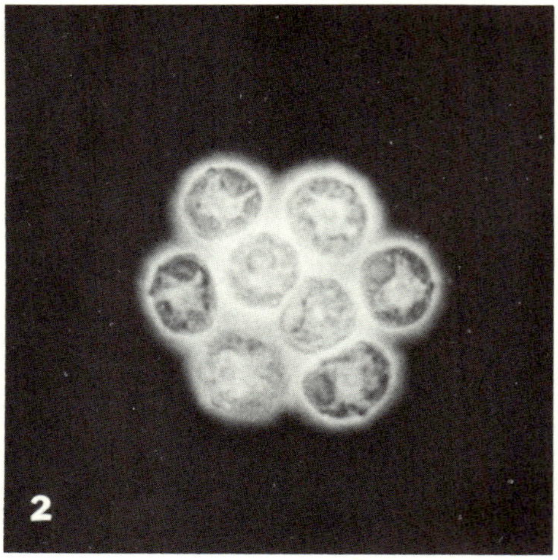

Fig. 1. *Gonium pectorale* (x690). Fig. 2. *Gonium octonarium* (x600).

lates in which the process has been studied, either cytologically or genetically, meiosis occurs at the time of zygote germinations. In *Gonium* all four products survive (Stein, 1958b) but in all other genera there is a random survival of only one of the four.

Pandorina Bory (Bory, 1824; Coleman, 1959; Coleman and Zollner, 1977), as a result of the inversion process during development, has globular colonies with the cells arranged in tiers. *Pandorina morum* Bory with colonies of 16 cells (Fig. 3, 4) tightly appressed at their posterior poles, and *P. charkowiensis* Korschikoff, with 16-32 cells separated from each other by matrix, reproduce sexually by the fusion of isogametes. This distinguishes *P. charkowiensis*

Fig. 3. *Pandorina morum.* Surface view of colony (x1260).

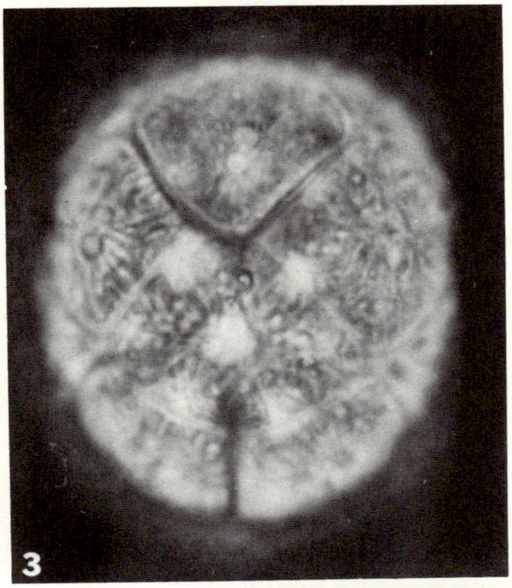

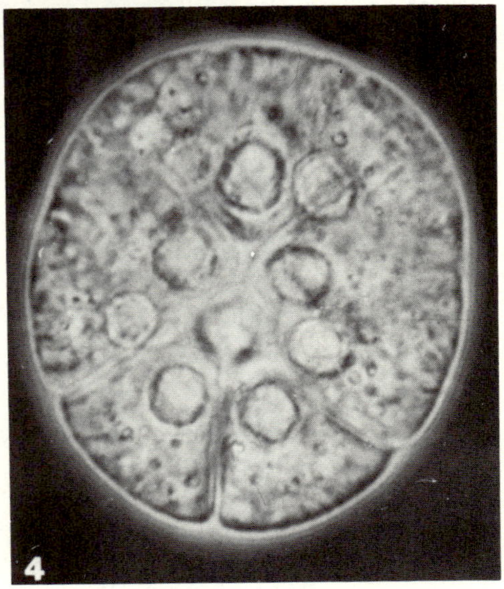

Fig. 4. *Pandorina morum.* Optical section of colony (x1260).

from species of similar colonial construction belonging to the genus *Eudorina*, with its anisogamous sexual process.

Eudorina Ehrenberg (1831) has spheroidal colonies of 16-32 cells (Fig. 5)

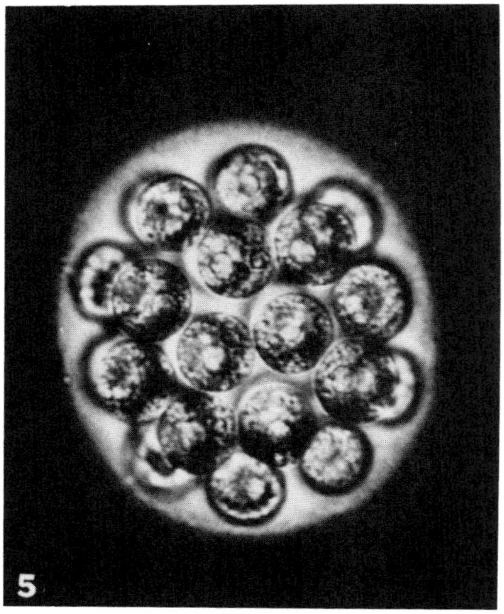

Fig. 5. *Eudorina elegans* (x520).

arranged in alternating tiers. In sexual reproduction packets of elongate sperm cells are differentiated, but the female gametes are transformed vegetative cells. In his monograph of the genus, Goldstein (1964) has merged the genus *Pleodorina* Kofoid with *Eudorina* based on the successful intercrossing between species of the two genera; on the other hand, Goldstein has retained within the same species in *Eudorina*, strains which do not intercross.

Pleodorina Shaw (1894) includes those colonial flagellates (Fig. 6, 7) in which the cells in the anterior tier or tiers are so reduced in size as to preclude their usual participation in either asexual or sexual reproduction. Sexual reproduction is anisogamous.

Volvox L. em. Ehrenberg (1832) has spheroidal colonies. Current researchers using this alga for studies of cellular differentiation (Starr, 1970; Kochert, 1975; Huskey *et al.*, 1979) prefer to interpret *Volvox* as a simple multicellular organism and thus they use the term "spheroid" rather than colony. The spheroids may have 1000-50,000 cells, only a few of which remain reproductive (Smith, 1944) (Fig. 8). Sexual reproduction is oogamous with the eggs and sperm formed in the same individual in some species, and in separate individuals in other species.

The other genera in the Volvocaceae are more rarely observed and are usually

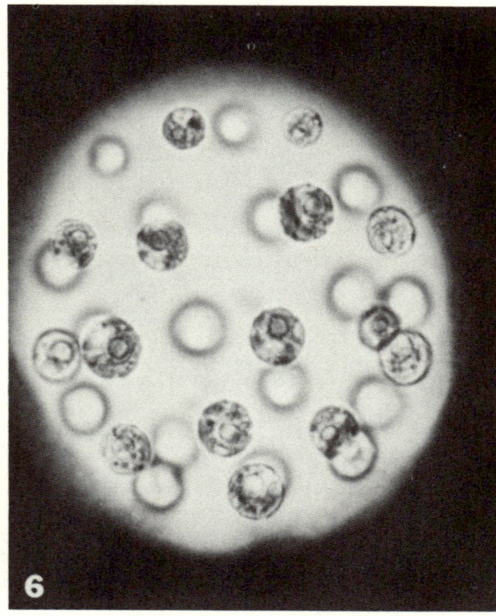

Fig. 6. *Pleodorina illinoisensis* (x530).

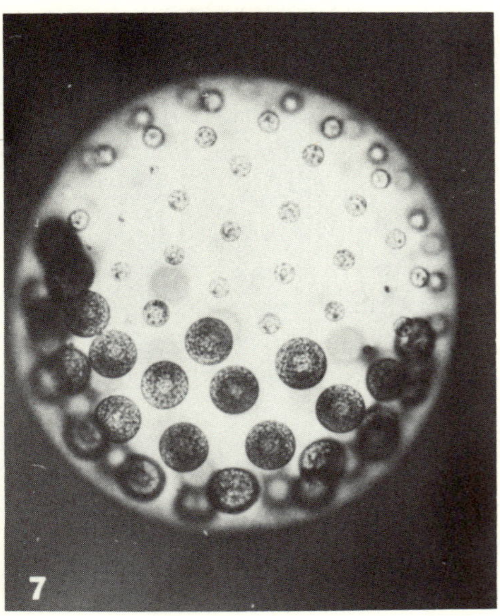

Fig. 7. *Pleodorina californica* (x170).

not included in the *Gonium-Volvox* series which is probably more artificial than real.

Platydorina Kofoid (Kofoid, 1899; Harris and Starr, 1969) may well have been evolved from a *Eudorina*-like ancestor in which the young embryos did not expand immediately after inversion but instead collapsed to form a secondarily flattened colony (Fig. 9) of 16-32 cells. The fact that the flagellated ends of

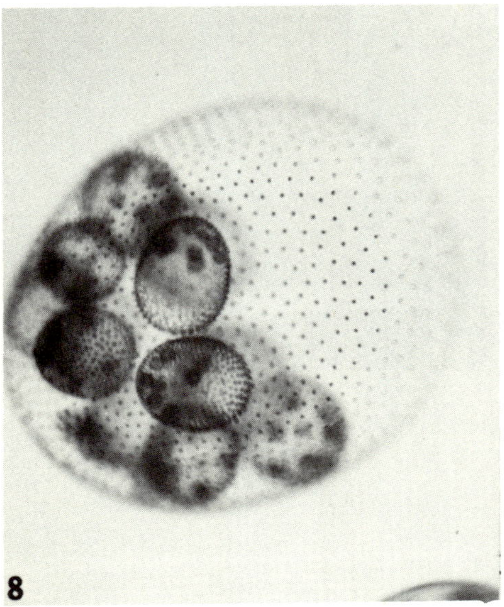

Fig. 8. *Volvox carteri* (x45).

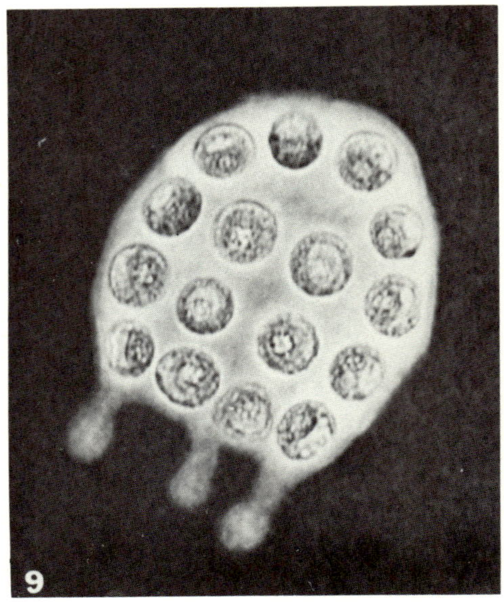

Fig. 9. *Platydorina caudata* (x 530).

two of the four cells in the middle of a 16-celled colony are facing one direction and the other two in the opposite direction remains as evidence of this secondary flattening. The pattern of swimming is that expected of a compressed globoid, with the curved edge of the colony directed forward. Eyespots on the cells of this curved edge reflect that these cells correspond to the cells in the anterior tier of a typical *Eudorina*-like colony. Sexual reproduction is anisogamous with packets of sperm being formed. This genus is known only from a few states in the Mississippi and Ohio drainage basins.

Volvulina Playfair (Playfair, 1915; Pocock, 1953) is distinctive because of its hemispherical cells, each surrounded by a nonconfluent sheath (Fig. 10). Sexual reproduction is through the fusion of isogametes. The most common species, *V. steinii* Playfair, occurs only in highly eutrophic situations and is known best from studies using cultured populations (Pocock, 1953; Stein, 1958a; Carefoot, 1966, 1967).

The genera *Corone* Fott (1949) and *Lundiella* Sarma et Shyam (1974) are less

Fig. 10. *Volvulina steinii* (x900).

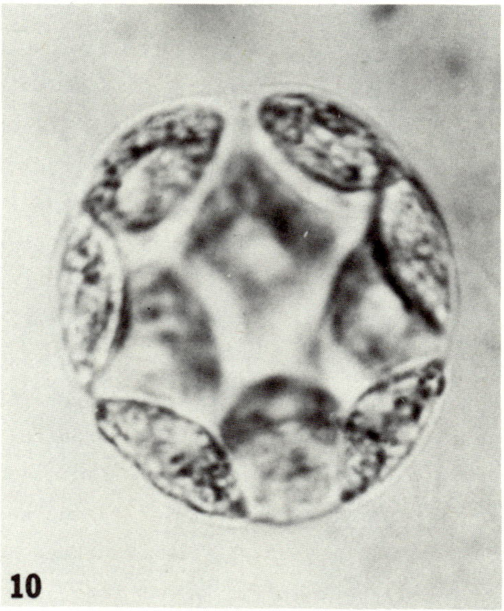

well-known colonial flagellates, having been observed only by the original workers. *Corone* is described as an 8-celled colony with the cells arranged in two tiers. The cells are connected by tough gelatinous strands but are not enclosed in a common mucilaginous matrix as is common to the family Volvocaceae. For this reason Fott erected a new family, the Coronaceae; Bourrelly (1966) places the genus in the family Spondylomoraceae but the writer feels it is more closely allied to the family Volvocaceae. *Lundiella* is an even more problematic genus, inasmuch as nothing is known about its reproduction. It resembles most closely a very small (64-celled) *Volvox* with angular cells, but no matrix was observed by Sarma and Shyam.

The genera *Stephanoon* Schewiakoff (1893) and *Mastigosphaera* Schewiakoff (1893), included in the Volvocaceae by Bourrelly (1966), are so little known as to be of doubtful existence. The 8-16 celled colonies of *Stephanoon* with cells in two tiers described by various workers were considered by Fott (1949) "as embracing elements altogether incoherent and becoming a permanent source of confusion or error" and he proposed the abolishment of this genus. *Mastigosphaera*, most closely resembling *Pandorina* but with only a single flagellum on each cell, may well have been based on an incorrect observation of *Pandorina*.

The Colonial Envelope. The matrix common to the genera of the Volvocaceae has yet to be the subject of a comparative detailed investigation of its chemical properties. Lang (1963), in an electron-microscopic investigation of various genera, showed that the matrix was not a single amorphous secretion of the cells. In most of the genera the matrix has at least two layers, the inner bounded by a tripartite structure consisting of two electron-dense layers separated by an electron-transparent layer. The chemical nature of the matrix in *Volvox* has been studied by various investigators, but its exact nature is still unknown. Protein and carbohydrates are major constituents. Kochert (1968), using the proteolytic enzyme pronase, was able to solubilize the matrix. Kochert and Yates (1974) reported (via a personal communication from Lamport) the amino-acid composition of the matrix glycoprotein from *Volvox carteri*, but the high yield of glycine in the analysis would indicate the possible contamination with the medium in which the alga had been grown. McCracken and Barcellona (1976), using electron histochemistry, postulated that the internal matrix contains glycoproteins or a mixture of acid mucosaccharides and glycoproteins, while the external layer contains acid mucosaccharides alone. McCracken and Barcellona speculate that the carbohydrates in the matrix may be involved in determining cell permeability and surface specificity in *Volvox* as similar compounds do at the surfaces of many other cells.

Motility. In the globoidal colonies of the Volvocaceae large eyespots occur on the cells at one end, their size gradually diminishing in the cells of successively posterior tiers. When a colony is swimming, it is the pole with the large eyespots that, coincidentally or otherwise, is directed forward as the colony moves through the medium, rotating on its longitudinal axis. The positioning of the two flagella in relation to the eyespot in each cell is important in determining normal movement, or abnormal movement in the case of mutants which have been studied in *Volvox carteri* (Huskey, 1979). The positively phototactic movement in *Volvox* has been explained by the activation or cessation of movement of the flagella in response to light (Huth, 1970; Hand and Haupt, 1971; Schletz, 1976) mediated by a receptor in the eyespot and/or chloroplast (Dodge, 1969; Schletz, 1976). The fact that the cells of many species of the Volvocaceae, e.g., *Volvox carteri*, are not interconnected by cytoplasmic strands would preclude having a single area of the spheroid which would sense the direction of the light and direct the flagella action of the cells in other areas, thus the response of the colony to a unidirectional light source must be the sum of the responses of the individual cells.

Colonial Development. The globular colonies of the Volvocaceae develop in a similar manner whether they are the 16-celled *Pandorina* or *Volvox* with several thousand cells. Cleavage of the asexual reproductive cells occurs parallel to the longitudinal axis in a slightly oblique plane which alternates clockwise and counterclockwise in succeeding cleavages. Such a spiral cleavage is found in the development of certain animals such as gastropods and is significant in that it results in alternating tiers of cells, a characteristic of the cell arrangement in all the Volvocaceae. The details of the cleavage patterns in *Volvox* have been illustrated in the classic paper by Janet (1923).

The cleavage pattern results in the circumstance that at the end of cleavage the cells in the resulting curved plate or hollow sphere are positioned with their potentially flagellar ends pointing to the inside of the sphere. The process of inversion by which the position of the cells is reversed was described in *Pleodorina* by Merton (1908) and in *Volvox* by Powers (1908). In the latter instance, the sequence of events was not properly ordered, and it was not until 1922 that Kuschakewitsch described the process correctly. Pocock (1933) pointed out the changes in cell shape which appear in various regions of the inverting embryo; Kelland (1977) and Viamontes *et al.* (1979) related these changes to the morphogenetic movements by which inversion is effected. *Pandorina morum* (Fulton, 1978) and *Eudorina elegans* (Gottlieb and Goldstein, 1977) have also been the subjects of recent investigations of the inversion process.

The Sexual Process. From an experimental point of view the colonial green algae belonging to the family Volvocaceae are unique among the algal flagellates in the ease with which sexual strains may be isolated from natural populations and successfully mated in the laboratory. This may seem surprising in view of comments which the late Professor Gilbert Morgan Smith made to the author in a letter dated May 27, 1955, following the receipt of sexual cultures of *Pandorina morum* and *Eudorina elegans* which the author had sent to him:

> "I have collected *Pandorina* in abundance many times but only once, some thirty years ago, did I ever observe fruiting material. This was in a collection from an ornamental pool in a garden. The same thing holds for *Chlamydomonas* in which I have only seen gametic union twice in material taken from the field. Once was 25 years ago and once last year."

Although natural populations may not be exhibiting sexual phases at the time they are collected, clones isolated from such populations will, as a rule, produce populations which either alone (homothallism) or when mixed with another population (heterothallism) undergo sexual reproduction. More importantly, soil samples from the edge of ponds where such algae are growing can be air-dried and then later re-wet to produce populations in the laboratory at will from zygotes in the dried soil (Starr, 1973). In this way one is not limited to populations collected near the laboratory. The formulation of soilwater media by Pringsheim (1950) made it possible to isolate and grow vigorous clonal populations of the Volvocaceae. With the exception of *Volvox*, the sexual process can be evoked without difficulty in such soilwater populations. Although some species of *Volvox* will produce sexual individuals in this medium, the axenic medium compounded by Provasoli and Pintner (1959) has proved to be the most reliable.

Dr. Annette Coleman began her doctoral study of the mating behavior of *Pandorina* (1958) with the sexual clones which the author had isolated in 1955. Her investigation of these and other populations was followed by others concerned with the genera *Gonium* (Stein, 1958b), *Eudorina* (Goldstein, 1964), *Platydorina* (Harris and Starr, 1969), *Volvulina* (Stein, 1958a; Carefoot, 1966, 1967), and many species of *Volvox*, namely, *V. aureus* (Darden, 1966), *V. carteri* f. *weismannia* (Kochert, 1968), *V. carteri* f. *nagariensis* (Starr, 1969), *V. rousseletii* (McCracken and Starr, 1970), *V. gigas* (Vande Berg and Starr, 1971), *V. dissipatrix* (Starr, 1972), *V. obversus* (Karn *et al.*, 1974) and *V. capensis* Starr *et al.*, 1980). These basic studies have, in many instances, led to further investigations by these and other researchers.

A detailed, masterful account of "Sexuality in Colonial Green Flagellates" by Annette Coleman (1979) has recently appeared. In this 34-page review, based on 101 references, she discusses the sexual reaction, the hormonal control of differentiation in *Volvox*, cell-cycle events and cell differentiation, mating types and syngens, the genetics of sexuality, and other topics--all in greater detail than space here permits. The reader is referred to this review for a summary of the experimental results dealing with all aspects of the sexual process.

FAMILY ASTREPHOMENACEAE

This is a monotypic family with the single genus and species *Astrephomene gubernaculifera* Pocock (1953). Superficially, the 64-128 celled globoidal colonies (Fig. 11) resemble genera in the family Volvocaceae, but the matrix in this genus surrounds only each individual cell and does not encompass the colony as a whole. Furthermore, during the development of the embryos, the cleavage pattern is such that the embryo develops with the flagellar ends of the cells pointing toward the outside of the colony; therefore inversion does not occur.

The colonies of *Astrephomene* often have up to seven small "rudder" cells at the posterior end of the colony. In Fig. 11 the colony is seen from the posterior end and four small cells are evident. The flagella of these small cells

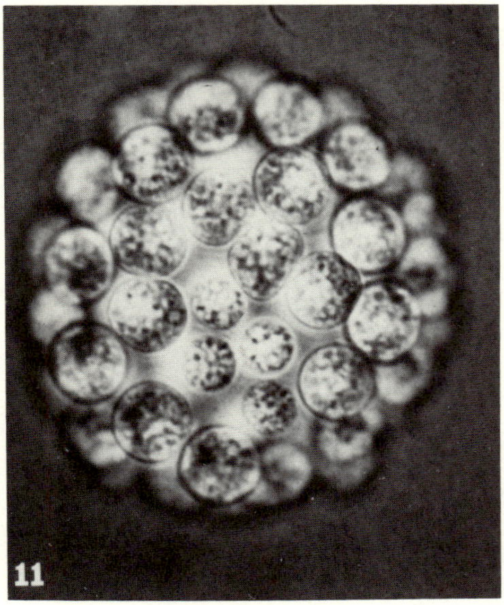

Fig. 11. *Astrephomene gubernaculifera* (x520).

extend without bending when the colony is swimming, giving the appearance of a rudder, hence the specific epithet *gubernaculifera*. These cells do not participate in the reproductive processes. Pyrenoids cannot be seen in the vegetative cells of *Astrephomene*, but, as in *Volvulina steinii*, following the fusion of isogametes, the zygote develops two pyrenoids (Stein, 1958a).

Natural populations of *Astrephomene* are rarely encountered possibly due to their need for highly eutrophic conditions to supply a carbon source. The alga is green, but cultures require an exogenous source of carbon. In axenic culture this is supplied as acetate; in bacterized soilwater culture, in the form of a rotting pea cotyledon. Soil cultures baited with a pea cotyledon have proved to be invaluable in getting multiple populations (Brooks, 1966, 1972).

In actively growing populations the cells of young colonies are hemispherical. In the plate illustrating Playfair's (1915) original description of the genus *Volvulina* (family Volvocaceae), some figures were of *Astrephomene*. Pocock's (1953) account of *Volvulina* and *Astrephomene* is a classic illustrating the best use of the literature, preserved collections and algal cultures to solve a problem of confused identity.

Astrephomene has served as experimental material in investigations of mating behavior and chromosome number by Stein (1958a) and Brooks (1966). Cave and Pocock (1956) had already demonstrated variable chromosome numbers in the single species of the genus, but this alone did not explain the degree of sexual isolation found by Brooks.

FAMILY HAEMATOCOCCACEAE

This family includes the unicellular genus *Haematococcus* and the colonial *Stephanosphaera*, each with protoplasts having extensions. The cells in *Stephanosphaera* are embedded in a mucilaginous matrix. This and the fact that the cells of certain species of *Volvox* have similar extensions have caused some authors (Bourrelly, 1966; Bold and Wynne, 1978) to include *Stephanosphaera* in the family Volvocaceae.

The colonies of *Stephanosphaera* (Fig. 12) are rarely encountered and then usually in pools formed in depressions of granitic rocks. When isolated in

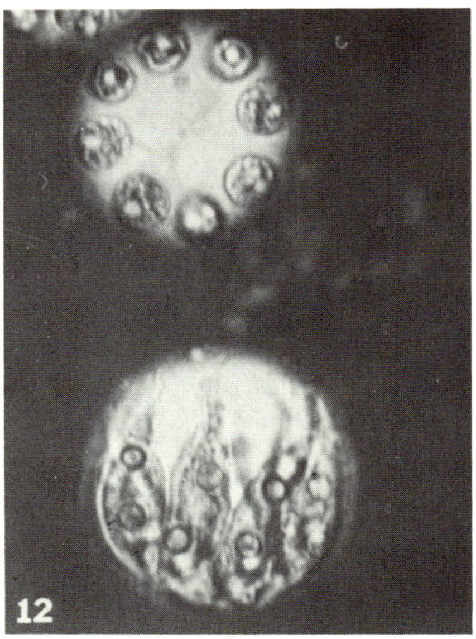

Fig. 12. *Stephanosphaera pluvialis* (x550).

cultures, however, the alga grows well in the same soilwater media in which the *Gonium-Volvox* series flourishes, even though these more commonly occur in farm ponds. Sexual reproduction occurs through the fusion of spindle-shaped isogametes.

FAMILY SPONDYLOMORACEAE

The cells in the colonies of this family are arranged in tiers which are loosely held together without the encompassing mucilaginous matrix common to the family Volvocaceae.

The family is based on *Spondylomorum* Ehrenberg (1848) as the type, and unfortunately there are doubts as to its existence (Pringsheim, 1960). The 8-16 cells in colonies of *Spondylomorum* are described as quadriflagellate. *Pascherina* Silva (1959) was formerly known as *Pascheriella* Korschikoff (1928), but the prior use of the epithet for a genus in the Chrysophyceae necessitated the name change. This 4-celled species is rarely encountered, but it is now maintained in the Culture Centre of Algae and Protozoa in Cambridge.

The only genus of the Spondylomoraceae commonly found throughout the world has colonies of 8-16 biflagellated cells (Fig. 13). This alga appears in the

Fig. 13. *Uva* sp. (x500).

literature under a variety of names. *Chlamydobotrys* Korschikoff (1924) was placed by Korschikoff (1938) as a synonym of *Pyrobotrys* Arnoldi (1916); more recently, Bourrelly (1962) has pointed out that *Uva* Playfair (1914) has priority over *Pyrobotrys*.

When one considers that this alga is commonly found in highly eutrophic waters such as sewage oxidation ponds, it is not surprising that it assimilates acetate as a carbon source more readily than it fixes carbon through photosynthesis (Wiessner, 1962).

Sexual reproduction by isogametes is known for both *Pascherina* and *Uva* (*Chlamydobotrys*, *Pyrobotrys*). In *Uva* the quadriflagellated zygote remains motile for 10 days or more. The question as to whether the unicellular flagellate *Chlorobrachis* Korschikoff (1925) is the motile zygote of *Uva* or an independent genus has been reviewed by Belcher and Swale (1961).

CULTURES

The ever increasing interest in the algae as materials for research in physiology, genetics, ecology, etc. has been fostered by the availability of algal cultures. Genera of the colonial green flagellates are especially well represented in the collections of algae at Cambridge (George, 1976), Göttingen (Koch, 1964), and Austin (Starr, 1978). The strains used by many of the researchers concerned with the sexual process in these flagellates are in the Austin collection (University of Texas, formerly at Indiana University), and with a few exceptions they still exhibit sexual activity when placed under the proper environmental protocol. The problem of the loss of fertility is especially pronounced in the strains of *Volvulina steinii* and *Astrephomene gubernaculifera*. Cultures of these two species have been maintained with ease for years through serial transfer (as is done with other genera), but it has been impossible to maintain their fertility. Brooks (1966) experienced this difficulty even during his study of *Astrephomene*, and the fertile clones of *Astrephomene* which he and Stein (1958a) deposited in the collections, as well as those of *Volvulina* deposited by Stein (1958a) and Carefoot (1966), have long since become infertile. Coleman (1975) has addressed this problem of longtime survival of fertile clones in the genus *Pandorina* and has discussed methods of preservation which may be used as an alternative to the usual method of maintenance through serial transfer of vegetative material.

REFERENCES

Arnoldi, W. 1916. Ein neuer Organismus aus der Volvokazeenordnung: *Pyrobotrys incurva*. Vol. Jubil. C. Timiriazeff, Moscow.

Belcher, J. H. and E. M. F. Swale. 1961. Some new and uncommon British Volvocales. Brit. Phycol. Bull. 2: 56-62.

Bold, H. C. and M. J. Wynne. 1978. *Introduction to the Algae*. Prentice-Hall, Englewood Cliffs, New Jersey. 706 pp.

Bory de St. Vincent, J. B. 1824. Encyclop. method. hist. natur. zoophyt. Vol. 2. Agasse, Paris. (Ref. not seen.)

Bourrelly, P. 1962. *Chlamydobotrys*, *Pyrobotrys* ou *Uva*? Rev. Algol. 6: 126-128.

Bourrelly, P. 1966. *Les Algues d'eau douce*. Vol. I. N. Boubée & Cie. Paris. pp. 1-511.

Brooks, A. E. 1966. The sexual cycle and intercrossing in the genus *Astrephomene*. J. Protozool. 13: 367-375.

Brooks, A. E. 1972. The physiology of *Astrephomene gubernaculifera*. J. Protozool. 19: 195-199.

Carefoot, J. R. 1966. Sexual reproduction and intercrossing in *Volvolina steinii*. J. Phycol. 2: 150-156.

Carefoot, J. R. 1967. Nutrition of *Volvolina* Playfair. J. Protozool. 14: 15-18.

Cave, M. S. and M. A. Pocock. 1956. The variable chromosome number in *Astrephomene gubernaculifera*. Amer. J. Bot. 43: 122-134.

Coleman, A. W. 1959. Sexual isolation in *Pandorina morum*. J. Protozool. 6: 249-264.
Coleman, A. W. 1975. Long-term maintenance of fertile algal clones: experience with *Pandorina* (Chlorophyceae). J. Phycol. 11: 282-286.
Coleman, A. W. 1979. Sexuality in the colonial flagellates. In Hutner, S. H. and M. Levandowsky [Eds.] *Physiology and Biochemistry of the Protozoa*, Vol. 1, 2nd Ed. Academic Press, New York, pp. 307-340.
Coleman, A. W. and J. Zollner. 1977. Cytogenetic polymorphism within the species *Pandorina morum*. Arch. Protistenk. 119: 224-232.
Darden, W. H. 1966. Sexual differentiation in *Volvox aureus*. J. Protozool. 13: 239-255.
Dobell, C. 1932. *Antony van Leeuwenhoek and His "Little Animals."* John Bale, Sons & Danielsson Ltd., London. Reprint, Dover Publ., New York, pp. 1-435.
Dodge, J. D. 1969. A review of the fine structure algal eyespots. Brit. Phycol. J. 4: 199-210.
Ehrenberg, C. G. 1831. Über der Entwicklung und Lebensdaur der Infusionthiere; nebst ferneren Beiträgen zu unser Vergleichung iher organischen Systeme. Monatsb. Akad. Wiss. Berlin. pp. 78 and 152, taf. II, figs. 10A-10D. (Ref. in Goldstein, 1964.)
Ehrenberg, C. G. 1832. Zur Erkenntniss der Organisation in der Richtung der kleinsten Raumes. Abhandl. kgl. Akad. Wiss. Berlin 1831: 1-154.
Ehrenberg, C. G. 1848. Boebachtung zweier generisch neuer Formen der Frühlingsgewässer bei Berlin. Monatsber. Konigl. Akad. Wiss. z. Berlin 1848: 233-247.
Fott, B. 1949. *Corone*, a new genus of colonial Volvocales. Mem. Soc. Roy. Letter. Scienc. de Boheme., CL. Scienc: 1-9.
Fulton, A. B. 1978. Colonial development in *Pandorina morum*. II. Colony morphogenesis and formation of the extracellular matrix. Dev. Biol. 64: 236-251.
George, E. A. 1976. List of Strains 1976 at Culture Centre of Algae and Protozoa. 36, Storey's Way, Cambridge, England. p. 120. (Privately printed)
Goldstein, M. 1964. Speciation and mating behavior in *Eudorina*. J. Protozool. 11: 317-344.
Gottlieb, B. and M. E. Goldstein. 1977. Colony development in *Eudorina elegans*. J. Phycol. 13: 358-364.
Hand, W. B. and W. Haupt. 1971. Flagellar activity of the colony members of *Volvox aureus* during light stimulation. J. Protozool. 18: 361-364.
Harris, D. O. and R. C. Starr. 1969. Life history and physiology of reproduction of *Platydorina caudata* Kofoid. Arch. Protistenk. 111: 138-155.
Huskey, R. J. 1979. Mutants affecting vegetative cell orientation in *Volvox carteri*. Dev. Biol. 72: 236-243.
Huskey, R. J., B. E. Griffin, P. O. Cecil and A. M. Callahan. 1979. A preliminary genetic investigation of *Volvox carteri*. Genetics 91: 229-244.
Huth, K. 1970. Bewegung und Orientierung bei *Volvox aureus* Ehrb. I. Mechanismus der phototaktischen Reaktion. Z. Pflanzenphysiol. 62: 436-450.
Janet, C. 1923. Le *Volvox*, Troisième Mémoire. pp. 1-179, plates 5-20.
Karn, R. C., R. C. Starr and G. A. Hudock. 1974. Sexual and asexual differentiation in *Volvox obversus* (Shaw) Printz, strains Wd3 and Wd7. Arch. Protistenk. 116: 142-148.
Kelland, J. L. 1977. Inversion in *Volvox* (Chlorophyceae). J. Phycol. 13: 373-378.
Koch, W. 1964. Verzeichnis der Sammlung von Algenkulturen am Pflanzenphysiologischen Institut der Universität Göttingen. Arch. Microbiol. 47: 402-432.
Kochert, G. 1968. Differentiation of reproductive cells in *Volvox carteri*. J. Protozool. 15: 438-452.
Kochert, G. 1975. Developmental mechanisms in *Volvox* reproduction. In Markert, C. L. and J. Papaconstantinou [Eds.] *The Developmental Biology of*

Reproduction. 33rd Symposium Soc. Develop. Biol., Academic Press, New York. pp. 55-90.

Kochert, G. and I. Yates. 1974. Purification and partial characterization of a glycoprotein sexual inducer from *Volvox carteri*. Proc. Nat. Acad. Sci. U.S. 71: 1211-1214.

Kofoid, C. A. 1899. Plankton studies. III. On *Platydorina*, a new genus of the family Volvocaceae from the plankton of the Illinois River. Bull. of the Illinois State Laboratory 5: 419-440.

Korschikoff, A. A. 1924. Zur Morphologie und Systematik der Volvocales. Arch. Russ. Protistol. 3: 153-197.

Korschikoff, A. A. 1925. Beitrage zur Morphologie und Systematik der Volvocales. Arch. Russ. Protistol. 4: 153-197. (Ref. in Belcher & Swale, 1961.)

Korschikoff, A. A. 1928. On two new Spondylomoraceae: *Pascheriella tetras* n. gen. et sp., and *Chlamydobotrys squarrosa* n. sp. Arch. Protistenk. 61: 223-238.

Korschikoff, A. A. 1938. Contribution to the algal flora of the Gorky district. Proc. Kharkov A. Gorky State Univ. Book 14. Proc. Bot. Inst. 3: 1-21. (Ref. in Smith, 1950.)

Kuschakewitsch, K. 1922. Zur Kenntnis der Entwicklungsgeschichte von *Volvox*. Bulletin de l'Academie des Sciences l'Oukraine 1: 31-36. (Reprinted in Arch. Protistenk. 73: 323-330, 1931.

Lang, N. J. 1963. Electron microscopy of the Volvocaceae and Astrephomenaceae. Amer. J. Bot. 50: 280-300.

Linnaeus, C. 1758. *Systema Naturae*. (10th ed.) 8°. 2 Vols. Holmiae. (Ref. in Dobell, 1932.)

McCracken, M. D. and W. J. Barcellona. 1976. Electron histochemistry and ultrastructural localization of carbohydrate containing substances in the sheath of *Volvox*. J. of Histochem. & Cytochem. 24: 668-673.

McCracken, M. D. and R. C. Starr. 1970. Induction and development of reproductive cells in the K-32 strains of *Volvox rousseletii*. Arch. Protistenk. 112: 262-282.

Merton, J. 1908. Über den Bau und die Fortpflanzung von *Pleodorina illinoisensis* Kofoid. Z. wiss. Zool. 90: 445-477.

Müller, O. 1773. *Vermium terr. et fluviatil, seu infus. historia*. Hauniae et Lipsiae. (Ref. not seen.)

Playfair, G. I. 1914. Contributions to a knowledge of the Biology of Richmond River. Proc. Linn. Soc. New South Wales 39: 93-151.

Playfair, G. I. 1915. Fresh-water algae of the Lismore district: with an appendix on the algae fungi and Schizomycetes. Proc. Linn. Soc. New South Wales 40: 310-362.

Pocock, M. A. 1933. *Volvox* in South Africa. Ann. South Afr. Mus. 16: 523-646.

Pocock, M. A. 1953. Two multicellular motile green algae, *Volvulina* Playfair and *Astrephomene*, a new genus. Trans. Roy. Soc. S. Africa. XXXIV (1): 103-127.

Powers, J. H. 1908. Further studies in *Volvox*, with descriptions of three new species. Trans. Amer. Microsc. Soc. 27: 123-149.

Pringsheim, E. G. 1950. The soil-water culture technique for growing algae. In Brunel, J., G. W. Prescott and L. H. Tiffany [Eds.] *The Culturing of Algae*. The Charles F. Kettering Foundation, Yellow Springs, Ohio. pp. 19-26.

Pringsheim, E. G. 1960. Zur Systematik und Physiologie der Spondylomoraceen. Osterr. Botan. Z. 107: 425-438.

Provasoli, L. and I. J. Pintner. 1959. Artificial media for freshwater algae; problems and suggestions. In Tryon, C. A. and R. T. Hartman [Eds.] *The Ecology of Algae*. Spec. Publ. No. 2, Pymatuning Lab. of Field Biology, University of Pittsburgh. pp. 84-96.

Sarma, Y. S. R. K. and R. Shyam. 1974. *Lundiella indica* gen. et sp. nov., a new member of colonial Volvocales (Chlorophyceae) from India. Brit.

Phycol. J. 9: 307-311.
Schewiakoff, W. 1893. Ueber geographische Verbreitung der Süsswasserprotozoen. Mem. Acad. Imp. Sci. St. Petersbourg VII, 41, No. 8.
Schletz, K. 1976. Phototaxis in *Volvox*--Pigments involved in the perception of light direction. Z. Pflanzenphysiol. 77: 189-211.
Shaw, W. R. 1894. *Pleodorina*, a new genus of the Volvocineae. Bot. Gaz. 19: 279-283.
Silva, P. C. 1959. Remarks on algal nomenclature II. Taxon 8: 60-64.
Smith, G. M. 1944. A comparative study of the species of *Volvox*. Trans. Amer. Microsc. Soc. 63: 265-310.
Smith, G. M. 1950. *The Fresh-water Algae of the United States*, 2nd Ed. McGraw-Hill, New York. 719 pp.
Starr, R. C. 1969. Structure, reproduction and differentiation in *Volvox carteri* f. *nagariensis* Iyengar, strains HK9 and HK10. Arch. Protistenk. 111: 204-222.
Starr, R. C. 1970. Control of Differentiation in *Volvox*. Develop. Biol. Suppl. 4: 59-100.
Starr, R. C. 1972. Sexual reproduction in *Volvox dissipatrix*. Br. Phycol. J. 7: 284 (Abstr.)
Starr, R. C. 1973. Special methods--dry soil samples. In Stein, J. R. [Ed.] *Handbook of Phycological Methods* Vol. 1. Cambridge University Press, Cambridge. pp. 161-167.
Starr, R. C. 1978. The Culture Collection of Algae at the University of Texas at Austin. J. Phycol. 14 (suppl.): 47-100.
Starr, R. C., R. M. O'Neil and C. M. Miller, III. 1980. L-Glutamic acid as a mediator of sexual morphogenesis in *Volvox capensis*. Proc. Nat. Acad. Sci. U.S. 77 (2). (paper in proof)
Stein, J. R. 1958a. A morphological study of *Astrephomene gubernaculifera* and *Volvulina steinii*. Amer. J. Bot. 45: 388-397.
Stein, J. R. 1958b. A morphological and genetic study of *Gonium pectorale*. Amer. J. Bot. 45: 664-672.
Vande Berg, W. J. and R. C. Starr. 1971. Structure, reproduction and differentiation in *Volvox gigas* and *Volvox powersii*. Arch. Protistenk. 113: 195-219.
Viamontes, G. I., L. J. Fochtmann and D. L. Kirk. 1979. Morphogenesis in *Volvox*: Analysis of critical variable. Cell 17: 537-550.
Wiessner, W. 1962. Kohlenstoffassimilation von *Chlamydobotrys* (Volvocales). Arch. für Mikrobiol. 43: 402-411.

Copyright 1980 by Elsevier North Holland, Inc.

EUGLENOID FLAGELLATES

PATRICIA L. WALNE
Department of Botany, University of Tennessee, Knoxville, Tennessee 37916, U.S.A.

ABSTRACT

Since considerable information on *Euglena* and a number of other euglenoid flagellates is summarized in other sources, this chapter focuses primarily on some very new data derived from current state-of-the-art investigations on a few of the organelles and structures, especially chloroplasts, flagellar mastigonemes, stigma (eyespot) and envelope (lorica). Some attention is given to recent data on the nucleus and mitosis, pellicle, ingestion organelles and trichocysts. A brief overview of green and colorless forms is provided, together with a summary of some of the hypotheses on chloroplast evolution and the phylogeny of euglenoid organisms.

INTRODUCTION

Several rather well-defined diagnostic features characterize the euglenoid flagellates, but despite their basic similarities, they manifest considerable diversities in form and function that pose intriguing biochemical, systematic, phylogenetic, developmental and other problems at nearly every taxonomic level. Traditional criteria that are used to infer relationships among euglenophytes include flagellar number, arrangement and structure, pellicle structure and elasticity, cell shape and type of symmetry, and the presence and nature of envelopes. Details of these and other diagnostic and taxonomically important features are presented by Leedale (1967) and Buetow (1968).

The apparent lack of sexual reproduction in this group complicates its systematic treatment since the interbreeding criterion cannot be used in delimiting species. Their demarcation has thus been based largely on morphology, which is capricious at best, resulting in the description of an inordinate number of species and the erection of a number of artificial phylogenetic schemes. Klebs (1883), for example, complicated the issue for many years by his separation of green and colorless organisms, when in fact many of them subsequently have been shown to be closely related. Hollande (1942) introduced a more natural scheme with his emphasis on flagellar organization, and more recently, Leedale (1967, 1978) has intensified efforts to devise a natural classification system for euglenoid flagellates by increasing the number of special attributes that must be considered.

In contrast to some of the other phytoflagellate groups for which there have been no general reviews or central sources of information, there are several excellent treatises and monographs on *Euglena* and euglenoid flagellates (*e.g.*, Gojdics, 1953; Pringsheim, 1956; Leedale, 1967; Wolken, 1967; Buetow, 1968), and the reader is referred to them for basic information on the general characteristics of these organisms, their life histories, habitats, nutrition, culture methods and conditions, and many biophysical and biochemical processes special to the Euglenophyta. In addition, ultrastructural details on various organelles are compiled by Dodge (1973); a short monograph summarizing the major biological and structural features of euglenoids in general, together with selected areas of research on the genus *Euglena*, is presented by Leedale (1971); and a brief commentary and bibliography highlighting research on the euglenoids through 1970 are provided by Leedale and Walne (1971).

The application of many new and exciting analytical techniques to old and tenacious problems is permitting some new insights into various aspects of form, function and phylogenetic relationships in the euglenoid flagellates. It is impossible, of course, to include all of the noteworthy contributions of many scientists, and regrettably, limitations of space dictate a very superficial treatment for those few that are considered, with the understanding that the original sources in the literature must be consulted for additional details. This chapter, then, is intended to complement the published works cited above by summarizing some of the current 'state-of-the-art' research on euglenoid organisms, as well as some earlier information, so as to provide a little background and perspective.

OVERVIEW OF GREEN AND COLORLESS ORGANISMS

Although most members of the Euglenophyta are flagellated, except when they are encysted or in a transitory non-motile or palmelloid phase, there are some exceptions, *e.g.*, stalked *Colacium*, *Euglena obtusa*. Most euglenoids are green and phototrophic, having chlorophylls *a* and *b* in their chloroplasts, as do the Chlorophyta and Charophyta, but they differ from them in most respects, including the chemical nature of the reserve photosynthate, which is a β-1:3 glucan termed 'paramylon' and not the typical α-1:4-linked glucose polymer, starch. Despite their photosynthetic capabilities, the chlorophyll-containing euglenoids may require vitamins and are thus considered photoauxotrophic. In their studies of *Euglena*, Kivic and Vesk (1974c) reported the uptake of proteins by pinocytosis from the reservoir, an anterior invagination, which lies adjacent to a large contractile vacuole and which has been shown in several euglenoid genera to contain abundant hairs (*e.g.*, Dynesius and Walne, 1975).

Some euglenoids are naturally colorless and osmotrophic (*e.g.*, *Astasia*, *Distigma*) or phagotrophic (*e.g.*, *Peranema*, *Entosiphon*), while some strains of green, wild-type *Euglena gracilis* have been bleached experimentally by streptomycin (Provasoli *et al.*, 1948), high temperature (Pringsheim and Pringsheim, 1952), UV light (Lyman *et al.*, 1961), porfiromycin (Kronestadt and Walles, 1975) or prolonged darkness (Siegesmund *et al.*, 1962; Palisano and Walne, 1976).

Ultrastructural changes in *Euglena* after UV irradiation were reported by Michaels and Gibor (1973), and Palisano and Walne (1976) described the ultrastructural and fluorescence properties of two permanently bleached strains of *E. gracilis*, induced by UV irradiation or by darkness. Structural degeneration of chloroplasts of *E. gracilis* Z during heterotrophic growth in darkness was shown by Ehara *et al.* (1975). The effects of darkness and streptomycin on the ultrastructure of *E. gracilis* were reported by Siegesmund *et al.* (1962). Mutants of *Euglena* resistant to streptomycin were isolated and characterized by Diamond and Schiff (1974). The comparative ultrastructure of *E. gracilis*, bleached *Euglena* and *Astasia longa* was investigated by Rogers *et al.* (1972); and Blum *et al.* (1965) compared *A. longa* and a bleached *Euglena gracilis* with respect to morphology, cytology and biochemistry. Ultrastructural studies on colorless, phagotrophic euglenoids have been carried out on *Entosiphon* (Mignot, 1963, 1967) and on *Peranema* (Roth, 1959, Mignot, 1966; Nisbet, 1974; Hilenski and Walne, 1979).

Encysted cells of *Euglena gracilis* strain Z, resulting from growth in darkness for 48-72 hr in a medium lacking nitrogen, showed a decrease in number of Golgi cisternae and their concomitant dilation with mucilage, subsequently secreted into the reservoir and onto the cell surface, as the encysting cells rotated during cyst formation. The mucilaginous product in the dictyosomes and on the cell surface reacted identically to staining with PAS/silver methenamine reagent; no mucilage was observed in the subpellicular ER complex (Triemer, 1980).

Cytolysome formation, nuclear cytology and cytological changes have been described in cells from carbon-starved and ageing cultures of *Euglena* (Gomez *et al.*, 1974a, 1974b; Malkoff and Buetow, 1974; Leedale, 1975b; Leedale and Buetow, 1976). Cytochemical localization of acid phosphatase activity in *E. gracilis* and *E. granulata* has been reported by Sommer and Blum (1965), Palisano and Walne (1972) and Gomez *et al.* (1974a, 1974b).

In terms of phylogenetic considerations, Leedale's well-known scheme (1967) had assumed a green euglenoid line as a point of origin for extant genera. However, in keeping with a theory of symbiotic origin of chloroplasts, he has recently proposed a scheme that assumes a colorless ancestral line, in which green forms were introduced later by phagocytosis and the subsequent symbiosis of a prokaryote protochloroplast (Leedale, 1978). Klein *et al.*(1972) hypothesized that the original symbiont was a simple blue-green algal cell with a structure similar to a proplastid in *Euglena* and that as they evolved, they acquired the ability to fuse thylakoids and to convert girdle thylakoids to straight thylakoids. They further theorized that since plastid development is light dependent, the proplastid in *Euglena* may represent the mature structure of the initial blue-green algal invader or symbiont (Klein *et al.*, 1972). However, Stewart and Mattox (1975) and Gibbs (1978) proposed that the chloroplasts of *Euglena* may have evolved from the progressive reduction of endosymbiotic green algae, and Gibbs (1978) offered ultrastructural and experimental evidence in support of her theory, which has important biochemical implications.

CHLOROPLASTS

Considerable information on chloroplast structure and biochemistry is summarized elsewhere (Leedale, 1967; Buetow, 1968; Gibbs, 1970; Dodge, 1973; Bisalputra, 1974). For some time, it has been reported that euglenoid chloroplasts are surrounded by 3 membranes, which have been interpreted to represent the chloroplast envelope-chloroplast ER complex. Gibbs (1978), however, suggests that although euglenoid chloroplasts are indeed surrounded by 3 membranes, their significance has been misinterpreted, and that the third membrane is not ER. She proposes instead that the inner two membranes are the chloroplast envelope and that the outer membrane is derived from the plasmalemma of a green algal symbiont. Evidence against an ER derivation for any of the membranes that surround the chloroplast includes: 1) contrary to earlier reports, the outer membrane is devoid of ribosomes, although they may be in close positional relationship to it, 2) there are no connections or attachments between the plastid membranes and either the outer membrane of the nuclear envelope or cytoplasmic rough ER, and 3) there are no vesicles between any of the 3 surrounding membranes, suggesting that no synthesis or transport of locally manufactured proteins is occurring in these membrane systems (Gibbs, 1978). Gibbs also presents other evidence based on naturally occurring colorless species, experimentally induced chloroplast loss, present-day symbiotic relationships, ultrastructural similarity to green algal chloroplasts, biochemical implications, and precedents in other algal groups in support of her theory that chloroplasts of *Euglena* may have evolved from symbiotic green algae.

Although most green euglenoid cells contain few-to-many chloroplasts of diverse morphologies, *Cryptoglena pigra* is unusual in its possession of only one rather large chloroplast, contrary to reports of two in the literature (Rosowski and Lee, 1978). Of previously uncertain affinities with the Euglenophyta because of the difficulty of resolving a number of cellular details with light microscopy alone, *C. pigra* has more recently been shown by a combination of studies with TEM, SEM and optical microscopy, to have many features typical of the euglenoids, thus opening the door for future comparative investigations with other euglenoid genera of questionable taxonomic affinities (Rosowski and Lee, 1978).

Many of the developmental and biochemical studies on euglenoid flagellates have been carried out on the chloroplasts (Schiff and Epstein, 1968; Schiff, 1971; Freyssinet *et al.*, 1979). In a long and informative series of ultrastructural, biochemical and other experimental studies, Schiff and his co-workers have characterized diverse aspects of chloroplast replication, development and differentiation in *Euglena* (e.g., Epstein and Schiff, 1961; Schiff *et al.*, 1961; Ben-Shaul *et al.*, 1964, 1965; Stern *et al.*, 1964; Schiff and Epstein, 1968; Holowinsky and Schiff, 1970; Schiff, 1971; Freyssinet *et al.*, 1979). A particularly valuable approach in these and related studies of chloroplast synthesis and replication has been the use of wild-type and mutant strains of *E. gracilis* (Schiff *et al.*, 1971; Schiff, 1973; Schmidt and Lyman, 1976; Lyman and Traverse, 1980). Many of the mutant strains described by Schiff *et al.* (1971) are aplastidic due to bleaching or have structural and/or functional deficiencies or lesions of the chloroplasts.

Moreover, since chloroplast replication or synthesis may be readily and specifically inhibited, *Euglena* is an excellent system for studying regulatory mechanisms and nuclear/organellar genome interactions (Lyman and Traverse, 1980). Other advantages and considerations for studies involving mutations, chloroplast bleaching and differentiation in *Euglena* are summarized by Lyman and Traverse (1980), who also provide detailed information on laboratory procedures and methods for carrying out some of those investigations.

In his article on photocontrol of chloroplast development in *Euglena*, Schiff (1978) reviewed many of the contributions from his laboratory, as well as those of other investigators. In his summary, he showed the usefulness of mutants and inhibitors in studying chloroplast function, and suggested nuclear-coded proteins and chloroplast-coded proteins as two major sources of genetic information for plastid development. During the first 12 hours of development, there is a lag in the formation of chlorophyll and other photosynthetic parameters, and apparently a blue-light photoreceptor localized outside the plastid controls the non-plastid contributions to plastid development. As the plastid becomes active, intra-plastid synthesis is apparently controlled by a blue-red photoreceptor that is similar to or identical with protochlorophyllide. Thus, the complex process of the development of mature chloroplasts involves highly coordinated substrate induction and the functional coordination of plastid and non-plastid cellular compartments by at least two different photoreceptors (Schiff, 1978).

Edmunds (1965) developed a procedure for attaining and characterizing rhythmic cell division in *Euglena*, and Boasson and Gibbs (1973) reported on chloroplast replication in synchronously dividing cells of *E. gracilis*. Brody *et al.* (1965) monitored fluorescence changes during chlorophyll formation in *E. gracilis* and estimated lamellar area as a function of age. Using freeze-etch techniques, Holt and Stern (1970) concluded that chloroplast thylakoids of *Euglena* are particulate and similar to those of higher plants.

Carrell *et al.* (1970) found that extracts of *Euglena* contain at least 2 DNases: one with a pH optimum of 9.4, '*Euglena* alkaline DNase', is absent or present in low activity or inhibited in dark-adapted cells, and its activity increases during light induction of chloroplast development. The authors hypothesized that the alkaline DNase is a chloroplast-related enzyme. Studies of phosphate uptake kinetics in *E. gracilis* Z grown in light/dark cycles showed that incorporation of P is non-linear with time and is completely inhibited by darkness; in PO_4-limited cultures, uptake follows Michaelis-Menten kinetics (Chisholm and Stross, 1976).

A number of recent research efforts have dealt with the problems in general of isolating functionally competent proplastids and mature chloroplasts that can be used in subsequent biochemical investigations, such as enzyme localization or for studies on thylakoid membrane structure and development. There are continuing efforts to perfect techniques for the isolation of pure fractions of chloroplast envelopes and thylakoid membrane systems.

Price (1978) addressed the general questions and problems concerning procedures for isolating chloroplasts from *Euglena* and outlined a technique involving isopycnic sedimentation that yields structurally intact plastids. Although the gradient zone containing them is rich in ribulose-bis-phosphate carboxylase (RuBPC) (Salisbury *et al.*, 1975), the isolated chloroplasts apparently lack the capacity for CO_2 fixation or for CO_2-dependent O_2 evolution at rates that are physiologically significant, even though they show Hill reaction activity (Price, 1978). However, the isolated *Euglena* plastids do have the capacity to incorporate exogenous amino acids and proteins (Vasconcelos, 1976), and analyses of the polypeptides of chloroplast subfractions show a qualitative correspondence between them and those observed in higher plant chloroplasts (Vasconcelos *et al.*, 1976).

Knight and Price (1968) measured the physical characteristics of developing *Euglena* chloroplasts using high molar sucrose gradients. They observed a lag in S* values (S* is defined as the sedimentation coefficient of an ideal, noninteracting particle) during early stages of plastid development and attributed the initial drop in sedimentation coefficients to the laying down of newly synthesized lipids. More recently, Dockerty and Merrett (1979) found that the buoyant density of *Euglena* plastids dropped from 1.24 g/cm^3 in dark-grown cells to 1.18 g/cm^3 after regreening for 15 hr and to 1.17 g/cm^3 for mature chloroplasts. They explained the decrease in density as coinciding with the formation of chlorophyll. Palisano *et al.* (in press) devised a double-sigmoid, sucrose-sorbitol gradient for the isolation and separation of intact developing *Euglena* plastids that differed greatly in size and density, and found that the radii and densities of the developing plastids increased with time but that the sedimentation coefficients did not increase for the first 24 hr after exposure to light and thereafter increased substantially (Table 1). They observed with electron microscopy that the early proplastids were extremely convoluted compared to later periods when they were more typically discoidal in shape. Since other studies have shown similar convoluted plastid structures in three-dimensional models based on serial sections of *in situ* plastids of *Euglena* (Ehara *et al.*, 1975), Palisano *et al.* (in press) concluded that the highly convoluted structures are not the result of cell fractionation or isolation procedures. Whether those structures can be correlated to lipid or to any other synthesis remains to be determined. Palisano *et al.* suggested that a change in the frictional coefficients of the developing plastids might account for the fact that the S* values did not increase during initial periods of their development. Configurational irregularities would cause them to tumble in the gradient, resulting in their slower movement during centrifugation compared to an ideal spherical particle of similar size; thus, the delayed increase in S* is probably a function of the convoluted shapes of the organelles.

Using Ficoll, Ludox, silica sols, sucrose or Urografin as gradient materials, several other investigators have encounted osmotic problems that result in structural damage to isolated organelles (e.g., Vasconcelos *et al.*, 1971;

Table 1. Densities, Sedimentation Coefficients and Mean Plastid Radii of Developing Plastids Isolated from *Euglena gracilis* var. *bacillaris*[1]

Hr Illumination	Density (g/cm^3)	Sedimentation Coefficient (S*)[†]	Mean Particle Radius (R)[x]
0	1.13[o]	1.6×10^5[o]	1.0 ± 0.2 μm
12[+]	1.14	$3.5 \times 10^4 \pm 0.5 \times 10^4$	1.5 ± 0.3 μm
24[+]	1.15	$1.5 \times 10^4 \pm 0.5 \times 10^4$	2.0 ± 0.5 μm
48[+]	1.17	$1.5 \times 10^5 \pm 0.1 \times 10^5$	3.0 ± 0.5 μm
72[+]	1.165	$3.75 \times 10^5 \pm 0.25 \times 10^5$	4.0 ± 1.0 μm

[o]Computed from the plot of the measured data according to the equations cited in the text.

[+]Measured data.

[x]Based on measurements obtained with the use of fluorescence optics.

[†]S* is defined as the sedimentation coefficient of an ideal, non-interacting particle.

[1]From Palisano *et al.* (in press)

Preston *et al.*, 1972; Ophir and Ben-Shaul, 1973, 1974; Price *et al.*,1973; Salisbury *et al.*, 1975). Salisbury *et al.* (1975) cautioned that high sucrose concentrations are hypertonic and lead to dehydration of plastids in *Euglena*. The recent work of Reardon *et al.* (1978) has shown that Ludox is the present gradient material of choice, but regardless of the isolation materials or procedures used, the plastids have been found to be leaky; large amounts of small proteins apparently are lost during initial cell breakage procedures and subsequent purification. For example, isolated plastids are reported to lack 95% of the cytochrome C-552. Such loss, together with loss of other molecules, may help to explain why the isolated chloroplasts are not photosynthetically active (Freyssinet *et al.*, 1979). In outlining their procedure for chloroplast isolation from *Euglena*, Lyman and Traverse (1980) also indicated that most cytochrome 552, ferredoxin, many soluble Calvin cycle enzymes and other soluble components may be lost but that DNA, RNA, ribosomes and membrane components are retained. It is clear that more work is needed, but with appropriate modifications of gradient materials and isolation procedures, it should be possible to isolate functionally competent, intact plastids, an important technical advance that would facilitate future biochemical and biophysical characterizations of developing and mature chloroplast structures and functions.

NUCLEUS AND MITOSIS

Euglenoid nuclei have been characterized in several genera, but especially in *Euglena* (e.g., Leedale, 1968). During mitosis, which is nonclassical, the nuclear envelope and the endosome (nucleolus) persist; the latter, which may be multiple or lobed, elongates and divides along the plane of division axis. Chromosomes remain condensed during interphase (Leedale, 1958, 1968, 1970; Dodge, 1973). Apparently, the timing of chromosome duplication and condensation is quite variable among euglenoid organisms, occurring variously in preprophase, prometaphase and telophase; and the chromatin, although remaining continuously condensed, does undergo discernible changes. Chaley *et al.*, (1977) described 3 discernible interphase nuclear states in *Astasia longa*, with condensed chromatin in every case; they also reported that the chromatin may be rendered in reversed contrast in cells fixed with OsO_4 without prefixation in aldehyde. In their study of mitosis in *Euglena gracilis* var. *bacillaris* and *Phacus longicaudus*, Pickett-Heaps and Weik (1977) found that the chromatin becomes less distinct and more diffuse during mitosis than during interphase. They did not observe chromosome attachment to the nuclear envelope; rather, chromosomal microtubules extend toward the poles but stop short of the nuclear envelope.

Moyne *et al.* (1975) found that thin fibrils of DNA link some chromosomes to one another in *Euglena gracilis* Z. DNA content of the nucleus in *E. gracilis* is reported to be constant throughout log and stationary phases of growth, whereas RNA content decreases from log to stationary phase; sedimentation values of RNA extracted from isolated nuclei are 19, 11 and 6 S (Paranti *et al.*, 1969). In addressing the general problems of isolating intact, functional nuclei from *Euglena*, Buetow (1978) provided a brief historical account and some of the advantages and disadvantages of a number of isolation techniques. He also detailed procedures for the isolation of *Euglena* nuclei that are enzymatically active and ultrastructurally intact (cf. Lynch and Buetow, 1975) and for the isolation and fractionation of chromatin from those nuclei (cf. Lynch *et al.*, 1975).

In his discussion of phylogenetic aspects of nuclear cytology, Leedale (1970) suggested that, based on similarities in mitotic processes, flagellar insertion and morphology, the euglenoid flagellates show possible affinities with trypanosomes and the lower fungi.

FLAGELLA AND RELATED STRUCTURES

Flagellar number, morphology and form of insertion are important indicators of phylogenetic relationships in euglenoid flagellates in the schemes proposed by Leedale (1967, 1978). Inserted just below the posterior or lateral wall of the reservoir, the flagella may vary in number from one to several. The emergent flagella are typically coated by a 'felt' of fibrous material and bear numerous fine hairlike appendages (Leedale, 1967; Buetow, 1968). The terminology surrounding the diverse hairlike structures or mastigonemes associated with the flagella of these and other flagellates is controversial. Considered by some investigators to be synonymous with 'Flimmer' and 'flagellar hairs' the term 'mastigoneme' is frequently used to denote essentially filamentous appendages that are distinct from the flagellar membrane (Bouck, 1972). Other investigators prefer a more restrictive usage of the term.

Mignot (1966) demonstrated that the hairs or mastigonemes are attached in a spiral band. Piccini *et al.* (1975) localized ATPase activity in the flagella of *Euglena* and also reported that digitonin-effected dissociation of the flagellar membrane revealed a network on its surface.

In their recent studies of surface organization and composition of *Euglena*, Bouck *et al.* (1978) reported the flagellar surface to be coated with *ca.* 30,000 fine, nontubular fibrils of 2 types--longer (*ca.* 3 μm) mastigonemes that apparently are attached to the paraflagellar rod, and shorter (*ca.* 1.5 μm) mastigonemes that constitute the terminal or centrifugal portions of a larger complex of laterally aligned, longitudinally overlapping units outside the flagellar membrane. Similar structural arrays have been observed in the anterior flagellum of *Peranema* (Figs. 40, 41) (Hilenski and Walne, 1979). Long mastigoneme filaments appear to arise from small tufts arranged helically along the flagellar membrane, and 2 lateral complexes of shorter mastigonemes seem to form a network on the flagellar membrane surface. Early on, Pitelka and Schooley (1955) observed detachable mastigonemes in shadowcast preparations of *Peranema*.

Removal of the flagellar membrane of *Euglena gracilis* with neutral detergents reveals the persistent attachment of the shorter complexes to the axoneme microtubules (Bouck *et al.*, 1978). SDS-acrylamide gel electrophoresis of isolated whole flagella yields *ca.* 30 polypeptides, with 2 giving strong positive PAS staining reactions; one of those 2 glycoprotein bands can be correlated to the surface mastigonemes by parallel analysis of isolated, purified mastigonemes. TLC analyses of hydrolysates of whole flagella and of mastigonemes suggest that the major sugar in both constituents is xylose. These data, together with future studies, should be helpful in resolving questions on synthesis and maintenance of segregated cell surfaces. With respect to phylogeny, Bouck (1977) hypothesized that euglenoid mastigoneme units might be homologous with the tubular mastigonemes of heterokont flagellates.

STIGMA (EYESPOT) AND PARAFLAGELLAR BODY

Cells of most euglenoid flagellates possess a photoreceptor/photoresponse system that can detect the direction, intensity and quality of light. The apparatus is generally believed to consist of two major components: a cytoplasmic stigma (eyespot) and a paraflagellar body (PFB) on the emergent flagellum. While there is still some controversy over their respective functions, a widely accepted theory of photoreception in euglenoid organisms assigns a shading or filtering role to the stigma and designates the PFB as the main photoreceptor (Diehn, 1969; Diehn and Kint, 1970). More recently, however, Creutz and Diehn (1976) (see below) provided an alternative to the shading hypothesis in their proposed new model for positive phototaxis that is based on the behavior of *Euglena* in polarized light.

Ultrastructure. Ultrastructural characterization of stigmata and determination of their relationship to other organelles have been carried out for a number of euglenoid flagellates (Leedale, 1967; Walne and Arnott, 1967; Walne and Lembi, 1969; Walne, 1971; Kivic and Vesk, 1972; West *et al.*, 1973; Dynesius and Walne, 1974, 1975). The extrachloroplastidic stigma, unique to the Euglenophyceae and Eustigmatophyceae, is composed of orange-red pigmented granules that are individually membrane bound or in packets, located near the base or sides of the reservoir, in line with the PFB. The component stigma granules move freely and apparently are not surrounded by a common membrane, although one report suggests that they are (Kivic and Vesk, 1972). In all cases, close positional relationships between eyespot granules and microtubules have been noted (e.g., Walne, 1971). In *Trachelomonas*, *Phacus*, *Hyalophacus* and *Euglena* a complex array of microtubules and ER in the reservoir region encompasses the stigma, but a clear functional relationship, if any, to the latter has not been demonstrated. In *Phacus* and *Hyalophacus* (Dynesius and Walne, 1974, 1975), multitubular structures (MTS) are always observed in a particular positional relationship to the stigma and PFB. Wolken (1977) suggested that twisted lamellae in association with PFBs are consistent with a model of rods packed in a helical array.

Comparative studies of the UV-induced W_3BUL colorless mutant and of a dark-induced non-chlorophyllous form of *Euglena gracilis* reveal quite different stigmatal structures. Stigmata of the former are comparable to those in green, wild-type cells in structure and electron density, whereas those in the latter are generally reduced in electron density and number of component granules and in structural integrity of the organelle (Palisano and Walne, 1976). Smaller stigmata with varying electron densities were reported in all strains of bleached *Euglena* that were examined by Kivic and Vesk (1974a, 1974b), who also found no eyespot or PFB in the naturally occurring colorless *Astasia longa*. Kronestadt and Walles (1975), however, reported the presence of both a PFB and a stigma of normal appearance in porfiromycin-bleached cells of *Euglena gracilis*. Changes in stigmata during different parts of the cell cycle have also been noted (Palisano and Walne, 1972; Gomez *et al.*, 1974a, 1974b).

Isolation and Chemical Characterization. In attempting to understand more about the structure and composition of the euglenoid photoreceptor apparatus and thus to relate structure with function, it is important to be able to work with pure fractions of isolated organelles, especially for subsequent chemical, biochemical and biophysical characterizations. Stigmata were first isolated from *Euglena* by Batra and Tollin (1964), who used conventional centrifugation techniques and demonstrated a close correlation between the action spectrum of phototaxis and the absorption spectrum of the isolated granules. The large-scale procedure for isolation of euglenoid stigmata developed by Bartlett *et al.* (1972) has permitted more thorough and precise compositional analyses. Additionally, other types of analytical centrifugation, employing reorienting gradients, have yielded confirmatory data on particle density and sedimentation coefficients (Walne and Breillatt, 1974).

Assays for nucleic acids and proteins indicate no DNA or RNA in isolated stigmata; protein concentrations are reported to decrease with increasing purity of stigma fractions (Sperling *et al.*, 1974).

Flavins. Microspectrophotometric analyses of stigmata in *Euglena acus* and *E. gracilis* suggest the presence of a flavin or a flavoprotein (Pagni *et al.*, 1975). Flavins were suggested as photoreceptor pigments (Diehn, 1969; Tollin and Robinson, 1969; Diehn and Kint, 1970; Froehlich and Diehn, 1974; Song and Moore, 1974) and were reported in the PFBs of living cells of *Euglena* by Benedetti and Checcucci (1975) utilizing fluorescence microscopy, and by Benedetti and Lenci (1977), utilizing fast-scanning microspectrofluorometry (described by Benedetti *et al.*, 1976a). On the basis of evidence obtained from action spectra for photoaccumulation of green *Euglena* with PFB, stigma and chloroplasts and of dark-grown *Euglena* with PFB and stigma, and for photodispersal of streptomycin-bleached cells with PFBs, Checcucci *et al.* (1976) also hypothesized a flavin nature for the photoreceptor, since all the spectra had major peaks at 340 and 450 nm and were similar to the absorption spectrum of a flavoprotein. Fluorometric analyses of isolated stigmata of *E. gracilis* var. *bacillaris* also suggest the presence of flavins (Pagni *et al.*, 1976). Extracts from isolated stigmata were shown to have excitation and emission spectra similar to riboflavin, and the results of photoreduction experiments with EDTA and sodium dithionite suggest that the flavin molecule may be complexed to a protein in the stigma. Estimates of the flavin content of stigma granules, obtained by fluorescence measurements, yield values of ca. 0.5 ng flavin/ml stigma sample.

Lipids. Aliphatic chains of esterified lipids have also been characterized from isolated stigmata of *Euglena* (Hilenski *et al.*, 1976). Other analyses of lipids from euglenoid organisms (Korn, 1964; Guehler *et al.*, 1964; Rosenberg, 1967; Erwin, 1968; Khan and Kolattukudy, 1973) have not dealt exclusively with stigmatal lipids. Hilenski *et al.* (1976) separated 4 major fractions of stigmatal lipid extracts: wax esters, triacylglycerols, free fatty acids and phospholipids. Methanolysis of each fraction yields 27-29 fatty acids, ranging from 12:0-22:6. Acetates of fatty alcohols of the wax fraction consist of 11:0-18:0 carbon chains, with 14:0 as the major component. It is also the major component in waxes from whole cells of *Euglena* grown on ethanol (Guehler *et al.*, 1964) and in the dark (Rosenberg, 1963) and in waxes from mitochondria of etiolated *Euglena* (Khan and Kolattukudy, 1973). Unsaturated alcohols were not detected by Hilenski *et al.* (1976), who also reported that the isolated stigma granules contain *ca.* 6.2 per cent lipids based on proteins.

Carotenoid pigments. Although all of the photoreceptor pigments have not been identified, many investigators assumed that the euglenoid stigma contained carotenoid pigments, specifically astaxanthin, because of the orange color of the organelle and the similarity of the action spectrum of phototaxis to the absorption spectrum of carotenoids (Diehn, 1969; Wolken, 1977). The actual presence of carotenoids in the stigma was not confirmed, however, until Batra and Tollin (1964) separated 4 carotenoid pigments from isolated stigmata of *Euglena* and reported the presence of lutein, β-carotene, cryptoxanthin and an unknown pigment. More recently, using fast-scanning microspectrophotometry, Benedetti *et al.* (1976b) obtained absorption spectra on *in vivo* and isolated stigma granules of green and dark-grown cells of *Euglena* and reported peaks at *ca.* 410, 450 and 490 nm, thus confirming the presence of carotenoid pigments in the stigma.

With thin layer chromatography, Heelis *et al.* (1979) resolved 29 compounds in isolated stigmata of *E. gracilis* strain Z, 16 of which could be eluted and their absorption spectra recorded. Seven of these were identified, including the major carotenoids β-carotene, diatoxanthin and diadinoxanthin, that together

Table 2. Carotenoid pigments separated from stigmata of *Euglena gracilis* var. *bacillaris*.[1]

Thick-Layer Pigment Band	Rf	Pigment Fraction Eluted from the Deactivated Alumina Column	Maxima in nm*			Solvent in which the Absorption Spectrum was Recorded	Pigment Identification
A	0.95-0.98	Petroleum Ether	423	446	474	Hexane	α-Carotene
			430	457	485	Chloroform	
			(445)	475	503	Carbon Disulfide	
			422	450	476	Acetone	
			427	451	478	Ethanol	
			427	451	479	Ethanol + HCl	
B	0.84-0.92	4% Acetone in Petroleum Ether	420	447	476	Hexane	α-Cryptoxanthin
			434	457	487	Chloroform	
			449	473	507	Carbon Disulfide	
			426	449	475	Acetone	
			427	450	476	Ethanol	
			428	450	476	Ethanol + HCl	
			421	449	476	Ethanol + $NaBH_4$	
			422	447	475	Ether	
C	0.59-0.64	2% Acetone in Petroleum Ether	423	450	478	Hexane	Zeaxanthin
			431	460	489	Chloroform	
				476	507	Carbon Disulfide	
			428	451	480	Acetone	
			428	451	480	Ethanol	
			427	452	479	Ethanol + HCl	
			422	448	479	Ethanol + $NaBH_4$	
D	0.36-0.42	95% Ethanol (?)		400		Hexane	Unknown[†]
				406	435	Chloroform	
				417	(455)	Carbon Disulfide	
				404	(430)	Acetone	
			415	441	469	Ethanol	
			405	428	455	Ethanol + HCl	

E	0.34	Second Fraction Eluted with 4% Acetone in Petroleum Ether	420	444	471	Hexane	Lutein
			429	455	484	Chloroform	
			442	472	501	Carbon Disulfide	
			(420)	445	473	Ether	
F	0.26	8% Acetone in Petroleum Ether	413	440	(470)	Hexane	Taraxanthin
			(427)	453	487	Chloroform	
			429	469	502	Carbon Disulfide	
			(412)	440	(470)	Ether	
G	0.10–0.15	40% Acetone in Petroleum Ether	418	438	469	Hexane	Neoxanthin
			425	448	480	Chloroform	
			(416)	439	470	Ethanol	
			(400)	425	452	Ethanol + HCl	
H	0.05	16% Acetone in Petroleum Ether		434	472	Hexane	Unknown
			430	450	(480)	Chloroform	
(D?)		95% Ethanol					Unknown†

*The wavelengths in parentheses are shoulders in the absorption spectra.

†The D Band pigment and the pigment fraction eluted in 95% Ethanol may be the same.

[1]From Pagni (1978).

comprised ca. 60% of the stigmatal pigments. Significant amounts of canthaxanthin, echinenone and cryptoxanthin also were isolated, and a possible carotenoid ester was detected (Heelis *et al.*, 1979). Pagni (1978) (Table 2) separated 8 carotenoid pigments from isolated stigmata of *E. gracilis* var. *bacillaris*, including the major carotenoids α-cryptoxanthin and zeaxanthin. Taraxanthin, neoxanthin, lutein and α-carotene also were identified; two 5,6-monoepoxide unknowns were detected, but no ketoxanthophylls were found (Pagni (1978).

Krinsky and Goldsmith (1960) (Table 3) studied the carotenoids of whole cells of *Euglena gracilis* var. *bacillaris* and *E. gracilis* strain Z and reported the major pigments in both strains to be β-carotene, antheraxanthin and neoxanthin, but in comparing their data with those of Heelis *et al.* (1979) and Pagni are (1978) it must be remembered that Krinsky and Goldsmith were working with extracts of whole cells, whereas the data of Heelis *et al.* and Pagni based on isolated stigmata. Since the ketocarotenoids echinenone, euglenanone and hydroxyechinenone were present in trace amounts in their extracts, Krinsky and Goldsmith (1960) suggested that they might be stigmatic pigments. The recent work of Pagni (1978) does not substantiate their suggestion, but Heelis *et al.* (1979) did report the presence of echinenone in their stigma preparations. Astaxanthin, sometimes assumed to be a stigmatal pigment, has not been identified in isolated stigmata. Table 3 compares the carotenoid pigments identified in extracts of stigmata isolated from *E. gracilis* and in whole-cell extracts of *E. gracilis* and *Astasia ocellata* based on several investigations.

Diehn (1969) hypothesized two perpendicularly oriented photoreceptor systems in the stigma--one based on reported carotenoid pigments and a second based on flavins which might be bound in a lipid matrix (Diehn and Kint, 1970; Froehlich and Diehn, 1974). The reaction of the stigma samples with sodium dithionite suggests that the flavins may be complexed with a protein (Pagni *et al.*, 1976), and the work of Hilenski *et al.* (1976) demonstrated the presence of lipids in which they might be bound. In general, our investigations lend credence to Diehn's 1969 hypothesis, since all those constituents--carotenoids, flavins, lipids and proteins--have now been confirmed in isolated stigmata. From their studies of the effect of KI on phototaxis, Mikolyjczyk and Diehn (1975) also concluded that there are two photoreceptor systems in *Euglena*, although not necessarily restricted to the stigma; the system for "step-up" ("inverse") responses may be mediated by a flavin, whereas the direct phobic response is not.

The presence of a flavin pigment in the stigma would increase the effectiveness of the shading or filtering mechanism by broadening the range of light

Table 3. Comparison of Carotenoid Pigments Identified in Extracts of Stigmata Isolated from *Euglena gracilis*, in Whole Cells of *Euglena gracilis* and in *Astasia ocellata*.[1]

	Pigments Present in Extracts of Isolated Stigmata of *Euglena gracilis*		Pigments Present in Extracts of Whole Cells: *Astasia ocellata*	Pigments Present in Extracts of Whole Cells: *Euglena gracilis*	
Heelis *et al.* (1979)	Pagni (1978)	Batra & Tollin (1964)	Thomas *et al.* (1967)	Krinsky & Goldsmith (1960)	Goodwin & Jamikorn (1954)
β-carotene	α-carotene		α-carotene		
		β-carotene		β-carotene	β-carotene
				γ-carotene	
cryptoxanthin			ε-carotene		
	α-cryptoxanthin				
		cryptoxanthin		cryptoxanthin	
	zeaxanthin				
	lutein	lutein			lutein
	taraxanthin				
	neoxanthin			neoxanthin	neoxanthin
				antheraxanthin	
	unknown (D Band)			unknown	
	unknown (H Band)				
		unknown			
diatoxanthin			unknown		
diadinoxanthin			echinenone	echinenone	
echinenone			4-keto-α-carotene		
canthaxanthin			canthaxanthin		
carotenoid ester			polytomaxanthin		
				euglenanone	
				hydroxyechinenone	

[1]Modified from Pagni 1978

absorbed by the carotenoids. Whether the flavin compounds in the stigma and PFB are similar or different has not yet been determined, but all flavins have similar absorption spectra and would thus function in a similar way in extending the shading range of the stigma (Pagni *et al.*, 1976). Circular dichroism analyses and studies with polarized light would be helpful in determining the ordered or non-ordered arrangement of the pigments in the stigma granules.

With respect to the PFBs, the findings of Creutz and Diehn (1976) suggest that the photoreceptor molecules (perhaps only flavins since Benedetti *et al.*, [1976] found no evidence of the presence of other pigments) are oriented with their transition moments perpendicular to the long axis of the cell. Creutz and Diehn (1976) noted that when *Euglena* cells are subjected to a beam of polarized light, they align perpendicular to it, apparently thus allowing for maximum light absorption. If so, then other adjustments in cell position relative to the direction of light would tend to affect thé amount of light absorbed by the photoreceptors. There is also speculation that a geosensory system, as well as a photosensory system, may affect cellular orientation by interacting with the locomotory apparatus (Creutz and Diehn, 1976).

High-speed cinematographic studies showed that *Euglena* cells react to high light intensity by turning toward the dorsal side of the cell (Diehn *et al.*, 1975). Ultrastructural examination of cells of *E. gracilis* treated with the cationic detergent CTAB revealed that the reservoir and flagellar membranes were selectively destroyed; correlative studies on photophobic behavioral responses showed that the treatment inhibited step-down photophobic response (i.e., the ability of the cells to respond with flagellar reorientation to a sudden decrease of light intensity), whereas there was no effect on step-up photophobic response (i.e., to an increase in light intensity) (Mikolajczyk and Diehn, 1978).

In a series of experiments on how cells monitor light intensity changes, Creutz *et al.* (1978) investigated the kinetics of photoreceptor pigment interconversions and photophobic behavioral responses of *Euglena* in a light intensity gradient and showed that the rate of accumulation of cells in small illuminated regions called light traps is a function of both the light intensity within the trap and change of intensity at the trap boundary. They concluded that the strength of behavioral response in a single cell is proportional to the rate of change in the amount of photochemically active form of a photoreceptor pigment molecule which can exist in two forms, and postulated that a mechanism involving a change in membrane potential in one direction but not in the other might cause cellular response (Creutz *et al.*, 1978).

Using a video system coupled to an optical microscope, Barghigiani *et al.*, (1979) investigated light-induced behavioral responses in single cells of *E. gracilis*. From light intensity-effect curves at different wavelengths in the near UV and visible ranges, they calculated the action spectrum for the step-down photophobic response of *Euglena* and concluded that a single photomotile reaction is responsible for cell accumulation.

Miller and Diehn (1978) reported the selective inhibition of chemoaccumulation of *E. gracilis* toward oxygen, without concomitant effects on cellular motility, by carbon monoxide and cyanide; azide had no effect on chemoaccumulation, even at concentrations that inhibit cell motility. Since the observed effects did not reflect a generalized inhibition of sensory transduction in *E. gracilis*, they concluded that the cytochrome a_3 moiety of cytochrome c oxidase serves as the receptor molecule for oxygen-mediated chemosensory behavioral responses in *Euglena* (Miller and Diehn, 1978).

Earlier information on euglenoid stigmata and on the historical controversy over photoreception in these flagellates is summarized by Leedale (1967), Walne and Arnott (1967), Buetow (1968), Walne (1971) and Wolken (1977). The general subject of photomovement is reviewed by Hand (1977), and a collection of lectures on the biophysics of photoreceptors and photobehavior in microorganisms is edited by Colombetti (1976). A biennial review of publications on photomovement in microorganisms, including *Euglena*, and a summary of terminology and response strategies for such photomovement are provided by Feinleib (1977, 1978).

CELL COVERINGS--PELLICLE AND ENVELOPE (LORICA)

<u>Pellicle</u>. Cells of euglenoid flagellates are delimited by a plasmalemma immediately subtended by a pellicle composed of helically arranged, articulating strips that converge by tucking under (Guttman and Ziegler, 1974) or by fusing together (Leedale, 1967) at the anterior and posterior ends of the cell. Variously ornamented, the pellicles of different species may be highly elastic to quite rigid and have been implicated in the typical euglenoid movements, with the contortions made possible by the sliding and positional modifications of the striae (Guttman and Ziegler, 1974). In close association with the pellicle are arrays of regularly arranged microtubules and sometimes rows of muciferous bodies that are continuous with the cell exterior via narrow channels or pores, through which mucilage can be secreted (Leedale, 1964, 1967; Arnott and Walne, 1967; Buetow, 1968; Guttman and Ziegler, 1974; Hofmann and Bouck, 1976), al-

though mucilage can apparently also be secreted via Golgi-derived vesicles into the reservoir of encysting cells of *Euglena* (Triemer, 1980).

Hofmann and Bouck (1976) suggested that pellicle form in *E. gracilis* is developmentally related to the distribution of microtubules, even though later they are not required, since strips of isolated pellicle retain their characteristic shape in the absence of associated microtubules or ER cisternae.

Freeze-etch investigations of the pellicle of *E. gracilis* reveal a bilayered structure--a particulate outer layer with particles 150Å in diameter, and a striated inner layer with striations of *ca.* 50Å width separated by *ca.* 34Å and showing a repeating pattern with a periodicity of *ca.* 450A (Schwelitz *et al.*, 1970). Miller and Miller (1978) suggested that in *E. gracilis* the pellicle membrane is organized along the lines of a two-dimensional crystal but that such a pattern is restricted to the cortical region. Freeze fracture of the membrane in this region reveals 2 non-complementary fracture faces. One face (P) is covered with particles with an average diameter of 110Å, and the other face (E) shows a complex series of striations that occur at regular angles of the pellicle ridges. Under certain conditions, striations visible on the P face are exactly complementary to those seen on the E face. As the membrane invaginates at the anterior end of the cell and forms the reservoir, the bounding membrane is more fluid in structure; the fracture faces change abruptly and reveal an organization more typical of other cell membranes, results suggesting that the cell membrane in *Euglena* is divided into 2 distinct but continuous regions with specialized structures and functions (Miller and Miller, 1978).

Chemically, isolated pellicles of *E. gracilis* var. *bacillaris* have been reported to consist largely of protein (*ca.* 80%), and of lesser amounts of lipid (*ca.* 11%) and carbohydrate (*ca.* 17%) (Barras and Stone, 1965). More recent investigations (Hofmann and Bouck, 1976) of isolated pellicles of *E. gracilis* strain Z, employing SDS acrylamide gels, have shown *ca.* 10 polypeptide bands, 3 of which are apparently major pellicle constituents, comprising *ca.* 50% of the total polypeptides present. One of the major polypeptides is apparently localized to the surface membrane, and one or more may be identical with a submembrane layer present in intact and isolated pellicles. Protein to lipid ratios were never less than 3, values suggesting a protein composition for much of the submembrane material.

Hofmann and Bouck (1976) also presented immunological evidence for intussusceptive surface growth and replication. The use of ferritin, fluorescein and latex spheres in TEM, fluorescence optical microscopy and SEM, respectively, permit localization of pellicle surface antigens and show that antigenic sites are immobile and that new antigenic sites (i.e., daughter strips) are apparently formed *de novo* and are inserted between the old strips in dividing cells.

The complexity and rigidity of the pellicle seem to be largely a function of the extensiveness and organization of a proteinaceous layer, which may be elaborated and articulated as in *Euglena spirogyra* (Leedale *et al.*, 1965) or fused with adjacent strips as in *Rhabdomonas* and *Menoidium* (Leedale and Hibberd, 1974). An evolutionary trend toward increased pellicle rigidity has been discussed by Leedale (1967) and by Leedale and Hibberd (1974), who suggested that rigid pellicles such as those of *Menoidium* and *Rhabdomonas* arose as a further development beyond the articulated type exhibited by *E. spirogyra* and that evolution of rigid pellicles has occurred along several phylogenetic lines. For example, the rigid pellicle of *Phacus* is thought to result from a cementing together of highly articulated strips rather than from fusion. Dodge (1975) suggested that the simplified pellicle of *Trachelomonas* arose through reduction, in the absence of the selection pressures exerted on non-loricate genera. A comparison of the pellicle of *T. hispida* var. *coronata* (West, 1977), *T. bulla*, *T. lefevrei* and *T. zorensis* (Donnelly, 1979) with that of *T. volvocina* (Dodge, 1975) confirms the general simplicity of the trachelomonad pellicle. Thus, there may be a phylogenetic trend toward pellicle reduction in enveloped genera; also, an evolutionary trend toward increased pellicle rigidity in some other genera might have been replaced in part by the protective envelope in *Trachelomonas*.

Guttman and Ziegler (1974) suggested a possible continuity between pericanal microtubules and some of the pellicle microtubules in *Euglena gracilis*. In *Menoidium* and *Rhabdomonas*, microtubules and amorphous materials reportedly articulate in a complex scroll fashion to form peri-reservoir/canal skeletal structures; a lateral displacement of some of those interconnected microtubules and associated materials might be visualized as a step toward the evolution of ingestion organelles (Leedale and Hibberd, 1974). Clearly, more comparative studies of pellicle ultrastructure, biochemical composition and function in euglenoid organisms are needed.

Other basic information and controversies on pellicle structure, function and composition are summarized by Leedale (1967), Buetow (1968), Guttman and Zeigler, 1974, and Hofmann and Bouck (1976).

Envelope (Lorica). Cells of most euglenoid flagellates are naked except for those that are enclosed partially or nearly completely by an envelope composed largely of mineral-impregnated mucilage (e.g., *Trachelomonas*, *Ascoglena*, *Strombomonas*). Cells of the colonial, stalk-producing *Colacium* are enclosed in a thin mucilaginous envelope and secrete mucilage that may subsequently mineralize in the stalk. The organism thus shares some developmental, structural and other attributes with *Trachelomonas*, but it is not considered in detail here. Historical summaries and descriptive, histochemical and experimental studies on *Colacium*, its stalk and attachment material are provided by Rosowski and Kugrens (1973), Rosowski and Willey (1975, 1977) and Willey *et al.* (1977).

A basic difference between *Euglena* and loricate euglenoids such as *Trachelomonas* may be the production of a primary skin by the latter (Figs. 1, 10, 11). The skin may serve to retain subsequent mucilaginous secretions that tend to become mineralized and may also delimit an external microenvironment in which the cell might exert physico-chemical influences, thereby promoting conditions favorable to enhanced mineralization (Leedale, 1975a). Mucilage retention alone, however, does not necessarily result in mineral deposition. For example, palmelloid stages in *Euglena* do not always become mineralized simply because the mucilage remains around the cell. Moreover, although *Colacium* stalks consist of mucilage, the extent of their mineralization may vary (Rosowski and Kugrens, 1973; Rosowski and Willey, 1977). Thus, mineralizable mucilage may be different chemically with regard to binding affinities from non-mineralizing mucilage, or some other factors may influence the advent of mineralization, such as the provision of a special mineralizing-conducive microenvironment by the primary skin. Perhaps both function in mineralization, such that a first step in mineralized-envelope evolution could have been the development of mucilages with mineral affinities. Then, greater efficiency could have been added by compartmentalization of those mucilages, yielding greater mineralization potentials (West and Walne, 1977; West *et al.*, 1978; West *et al.*, submitted).

Although envelope production in *Trachelomonas* can be influenced by the availability of reduced Fe and Mn in the medium or environment, mineralization potential has also been shown to be somewhat species specific. Pringsheim (1953) reported that some species (e.g., *T. hispida*) did not darken as well as others, perhaps due to a lower binding affinity of the mucilaginous surface. Singh (1956) reported an inability to promote mineralization of envelopes of *T. grandis* even in mineral-supplemented medium and suggested that Fe and Mn were not the only requisites to envelope mineralization. The data on *T. hispida* var. *coronata* indicate that in accord with a generally accepted theory on biological mineralization, envelope mineralization is directed, the cells produce first an organic matrix and deposition of the inorganic phase follows (West and Walne, 1977; West *et al.*, submitted). Leedale (1975a) reported a similar developmental sequence in envelopes of *T. oblonga* var. *punctata*, wherein the immature envelope consists of a transparent, mucilaginous band (skin) secreted by the cell. Later, electron-dense ferric hydroxide and manganese compounds are precipitated on the inner layer of the immature envelope, providing species-specific size, shape, ornamentation and collar characteristics in the mature envelopes. Some very recent data suggest that the mineralizing elements are precipitated on the outer layer of the immature envelope in some species (Fig. 11) (Donnelly and Walne, 1979, and unpublished data).

Our own studies and those of Rosowski *et al.* (1975a, 1975b) have shown that there can be marked differences in morphology and color, especially in envelopes, in clonal populations of *Trachelomonas* (Figs. 3, 4). Such variability may often be related to nutritional and culture conditions. Many of the species described in the literature are undoubtedly synonymous, the variations in cell size, envelope ornamentation, coloration etc., being due to varying environmental conditions.

Envelope structure and morphology have been characterized to some extent with the light microscope (Pringsheim, 1953; Singh, 1956; Leedale, 1967, 1975a), with TEM (Ueda, 1960; Leedale, 1967, 1975a; West *et al.*, 1973; West and Walne, 1973, 1977, and submitted; Dodge, 1975; Donnelly and Walne, 1979) and with SEM (Rosowski *et al.*, 1975; West and Walne, 1977; West *et al.*, 1978, and submitted; Donnelly and Walne, 1979). SEM investigations reveal that the less-developed, less-mineralized, flexible immature envelopes collapse readily during preparative procedures (Figs. 1, 2), whereas the more mineralized and brittle mature envelopes do not (Figs. 3, 4) (Rosowski *et al.*, 1975a; West, 1977; Donnelly, 1979; Donnelly and Walne, 1979).

Leedale (1975a) and we (West and Walne, 1977 and submitted; Donnelly and Walne, 1979) have demonstrated at least two general classes of envelopes in *Trachelomonas* on the basis of envelope microarchitecture: those with a granular texture (type 1) (Figs. 10, 13, 14) and those with a fibrillar or 'fibrous mat' texture (type 2) (Figs. 7, 16, 18). Leedale concluded that gross ultrastructural envelope patterns, such as granular or fibrillar textures, do not support the cytology-based phylogenetic schemes of Pringsheim. Leedale implied, however, that those basic features remain constant within a given species. Our studies (West and Walne, 1977, and submitted; Donnelly and Walne, 1979) on *Trachelomonas* envelopes indicate that both granular and fibrillar textures may be found in a single species, occasionally in a single envelope (Fig. 5). Moreover, we find that texture is not exclusively a function of envelope thickness (Figs. 15-18). In fact, some of our data suggest that texture may be related to the chemical composition of the envelope. Granular envelopes seem to be correlated with higher concentrations of Fe, while fibrillar or woven envelopes may be correlated with higher concentrations of Mn (*cf.* Figs. 21-32) (Donnelly and Walne, 1979; West and Walne, submitted).

TEM investigations of envelope structure in *Trachelomonas hispida* var. *coronata* reveal a complex arrangement of anastomosing fibrillar constituents (Figs. 6, 7, 16, 18). Glancing sections and face views of envelopes show an inner reticulum of fine fibrils that underlie the outer, coarser, ornamented meshwork. Optical sections of the surface matrix and ornamentations reveal single and folded mucilage strands and papillae that contain fine deposits and needle-like electron-dense regions of 40-80 Å in width (Fig. 5). Since the needle-shaped constituents appear electron dense in envelopes prepared without osmication, we conclude that they are composed of elements of greater atomic weight than those of the mucilage, and we suggest that they are the mineral deposits of the envelope. Micro micro electron diffraction (μμED) of the needle-shaped mineral deposits yields a pattern of discrete spots, a result demonstrating that the deposits are crystalline (Fig. 8). Thus, the crystalline mineral deposits in these envelopes occur as needle-shaped microcrystallites *ca.* 40-90 Å thick. The inner network of the envelopes can be seen readily through punctae (Figs.

Figs. 1-9. Figs 1-4, SEM, *Trachelomonas zorensis* Defl. Fig. 1. Pellicle ridges on protoplast are seen through the wispy, hyaline, skin-like immature envelope (x1650). Fig. 2. Immature envelope with open reticulate surface is collapsed around protoplast (x2000). Fig. 3. Variation in surface pattern is evident in three mature envelopes (x1200). Fig. 4. Delicate mucilaginous material covers punctate envelope surface. Portions of envelopes with markedly different surface patterns are also shown (x1650). Figs. 5-9, *Trachelomonas hispida* var. *coronata* Lemm. Fig. 5. Optical section shows single, folded and agglutinated mucilage strands (arrowheads) and papillae (arrow) of *ca.* 200Å thickness that contain fibrous deposits and finely granular electron-dense regions of *ca.* 50Å width. Unfixed, CPD uncoated TEM whole mount, negative image (x83,000). Fig. 6. Face view of envelope showing inner reticulum (arrow) that underlies the fibrous meshwork. Lateral adherence of several individual strands gives rise to thicker strands that are readily visualized. Unfixed, CPD TEM whole mount (x13,000). Fig. 7. Face view of punctate envelope with 'fibrous mat' substructure. Fibrous or needle-like inclusions are twisted and folded within single and anastomosed mucilage strands to produce fibrous mat appearance. Unfixed, CPD uncoated TEM whole mount, negative image (x37,500). Fig. 8. A μμED pattern from peripheral fibril *ca.* 75Å in width (*cf.* Fig. 5). Discrete spots demonstrate the presence of a crystal in the path of the electron beam. Fig. 9. Degenerating envelope. Large ropelike structures and elements of underlying inner reticulum are composed of fine mucilage strands. Degeneration of ropelike interpunctal mucilage results in large openings which expose the subtending inner reticulum (x4000).

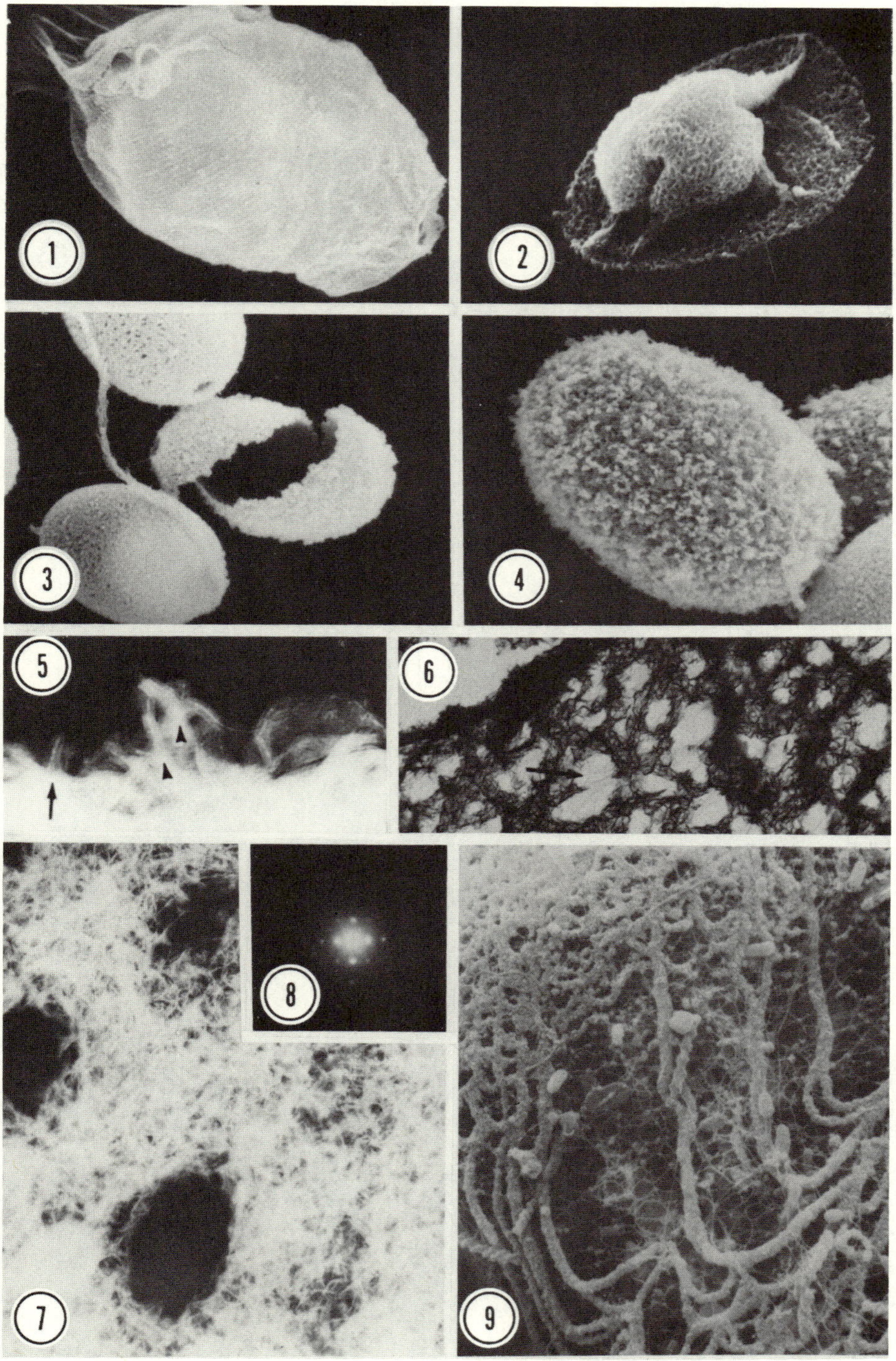
1
2
3
4
5
6
7
8
9

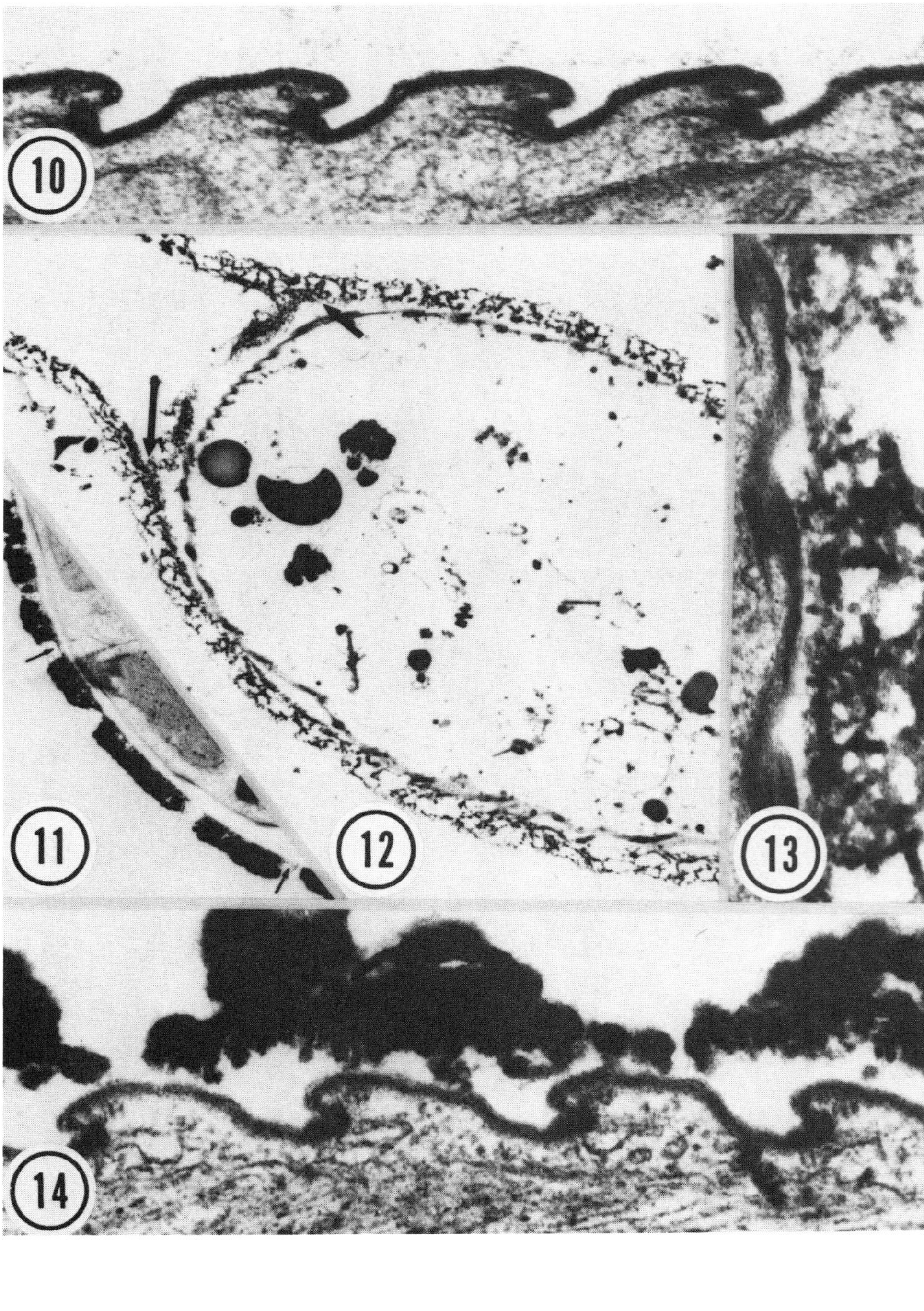
10
11
12
13
14

Figs. 10-14. TEM, *Trachelomonas bulla* Stein. Fig. 10. Pellicle is covered by very delicate, granular envelope (x76,000). Fig. 11. 'Skin' (arrow) is evident between and below very dense region of envelope (x7300). Fig. 12. Reticulate envelope of degenerated cell shows oblique neck and apical pore; arrows denote delicate inner envelope layer that extends anteriorly to form the pore (x7300). Fig. 13. Reticulate envelope with finely granular substructure (x56,000). Fig. 14. Higher magnification (*cf*. Fig. 13) showing portion of very dense envelope with finely granular substructure (x77,000).

6, 7) and through lesions resulting from the degeneration or disruption of thick bundles of mineralized mucilage organized into ropelike structures (Fig. 9). The unit mucilage strands of the inner reticulum are of comparable thickness to those in other envelope layers, but they do not contain mineral microcrystallites and are thus electron lucent in thin section (West *et al.*, 1978; West and Walne, submitted; West *et al.*, submitted).

Compositionally, *Trachelomonas* envelopes have been reported to react positively to PAS and Alcian Blue tests for polysaccharides (e.g., glycosaminoglycans) and to the Prussian Blue test for iron (Pringsheim, 1953; Singh, 1956; Moss,

Table 4. SEM-EDX Semi-quantitative Elemental Analyses of Air-dried Envelopes of *Trachelomonas hispida* var. *coronata* Mounted on Solid Graphite Stubs[1].

	Characteristic Fluorescent X-rays Counted (Peak Channel)	
Element (Kα)	Light Envelope	Dark Envelope
Na	393	109
Si	1949	607
S	588	207
Cl	136	60
K	81	--[+]
Ca	615	384
Mn	919	1827

[1]Modified from West *et al.*, submitted

[+]Not detected

1964; West and Walne, 1977). In agreement with Singh (1956), Leedale (1975a) reported no spectral peaks for Mn in 8 species of *Trachelomonas* analyzed with TEM-EDX. To the contrary, in our investigations of *T. hispida* var. *coronata*, using Energy-Dispersive X-ray Analysis (STEM-EDX, SEM-EDX), Atomic Absorption Spectrophotometry (AA) and Electron Probe Microanalysis (EPM), we find Mn rather than Fe, as the main mineralization element in most envelopes, with an occasional exception in some light envelopes where Si (Fig. 19, Table 4) is the main element (Figs. 19, 20, 21-26, 33-35, Table 4). When detectable, the small concentrations of Fe are higher in the light envelopes than in the intermediate or dark envelopes, where it is barely recorded (Table 5) (West, 1977; West and Walne, 1977; West *et al.*, submitted). More recently, Leedale reports that he, too, finds Mn in the envelopes (Leedale, personal communication). Preliminary quantitative analyses by AA yield estimates of *ca.* 20-30 wt % Mn. Quantitation with EPM yields estimates of *ca.* 23 wt % Mn, with other elements yielding estimates of less than 1 wt %. STEM-EDX analysis of thin specimens yields an average of *ca.* 36 wt % Mn; other elements yield concentrations of less than 5 wt % (Figs. 21-26; 33-35) (Table 5) (West, 1977; West and Walne, 1977; West *et al.*, 1978 and submitted).

Our very recent data suggest that the chemical composition of envelopes may be related not only to the chemical composition of the medium but also to development or mineralization state of the envelope (Donnelly and Walne, unpublished data). Comparative TEM, SEM and SEM-EDX studies of envelopes of *Trachelomonas bulla*, which do not become dark, with *T. lefevrei* and *T. zorensis*, which do darken markedly, reveal that Fe is the major mineralizing element in hyaline to very lightly colored envelopes in all 3 species. Mn concentrations increase as envelopes darken in *T. lefevrei* and *T. zorensis*. In very dark envelopes, Fe has not been detected with SEM-EDX, but more precise analyses, such as those afforded by STEM-EDX for example, may eventually reveal Fe in those envelopes.

Figs. 15-20. *Trachelomonas hispida* var. *coronata* Lemm. Fig. 15. Thin section of a pale-yellow envelope, *ca.* 250-350 nm in thickness (x1300). Fig. 16. Higher magnification of thin envelope (*cf.* Fig. 15) showing 'fibrous mat' substructure. Fibrillar (arrow) and linear globular (arrowhead) arrays are evident; fibrillar constituents are *ca.* 45-95Å thick (x50,500). Fig. 17. Thin section of medium/dark envelope (x1300). Fig. 18. Higher magnification of dark envelope (*cf.* Fig. 17), showing again the 'fibrous mat' substructure, with fibrillar and irregular linear (arrowhead) and parallel (arrow) arrays; the latter impart a flake-like appearance (x52,500). Figs. 19-20. SEM-EDX spectra obtained from populations of light (Fig. 19) and dark (Fig. 20) envelopes. The major element identified in the light envelopes (Fig. 19) is Si, with lesser amounts of S, Ca and Mn. In the dark envelopes (Fig. 20), the major element identified is Mn, with much-reduced peaks for Si, S and Ca.

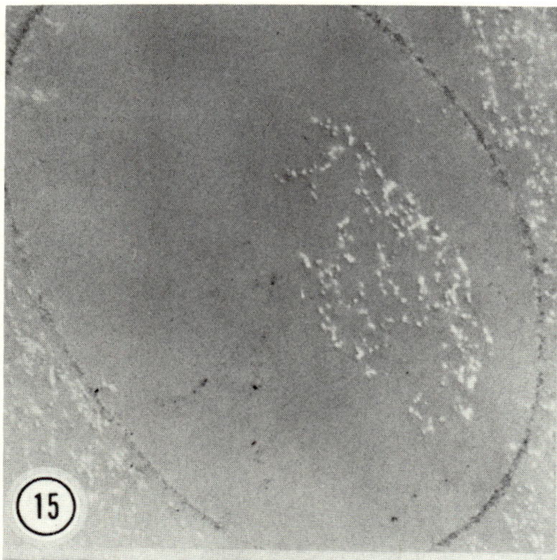
15

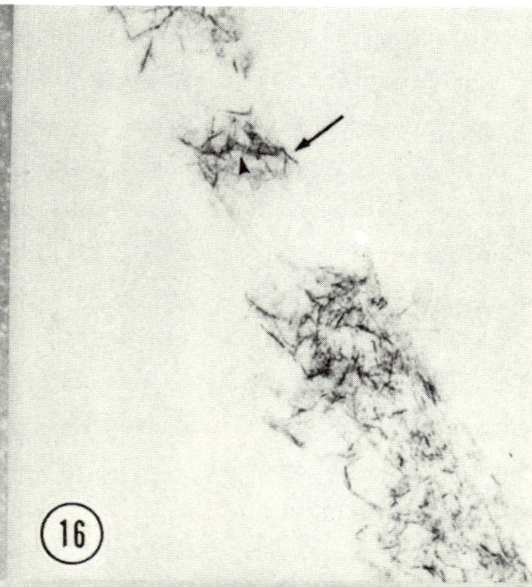
16

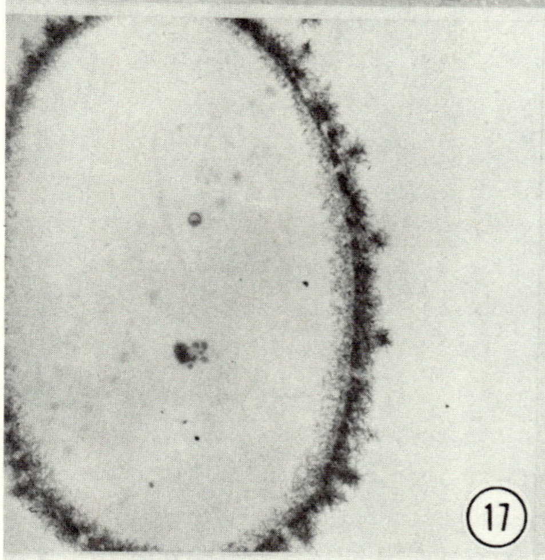
17

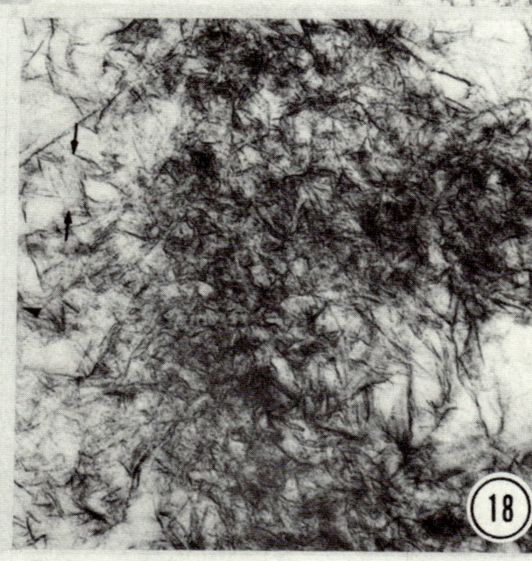
18

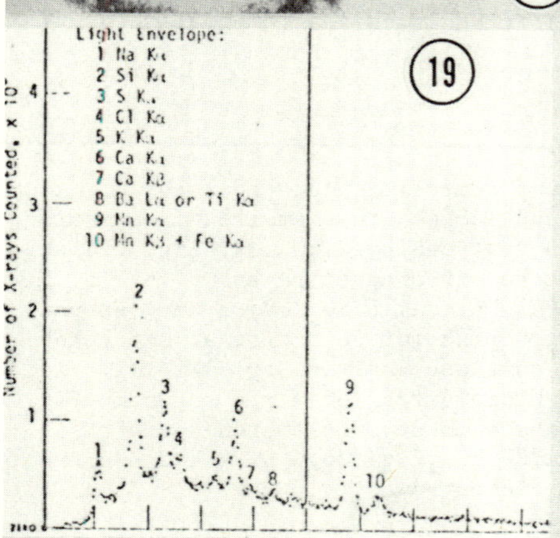
Light Envelope:
1 Na Kα
2 Si Kα
3 S Kα
4 Cl Kα
5 K Kα
6 Ca Kα
7 Ca Kβ
8 Ba Lα or Ti Kα
9 Mn Kα
10 Mn Kβ + Fe Kα
19
Number of X-rays Counted, x 10³
4
3
2
1
1
2
3
4
5
6
7
8
9
10

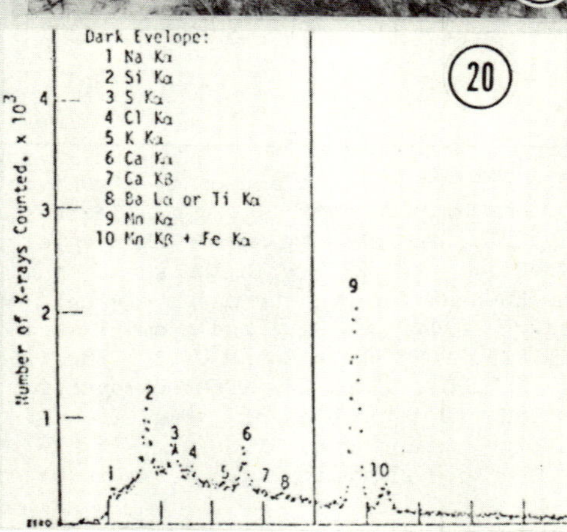
Dark Evelope:
1 Na Kα
2 Si Kα
3 S Kα
4 Cl Kα
5 K Kα
6 Ca Kα
7 Ca Kβ
8 Ba Lα or Ti Kα
9 Mn Kα
10 Mn Kβ + Fe Kα
20
Number of X-rays Counted, x 10³
4
3
2
1
1
2
3
4
5
6
7
8
9
10

The relative concentrations of other elements (e.g., Si, Ca) also decrease as Mn concentrations increase (Figs. 27-32; *cf.* Figs. 21-26, 33-35). Thus, there are apparently shifting ratios of Fe/Mn in light, intermediate and dark envelopes. Spectra of mixed populations (light and dark) of envelopes of *T. lefevrei* and *T. zorensis* however, show Mn as the major peak with very little to no detectable Fe (Donnelly, 1979; Donnelly and Walne, 1979 and unpublished data).

In summary, envelopes of *Trachelomonas hispida* var. *coronata* are composed of mucilage strands anastomosed into fibrous mat-like structures, sometimes ornamented with spines and punctae. The mucilage strands of the matrix are impregnated with microcrystallites composed largely of Mn, of up to *ca.* 36-40 wt % Mn, with traces of other elements such as Si and Ca and only small amounts of Fe, if any. Non-mineralized mucilage may form an inner reticulum between the pellicle and the outer envelope material. Envelopes of *T. bulla*, *T. lefevrei* and *T. zorensis* show shifting ratios of Fe/Mn as they darken, and their texture seems to be correlated with their chemical composition--granular envelopes with Fe, fibrillar envelopes with Mn, and some of intermediate texture when the element composition is more balanced.

These data thus provide confirmatory evidence for the historical assumptions that Mn and Fe are major elements involved in envelope biomineralization, which is now understood as very complex, involving also several other elements. The data also support Pringsheim's suggestion that Mn is responsible for darkening of envelopes. Finally, the data raise interesting questions and point to the need for additional and ever-more-definitive studies. For example, do the differences in major mineralizing elements reflect real species differences, or do they perhaps reflect subtle differences in nutrition, development or instrumentation? It is noteworthy that differences in instrumentation and

---

Figs. 21-26. STEM analyses of structure and composition of envelopes of *Trachelomonas hispida* var. *coronata*. See text and Table 4 for detailed information. Fig. 21. TEM whole mount of thin envelope analyzed by STEM (*cf.* Fig. 22) (x3500). Fig. 22. STEM-EDX spectrum obtained from envelope shown in Fig. 21. The Kα peak for Mn, the major element, occurs at *ca.* 5.89 KeV; a Mn + Fe peak occurs at *ca.* 6.4 KeV; and a small secondary peak for Fe occurs at *ca.* 7 KeV. Fig. 23. TEM whole mount of intermediate envelope analyzed by STEM (*cf.* Fig. 24) (x2200). Fig. 24. STEM-EDX spectrum obtained from envelope shown in Fig. 23. Mn is major element. Fig. 25. TEM whole mount of electron-dense thick envelope analyzed by STEM (*cf.* Fig. 26) (x3000). Fig. 26. STEM-EDX spectrum of envelope shown in Fig. 25. Mn is major element.

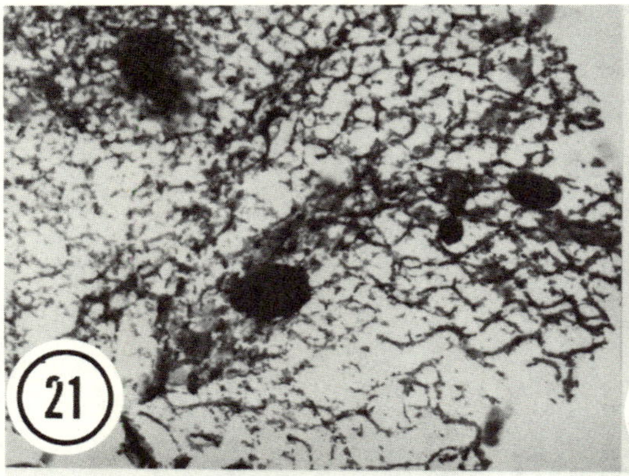
21

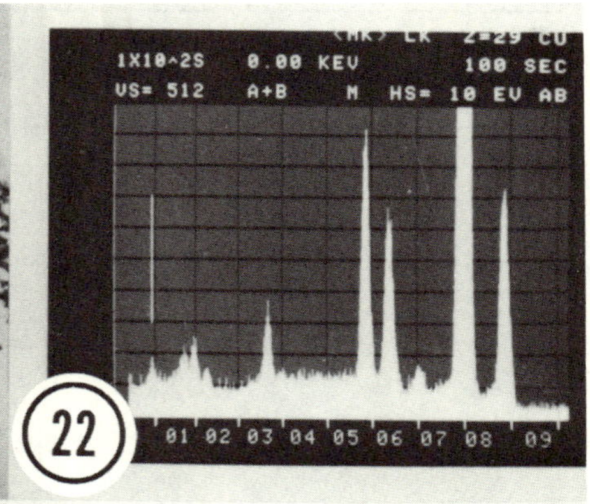

22

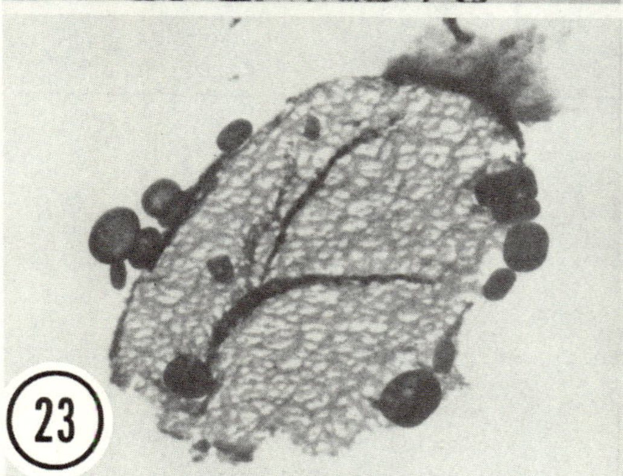
23

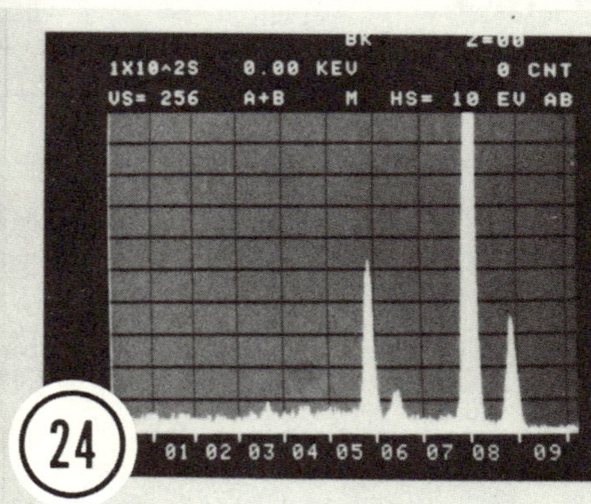

24

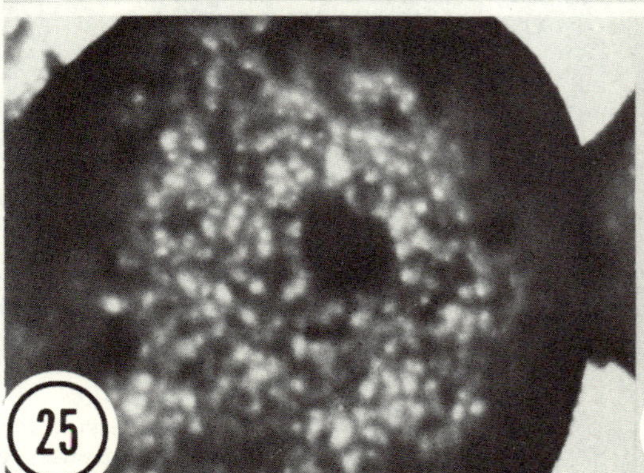
25

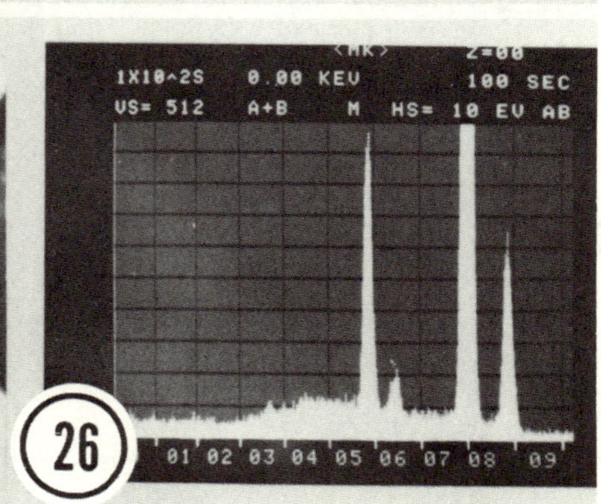

26

Table 5. TEM-EDX Analyses of Three Envelopes of *Trachelomonas hispida* var. *coronata*[1,2]

Element	Characteristic Fluorescence Intensities (I_Z)* & Mass Fractions (I_Z/I_W)+ e⁻-lucent envelope (light envelope)[a]		Medium e⁻-dense envelope (intermediate envelope)[b]		e⁻-dense envelope (dark envelope)[c]	
Z	I_Z	I_Z/I_W	I_Z	I_Z/I_W	I_Z	I_Z/I_W
Si	.73	.19	--	--	--	--
P	.84	.22	--	--	--	--
S	.36	.09	--	--	--	--
Ca	1.24	.32	3.3	0.11	8.8	.07
Mn	5.62	1.46	65.6	2.12	228.0	1.82
Fe†	3.44	.89	2.0	0.60	25.0	.20
"White" (I_W)	3.86		30.9		125.0	

[1]Modified from West *et al.* (Submitted)

[2]Envelopes unfixed, uncoated, air-dried, mounted on a formvar-coated grid and examined in a JEM-100 C Analytical Electron Microscope (AEM) with a beryllium specimen holder.

*Characteristic fluorescence intensity, I_Z, (cps/nA), corrected for inter-element relative fluorescence yields using values published by Cliff and Lorimer (1975).

+Corrected fluorescence intensities (I_Z) divided by the continuum ("white") radiation (I_W), collected in the range of 4.2-5.2 Kev. I_Z/I_W is sometimes termed a "mass fraction" and is proportional to concentration.

†The values for FeK_α were calculated by difference from the combined MnK_α and FeK_α radiation. MnK_β was calculated as 15% of the MnK_α radiation, then subtracted to leave FeK_α value.

[a]*cf.* Fig. 21, [b]*cf.* Fig. 23, [c]*cf.* Fig. 25.

Figs. 27-32. TEM and SEM-EDX analyses of structure and composition of envelopes of *Trachelomonas lefevrei* Defl. Fig. 27. TEM whole mount of thin, hyaline envelope reveals its porous nature (*cf.* Fig. 28) (x3000). Fig. 28. SEM-EDX spectrum of hyaline envelope (*cf.* Fig. 27), showing P and Fe as the major elements. Fig. 29. TEM whole mount of intermediate, golden envelope, which is slightly porous (*cf.* Figs. 27, 30, 31) (x 2800). Fig. 30. SEM-EDX spectrum of golden envelope (*cf.* Fig. 29); Mn is the major element; only a small Fe peak is identified. Fig. 31. TEM whole mount of brown envelope showing a non-porous surface (*cf.* Figs. 27, 29, 32) (x2000). Fig. 32. SEM-EDX spectrum of brown envelope (*cf.* Fig. 31), with Mn as the major element; no Fe is recorded.

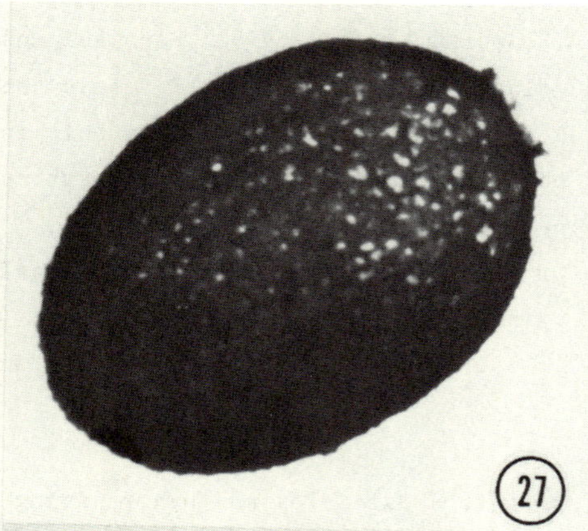
27

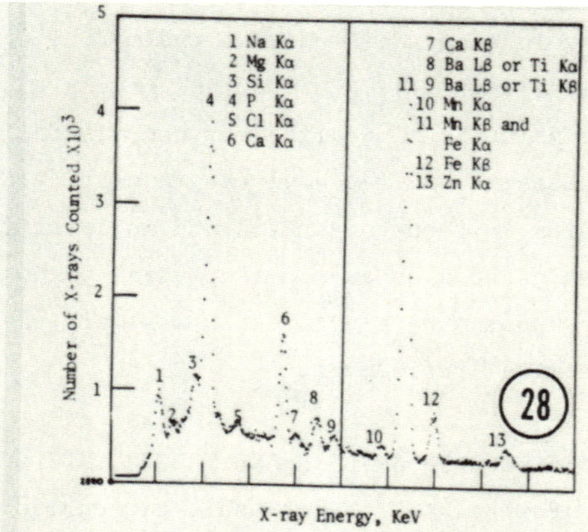

28

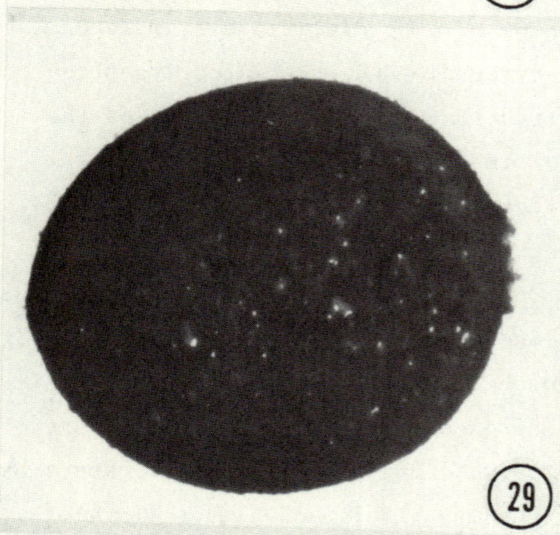
29

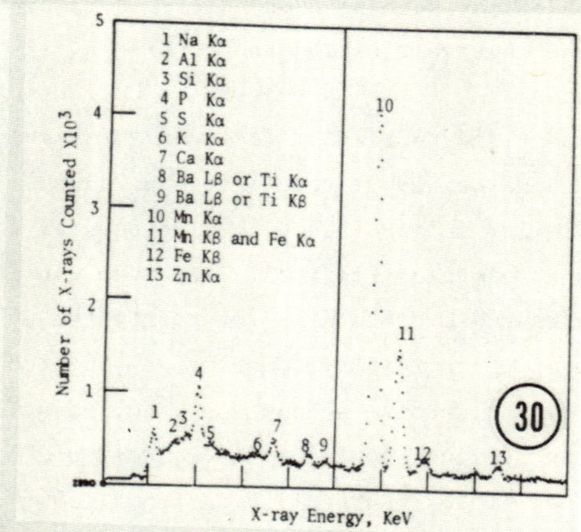

30

31

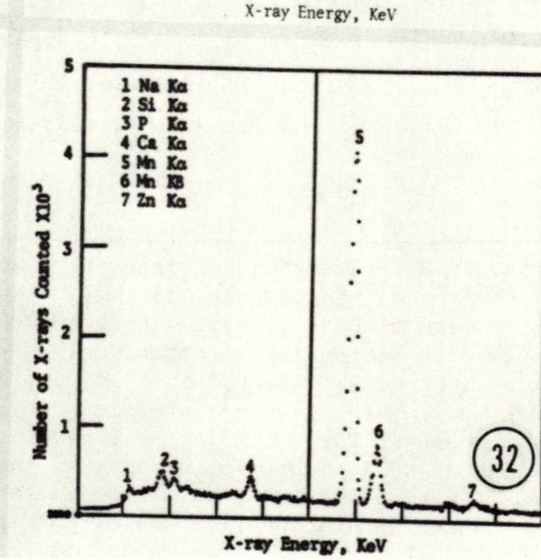

32

in preparative procedures for SEM have been reported to have a marked effect on envelope surface features and morphology (Rosowski *et al.*, 1975b; Rosowski and Glider, 1977). A comparison of euglenoid flagellates that produce, retain and/or maintain mineralized or non-mineralized mucilaginous structures may help in assessing their usefulness in clarifying phylogenetic trends and interrelationships in these organisms and in determining whether the euglenoid envelopes represent a unique developmental pathway or whether envelopes in several algal groups may have evolved from a more basic type by structural and physico-chemical refinement.

INGESTION ORGANELLES AND TRICHOCYSTS

Some of the phagotrophic euglenoids ingest food particles and other living organisms such as *Euglena* or yeast, by means of a specialized system lateral to the reservoir region and consisting of a cytostome and associated ingestion apparatus termed a rodorgan. Current ultrastructural investigations of the colorless, voracious *Peranema trichophorum* (Ehr.) Stein (Fig. 36) (Hilenski and Walne, 1979) reveal several unusual features, including striated fibrils, similar in appearance to kinetodesmal fibers that interconnect the basal bodies and rodorgan (Fig. 43). The rodorgan is composed of an array of regularly arranged microtubules interconnected by fine threads or fibrils (Figs. 42, 43); 2 microtubules may sometimes be interconnected by more than one fibril (Fig. 42). Extensive accounts of the ingestion process and of the ultrastructure of the feeding apparatus in *Peranema* and *Entosiphon* may be found in the papers by Chen (1950), Roth (1959), Mignot (1966), Leedale (1967) and Nisbet (1974).

Trichocysts, a type of ejectile organelle, have been shown in *Entosiphon* (Mignot, 1966; Mignot and Hovasse, 1973) and more recently in *Peranema* (Figs. 37-39) (Hilenski and Walne, 1979 and unpublished data). In the latter, tri-

---

Figs. 33-35. STEM-EDX analyses for elemental microdistribution in envelopes of *Trachelomonas hispida* var. *coronata*. See text for details. Fig. 33. STEM whole mount. Scan line illustrates electron beam path followed across 3 envelopes with increasing electron density--thin, intermediate, and thick, to produce line scans presented in Figs. 34, 35 (x1150). Fig. 34. STEM-EDX line scan showing Mn K_α X-ray intensity as a function of distance along the line scan shown in Fig. 33. In the 3 envelopes shown there, Mn is shown by this line scan to increase in quantity as the beam crosses envelopes of increasing density, thus verifying Mn mineralization. Fig. 35. STEM-EDX line scan showing 'white' radiation line scan from Fig. 33. Relative differences in 'white' radiation reflect differences in mass.

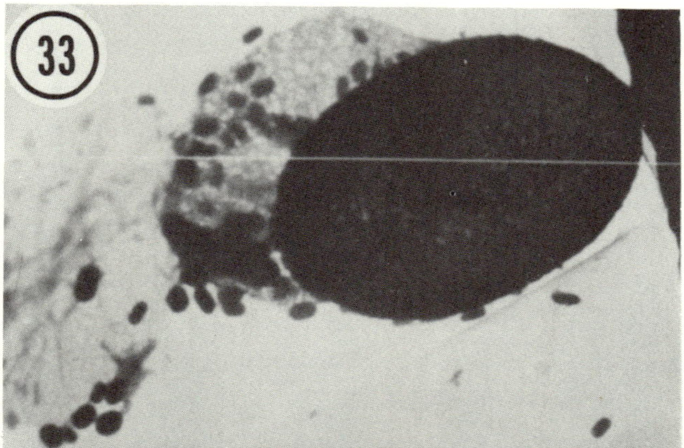
33

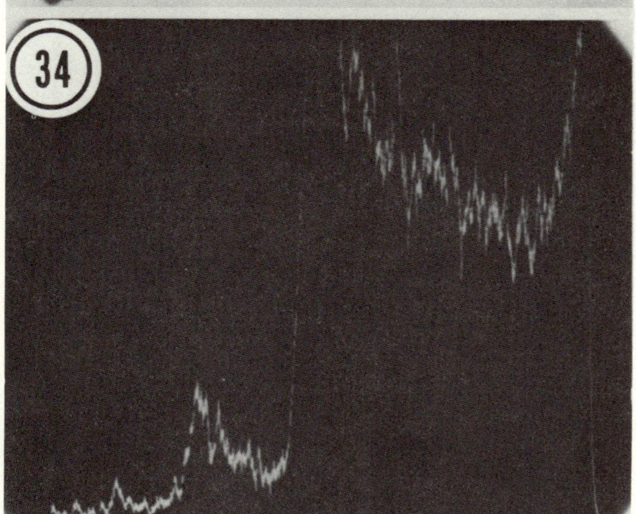
34

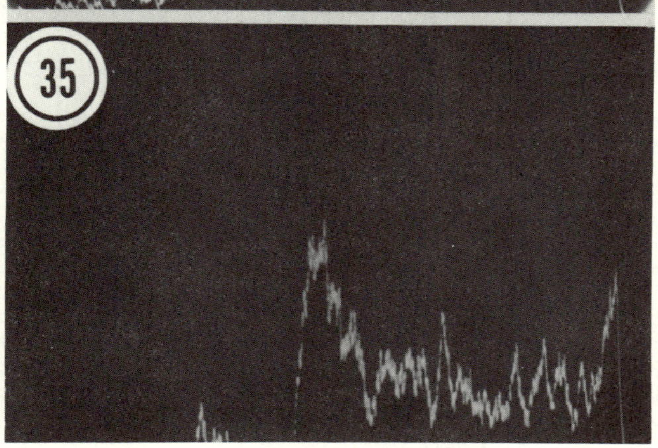
35

chocysts can be seen to protrude through the pellicle and eject mucilaginous material (Fig. 37); sometimes the entire trichocyst may be ejected (Fig. 39). In the undischarged condition, they are located at the cell periphery just below and at *ca.* right angles to the pellicular grooves (Fig. 38) (Hilenski and Walne, 1979). Our preliminary data (Hilenski and Walne, unpublished data) suggest that trichocysts are initiated in Golgi-associated vesicles, which migrate to subpellicular positions where development continues. This is similar to trichocyst ontogeny suggested in marine dinoflagellates by Bouck and Sweeney (1966), and to a close association of presumptive trichocyst precursors with the Golgi apparatus, as described in *Entosiphon* by Mignot and Hovasse (1973). Immature trichocysts in *Peranema* are seen as dispersed amorphous material within a vesicle; more mature, vesicle-enclosed trichocysts are tubular, elongate, fusiform and slightly hourglass shaped (Fig. 38) (Hilenski and Walne, 1979). Negatively stained (Fig. 39) or shadowcast preparations of discharged trichocysts, *ca.* 1,000 nm long and *ca.* 200 nm in diameter, show a distinctive dark band on the outer 'jacket' or tube, which has a complex diamond or latticework pattern composed of tightly coiled helical elements; helical striae are also evident on the inner tube (Hilenski and Walne, 1979). Tubular trichocysts rich

---

Figs. 36-43. *Peranema trichophorum* (Ehr.) Stein. Fig. 36. Specimen of *P. trichophorum* showing anteriorly directed flagellum, and trailing flagellum appressed in a groove on the surface of the cell. Pellicle and flagellar surfaces are coated with mucilaginous material. SEM preparation (x4200). Fig. 37. Longitudinal section through peripheral trichocyst discharging mucilage (x55,000). Fig. 38. Longitudinal section through 2 subpellicular trichocysts in different stages of development. Immature trichocyst, on left, is seen as dispersed amorphous material within a vesicle; more mature, vesicle-enclosed trichocyst, on right, is elongate and fusiform, with slightly hourglass shape (x26,000). Fig. 39. Negatively stained preparation of discharged trichocyst, *ca.* 1,000 nm long and *ca.* 200 nm in diameter. Distinctive dark central band is seen on the outer 'jacket' or tube, which has a diamond pattern composed of tightly coiled helical elements. Regularly arranged helical striae are evident at one end of the inner tube (arrow). Uranyl acetate stain (x67,000). Fig. 40. Tangential section of anterior flagellum. Surface of mastigoneme ribbon is seen at the top, with progressively more underlying portions of the flagellum revealed toward the bottom (x32,000). Fig. 41. Transverse section of anterior flagellum showing the axoneme and 2 ribbonlike complexes of mastigonemes (x62,000). Fig. 42. Transverse section of portion of rodorgan showing microtubules interconnected by fine threads; 2 microtubules may be interconnected by more than 1 fibril (arrow) (x65,500). Fig. 43. Portion of rodorgan and striated fibril that is similar to a kinetodesmal fibril. Part of the ingestion apparatus, the rodorgan is composed of an array of regularly arranged microtubules interconnected by fine threads or fibrils (x62,000).

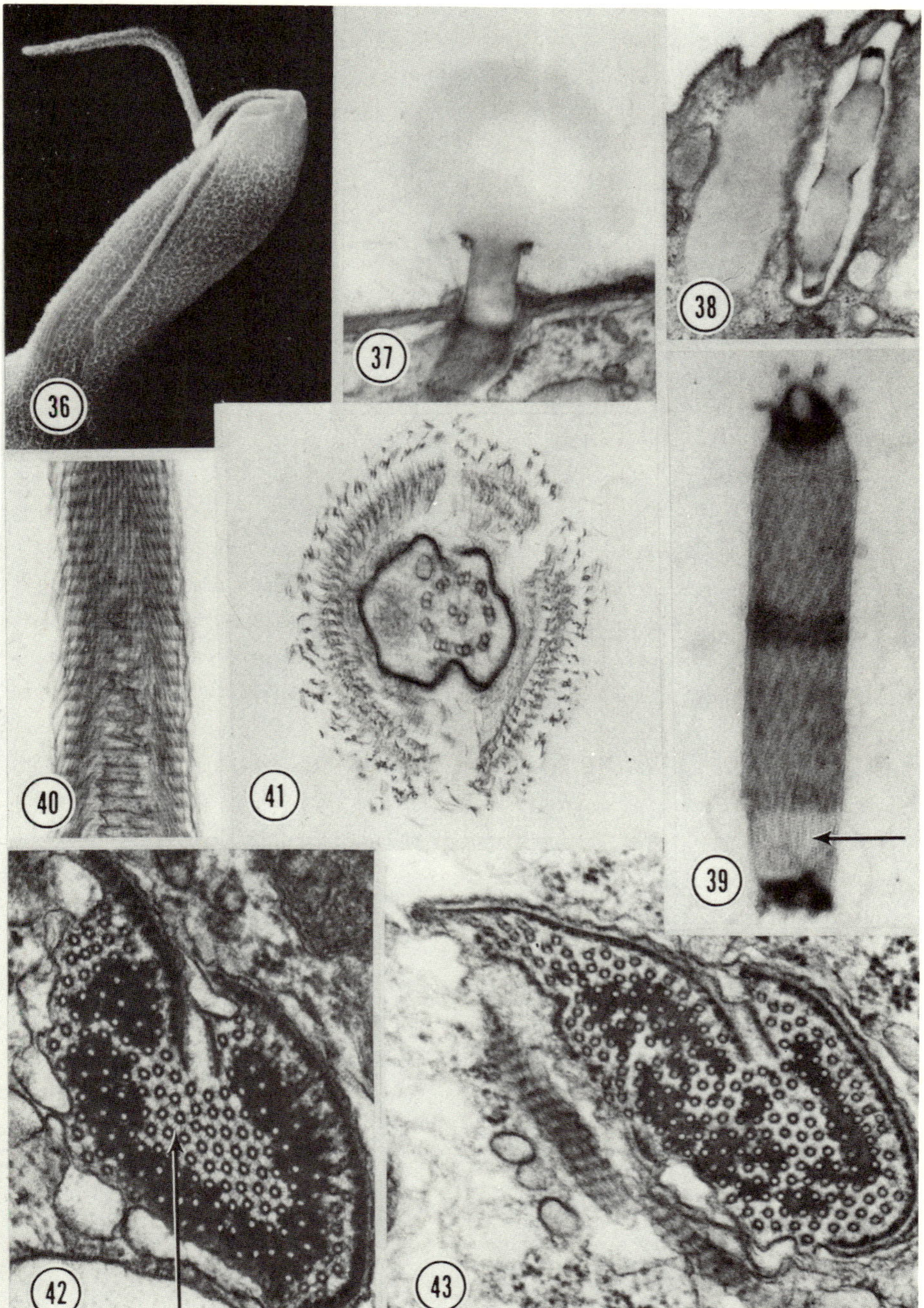
36
37
38
39
40
41
42
43

in polysaccharides and with walls consisting of 2 complexes of spiralling filaments that form a characteristic latticework were also reported in *Entosiphon* Mignot and Hovasse, 1973). There are no data available on the composition or the triggering mechanisms for discharge of the trichocysts.

FUTURE RESEARCH

From a perusal of this chapter and the literature on euglenoids at large, it is quite apparent that, with few exceptions, most of the in-depth investigations on euglenoid flagellates have dealt with *Euglena*, and especially with *E. gracilis*, since it can be maintained and manipulated more readily in axenic culture than can most other euglenoid organisms. Thus, basic studies on almost all of the other euglenoids are needed, with respect to such considerations as culture conditions, nutrition, growth and development, mitosis, ultrastructural, biochemical and biophysical characterizations of cellular, subcellular and molecular events and constituents. Concerning the last, more efficient techniques need to be devised for the successful isolation of pure fractions of functionally competent subcellular organelles and other components.

Some euglenoids are well suited to serve as model membrane systems with unusual and unique properties, manifested in *Euglena*, for example, by the distinctively stable pellicle and by the contiguous but segregated regions of pellicle, reservoir and flagellum. These pose intriguing questions on the synthesis, development and maintenance of cell surfaces. The pathways of glycoprotein synthesis and insertion into different surface areas may be investigated with the use of such specialized tools as immunospecific markers or specific lectins, as suggested by Bouck *et al.* (1973), since we now know that different carbohydrate sugars occur on the flagellum and cell membrane.

Envelopes and other structures and secretions as stalks and warts in such organisms as *Trachelomonas*, *Colacium* and *Euglena spirogyra* are prime materials for the study of fundamental questions on synthesis, development and maintenance of extracellular matrices and surfaces. With regard to *Trachelomonas* and other enveloped genera, additional studies with μμED, stereoscopic TEM, ESCA (Electron Spectroscopy for Chemical Analysis) or AES (Auger Electron Spectroscopy) on untreated and demineralized envelopes are needed to extend present investigations on microcrystallite formation. Studies with diverse inhibitors and comparative studies of development in other mucilage-secreting organisms (e.g., stalk production in *Colacium* and wart production in *Euglena spirogyra*) would help to provide new insights into envelope production. *In vitro* incubation experiments with isolated envelopes and limnological studies of environmental parameters could add information on envelope function.

With respect to cellular euglenoid movements, intranuclear mitoses, and the mechanisms and control of cell duplication, there is much that we still do not understand, and the possibility of actin-type interactions in these areas should not be overlooked in future investigations.

Concerning photoreception in *Euglena* and other euglenoids, a most important contribution would involve the redoubtable task of developing a procedure for isolating the paraflagellar bodies (PFBs), since they are considered to be the primary photoreceptors. Chemical assays could then be carried out, which could lead to the precise identification of the photoreceptor pigments. The identification of the stigmatal carotenoid pigments is tentative; larger quantities of some pigments are required for their precise identification by such techniques as absorption, mass, IR and NMR spectroscopy in determining the actual maxima, compound weights, pigment functional groups and their positions, respectively. ESR analysis (Electron Spin Resonance), a technique which detects and identifies free radicals, may prove useful in future analyses. Since flavins are known to be involved in electron transport, perhaps such a process could be photochemically induced in the stigma and/or PFB, which could be analyzed in darkness and under illumination for the presence of radicals. If, for example, ESR analyses were characteristic of an anisotropic environment for the radical, a highly ordered 'liquid crystal' structure would be suggested for the stigma granules. Comparative studies of the ultrastructure and chemical composition of isolated stigmata of other species of *Euglena*, induced colorless mutants, and other genera would also prove interesting.

Other biophysical and biochemical approaches to the study of photoreception, phototaxis and photomovement in *Euglena* include investigations on the molecular bases of sensory transduction, such as those studies being carried out by Lenci, Benedetti, Colombetti, Diehn and others.

Questions on the origin of chloroplasts in euglenoids persist and may be addressed in future studies by the kinds of experiments outlined by Schiff (1973), Gibbs (1978) and others. For example, investigations involving induced chloroplast loss, comparative ultrastructural and biochemical studies of naturally occurring and induced colorless organisms, and careful examinations of extant symbiotic relationships, among other experimental approaches, might provide us with some unexpected and exciting new insights into chloroplast origins and euglenoid phylogeny.

These are but a few of the many provocative questions that might be addressed in future studies, for obviously it is impossible to suggest here all of the kinds of investigations that could or should be carried out on this fascinating group of organisms.

ACKNOWLEDGEMENTS

The author acknowledges research support for some of the studies cited in the text from NSF Grants PCM-7519872, PCM-7702719 and DEB-7820210. I thank my students and colleagues, Drs. J. Bentley, ORNL; Linda Sue Donnelly, Larry K. West, Ms. Lu L. Hilenski and Mr. John Dunlap, University of Tennessee, Knoxville, for their kindness in allowing me the use of some of their electron micrographs and analytical data as follows: Figures 8, 21-26, 33-35 (Bentley); Figures 1-4, 10-14, 28, 30, 32 (Donnelly); Figures 27, 29, 31 (Dunlap); Figures 5-7, 9, 15-20 (West); Figures 36-43 (Hilenski). I am grateful to Mr. David A. Gerard for excellent and gracious technical assistance, and to Mr. Thomas E. Byrne for clerical assistance. I thank Dr. L. K. West for stimulating ideas and discussions, especially on envelope biomineralization, and I express sincere appreciation to him and to Drs. G. F. Leedale, G. B. Bouck, J. R. Rosowski, P. Benedetti, G. Colombette, F. Lenci and B. Diehn for their helpful suggestions on the manuscript. Contribution No. 513 from the Department of Botany, University of Tennessee, Knoxville.

REFERENCES

Arnott, H. J. and P. L. Walne. 1967. Observations on the fine structure of pellicle pores of *Euglena granulata*. Protoplasma 64: 330-344.

Barghigiani, C., G. Colombetti, B. Franchini and F. Lenci. 1979. Photobehavior of *Euglena gracilis*: Action spectrum for the step-down photophobic response of individual cells. Photochem. Photobiol. 29: 1015-1019.

Barras, D. R. and B. A. Stone. 1965. The chemical composition of the pellicle of *Euglena gracilis* var. *bacillaris*. Biochem. Journ. 97: 14-15.

Bartlett, C. J., P. L. Walne, O. J. Schwarz and D. H. Brown. 1972. Large-scale isolation and purification of eyespot granules from *Euglena gracilis* var. *bacillaris*. Pl. Physiol. 49: 881-885.

Batra, P. and G. Tollin. 1964. Phototaxis in *Euglena*. I. Isolation of the eye-spot granules and identification of the eye-spot pigments. Biochem. Biophys. Acta 79: 371-378.

Benedetti, A. and A. Checcucci. 1975. Paraflagellar body (PFB) pigments studied by fluorescence microscopy in *Euglena gracilis*. Plant Sci. Let. 4: 47-51.

Benedetti, P. A. and F. Lenci. 1977. *In vivo* microspectrofluorometry of photoreceptor pigments in *Euglena gracilis*. Photochem. Photobiol. 26: 315-318.

Benedetti, P. A., G. Bianchini and G. Chiti. 1976a. Fast scanning microspectroscopy: an electrodynamic moving-condenser method. Appl. Optics 15: 2554-2557.

Benedetti, P. A., G. Bianchini, A. Checcucci, R. Ferrara, S. Grassi and D. Percival. 1976b. Spectroscopic properties and related functions of the stigma measured in living cells of *Euglena gracilis*. Arch. Mikrobiol. 111: 73-76.

Ben-Shaul, Y., J. A. Schiff and H. T. Epstein. 1964. Studies of chloroplast development in *Euglena*. VII. Fine structure of the developing plastid. 39: 231-240.

Ben-Shaul, Y., H. T. Epstein and J. A. Schiff. 1965. Studies of chloroplast development in *Euglena*. 10. The return of the chloroplast to the proplastid condition during dark adaptation. Canad. Journ. Bot. 43: 129-136.

Bisalputra, T. 1974. Plastids. In Stewart, W. D. P. [Ed.] *Algal Physiology and Biochemistry*. University of California Press, Berkeley. pp. 124-160.

Blum, J. J. 1965. Observations on the acid phosphatases of *Euglena gracilis*. Jour. Cell Biol. 24: 223-234.

Blum, J. J., J. R. Sommer and V. Kahn. 1965. Some biochemical, cytological and morphogenetic comparisons between *Astasia longa* and a bleached *Euglena gracilis*. Jour. Protozool. 12: 202-209.

Boasson, R. and S. P. Gibbs. 1973. Chloroplast replication in synchronously dividing *Euglena gracilis*. Planta 115: 125-134.

Bouck, G. B. 1972. Architecture and assembly of mastigonemes. In *Advances in Cell and Molecular Biology*, Vol. 2. Academic Press, NY. pp. 237-271.

Bouck, G. B. 1977. Flagellar mastigonemes may provide clues to algal systematics and evolution. V Int. Congr. Protozool. Abstr. 459a.

Bouck, G. B. and B. M. Sweeney. 1966. The fine structure and ontogeny of trichocysts in marine dinoflagellates. Protoplasma 61: 205-223.

Bouck, G. B., A. Rogalski and A. Valaitis. 1978. Surface organization and composition of *Euglena*. II. Flagellar mastigonemes. Jour. Cell Biol. 77: 805-826.

Brody, M., S. S. Brody and J. H. Levine. 1965. Fluorescence changes during chlorophyll formation in *Euglena gracilis* (and other organisms) and an estimate of lamellar area as a function of age. Jour. Protozool. 12: 465-476.

Buetow, D. E. 1968. *The Biology of Euglena*. Vols. I and II. Academic Press, NY. 361 pp and 417 pp.

Buetow, D. E. 1978. Nuclei and chromatin from *Euglena gracilis*. In Hellebust, J. A. and J. S. Craigie [Eds.] *Handbook of Phycological Methods. Physiological and Biochemical Methods*. Cambridge Univ. Press, Cambridge. pp. 15-23.

Carell, E. F., J. M. Egan and E. A. Pratt. 1970. Studies on chloroplast development and replication in *Euglena*. II. Identification of two different deoxyribonucleases. Arch. Biochem. Biophys. 138: 26-31.

Chaley, N., A. Lord and F. G. Fontaine. 1977. A light and electron microscope study of nuclear structure throughout the cell cycle in the euglenoid *Astasia longa* (Jahn). Jour. Cell Sci. 27: 23-45.

Checcucci, A., G. Colombetti, R. Ferrara and F. Lenci. 1976. Action spectra for photo-accumulation of green and colorless *Euglena*: Evidence for identification of receptor pigments. Photochem. Photobiol. 23: 51-54.

Chen, Y. T. 1950. Investigations of the biology of *Peranema trichophorum* (Euglenineae). Quart. J. Microscop. Sci. 91: 279-308.

Chisholm, S. W. and R. G. Stross. 1976. Phosphate uptake kinetics in *Euglena gracilis* Z grown in light/dark cycles. I. Synchronized batch cultures. II. Phased PO_4-limited cultures. J. Phycology 12: 210-217 and 217-222.

Cliff, G. and G. W. Lorimer. 1975. The quantitative analysis of thin specimens. J. Micros. 103: 203-207.

Colombetti, G. [Ed.]. 1976. *Proceedings of the International School on: Biophysics of Photoreceptors and Photobehavior of Microorganisms*. Badia Fiesolana, Italy. National Research Council, Pisa, Italy.

Creutz, C. and B. Diehn. 1976. Motor response to polarized light and gravity sensing in *Euglena*. J. Protozool. 23: 552-556.

Creutz, C., G. Colombetti and B. Diehn. 1978. Photophobic behavioral responses of *Euglena* in a light intensity gradient and the kinetics of photoreceptor pigment interconversions. Photochem. Photobiol. 27: 611-616.

Diamond, J. and J. A. Schiff. 1974. Isolation and characterization of mutants of *Euglena* resistant to streptomycin. Plant Sci. Let. 3: 259-295.

Diehn, B. 1969. Two perpendicularly oriented pigment systems involved in phototaxis of *Euglena*. Nature 221: 336.

Diehn, B. and B. Kint. 1970. The flavin nature of the photoreceptor molecule for phototaxis in *Euglena*. Physiol. Chem. and Phys. 2: 483-488.

Diehn, B., J. R. Fonseca and T. L. Jahn. 1975. High speed cinematography of the direct photophobic response of *Euglena* and the mechanism of negative phototaxis. J. Protozool. 22: 492-494.

Dockerty, A. and M. J. Merrett. 1979. Isolation and enzymic characterization of *Euglena* proplastids. Plant Physiol. 63: 468-473.

Dodge, J. D. 1973. The Fine Structure of Algal Cells. Academic Press, NY. 261 pp.

Dodge, J. D. 1975. The fine structure of *Trachelomonas* (Euglenophyceae). Arch. Protistenk. 117: 65-77.

Donnelly, L. S. 1979. Comparative cellular ultrastructure and chemical characterization of envelopes of three species of *Trachelomonas* (Euglenophyta). Ph.D. Dissertation. University of Tennessee, Knoxville. 194 pp.

Donnelly, L. S. and P. L. Walne. 1979. Comparative ultrastructure of three species of *Trachelomonas* (Euglenophyceae). ASB Bull. 26: 46.

Doughty, M. J. and B. Diehn. 1979. Photosensory transduction in the flagellated alga, *Euglena gracilis*. I. Action of divalent cations, Ca^{2+} antagonists and Ca^{2+} ionophore on motility and photobehavior. Biochem. Biophys. Acta. 588: 148-168.

Dynesius, R. A. and P. L. Walne. 1974. Ultrastructural and cytochemical comparisons of the euglenoid flagellates, *Phacus pleuronectes* and *Hyalophacus ocellatus*. J. Phycol. 10: 16-17.

Dynesius, R. A. and P. L. Walne. 1975. Ultrastructural aspects of *Phacus pleuronectes*. J. Phycol. 11: 125-130.

Edmunds, L. N. 1965. Studies on synchronously dividing cultures of *Euglena gracilis* Klebs (strain Z). I. Attainment and characterization of rhythmic cell division. J. Cell Comp. Physiol. 66: 147-158.

Ehara, T., I. Shihira-Ishikawa, T. Osafune, E. Hase and I. Ohkuro. 1975. Some structural characteristics of chloroplast degeneration in cells of *Euglena gracilis* Z during their heterotrophic growth in darkness. J. Electron Micros. 24: 253-261.

Epstein, H. T. and J. A. Schiff. 1961. Studies of chloroplast development in *Euglena*. 4. Electron and fluorescence microscopy of the proplastid and its development into a mature chloroplast. J. Protozool. 8: 427-432.

Erwin, J. 1968. Lipid metabolism. In Buetow, D. E. [Ed.] *The Biology of Euglena*. Academic Press, NY. pp. 133-148.

Feinleib, M. E. 1977. Photomovement in microorganisms: Strategies of response. In Castellani, A. [Ed.] *Research in Photobiology*. Plenum, London. pp. 71-84.

Feinleib, M. E. 1978. Photomovement of microorganisms. Photochem. Photobiol. 27: 849-854.

Freyssinet, G., G. C. Harris, M. Nasatir and J. A. Schiff. 1979. Events surrounding the early development of *Euglena* chloroplasts. 14. Biosynthesis of cytochrome c-552 in wild type and mutant cells. Plant Physiol. 63: 908-915.

Froehlich, O. and B. Diehn. 1974. Photoeffects in a flavin-containing lipid bilayer membrane and implications for algal phototaxis. Nature 248: 802-804.

Gibbs, S. P. 1970. The comparative ultrastructure of the algal chloroplast. Ann. N.Y. Acad. Sci. 175: 454-473.

Gibbs, S. P. 1978. The chloroplasts of *Euglena* may have evolved from symbiotic green algae. Canad. Journ. Bot. 56: 2883-2889.

Gojdics, M. 1953. The Genus *Euglena*. Univ. Wisconsin Press, Madison. 268 pp.

Gomez, M. P., J. B. Harris and P. L. Walne. 1974a. Studies of *Euglena gracilis* in ageing cultures. I. Light microscopy and cytochemistry. Brit. Phycol. J. 9: 163-174.

Gomez, M. P., J. B. Harris and P. L. Walne. 1974b. Studies of *Euglena gracilis* in ageing cultures. II. Ultrastructure. Brit. Phycol. J. 9: 175-193.

Goodwin, T. and M. Jamikorn. 1954. Studies on carotenogenesis. Some observations on carotenoid synthesis in two varieties of *Euglena gracilis*. J. Protozool. 1: 216-219.

Guehler, P. F., L. Peterson, H. M. Tschiya and R. M. Dodson. 1964. Composition of the wax formed by *Euglena gracilis*. Arch. Biochem. Biophys. 106: 294-298.

Guttman, H. N. and H. Ziegler. 1974. Clarification of structures related to function in *Euglena gracilis*. Cytobiol. 9: 10-22.

Hand, G. W. 1977. Photomovement. In Smith, K. [Ed.] *The Science of Photobiology*. Plenum, NY. pp. 313-328.

Heelis, D. V., W. Kernick, G. O. Phillips and K. Davies. 1979. Separation and identification of the carotenoid pigments of stigmata isolated from light-grown cells of *Euglena gracilis* strain Z. Arch. Microbiol. 121: 207-211.

Hilenski, L. L. and P. L. Walne. 1979. Some cytological features of the phagotrophic euglenoid flagellate *Peranema*. J. Phycol. 15 (Suppl.): 16.

Hilenski, L. L., P. L. Walne and F. Synder. 1976. Aliphatic chains of esterified lipids in isolated eyespots of *Euglena gracilis* var. *bacillaris*. Plant Physiol. 57: 645-646.

Hofmann, C. and G. B. Bouck. 1976. Immunological and structural evidence for patterned intussusceptive surface growth in a unicellular organism. J. Cell Biol. 69: 693-715.

Hollande, A. 1942. Étude cytologique et biologique de quelques flagellés libres. Volvocales, Cryptomonadines, Eugléniens, Protomastigines. Arch. Zool. Exp. Gen. 83: 1-268.

Holowinsky, A. W. and J. A. Schiff. 1970. Events surrounding the early development of *Euglena* chloroplasts. I. Induction by preillumination. Plant Physiol. 45: 339-397.

Holt, S. C. and A. I. Stern. 1970. The effect of 3-(3,4-dichlorophenyl)-1,1-dimethylurea on chloroplast development and maintenance in *Euglena gracilis*. I. Ultrastructural characterization of light-grown cells by the techniques of thin sectioning and freeze-etching. Plant Physiol. 45: 475-483.

Khan, A. A. and P. E. Kolattukudy. 1973. Control of synthesis and distribution of acyl moieties in etiolated *Euglena gracilis*. Biochem. 12: 1939-1948.

Kivic, P. A. and M. Vesk. 1972. Structure and function of the euglenoid eyespot. J. Exp. Bot. 23: 1070-1075.

Kivic, P. A. and M. Vesk. 1974a. An electron microscope search for plastids in bleached *Euglena gracilis* and in *Astasia longa*. Canad. J. Bot. 52: 695-699.

Kivic, P. A. and M. Vesk. 1974b. The structure of the eyespot apparatus in bleached strains of *Euglena gracilis*. Cytobiol. 10: 88-101.

Kivic, P. A. and M. Vesk. 1974c. Pinocytotic uptake of protein from the reservoir of *Euglena*. Arch. Mikrobiol. 96: 155-159.

Klebs, G. 1883. Über die Organisation einiger Flagellaten-Gruppe und ihre Beziehungen zu Algen und Infusorien. Untersuch. Bot. Inst. Tübingen 1: 233-262.

Klein, S., J. A. Schiff and A. W. Holowinsky. 1972. Events surrounding the early development of *Euglena* chloroplasts. II. Normal development of fine structure and the consequences of preillumination. Dev. Biol. 28: 253-273.

Knight, G. J. and C. A. Price. 1968. Measurements of S-ρ coordinates during the development of *Euglena* chloroplasts. Biochem. Biophys. Acta 158: 233-285.

Korn, E. D. 1964. The fatty acids of *Euglena gracilis*. J. Lipid Res. 5: 352-362.

Krinsky, N. I. and T. H. Goldsmith. 1960. The carotenoids of the flagellated alga, *Euglena gracilis*. Archiv. Biochem. Biophys. 91: 271-279.

Kronestadt, B. and E. Walles. 1975. On the presence of plastids and the eye-spot apparatus in a porfiromycin-bleached strain of *Euglena gracilis*. Protoplasma 84: 75-82.

Leedale, G. F. 1958. Nuclear structure and mitosis in the Euglenineae. Arch. Mikrobiol. 32: 32-64.

Leedale, G. F. 1964. Pellicle structure in *Euglena*. Brit. Phycol. Bull. 2: 291-306.

Leedale, G. F. 1967. *Euglenoid Flagellates*. Prentice-Hall, Inc., Engelwood Cliffs, NJ, 242 pp.

Leedale, G. F. 1968. The nucleus in *Euglena*. In Buetow, D. E. [Ed.] *The Biology of Euglena*. Academic Press, NY, pp. 185-242.

Leedale, G. F. 1970. Phylogenetic aspects of nuclear cytology in the algae. Ann. NY Acad. Sci. 175: 429-453.

Leedale, G. F. 1971. The Euglenoids. In Head, J. J. and O. E. Lowenstein [Eds.] *Oxford Biology Readers*. Oxford Univ. Press, Oxford. No. 5, 16 pp.

Leedale, G. F. 1975a. Envelope formation and structure in the euglenoid genus *Trachelomonas*. Brit. Phycol. J. 10: 17-41.

Leedale, G. F. 1975b. Preliminary observations on nuclear cytology and ultrastructure in carbon-starved streptomycin-bleached *Euglena gracilis*. Colloq. Intern. CNRS 240: 285-290.

Leedale, G. F. 1978. Phylogenetic criteria in euglenoid flagellates. Biosys. 10: 183-187.

Leedale, G. F. and D. E. Buetow. 1976. Observations on cytolysome formation and other cytological phenomena in carbon-starved *Euglena gracilis*. J. Micros. Biol. Cell. 25: 149-154.

Leedale, G. F. and D. J. Hibberd. 1974. Observations on the cytology and fine structure of the euglenoid genera *Menoidium* Perty and *Rhabdomonas* Fresenius. Arch. Protistenk. 116: 319-345.

Leedale, G. F., B. J. D. Meeuse and E. G. Pringsheim. 1965. Structure and physiology of *Euglena spirogyra*. I. and II. Arch. Mikrobiol. 50: 68-102.

Leedale, G. F. and P. L. Walne. 1971. Bibliography on Euglenophyta. In Parker, B. C. and J. R. Rosowski [Eds.] *Selected Papers in Phycology*. University of Nebraska Press, Lincoln. pp. 797-802.

Lenci, F. and G. Colombetti. 1978. Photobehavior of micro-organisms: A biophysical approach. Ann. Rev. Biophys. Bioeng. 7: 347-361.

Lyman, H. and K. Traverse. 1980. *Euglena*: Mutations, chloroplast 'bleaching' and differentiation. In Gantt, E. [Ed.] *Handbook of Phycological Methods. Developmental and Cytological*. Cambridge University Press, Cambridge. pp. 107-141.

Lyman, H., H. T. Epstein and J. A. Schiff. 1961. Studies of chloroplast development in *Euglena*. I. Inactivation of green colony formation by UV light. Biochim. Biophys. Acta. 50: 301-309.

Lynch, M. J. and D. E. Buetow. 1975. Isolation of intact nuclei from *Euglena gracilis*. Exp. Cell Res. 91: 344-348.

Lynch, M. J., R. E. Leake, K. M. O'Connell and D. E. Buetow. 1975. Isolation, fractionation and template activity of the continuously condensed chromatin of *Euglena gracilis*. Exp. Cell Res. 91: 349-357.

Malkoff, D. B. and D. E. Buetow. 1974. Ultrastructural changes during carbon starvation in *Euglena gracilis*. Exp. Cell Res. 35: 58-63.

Michaels, A. and A. Gibor. 1973. Ultrastructural changes in *Euglena* after ultraviolet irradiation. J. Cell Sci. 13: 799-809.

Mignot, J. P. 1963. Quelques particularités de l'ultrastructure d'*Entosiphon sulcatum* (Dujardin) Stein, Flagellé, Euglenien. C. R. Acad. Sci. 257: 2530-2533.

Mignot, J. P. 1966. Structure et ultrastructure de quelques Euglenomonadines. Protistologica 2: 51-117.

Mignot, J. P. 1967. Affininités des Euglénomonadines et des Chloromonadines Remarques sur la Systématique des Euglénida. Protistologica 3: 25-60.

Mignot, J. P. and R. Hovasse. 1973. Nouvelle contribution a la connaissance des trichocystes: Les organites grillagés d'*Entosiphon sulcatum* (Flagellata, Euglenida). Protistologica 9: 373-391.

Mikolyjczyk, E. and B. Diehn. 1975. The effect of potassium iodide on photophobic responses in *Euglena*: evidence for two photoreceptor pigments. Photochem. Photobiol. 22: 269-271.

Mikolyjczyk, E. and B. Diehn. 1978. Morphological alterations in *Euglena gracilis* induced by treatment with CTAB (Cetyltrimethylammonium Bromide) and Triton X-100: Correlation with effects on photophobic behavioral responses. J. Protozool. 25: 161-170.

Miller, S. and B. Diehn. 1978. Cytochrome c oxidase as the receptor molecule for chemoaccumulation (chemotaxis) of *Euglena* toward oxygen. Science 200: 548-549.

Miller, K. R. and G. J. Miller. 1978. Organization of the cell membrane in *Euglena*. Protoplasma 95: 11-24.

Moss, M. L. 1964. The phylogeny of mineralized tissues. Intl. Rev. Gen. Exp. Zool. 1: 297-331.

Moyne, G., O. Bertaux and E. Puvion. 1975. The nucleus of *Euglena*. 1. An ultracytochemical study of the nucleic acids and nucleoproteins of synchronized *Euglena gracilis* Z. J. Ultrastr. Res. 52: 362-376.

Nisbet, B. 1974. An ultrastructural study of the feeding apparatus of *Peranema trichophorum*. J. Protozool. 21: 39-48.

Ophir, I. and Y. Ben-Shaul. 1973. Separation and ultrastructure of proplastids from dark-grown *Euglena* cells. Plant Physiol. 51: 1109-1116.

Ophir, I. and Y. Ben-Shaul. 1974. Structural organization of developing chloroplasts in *Euglena*. Protoplasma 80: 109-127.

Pagni, P. G. S. 1978. Ultrastructure and Chemical Characterization of Euglenoid Stigmata. Ph.D. Dissertation, University of Tennessee, Knoxville, Tenn.

Pagni, P. G. S., P. L. Walne and S. W. Perdue. 1975. Microspectrophotometric analysis of the eyespots of *Euglena gracilis* var. *bacillaris* and *Euglena acus* (Euglenophyta). ASB Bull. 22: 71.

Pagni, P. G. S., P. L. Walne and E. Wehry. 1976. Fluorometric evidence for flavins in isolated eyespots of *Euglena gracilis* var. *bacillaris*. Photochem. Photobiol. 24: 373-375.

Palisano, J. R. and P. L. Walne. 1972. Acid phosphatase activity and ultrastructure of aged cells of *Euglena gracilis*. J. Phycol. 8: 81-88.

Palisano, J. R. and P. L. Walne. 1976. Light and electron microscopy of two permanently bleached cell lines of *Euglena gracilis* (Euglenophyceae). Nova Hedwig. 27: 455-482.

Palisano, J. R., P. L. Walne and D. H. Brown. 1980. Ultrastructure and physical characteristics of proplastids and chloroplasts isolated by zonal contrifugation from *Euglena gracilis* var. *bacillaris*. Euro. J. Cell Biol. In press.

Parenti, F., G. Brawerman, J. F. Preston and J. M. Eisenstadt. 1969. Isolation of nuclei from *Euglena gracilis*. Biochim. Biophys. Acta 195: 234-243.

Picinni, E., V. Albergoni and O. Coppellotti. 1975. ATPase activity in flagella from *Euglena gracilis*. Localization of the enzyme and effects of detergents. J. Protozool. 22: 331-335.

Pickett-Heaps, J. D. and K. Weik. 1977. Cell division in *Euglena* and *Phacus*. I. Mitosis. In Rost, L. and E. M. Gifford, Jr. [Eds.] *Mechanisms of Control of Cell Division*. Dowden, Hutchinson and Ross, Inc., Stroudsburg, PA.
Pitelka, D. R. and C. N. Schooley. 1955. Comparative morphology of some protistan flagella. Univ. Calif. Publ. Zool. 61: 79-128.
Preston, J. F., F. Parenti and J. M. Eisenstadt. 1972. Studies on the isolation and purification of chloroplasts from *Euglena gracilis*. Planta 107: 351-367.
Price, C. A. 1978. Chloroplasts from *Euglena gracilis*. In Hellebust, J. A. and J. S. Craigie [Eds.] *Handbook of Phycological Methods. Physiological and Biochemical Methods*. Cambridge Univ. Pr., Cambridge. pp. 5-13.
Price, C. A., E. N. Breden and A. Vasconcelos. 1973. Isolation of intact spinach chloroplasts in the CF-6 continuous-flow zonal rotor; implications for membrane-bound organelles. Analyt. Biochem. 54: 239-246.
Pringsheim, E. G. 1953. Observations on some species of *Trachelomonas* grown in culture. New Phytol. 52: 238-266.
Pringsheim, E. G. 1956. Contributions toward a monograph of the genus *Euglena*. Nova Acta Leopoldina 18: 1-168.
Pringsheim, E. G. and O. Pringsheim. 1952. Experimental elimination of chromatophores and eye-spot in *Euglena gracilis*. New Phyto. 51: 65-76.
Provosoli, L., S. H. Hutner and A. Schatz. 1948. Streptomycin-induced chlorophyll-less races of *Euglena*. Proc. Soc. Exp. Biol. and Med. 69: 279-282.
Reardon, E. M., M. Barltof, W. Ortiz, D. Santaro, R. Zielinski and C. A. Price. 1978. Isolation of chloroplasts active in protein synthesis from spinach and *Euglena*. In Akoyunoglou, G. and J. H. A. Akonyunoglou [Eds.] *Developments in Plant Biology. Vol. 2. Chloroplast Development*. Elsevier-North Holland Biomedical Pr., NY. pp. 277-282.
Rogers, T. D., V. E. Scholes and H. E. Schlichting. 1972. An ultrastructural comparison of *Euglena gracilis* Klebs, bleached *Euglena*, and *Astasia longa* Pringsheim. J. Protozool. 19: 133-139.
Rosenberg, A. 1963. A comparison of lipid patterns in photosynthesizing cells of *Euglena gracilis*. Biochem. 2: 1148-1154.
Rosenberg, A. 1967. *Euglena gracilis*: A novel lipid energy reserve and arachidonic acid enrichment during fasting. Science 157: 1189-1191.
Rosowski, J. R. and W. V. Glider. 1977. Comparative effects of metal coating by sputtering and by vacuum evaporation on euglenoid flagellates. Scanning Electr. Micros. II: 471-480.
Rosowski, J. R. and P. Kugrens. 1973. Observations on the euglenoid *Colacium* with special reference to the formation and morphology of attachment material. J. Phycol. 9: 370-383.
Rosowski, J. R. and K. W. Lee. 1978. *Cryptoglena pigra*: A euglenoid with one chloroplast. J. Phycol. 14: 160-166.
Rosowski, J. R. and R. L. Willey. 1975. *Colacium libellae* sp. nov. (Euglenophyceae), a photosynthetic inhabitant of the larval damselfly rectum. J. Phycol. 11: 310-315.
Rosowski, J. R. and R. L. Willey. 1977. Development of mucilaginous surfaces in euglenoids. I. Stalk morphology of *Colacium mucronatum*. J. Phycol. 13: 16-21.
Rosowski, J. R., R. L. Vadas and P. Kugrens. 1975a. Surface configuration of the lorica of the euglenoid *Trachelomonas* as revealed with scanning electron microscopy. Amer. J. Bot. 62: 48-57.
Rosowski, J. R., P. L. Walne and L. K. West. 1975b. Comparative effects of critical point and air-drying on the morphology of the rigid mucilaginous coating (lorica) of *Trachelomonas* (Euglenophyceae). Micron 5: 321-339.
Roth, L. E. 1959. An electron-microscope study of the cytology of the protozoan *Peranema trichophorum*. J. Protozool. 6: 107-116.

Salisbury, J. L., A. C. Vasconcelos and G. L. Floyd. 1975. Isolation of intact chloroplasts of *Euglena gracilis* by isopycnic sedimentation in gradients of silica. Plant Physiol. 56: 399-403.

Schiff, J. A. 1971. Developmental interaction along cellular compartments in *Euglena*. In Boardman, N. K., A. W. Lonnane and R. M. Smillie [Eds.] *Autonomy and Biogenesis of Mitochondria and Chloroplasts*. North Holland, Amsterdam. pp. 98-118.

Schiff, J. A. 1973. The development, inheritance and origin of the plastid in *Euglena*. In Abercrombie, M. and J. Brachet [Eds.] *Advances in Morphogenesis*, Vol. 10. Academic Pr., NY. pp. 265-309.

Schiff, J. A. 1978. Photocontrol of chloroplast development in *Euglena*. In Akoyunoglou, G. and J. H. Argyroudi-Akoyunoglou [Eds.] *Chloroplast Development*. Elsevier-North Holland Biomedical Pr., NY. 888 pp.

Schiff, J. A. and H. T. Epstein. 1968. The continuity of the chloroplast in *Euglena*. In Buetow, D. E. [Ed.] *The Biology of Euglena. Vol II. Biochemistry.* Academic Press, NY. pp. 285-333.

Schiff, J. A., H. Lyman and H. T. Epstein. 1961. Studies of chloroplast development in *Euglena*. II. Photoreversal of the U.V. inhibition of green colony formation. Biochim. Biophys. Acta 50: 310-318.

Schiff, J. A., H. Lyman and G. K. Russell. 1971. Isolation of mutants from *Euglena gracilis*. In San Pietro, A. [Ed.] *Methods of Enzymology, Vol. 23A*. Academic Press, NY. pp. 143-162.

Schmidt, G. W. and H. Lyman. 1976. Inheritance and synthesis of chloroplasts and mitochondria of *Euglena gracilis*. In Lewin, R. [Ed.] *The Genetics of Algae*. Blackwell Scientific, Oxford. pp. 257-299.

Schwelitz, F., W. R. Evans and H. H. Mollenhauer. 1970. The fine structure of the pellicle of *Euglena gracilis* as revealed by freeze-etching. Protoplasma 69: 341-349.

Siegesmund, K. A., W. G. Rosen and S. R. Gawlik. 1962. Effects of darkness and of streptomycin on the fine structure of *Euglena gracilis*. Amer. J. Bot. 49: 137-145.

Singh, K. P. 1956. Studies in the genus *Trachelomonas*. I., II. Amer. J. Bot. 43: 258-266, 274-280.

Sommer, J. R. and J. J. Blum. 1965. Cytochemical localization of acid phosphatase in *Euglena gracilis*. J. Cell Biol. 24: 235-251.

Song, P. S. and T. A. Moore. 1974. On the photoreceptor pigment for phototropism and phototaxis: Is a carotenoid the most likely candidate? Photochem. Photobiol. 19: 435-441.

Sperling, P. G., P. L. Walne and L. L. Triplett. 1974. Preliminary characterization of non-lipid components from isolated eyespots of euglenoid flagellates. J. Phycol. 10 (Suppl.): 15.

Stern, A. I., J. A. Schiff and H. T. Epstein. 1964. Studies of chloroplast development in *Euglena*. V. Pigment biosynthesis, photosynthetic oxygen evolution and carbon dioxide fixation during chloroplast development. Plant Physiol. 39: 220-226.

Stewart, K. D. and K. R. Mattox. 1975. Comparative cytology, evolution, and classification of the green algae with some consideration of the origin of other organisms with chlorophylls *a* and *b*. Bot. Rev. 41: 104-135.

Thomas, D. M., T. W. Goodwin and J. F. Ryley. 1967. Nature and distribution of carotenoid pigments in *Astasia ocellata*. J. Protozool. 14: 654-657.

Tollin, G. and M. I. Robinson. 1969. Phototaxis in *Euglena*. V. Photosuppression of phototactic activity by blue light. Photochem. Photobiol. 9: 411-416.

Triemer, R. E. 1980. Role of the Golgi apparatus in mucilage production and cyst formation in *Euglena gracilis* (Euglenophyta). J. Phycol. In Press.

Ueda, K. 1960. Structure of plant cells with special reference to lower plants. IV. Structure of *Trachelomonas* sp. Cytologia 25: 8-16.

Vasconcelos, A. C. 1976. Synthesis of proteins by isolated *Euglena gracilis* chloroplasts. Plant Physiol. 58: 719-721.

Vasconcelos, A. C., L. R. Mendiola-Morgenthaler, G. L. Floyd and J. L. Salisbury. 1976. Fractionation and analysis of polypeptides of *Euglena gracilis* chloroplasts. Plant Physiol. 58: 87-90.

Vasconcelos, A. C., M. Pollack, L. R. Mendiola, H. P. Hoffmann, D. H. Brown and C. A. Price. 1971. Isolation of intact chloroplasts from *Euglena gracilis* by zonal centrifugation. Plant Physiol. 47: 217-221.

Walne, P. L. 1971. Comparative ultrastructure of eyespots in selected euglenoid flagellates. In Parker, B. C. and R. M. Brown, Jr. [Eds.] *Contributions in Phycology*. Allen Press, Lawrence, Kans. pp. 107-120.

Walne, P. L. and H. J. Arnott. 1967. The comparative ultrastructure and possible function of eyespots: *Euglena granulata* and *Chlamydomonas eugametos*. Planta 77: 325-354.

Walne, P. L. and J. P. Breillatt. 1974. Isolation of *Euglena gracilis* eyespots by zonal flotation with continuous fraction collection. J. Phycol. 10 (Suppl.): 16.

Walne, P. L. and C. A. Lembi. 1969. Ultrastructural and time-lapse studies of *Trachelomonas*, *Lepocinclis* and *Euglena* with special reference to eyespots. ASB Bull. 16: 71.

West, L. K. 1977. Cellular organization and biomineralization of the envelope of *Trachelomonas hispida* var. *coronata*. Ph.D. Dissertation. University of Tennessee, Knoxville. 245 pp.

West, L. K. and P. L. Walne. 1973. Light and electron microscope studies of the euglenoid flagellate, *Trachelomonas*. ASB Bull. 20: 90.

West, L. K. and P. L. Walne. 1977. Manganese biomineralization in cell envelopes of *Trachelomonas* (Euglenophyceae). J. Cell Biol. 75: 409a.

West, L. K. and P. L. Walne. *Trachelomonas hispida* var. *coronata* (Euglenophyceae): II. Envelope substructure. J. Phycol. Submitted.

West, L. K., P. L. Walne and J. Bentley. 1978. Substructure and composition of cell envelopes (loricae) of *Trachelomonas hispida* var. *coronata*. ASB Bull. 25: 94.

West, L. K., P. L. Walne and J. Bentley. *Trachelomonas hispida* var. *coronata* (Euglenophyceae): III. Envelope elemental composition and mineralization. J. Phycol. Submitted.

West, L. K., P. L. Walne and J. R. Rosowski. *Trachelomonas hispida* var. *coronata* (Euglenophyceae): I. Ultrastructure of cytoskeletal and flagellar systems. J. Phycol. Submitted.

West, L. K., P. L. Walne, R. W. Pearman, J. C. Pendland, H. C. Aldrich, J. R. Rosowski and P. Kugrens. 1973. Ultrastructural observations on vegetative cells of *Trachelomonas*. J. Phycol. 9 (Suppl.): 17-18.

Willey, R. L., K. Ward, W. Russin and R. Wibel. 1977. Histochemical studies of the extracellular carbohydrate of *Colacium mucronatum* (Euglenophyceae). J. Phycol. 13: 349-353.

Wolken, J. J. 1967. *Euglena*. Appleton-Century-Crofts, NY. 204 pp.

Wolken, J. J. 1977. *Euglena*: The photoreceptor system for phototaxis. J. Protozool. 24: 518-522.

Copyright 1980 by Elsevier North Holland, Inc.
Cox, ed. Phytoflagellates

CHRYSOPHYTES

RICHARD N. PIENAAR
Department of Botany, University of Natal, P.O. Box 375, Pietermaritzburg 3200, Natal, South Africa.

INTRODUCTION

In 1914 Pascher erected the class Chrysophyceae to include a group of golden brown flagellates which Klebs (1892) had placed in the family Chrysomonadina. Since this date the class has been revised on a number of occasions. There are now excellent reviews that provide an up-to-date account of the current status of the class Chrysophyceae (Bourrelly, 1957, 1965, 1968; Hibberd, 1976; Bold and Wynne, 1978).

In the light of recent work it is generally accepted that some of the original orders grouped in the Chrysophyceae should be placed in other classes of the algae; e.g., the Isochrysidales now belong in the Prymnesiophyceae and the Craspedomonadales have been removed as they are no longer regarded as plants (Hibberd, 1976).

In this contribution the classification system will be the one followed by Bold and Wynne (1978) for the class Chrysophyceae, with the exception that information on the order Dictyochales (silicoflagellates) will be omitted as this group will be dealt with elsewhere in this book.

Bourrelly (1968) subdivided the Chrysophyceae into three series. Essentially these three series are based on whether the cells are non-flagellate or flagellate. If the cells are flagellate, then the number of flagella is of taxonomic importance. The three series are as follows:

1. Flagellate series - containing flagellate cells possessing two flagella of unequal length which exhibit heterodynamic movement. In this series the biflagellate condition may be predominant in the life history of the alga or it may be transitory, found only during the reproductive phase.
2. Uniflagellate series - containing representatives that possess motile cells with only one emergent flagellum of the pleuronematic type. This series should perhaps be re-evaluated because there is now evidence indicating that many of the genera which were once thought to possess a single flagellum possess a small rudimentary or vestigial second flagellum which can only be seen with the electron microscope.
3. Non-flagellate series - containing a number of genera which to date

have not been found to have flagellated cells. Reproduction in this group is thought to occur by means of amoeboid cells.

The following amended scheme of classification of the Chrysophyceae has been taken from Bold and Wynne (1978):

Biflagellate Series

Order Ochromonadales - monads, colonies or palmelloid masses.

Order Chrysapiales - coccoid, solitary or colonial.

Order Phaeothamniales - filamentous or thalloid.

Uniflagellate Series

Order Chromulinales - monads or colonies, rhizopodial, palmelloid or plasmodial aggregations.

Order Chrysophaerales - coccoid forms either solitary or colonial.

Order Thallochrysidales - filamentous or thalloid forms.

Order Dictyochales - monads with a siliceous skeleton surrounded by an amoeboid type protoplast.

Non-flagellate Series

Order Chrysococcales - palmelloid forms, the naked cells being embedded in mucilaginous envelopes.

Order Rhizochrysidales - rhizopodial forms, solitary or colonial, naked or loricate.

Order Stichogloeales - coccoid forms with a cell wall, solitary or colonial.

Order Phaeoplacales - filamentous or thalloid forms with a cell wall

GENERAL CHARACTERISTICS OF THE CHRYSOPHYCEAE

There is considerable variation in morphology within the Chrysophyceae. The range varies from the simple unicellular flagellate cells of the *Ochromonas* type through the colonial motile colonies (e.g. *Synura* Ehrenberg) to the thalloid and filamentous representatives (e.g. *Phaeoplaca* Chodat and *Chrysonema* Anand.). In this chapter more attention will be focused on the motile organisms and those representatives that possess motile stages as part of their life history.

It is now generally accepted by most phycologists that the basic organisation of most of the Chrysophyceae revolves around that found in the well-studied genus *Ochromonas* Wyssotzki. Bourrelly (1957, 1968) was the first to suggest that *Ochromonas* was probably the common ancestor of the class. His work was based almost entirely on information obtained with the light microscope. More recently, however, Hibberd (1976) has produced an excellent account of the

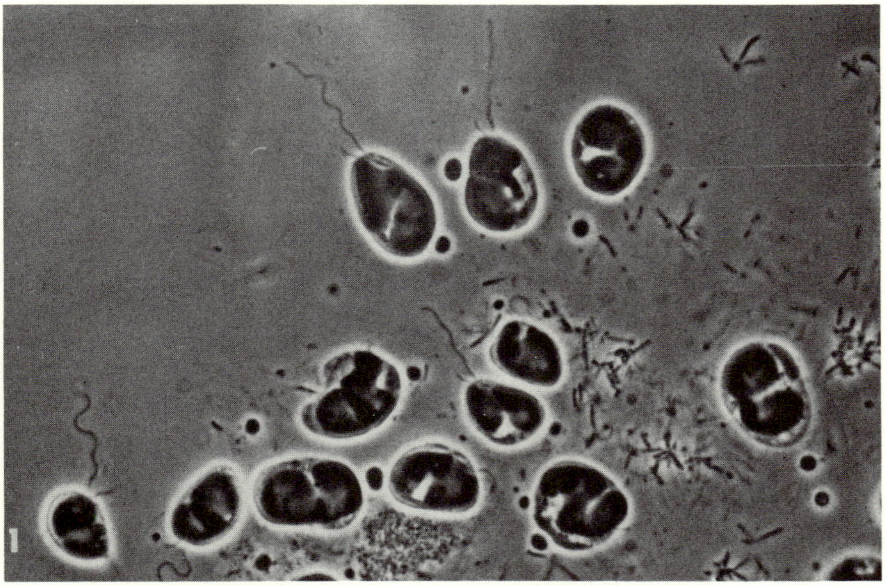

Fig. 1. Light micrograph of cells of *Ochromonas* sp. Note the well defined chloroplasts and the two heterodynamic flagella (x1000).

basic characteristics of the Chrysophyceae as seen with the electron microscope (Figs. 1, 2). He is also of the opinion that the *Ochromonas* type cell can be regarded as the basic type of cell within the class.

The motile representatives in the class usually have two laterally inserted unequal heterodynamic flagella. The two flagella are normally inserted at an oblique angle to each other. In a number of genera only one emergent flagellum can be distinguished with the light microscope (e.g. *Chromulina* Cienkowsky), but recent ultrastructural studies have revealed a reduced or vestidial second flagellum. Other cells (e.g. *Ochromonas*) exhibit two unequal flagella, while others (e.g. *Rhizochromulina* Hibberd and Chretiennot-Dinet) possess motile cells that are uniflagellate and do not possess a vestidial second flagellum (Hibberd and Chretiennot-Dinet, 1979). In those cells possessing two emergent flagella the larger of the two flagella is of the tinsel variety and possesses two rows of mastigonemes. The longer flagellum is usually directed forward when swimming and beats in an approximate sine wave. The shorter flagellum

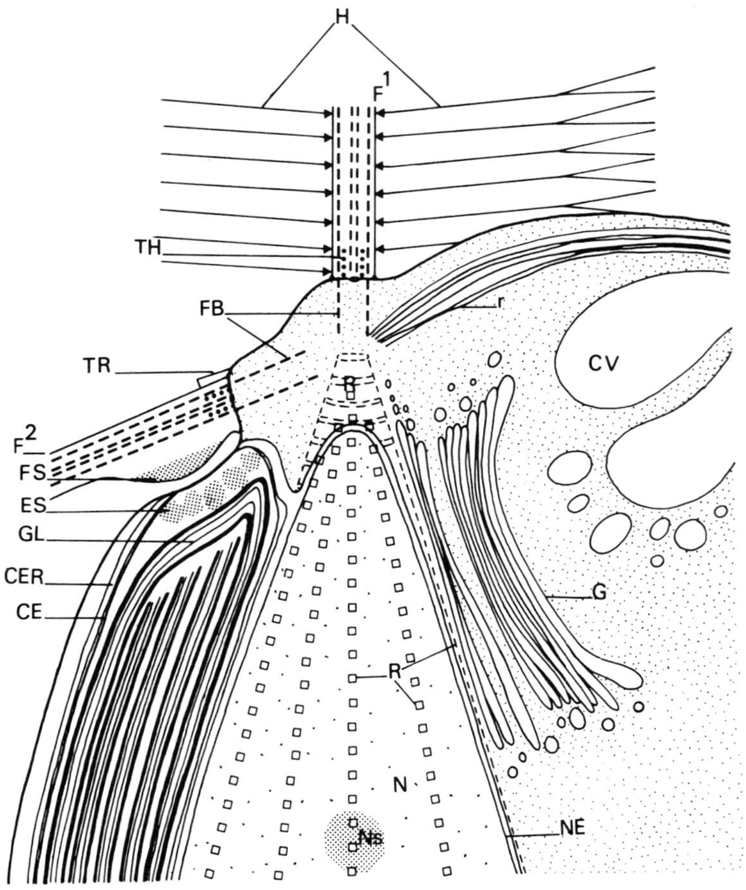

Fig. 2. Basic organization of the Chrysophyceae. Diagrammatic representation of the anterior end of the basic *Ochromonas* type of cell. CE, chloroplast envelope; CER, chloroplast endoplasmic reticulum; CV, contractile vacuole; ES, eyespot; F_1, anteriorly directed hairy flagellum; F_2, laterally directed smooth flagellum; FB, flagellar basal bodies; FS, flagellar swelling; G, Golgi body; GL, girdle lamella; H, flagellar hairs; N, nucleus; NS, nucleolus; NE, nuclear envelope; R, rhizoplast; r, microtubular flagellar root; TH, transitional helix; TR, transitional region. (Redrawn from Hibberd, 1976.)

lacks mastigonemes and when it is emergent it is directed towards the posterior of the cell. Some representatives have been demonstrated to possess a covering of scales over both the flagella (e.g. *Sphalaeromantis* Skuja).

A number of representatives of the class possess elaborate scale coverings over the cell surface (e.g. *Mallomonas* Perty, *Synura*, *Chrysosphaerella* Lauterborn, *Syncrypta* Ehrenberg) and some possess scale covering over both the cell and flagellar surface (e.g. *Sphalaeromantis* Pascher and *Synura*). Another form of cell covering is the lorica found in such genera as *Dinobryon* Ehrenberg, *Epipyxis* Ehrenberg and *Pseudokephyrion* Pascher.

Most members of the Chrysophyceae that have been studied with respect to their pigmentation have been demonstrated to possess chlorophyll *a* and some members possess chlorophyll *c* ($c_1 + c_2$). Chlorophyll *b* is entirely absent from this class (Ricketts, 1965; Meeks, 1974). Some controversy still exists about the presence of chlorophyll *c* in that some Chrysophyceae reported to contain chlorophyll *c* are now classified in the Prymnesiophyceae (Ricketts, 1965; Jeffrey, 1969). There is, however, one report of *Synura* possessing chlorophyll *c* (Guillard and Lorensen, 1972).

The most characteristic xanthophyll in the Chrysophyceae is fucoxanthin. Allen *et al.* (1960) have demonstrated that it represents 75% of the total pigment in *Ochromonas danica*. Other xanthophylls found in cells are diatoxanthin in *Sphalaeromantis* and *Ochromonas* and diadinoxanthin in *Sphalaeromantis* (Jeffrey, 1961). β carotene appears to be present in all members of the class. In some species trace amounts of α and γ carotenes can be found (Dales, 1960; Goodwin, 1974). Not all members of the Chrysophyceae are pigmented. There are a considerable number of genera that have colourless representatives which exhibit a heterotrophic form of nutrition (e.g. *Syncrypta* and *Anthophysa* Bory de Saint-Vincent).

Chrysophyceae store fat and chrysolaminarin, a β-1-3-linked glucan (Craigie, 1974). In *Ochromonas*, Krauss (1962) found that the chrysolaminarin became strongly radioactive during photosynthesis in $^{14}CO_2$ and it was deduced that chrysolaminarin was an assimilatory product.

Another characteristic of the Chrysophyceae is the ability of many of the cells to produce siliceous cysts termed statospores. Statospores are usually characterised by possessing an apical pore and a plug. In many instances the statospore has spines or other forms of ornamentation. These statospores are regarded as stages in the life history of members of this class.

ULTRASTRUCTURE OF DISTINCTIVE ORGANELLES

The Chrysophyceae have a number of organelles whose ultrastructure can be of some importance in delineating the chrysophytes from other phytoflagellates. Hibberd's review (1976) on the ultrastructural aspects of the chrysophycean type of cellular organisation has been used extensively for information in this section.

Cell covering and appendages. Many chrysophycean cells possess naked protoplasts while others have cell coverings of either scales, loricas or cell walls. The naked type of protoplast is simply bounded by a plasma membrane. Recent work on *Ochromonas danica* Pringsheim has revealed that the cell surface is unusually flexible and produces projections of varying shapes and sizes - large projections 340-360 nm and small projections 50-110 nm (Kahan *et al.*, 1978). Some of these projections were released into the medium while others were retained on the cell surface. Ruthenium red stained a layer outside the cell surface indicating that there was a layer of mucopolysaccharide external to the cell membrane. It is thought that this material is Golgi derived.

A recently isolated marine species of *Chromulina* has been found to possess small 'knob-like' projections arising from the plasma membrane (Pienaar, 1977). These structures bear superficial resemblance to the projections occurring on the cell surface of *Pavlova* Butcher, a member of the Prymnesiophyceae (Green and Manton, 1970).

Over the past decade a number of Chrysophycean genera and species have been described which possess an elaborate body covering made up of scales of various forms of architecture. *Mallomonas* and *Mallomonopsis* Matvienko are unicellular representatives of the order Ochromonadales that possess silicified scales and bristles. Many workers have regarded the scales as species specific and they are used extensively in identification. Harris (1970) has, however, drawn attention to the fact that under certain conditions *Mallomonas* produces "imperfect" scales. Such forms are found when populations are exhibiting a period of rapid cell division or "swarming". It is, therefore, important that considerable caution be exercised when relying entirely on scale architecture for the description of new species of these genera. *Synura*, the colonial flagellate, has also been found to possess silicified scales which cover the individual cells (Takahashi, 1972, 1975, 1978; Peterfi and Momeu, 1977). Other genera that have been reported to have a cell covering made up of scales include *Sphalaeromantis* (Harris, 1963; Manton and Harris, 1966; Pienaar, 1976), *Paraphysomonas* de Saedeleer (Manton and Leedale, 1961), *Phaeaster pascheri* Scherff. (Belcher and Swale, 1971), *Chrysosphaerella* (Bradley, 1964; Kristiansen, 1969a; Asmund, 1973) and the colourless flagellate *Syncrypta glomerifera* Clarke and Pennick (Clarke and Pennick, 1975). No work on the chemical composition of the scales in these last mentioned genera has been done. It is generally accepted that these scales are purely organic and are not of the silicified variety. It would, therefore, appear that the occurrence of body scales is common in a number of taxa of the Chrysophyceae from different orders.

The Golgi system appears to be of importance in the production of scales in some genera, e.g. *Sphalaeromantis* (Manton and Harris, 1966; Pienaar, 1976) and *Paraphysomonas* (Manton and Leedale, 1961). However, in other genera the production of the scales is not directly attributed to the dictyosomal cisternae. Schnepf and Deichraber (1969) have demonstrated that the silicified scales of *Synura petersenii* Korsh. are produced in vesicles that lie adjacent to the periplastidial reticulum. The shape of the scale vesicle and hence the shape of the scale is determined by a complicated process of membrane morphogenesis. The deposition of silica occurs intracellularly and the silicified scale is eventually released to the exterior of the cell. More recently Wujek and Kristiansen (1978) have demonstrated that the silica scales and bristles of *Mallomonas caudata* Ivanow emend. Krieger takes place in vesicles of the Golgi apparatus. Both the scale and the bristly initials are flat and a folding of varying degrees occurs during their formation. Scales and bristles are produced independently of each other and do not develop at the same time. It is not yet known how they are released, linked together, or arranged on the cell in a definite pattern.

Some chrysophycean cells have been shown to possess several appendages projecting from the cell surface. These include the problematic members of the family Pedinellaceae, e.g. *Pedinella hexacostata* Vysotskii and *Pseudopedinella ambisua* Bourrelly (Swale, 1969) and *Pseudopedinella pyriformis* Carter (Ostroff and Van Valkenburg, 1978). The cells of these species are characterised by unusual projections or tentacles - made up of fibrils and muciferous bodies which occur as a ring around the flagellar pole, and a well defined peduncle made up of microtubules among which are a number of circular profiles (Swale, 1969). Other chrysophycean cells possess fine filipodia or rhizopodia which radiate out from the cell, e.g. *Chrysamoeba radians* Klebs (Hibberd, 1971) and *Rhizochromulina marina* (Hibberd and Chretiennot-Dinet, 1979).

Certain Chrysophyceae are characterised by walls or loricas which can be of cellulosic-pectic nature, with or without a siliceous impregnation (Peterfi, 1969). In some genera such as *Chrysococcus* Klebs and *Pseudokephyrion* (Bourrelly, 1957, 1968; Belcher, 1969a), the loricas have been reported to be calcified while in *Pseudokephyrion pseudospirale* Bourrelly the fibrillar lorica is impregnated with iron and manganese (Belcher, 1968).

Franke and Heath (1973) are of the opinion that during the formation of the lorica in *Dinobryon sertularia* Ehrenberg new material is added to the lorica in the equatorial region by secretion in the plasma membrane attachment zones and is symmetrically distributed through rotational movements of the algal body which is basally anchored to the lorica bottom by a stalk. Similar hypotheses have been suggested by other workers (e.g. Karim and Round, 1967; Wujek, 1969; Kristiansen, 1969b). It has also been suggested that the fibril pattern of the lorica is species specific (Karim and Round, 1967; Kristiansen, 1969b, 1972).

Flagellar apparatus: Most flagellated cells possess two laterally inserted heterodynamic flagella; although there are examples of flagellate chrysophycean cells possessing only a single flagellum, e.g. zoospore of *Rhizochromulina marina* (Hibberd and Chretiennot-Dinet, 1979). The larger of the two flagella is often about the same length as the cell and is directed in a forward position when swimming, beating in an approximate sine wave (Hibberd, 1976). The longer flagellum is characterised by mastigonemes usually arranged in two unbalanced rows situated on opposite sides of the flagellum.

The mastigonemes have been studied in detail in *Ochromonas danica* (Bouck, 1971, 1972). According to Bouck (1971) each tubular mastigoneme consists of a 250 nm basal attachment region, a 1 μm microtubular-like shaft, a few relatively long (0.5 μm) terminal filaments, numerous short (40 nm) lateral filaments, and longer (200 nm) lateral filaments. A reconstruction of the mastigoneme is given in Fig. 3.

The production of the mastigonemes has been shown to be intracellular, usually associated with the perinuclear continuum (Leedale, Leadbeater and Massalski, 1970; Bouck, 1971). The mastigonemes are eventually released at the base of the flagellum. They are attached to the flagellar surface at a small disc-like region thought to be a modification of the flagellar membrane. The composition of the mastigonemes is probably a single major polypeptide and a number of carbohydrates.

Pedinella Wyssotzki has a single rather unusual emergent flagellum possessing one row of rather stiff hairs spiralling around the flagellum (Swale, 1969). This genus also possesses a winged flagellum (Swale, 1969).

The shorter flagellum when present is usually described as the smooth variety. However, Bouck (1971) has also observed very fine hairs on the supposedly smooth posterior flagellum of *Ochromonas* and Dodge (1973) has suggested that there may be no such thing as truly smooth flagella. Although this "smooth" flagellum may possess fine hairs, it lacks the typical mastigonemes found on the long flagellum. Associated with the short flagellum is a region containing electron dense material commonly referred to as the "flagellar swelling." This

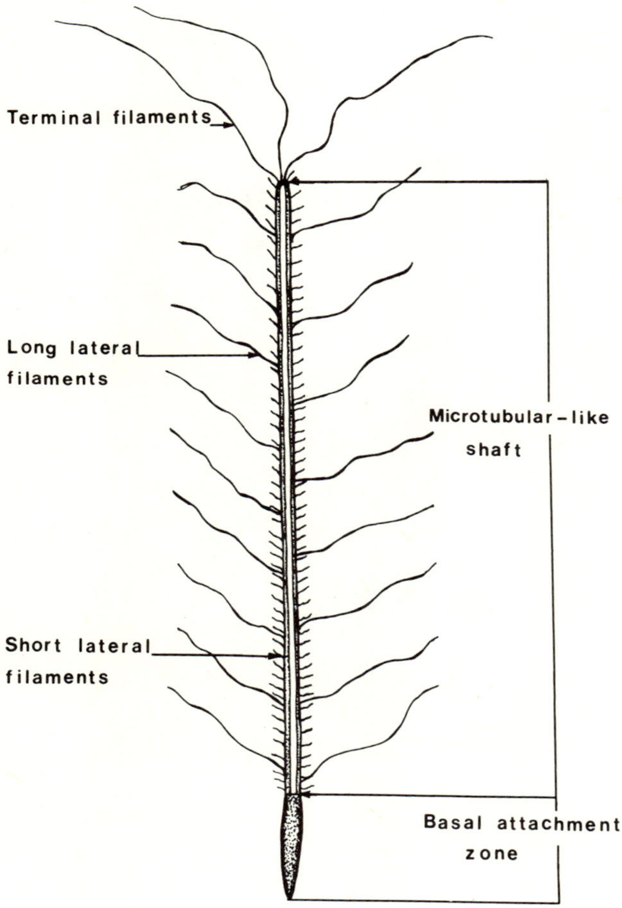

Fig. 3. Reconstruction of a mastigoneme of *Ochromonas danica* (after Bouck, 1971).

short flagellum is often found situated in a groove or small depression in the cell. There is also evidence of some fibrous material in the groove which may hold the flagellum within the groove. The swelling on the short flagellum is directed towards the plasma membrane, under which lies the eyespot associated with the chloroplast. This is illustrated in Figure 4a and 4b.

Both of the flagella possess transition zones which have a helix of dense material (referred to as the "transitional helix" just distal to the transverse portion which demarcates the flagellar shaft from the

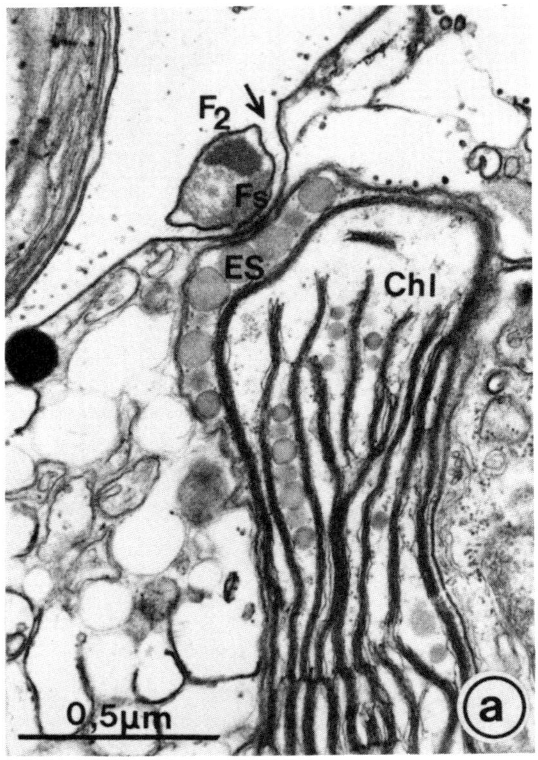

Fig. 4a. Anterior region of *Chromulina* sp. Transverse section through the short flagellum (F_2). Note the flagellar swelling (Fs); flagellar groove (arrow); the eyespot (ES) and chloroplast (Chl).

basal body (Hibberd, 1976).

A number of genera have been demonstrated to possess flagellar scales of varying architecture which cover the flagellar surfaces. In *Sphalaeromantis tetragona* (Harris, 1963; Manton and Harris, 1966) and *S. salina* Pienaar (Pienaar, 1976), two types of scales are found on both the flagellar surfaces. Hibberd (1973) has demonstrated flagellar scales on several species of *Synura*. In *S. echinulata* Korshikov, *S. sphagnicola* Korshikov (Korshikov), *S. spinosa* Korshikov, *S. uvella* Ehrenberg emend. Korshikov, *Mallomonopsis ouradion* (Harris and Bradley) Harris (Hibberd, 1973), *M. paxillata* (Bradley, 1966) and *Mallomonas striata* Asmund (Bradley, 1966), flagellar scales have been reported on the long flagellum. In *S. petersenii*, flagellar scales occur on both the short and the long flagellum (Hibberd, 1973). The architecture of the scales is

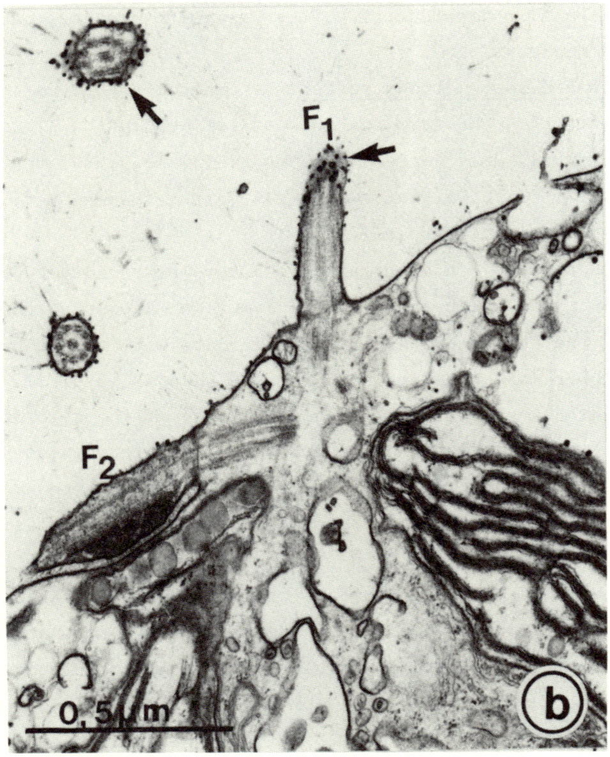

Fig. 4b. Anterior region of *Chromulina* sp. Longitudinal section through the anterior region of *Chromulina* sp. Note the long flagellum (F_1) with small knob-like projections (arrow); short flagellum (F_2).

quite variable. Pienaar (1977) has reported a marine species of *Chromulina* which has a series of small dense particles over the entire surface of the long tinsel flagellum. These structures are similar to the characteristic "knobs" found on the flagella of *Pavlova* (Green and Manton, 1970).

To date there is very little evidence as to the site of flagellar scale production. The flagellar scales of *Sphalaeromantis* may be produced in the dictyosome. Hibberd (1973) found that the flagellar scales in certain *Synura* and *Mallomonopsis* species accumulated and were probably produced in a large vesicle similar to the flagellar scale reservoir found in the genus *Pyramimonas* (Prasinophyceae).

The Chrysophyceae possess a rather elaborate flagellar root system. Hibberd (1976) is of the opinion that there are two types of flagellar roots; one consisting of small numbers of microtubules which run from near the basal bodies

JOHN RYLANDS
UNIVERSITY
LIBRARY OF
MANCHESTER

under the cell surface and a larger cross banded root or rhizoplast which extends from the basal bodies deep into the cell with branches ramifying over the surface of the nucleus.

Chloroplasts and associated structures: Generally there are one or two chloroplasts per cell. However, in the case of *Apedinella* (Throndsen) Throndsen (Throndsen, 1971) and *Pedinella* Wyssotzki (Swale, 1969) there are six chloroplasts per cell. In *Olisthodiscus luteus* Carter (Leadbeater, 1969) there are between 7-25 chloroplasts per cell.

In the Chrysophyceae, chloroplasts are always surrounded by a double chloroplast envelope inside which are arranged lamellae made up of three thylakoids. All members of the Chrysophyceae examined to date with the exception of *Rhizochromulina* (Hibberd and Chretiennot-Dinet, 1979) possess girdle lamellae. These are lamellae which are continuous around the edge of the chloroplasts and usually lie parallel to the chloroplast envelope. The chloroplasts are completely surrounded by a layer of endoplasmic reticulum which is often connected to the nuclear envelope. Frequently the outer surface of the endoplasmic reticulum is covered with ribosomes.

In some members of the Chrysophyceae there is a proliferation of the inner membrane of the chloroplast endoplasmic reticulum to form what has been termed the periplastidial reticulum. It is particularly noticeable over the region where pyrenoids are located and on the inner face of the chloroplast (Dodge, 1973; Hibberd, 1976).

The chloroplasts of Chrysophyceae have been found to possess chloroplast ribosomes, which in the case of *Ochromonas* (Gibbs, 1968) are 17-20 nm in size while cytoplasmic ribosomes are 21-23 nm in diameter. Chloroplast DNA is also present and in *Ochromonas* the DNA always seems to be located in a ring around the periphery of the chloroplast but enclosed in the girdle lamellae (Gibbs, 1968; Bisulputra, 1974).

Associated with the chloroplasts in the motile members of the Chrysophyceae is the eyespot or stigma which can easily be distinguished with the light microscope. The stigma is located towards the edge of the chloroplasts and is closely associated with the flagellar swelling on the shorter flagellum (Fig. 4a). The eyespot is made up of a number of osmiophyllic globules packed between the chloroplast envelope and the thylakoids. There is a close association between the plasma membrane and the chloroplast envelope in the region of the eyespot.

In *Sphalaeromantis tetragona* Skuja, Manton and Harris (1966) found a distinctly paracrystalline arrangement in the wedge-shaped flagellar swelling on the surface nearest to the eyespot. It is generally accepted that the flagellar swelling and the eyespot constitute the photoreceptor apparatus in the Chrysophyceae.

Kristiansen and Walne (1976, 1977) have demonstrated a structural connection between the flagellar base and stigma in *Dinobryon*. These workers have suggested that it might serve in impulse transfer between the two organelles or in mechanical support or possibly in both.

In *Chromulina psammobia* Fauré-Frémiet (Rouiller and Fauré-Frémiet, 1958) and *C. placentula* (Belcher and Swale, 1967), the eyespot is not situated in a superficial position, but is slightly embedded within the cell. Eyespots have not been detected in *Chrysamoeba radians* Klebs (Hibberd, 1971), *Mallomonas papillosa* Harris and Bradley (Belcher, 1969b), *Chrysococcus cordiformis* Naumann (Belcher and Swale, 1972b) and *Rhizochromulina marina* (Hibberd and Chretiennot-Dinet, 1979). Belcher and Swale (1972a) found a flagellar swelling and a simple eyespot located in a leucoplast in the non-pigmented genus *Anthophysa*.

Many members of the Chrysophyceae possess pyrenoids and at present they are usually of the immersed or semi-immersed variety (Hibberd, 1976). In most of the genera studied the pyrenoid is free of traversing thylakoids. According to Dodge (1973) members of the Chrysophyceae have been found to possess the following types of pyrenoids: internal simple pyrenoids with no lamellae, internal compound pyrenoids with random and parallel lamellae, and an invaginated pyrenoid in a non-starchy chloroplast. Stalked pyrenoids have not been reported in the Chrysophyceae. *Chrysococcus cordiformis* has unusual "pyrenoid areas" which resemble pyrenoids in that they have a denser matrix than the chloroplast stigma. They are found at the extremities of the chloroplasts and are surrounded by girdle lamellae and traversed by single, sinuously convoluted thylakoids (Belcher and Swale, 1972b).

Nucleus and Golgi body: The association between these two organelles appears to be reasonably constant in most of the Chrysophyceae studied to date. The single dictyosome making up the Golgi body lies against the nucleus towards the anterior end of the cell (Cole and Wynne, 1973; Hibberd, 1976). Cole and Wynne (1973) have reported a close association between the nuclear envelope and the forming face of the dictyosome. One edge of the dictyosome is directed towards the flagellar base. In the majority of chrysophytes the nucleus is pyriform and its narrow end is extended in the direction of the flagellar bases. There is a close association between the nuclear envelope and the chloroplast envelope.

Other Organelles: In *Ochromonas tuberculata* Hibberd (1970) described an unusual ejectile organelle, the "discobolocyst." These organelles have not been reported in other Chrysophyceae. They are located around the periphery of the cell and in the undischarged state are ovoid and protrude slightly above the surface of the cell. The discobolocyst is filled with fine fibrillar material thought to be a mucopolysaccharide and possesses a distal portion made up of an electron dense ring. When this organelle is discharged from the cell the dense ring is shot some distance from the cell leaving behind it a delicate cylinder of fibrous material. These unusual organelles are thought to be derived from the dictyosomal vesicles.

MOVEMENT PATTERNS

The majority of motile cells exhibit a positive phototactic response due to the presence of a photoreceptor. Most of the motile members move in a forward direction with a slight rotating movement. The role of the mastigonemes on the longer flagellum in the movement of the cells has been investigated by a number of workers (Jahn, Landman and Fonsea, 1964; Jahn, 1965; Bouck, 1971). Bouck (1971) is of the opinion that the distally propagated waves of the *Ochromonas* flagellum pull rather than push the cell through the medium. By current interpretations mastigonemes can perform this function passively provided they are relatively stiff, bend in some preferred direction, and are properly orientated relative to the planar undulations of the flagellum (Jahn *et al.*, 1964). For such a theory to be correct the mastigonemes must be arranged in two opposing rows in the same plane as the undulatory plane of the flagellum (Holwill and Sleigh, 1967). We know that the mastigonemes in *Ochromonas* are in fact attached to the flagellum in two opposing rows, but it is still not clear whether these rows are in the necessary flagellar plane.

Pedinella cells have been found to swim strongly and rotate rapidly on their axes with frequent changes in direction (Swale, 1969). The cells tend to move in circles or arcs. They have also been observed to swim for short periods and then attach themselves by the distal end of the peduncle. *Pedinella* cells moving slowly do not rotate, but move smoothly forward in one plane. At these slow speeds the large waves passing from the base to the apex of the flagellum are clearly visible. In a related genus *Apedinella*, Throndsen (1971) reported that the flagellum moved very rapidly in one plane and acted as a tractellum (i.e., it pulled the cell). Throndsen is of the opinion that the proximal part of the flagellum is not involved in the swimming activity and the cell does not show any kind of oscillation of the body as reported by Swale (1969) for *Pseudopedinella hexacosta* Vysotskii.

Throndsen (1973) studied the motility of some marine nanoplankton flagellates including *Olisthodiscus luteus*, *Apedinella spinifera*, *Pseudopedinella pyriformis* and *Ochromonas minima* using photomicrography techniques. The results he obtained for these species are summarized in Table 1.

Table 1: Swimming rates of certain chrysophycean flagellates modified from Throndsen (1973).

Species	Cell Volume μm^3	Gyration gyr/sec.	Swimming Speed µm/sec.	Migration Capacity metres/12 hours
Olisthodiscus luteus	600	1/2-1	140 (130-160)	7
Pseudopedinella pyriformis	500	1/2-1	105 (90-110)	4.5
Apedinella spinifera	450	1/2-1	110 (90-175)	5
Ochromonas minima	25	1/4-1 1/4	75	3

The slightly asymmetrical *Olisthodiscus luteus* exhibited one gyration or less per second, the swimming speed was a moderate 130-160 µm/second and the swimming direction variable. In the case of the more symmetrical species, (viz. *Pseudopedinella pyriformis* and *Apedinella spinifera*) rotation during swimming was also slow, probably due to the presence of the long trailing pseudopodium (peduncle) in *P. pyriformis* and the long protruding spines in *A. spinifera*. The rotation speed increased significantly in *P. pyriformis* when the pseudopodium (peduncle) was retracted. The diameter of the swimming path helix in *P. pyriformis* and *A. spinifera* was found to be more than twice the cell diameter. An interesting point emerging from this investigation is the fact that even small chrysophycean flagellates possess the ability to swim considerable distances over a 12 hour period and that this could be of some importance to the cells (Throndsen, 1973).

LIFE HISTORIES IN THE CHRYSOPHYCEAE

Sexual reproduction has been observed with certainty in relatively few genera. It has been observed and substantiated in the loricate members mainly due to the fact that zygote formation is easily recognizable (Bourrelly, 1957; Kristiansen, 1960, 1961, 1963a, 1963b; Fott, 1962). These workers have demonstrated that the loricate cells attach at their openings and the protoplasts migrate from the loricas to fuse and form the zygote. Sexual reproduction is invariably isogamous, although autogamous sexual reproduction has been described for *Ochromonas* (Gayral *et al.*, 1972) and for *Dinobryon* (Sandgren, 1978). It is generally believed that the vegetative phase of members of the Chrysophyceae is haploid with meiosis being zygotic (Bold and Wynne, 1978).

A unifying characteristic of members of the Chrysophyceae is the ability of the cells to produce endogenous siliceous cysts called statospores. The cyst contains a single pore which is sealed with a plug thought to be pectic (Bourrelly, 1957). The general process of statospore formation is well known as the result of a number of light microscopical studies (Pascher, 1932; Hollande, 1952; Bourrelly, 1957; Sheath *et al.*, 1975).

Important aspects of the encystment process include the following:

1. The number of nuclei within the cyst.
2. The development of the siliceous cyst wall.
3. The fate of the cytoplasm not initially included within the cyst wall.
4. The formation of the plug.

The mature cyst has been observed to contain a varying complement of cellular organelles as well as large storage reserves in the form of chrysolaminarin and lipids.

Until recently very little was known concerning the production of a statospore at the ultrastructural level. Hibberd (1977) studied the process of encystment in *Ochromonas tuberculata* Hibberd. Esser and Van Valkenburg (1977) studied cyst formation in *Uroglena volvox* Ehrenberg. Sandgren (1978) has undertaken an in depth study on cyst formation of a number of genera including *Dinobryon cylindricum* Imhoff, *Dinobryon divergens*, *Uroglena volvox*, *U. americana* and *Mallomonas caudata*. In *Ochromonas tuberculata* the cysts are produced within a cytoplasmic silica deposition vesicle whose membrane (silicalemma) appears to be formed by the coalescence of Golgi vesicles. The formation of the silica deposition vesicle is similar to that described during frustule formation in the Bacillariophyceae (Drum and Pankratz, 1964; Reimann, 1964; Darley, 1974). The pore is formed at the anterior end of the flagellate cell by the lack of deposition of silica over a small region of the silica deposition vesicle and the rupture of the silica deposition vesicle and other membranes crossing this region.

Once the cyst wall is complete, an extracystic plug is formed at the pore which results in the loss of some extracystic cytoplasm and the plasmalemma. The inner silica deposition membrane then becomes the functional plasmalemma. The plug develops first by coalescence with the cell membrane of Golgi derived vesicles containing dense but non-siliceous spicules surrounded by amorphous material. During the later stages of plug formation only fibrous material is deposited, some of which may be extracted through the pore. Hibberd (1977) found that on germination the plug completely disappears and a single cell similar to a normal vegetative cell emerges through the pore.

In *Dinobryon cylindricum* the first event in the developmental process of encystment is a pre-encystment nuclear division (Sandgren, 1978). This division is unusual in that it does not include the replication of the flagellar apparatus and the dictyosome. Only the plastid replicates prior to metaphase. This results in a lack of symmetry. The one pore is defined by the elongated electron dense rhizoplasts. The second diffuse spindle pore is occupied by the dictyosome and the short flagellum basal body. There is no cytokinesis following the pre-encystment nuclear division. The binucleate cell, instead of cleaving, continues to enlarge until it eventually fills the lorica. The basal bodies of the unreplicated flagellar apparatus are reported to reunite and migrate back to the normal subapical position. The cell increases in size and eventually bulges out of the lorica and an encystment vesicle is produced. The Golgi system is active during the early stages of encystment. These vesicles fuse and form the silica disposition vesicle which grows in a radiating manner and eventually encloses the majority of the cytoplasm. A pore is developed at the anterior of the cell - the only region where the silica deposition vesicle does not encapsulate the cytoplasm. The developing pore is identified by a ring of electron dense material. The portion of cytoplasm cut off from the main body of the cell contains many mitochondria, ribosomes and the contractile vacuole. Once the silica deposition vesicle is complete the deposition of the silica begins within its confines. Polymerization of silica occurs as a simultaneous event throughout the silica deposition vesicle and no nucleating sites are observed. Once the deposition of the silica has reached the maximum thickness there is a retraction of the cytoplasm external to the cyst through the pore (Fig. 5).

Small fragments of cytoplasm remaining outside the cyst wall contribute to the formation of an extension of the silica deposition vesicle inside which the collar of the cyst develops. A tripartite plug is now produced in the region of the pore (Figs. 6a, 6b).

As the cyst matures the cytoplasmic contents become more concentrated and the chrysolaminarin vacuole takes on a mottled appearance due to the fusion of dense granules. Once this stage has been reached there is a secondary phase of silicification which results in the production of the hooked extension of the cyst collar and the short body spines. It has been suggested that cytoskeletal microtubules may be involved in the retraction of the extracystic cytoplasm. As the cyst matures there is a general shrinking of the cytoplasm and reduction in the size of the plastids and associated organelles and an increase in chrysolaminarin and lipid.

Sandgren (1978) agrees with Geitler (1935) that the binucleate cyst of *Dinobryon* which results from the mitosis in the pre-encystment cell is most likely a vessel for autogamic recombination and is not a simple asexual resting stage.

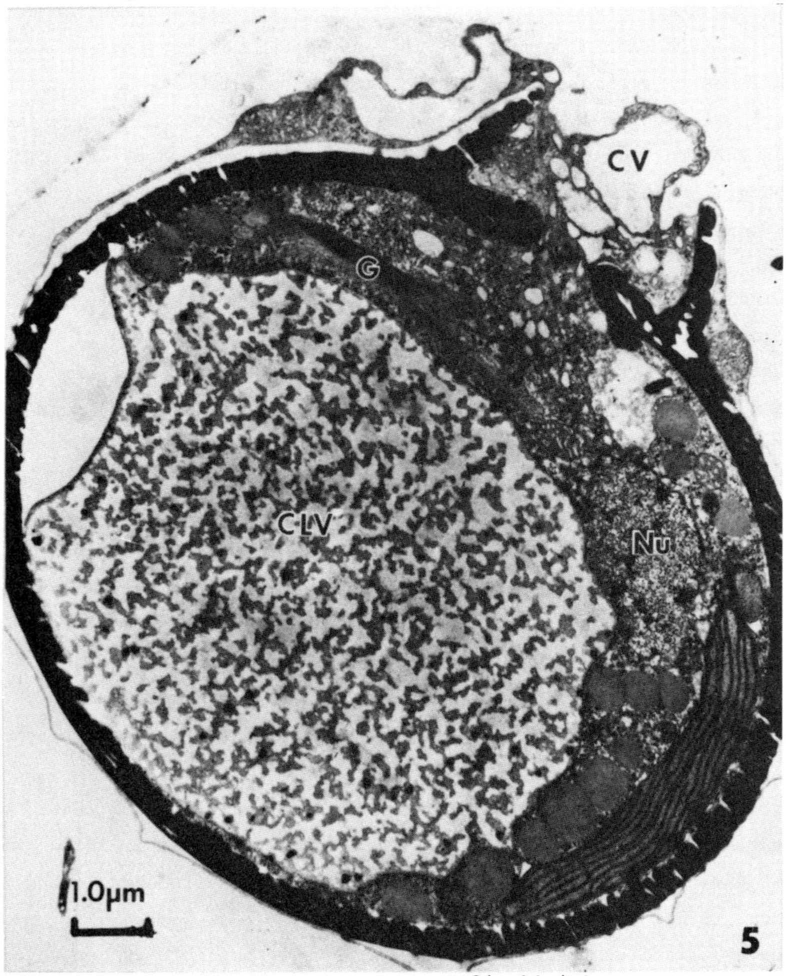

Fig. 5. Transverse section through a cyst of *D. cylindricum* with mature silica walls and basal portion of the collar. There is still some extracystic cytoplasm present which contains the contractile vacuole (CV). CVL, chrysolaminarin vacuole; G, Golgi body; Nu, Nucleus. (Micrograph kindly supplied by C. Sandgren.)

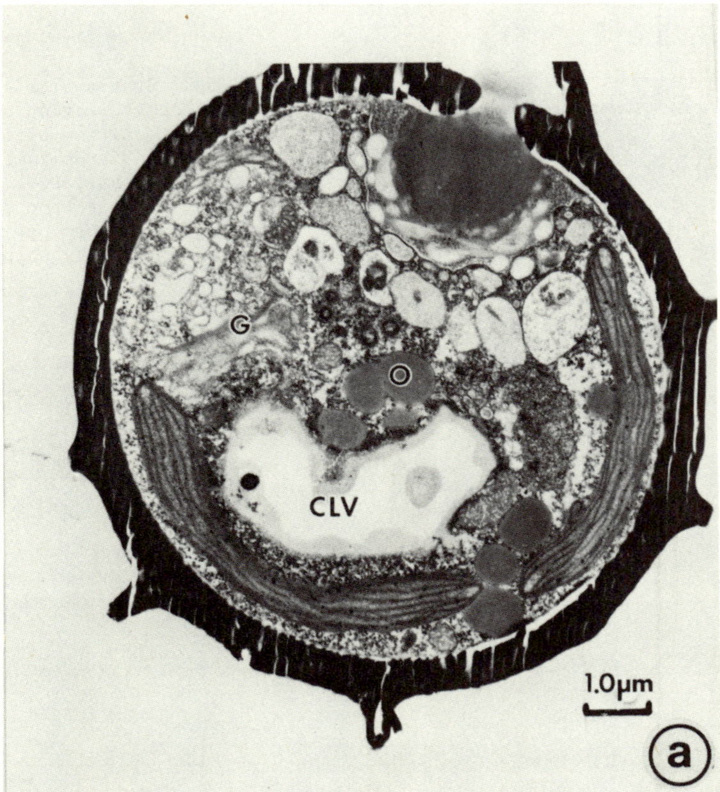

Fig. 6a. Statospores of *Dinobryon cylindricum*. Mature statospore of *D. cylindricum*. Note the well developed siliceous wall and collar. Body spines are present. G, Golgi; CLV, chrysolaminarin vacuole; O, oil droplets. (Micrograph kindly supplied by C. Sandgren.)

Sandgren found that cyst formation in *Dinobryon* and *Uroglena* was essentially the same, but that in *Mallomonas caudata* there are marked differences between this cyst and other chrysophyte cysts (Fig. 7). These differences include:

1. No conservation of extracystic cytoplasm prior to plug formation.
2. The cyst is enclosed within a scale layer.
3. The extracystic cytoplasm is rich in mitochondria and plastids.
4. The cyst contains a single plastid and nucleus.
5. The cyst wall consists of a basal lamina of silica from which an irregular array of reticulate elements project.

Sandgren (1978) thinks that there is sufficient diversity with regard to the

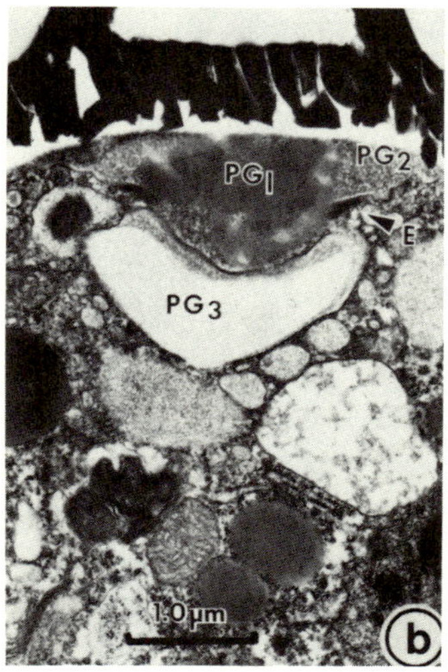

Fig. 6b. Statospores of *Dinobryon cylindricum*. Section showing the mature composition of the cyst plug of *D. cylindricum*. Three plug layers (PG_1, PG_2, PG_3) can be seen. E, electron dense material defining pore periphery. (Micrograph kindly supplied by C. Sandgren.)

key aspects of the encystment process to consider the use of this development sequence in determining phylogenetic relationships.

Sandgren (1978) has also been able to demonstrate the existence of sexually compatable mating types and a mating group in *Dinobryon cylindricum*. In contrast to the chlamydomonad mating system where gametogenesis must be induced in a special medium, sexually compatible clones of *D. cylindricum* function as gametes under normal culture conditions. There is either no special process of gametogenesis undergone by vegetating *Dinobryon* cells prior to gamete fusion or gametogenesis must occur after compatible clones are mixed. Sex in *D. cylindricum* has thus far been determined to be heterothallic and dioecious.

The production of self induced binucleate cysts by some clones might seem to suggest the existence of homothallic sexuality. But it should be remembered that these homothallic cysts become binucleate because there is no cytokinesis following a special mitotic division in the vegetative cyst mother cell. They are, therefore, thought to be autogamic.

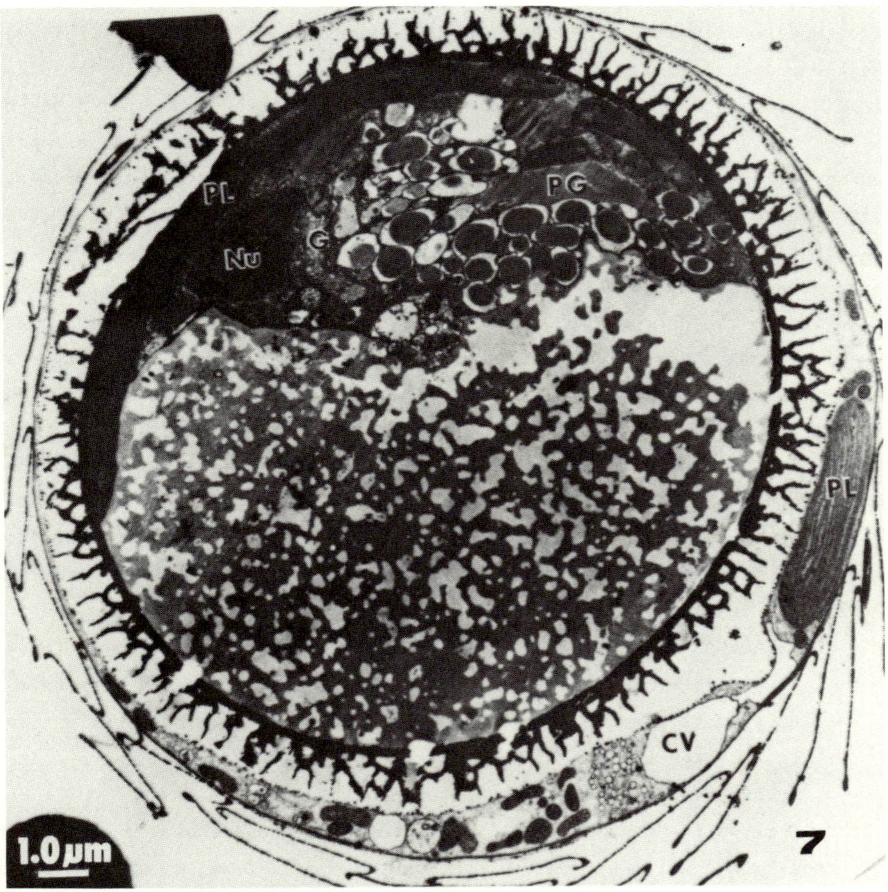

Fig. 7. Section lateral to the pore of a mature cyst of *Mallomonas caudata*. The cyst wall is composed of a basal lamina of silica from which reticulate elements arise. The extracystic cytoplasm contains several contractile vacuoles (CV), mitochondria and plastids (PL). All this is enclosed within the cell layer of siliceous scales. Lipid droplets are present around the large chrysolaminarin vacuole. G, Golgi body; PG, material comprising the plug; NU, nucleus. (Micrograph kindly supplied by C. Sandgren.)

It would appear that it is possible - in the case of *D. cylindricum* - to distinguish between autogamic and zygotic statospores based on the dynamics of encystment relative to the growth of the vegetative cell populations. This is indeed fortuitous because both classes of cysts are morphologically identical and both are binucleate. Sandgren found that autogamic cysts are produced in large numbers only after the growth of the vegetative population has slowed significantly and, thus, the production of these cysts seems to be dependent on stimuli which are extrinsic or environmental. Zygotic cysts on the other hand are produced during the exponential growth phase of the vegetating cell population so long as clones representing two compatible mating types are present. Environmental stimuli do not change the encystment dynamics of zygotic cysts. These static spores appear to be completely under intrinsic controls.

Sandgren was able to make two generalizations with regard to reproductive strategies in the Chrysophyceae from his culture work on *Dinobryon cylindricum*:

1. Generalist species tend to produce cysts at low frequencies by what is most likely to be an asexual mechanism. Specialist species tend to produce cysts at a high frequency and more commonly utilize the sexual mechanism of reproduction. High frequency intrinsic encystment which conclusively results from out crossing sexual reproduction on the basis of the *Dinobryon* culture experiments is restricted to the rapidly growing species.
2. The resulting cyst populations show great variation in size, but this is not correlated with either the mechanism of encystment or the growth rate of the species involved. The low frequency long term encystment common in generalist species appears to be as effective as the high frequency, short term encystment of specialist species in producing an equivalent population of cysts.

Other work on the reproductive life in terms of members of the Chrysophyceae are virtually non-existent. Recently *Olisthodiscus luteus* Carter, a problematic species, has been shown to produce a viable benthic stage which might be transitory in functioning as a vegetative propagule (Tomas, 1978a, 1978b).

MITOSIS AND MEIOSIS

There is very little published work on the ultrastructure of mitotically dividing cells in the Chrysophyceae. To date *Ochromonas danica* is the only species that has been studied in any detail. In this species the flagellar bases, chloroplast and dictyosome, divided during interphase. The disassociation of the nuclear envelope occurred, except for the regions closely associated with the chloroplasts, resulting in what could be termed an open spindle, with the spindle fibres inserted into the rhizoplasts. It is thought that the rhizoplast organises the microtubules in the spindle.

Sandgren (1978) studied the pre-encystment mitotic division in *Dinobryon cylindricum*. This nuclear division exhibits a number of differences when compared with the mitosis in vegetative cells of *Ochromonas danica*. An interesting feature in the pre-encystment division is that there is no replication of the flagellar apparatus or the Golgi complex. The only organelles that were observed to replicate were the single chloroplast which replicated prior to metaphase and the nucleus. This resulted in the metaphase spindle lacking in symmetry. One pore is defined by the elongated electron dense rhizoplast - which extends from the base of the long flagellum basal body. The second diffuse spindle pore is occupied by the dictyosome and the short flagellum basal body. After nuclear division is complete the basal bodies of the unreplicated flagellar apparatus reunite and migrate back to the normal subapical position.

Meiosis has not been studied at the ultrastructural level, but studies on the life cycle of members of the class indicate that it is zygotic.

ULTRASTRUCTURAL CHANGES DUE TO ENVIRONMENTAL CHANGES

There is an increasing awareness that considerable caution should be exercised when describing the ultrastructure of various algae that have been grown under different environmental conditions. Vesk and Jeffrey (1977) have demonstrated that the photosynthetic pigments and in some instances the structure of the chloroplast varied in unicellular marine algae exposed to blue-green light. *Olisthodiscus luteus* showed a 17% increase in the chlorophyll (chlorophyll *a* and *c*) when grown in 400 μ W. cm^{-2} white and blue-green light. The structure of the chloroplast, however, remained stable.

Bouck and Brown (1973) and Brown and Bouck (1973) examined the fine structure of the vegetating and dividing cells of *Ochromonas danica* and found that there was a close correlation between the distribution of cytoplasmic microtubules and the asymmetric cell shape and the association of these microtubules with two distinct sites in the anterior of the cell. These observations implied a cytoskeletal role for microtubules and further suggested that control of microtubule distribution and control of cell shape may be effected by restricting microtubule assembly to specific nucleating sites. It is known that various environmental factors (e.g. salinity) can markedly affect the shape.

In many of the scale-bearing representatives, the architecture of the scales is regarded as having taxonomic importance (Peterfi and Momeu, 1977). They are usually used to dilineate different species. However, considerable caution should be used in describing new species solely on the grounds of scale architecture as Harris (1970) has demonstrated that under different growth conditions *Mallomonas* produces, what she refers to as "imperfect forms", which have slight differences in scale structure. These differences could be regarded by some workers as sufficiently distinct to warrant the erection of a new species.

No work to date has been published which proves beyond a doubt that scale architecture is a stable and reliable criterion on which to base the systematics of various chrysophycean genera.

HABITAT

Members of the Chrysophyceae occur in fresh, brackish and marine water environments. By far the largest number of species occur in fresh water habitats. The fresh water forms seem to prefer unpolluted and cold waters, such as mountain lakes and streams. A number of genera occur in arctic and subarctic habitats and in cold waters of alpine regions.

Olisthodiscus luteus has been shown to be a euryhaline species being capable of surviving in media with salinities ranging from 2 to excess of 50 o/oo (Tomas, 1978a).

CULTURE MEDIA REFERENCES

A number of different media have been used in the cultivation of certain Chrysophyceae. The following are just a few that have been reported in the literature.

Marine Media:

1. Erdschreiber Medium (Starr, 1964).
2. 10% Modified Erdschreiber Medium used by Throndsen (1969) in the successful cultivation of marine nanoplankton species.
3. Provasoli's Enriched Seawater Media (Bold and Wynne, 1978) used extensively at different concentrations.
4. Modified Gold's E_2 Medium (Gold, 1964) used by Tomas (1978a) in the cultivation of *Olisthodiscus luteus*.
5. Aaronson's Medium used in the successful cultivation of various species of *Ochromonas* (Kahn *et al.*, 1978).

Freshwater Media:

1. Bold's Basal Medium (Bischoff and Bold, 1963).
2. Bold's Basal Medium supplemented with 10% leaf extract used by Hibberd (1977) to successfully culture *Ochromonas tuberculata*.

3. Pringsheim's Biphasic Soil Water Medium used by Belcher (1969a). A number of excellent references for the isolation and cultivation of microalgae can be found in Stein (1973).

BRIEF SUMMARY OF THE FOSSIL RECORD

Remarkably little is known about the geological range of the Chrysophyceae. Bradley (1946) reported the presence of remains of chrysophycean cells from coprolites found in the Bridger formation of Wyoming which are regarded as being Upper Eocene. These resembled the genera *Lagynion, Cyrtophora* and *Dinobryon*. There are also reports of cysts of chrysophycean algae from Miocean diatomites, e.g. *Cysta* sp. (in Prescott, 1969).

Chrysophyte cysts are commonly found in both marine and freshwater diatomites. Because in the early literature chrysophycean cysts were only described for the freshwater species of Chrysophyceae and marine fossils were placed in a separate family the Archaemonadaceae (Moore *et al.*, 1968). Recent discovery of abundant chrysophyte cysts in the plankton and in oceanic deposits in the Pacific invalidates their separation on an ecological basis (Moore *et al.*, 1968).

NEEDED RESEARCH

The Chrysophyceae have received a considerable amount of attention from phycologists over the past 25 years. This has been due primarily to the fact that there have been a number of excellent taxonomic works published (Bourrelly, 1957, 1965, 1968) on this very diverse class. Most of the work done on the Chrysophyceae has been concentrated in Europe, the Americas and the Far East. More work should be done on the chrysophyte flora of other parts of the world, particularly in Africa, Australia and the Arctic and Antarctic. Particular attention should also be paid, not only to the taxonomy of the group, but also to their ecology and seasonal distribution. More attention should be paid to the colourless representatives which, although they are very difficult to identify - particularly at the light microscope level - are undoubtedly important in both fresh water and marine ecosystems.

Because of the diversity of scale bearing representatives, some studies on the physiology and biochemistry of scale production is long overdue. The fact that many genera and species are identified by their scale architecture, should emphasise the need to investigate the reliability of scales and identification aids.

There still remains a great deal of work on the sexual reproduction of the Chrysophyceae and, in particular, the causative factors in triggering off this stage in the life cycle. Chromosome studies would also be most useful although probably difficult to undertake.

ACKNOWLEDGEMENTS

The author would like to thank his many colleagues for their help in the preparation of this paper and in particular to Dr. Craig Sandgren who very kindly allowed me to use information from his doctoral dissertation prior to it being published. This paper resulted out of work funded by the Council for Scientific and Industrial Research and the University of Natal Research Fund.

REFERENCES

Allen, M. B., L. Fries, T. W. Goodwin and S. Phagpolngarm. 1960. Carotenoid distribution in certain naturally occurring algae and some artificially induced mutants of *Chlorella pyrenoidosa*. J. Gen. Microbiol. 23: 93-103.

Asmund, B. 1973. Survey of the genus *Chrysosphaerella* (Chrysophyceae) as studied in the electron microscope with description of a new species. Bot. Tidsskr. 68: 132-139.

Belcher, J. H. 1968. Lorica construction in *Pseudokephyrion pseudospirale* Bourrelly. Br. Phycol. Bull. 3: 495-499.

Belcher, J. H. 1969a. A morphological study of the phytoflagellate *Chrysococcus refuscens* Klebs. in culture. Br. Phycol. J. 4: 105-117.

Belcher, J. H. 1969b. Some remarks upon *Mallomonas papillosa* Harris and Bradley and *M. calceolus* Bradley. Nova Hedwigia 18: 257-270.

Belcher, J. H. and E. M. F. Swale. 1967. *Chromulina placentula* sp. nov. (Chrysophyceae), a freshwater nanoplankton flagellate. Br. Phycol. Bull. 3: 257-267.

Belcher, J. H. and E. M. F. Swale. 1971. The microanatomy of *Phaeaster pascheri* Scherffel. (Chrysophyceae). Br. Phycol. J. 6: 157-169.

Belcher, J. H. and E. M. F. Swale. 1972a. The morphology and fine structure of the colourless colonial flagellate *Anthophysa vegetans* (O. F. Muller) Stein. Br. Phycol. J. 7: 335-346.

Belcher, J. H. and E. M. F. Swale. 1972b. Some features of the microanatomy of *Chrysococcus cordiformis* Naumann. Br. Phycol. J. 7: 53-59.

Bisalputra, T. 1974. Plastids. In Stewart, W. D. P. [Ed.] *Algal Physiology and Biochemistry*. Botanical Monographs, No. 10. Blackwell Scientific.

Bischoff, H. W. and H. C. Bold. 1963. Phycological Studies IV. Some algae from Enchanted Rock and related algal species. The University of Texas Pub. No. 6318. 95 pp.

Bold, H. C. and M. J. Wynne. 1978. *Introduction to the Algae*. Structure and Reproduction. Prentice Hall.

Bouck, G. B. 1971. The structure, origin, isolation and composition of the tubular mastigonemes of the *Ochromonas* flagellum. J. Cell Biol. 50: 362-384.

Bouck, G. B. 1972. Architecture and assembly of mastigonemes. Adv. Cell Molec. Biol. 2: 151-192.

Bouck, G. B. and D. L. Brown. 1973. Microtubule biogenesis and cell shape in *Ochromonas*. I. The distribution of cytoplasmic and mitotic microtubules. J. Cell. Biol. 56: 340-378

Bourrelly, P. 1957. Resérches sur les Chrysophycées. Revue Algol., Mem. Hors. S'er., 1: 1-412.

Bourrelly, P. 1965. La classification des Chrysophycées: ses problemes. Revue Algol., 1: 56-60.

Bourrelly, P. 1968. *Les algues d'eau douce* 2: Algues Jaunes et Brunes. Paris: Boubée et Cie. 438 pp.
Bradley, D. E. 1964. A study of the *Mallomonas, Synura* and *Chrysosphaerella* of Northern Iceland. J. Gen. Microbiol. 37: 321-333.
Bradley, D. E. 1966. The ultrastructure of the flagella of three chrysomonads with particular reference to the mastigonemes. Exp. Cell Res. 41: 162-173.
Brown, D. L. and G. B. Bouck. 1973. Microtubule biogenesis and cell shape in *Ochromonas* II. The role of nucleating sites in shape and development. J. Cell Biol. 56: 360-378.
Clarke, K. J. and N. C. Pennick. 1975. *Syncrypta glomerifera* sp. nov. A marine member of the Chrysophyceae bearing a new form of scale. Br. Phycol. J. 10: 363-370.
Cole, G. T. and M. J. Wynne. 1973. Nuclear pore arrangement and structure of the Golgi complex in *Ochromonas danica* (Chrysophyceae). Cytobios 8: 161-173.
Craigie, J. S. 1974. Storage Products. In Stewart, W. D. P. [Ed.] *Algal Physiology and Biochemistry*. Botanical Monographs, No. 10, Blackwell Scientific.
Dales, R. P. 1960. On the pigments of the Chrysophyceae. J. Mar. Biol. Assn. U.K. 39: 693-699.
Darley, W. M. 1974. Silicification and Calcification. In Stewart, W. D. P. [Ed.] *Algal Physiology and Biochemistry*. Botanical Monographs, No. 10. Blackwell Scientific.
Dodge, J. D. 1973. *The Fine Structure of Algal Cells*. Academic Press. London and New York.
Drum, R. W. and H. S. Pankratz. 1964. Post mitotic structure of *Gomphonema parvulum*. J. Ultra. Res. 10: 217-223.
Esser, S. C. and S. D. Van Valkenburg. 1977. The fine structure of vegetative and statospore forming cells of *Uroglena volvox* Ehrenberg. J. Phycol. (Suppl.) 13: 20.
Fott, B. 1962. Hologamic and agamic cyst formation in loricate Chrysomonads. Phykos 3: 15-18.
Franke, W. W. and W. Herth. 1973. Cell and lorica fine structure of the Chrysomonad alga, *Dinobryon sertularia* Ehr. (Chrysophyceae). Arch. Mikrobiol. 91: 323-344.
Gayral, P., C. Haas and H. Le Pailleur. 1972. Alternance morphologique de genérations et alternance de phases chez les Chrysophycées. Soc. Bot. Fr., Mem.: 215-230.
Geitler, L. 1935. Über zweikernige Cysten von *Dinobryon divergens*. Osterr. Bot. Z. 84: 282-286.
Gibbs, S. P. 1968. Autoradiographic evidence for the *in situ* synthesis of chloroplast and mitochondrial RNA. J. Cell Sci. 3: 327-340.
Gold, K. 1964. A microbiological assay for vitamin B_{12} in seawater using radiocarbon. Limnol. Oceanogr. 9: 343-347.
Goodwin, T. W. 1974. Carotenoids and biliproteins. In Stewart, W. D. P. [Ed.] *Algal Physiology and Biochemistry*. Botanical Monographs, No. 10. Blackwell Scientific.
Green, J. C. and I. Manton. 1970. Studies in the fine structure and taxonomy of flagellates in the genus *Pavlova gyrans*, the type species. J. Mar. Biol. Assn. U.K. 50: 1113-1130.
Guillard, R. R. L. and C. J. Lorenzen. 1972. Yellow-green algae with chlorophyllide C. J. Phycol. 8: 10-14.
Harris, K. 1963. Observations on *Sphaleromantis tetragona*. J. Gen. Microbiol. 33: 345-348.
Harris, K. 1970. Imperfect forms and the taxonomy of *Mallomonas*. J. Gen. Microbiol. 61: 73-76.

Hibberd, D. J. 1970. Observations on the cytology and ultrastructure of *Ochromonas tuberculatus* sp. nov. (Chrysophyceae) with special reference to the discobolocysts. Br. Phycol. J. 5: 119-143.
Hibberd, D. J. 1971. Observations on the cytology and ultrastructure of *Chrysamoeba radians* Klebs. (Chrysophyceae). Br. Phycol. J. 6: 207-223.
Hibberd, D. J. 1973. Observations on the ultrastructure of flagellar scales in the genus *Synura* (Chrysophyceae). Arch. Mikrobiol. 89: 291-304.
Hibberd, D. J. 1976. The ultrastructure and taxonomy of the Chrysophyceae and Prymnesiophyceae (Haptophyceae): a survey with some new observations on the ultrastructure of the Chrysophyceae. J. Linn. Soc. (Bot.) 72: 55-80.
Hibberd, D. J. 1977. Ultrastructure of cyst formation in *Ochromonas tuberculata* (Chrysophyceae). J. Phycol. 13: 309-320.
Hibberd, D. J. and M. J. Chretiennot-Dinet. 1979. The ultrastructure and taxonomy of *Rhizochromulina marina* gen. et sp. nov., an amoeboid marine chrysophyte. J. Mar. Biol. Assn. U.K. 59: 179-193.
Hollande, A. 1952. Classes des Chrysomonadines. In Grasse, P. [Ed.] *Traite de Zoologie*. Masson et Cie, Paris. pp. 471-570.
Holwill, M. E. J. and M. A. Sleigh. 1967. Propulsion in hispiol flagella. J. Exp. Biol. 47: 267.
Jahn, T. L. 1965. Hydrodynamic principles in the locomotion of microorganisms. Excerpta Med. Int. Congr. Ser. 91: 18.
Jahn, T. L., M. G. Landman and J. R. Fonsea. 1964. The mechanisms of locomotion of flagellates. II. Function of the mastigonemes of *Ochromonas*. J. Protozool. 11: 291.
Jeffrey, S. W. 1961. Paper chromatographic separation of chlorophylls and carotenoids from marine algae. Phytochem. 10: 3121-3127.
Jeffrey, S. W. 1969. Properties of two spectrally different components in chlorophyll *c* preparations. Biochem. Biophys. Acta 177: 456-467.
Kahan, D., R. Oren, S. Aaranson and U. Behras. 1978. Fine structure of the cell surface and Golgi apparatus of *Ochromonas*. J. Protozool. 25: 30-33.
Karim, A. G. A. and F. E. Round. 1967. Microfibrils in the lorica of the freshwater alga *Dinobryon*. New Phytol. 66: 409-412.
Kauss, H. 1962. Der einfluss von Thiamin - und Biotin - mangel auf den Stoffwechsel von *Ochromonas malhamensis*. Vortr. Gesamtbeg. Bot. No. 1: 129-132.
Klebs, G. 1892. Flagellaten studies. II. Z. Wiss. Zool. 55: 353-445.
Kristiansen, J. 1960. Some cases of sexuality in *Kephyriopsis* (Chrysophyceae). Bot. Tids. 56: 128-131.
Kristiansen, J. 1961. Sexual reproduction in *Mallomonas caudata*. Bot. Tids. 57: 306-309.
Kristiansen, J. 1963a. Sexual and asexual reproduction in *Kephyrion* and *Stenocalyx* (Chrysophyceae). Bot. Tids. 59: 244-254.
Kristiansen, J. 1963b. Observations on the structure and ecology of *Synura splendida*. Bot. Tids. 59: 280-289.
Kristiansen, J. 1969a. *Chrysosphaerella multispina* Bradley and some other remarkable Chrysophyceae from Lake Straken, Aneboda, Sweden. Osterr. Bot. Z. 116: 70-84.
Kristiansen, J. 1969b. Lorica structure in *Chrysolykos* (Chrysophyceae). Svensk. Bot. T. 64: 162-168.
Kristiansen, J. 1972. Studies on the lorica structure in Chrysophyceae. Svensk. Bot. T. 66: 184-190.
Kristiansen, J. and P. L. Walne. 1976. Structural connections between flagellar base and stigma in *Dinobryon*. Protoplasma 89: 371-374.
Kristiansen, J. and P. L. Walne. 1977. Fine structure of photokinetic systems in *Dinobryon cylindricum* var. *alpinum* (Chrysophyceae). Br. Phycol. J. 12: 329-341.

Leedale, G. F., B. S. C. Leadbeater and A. Massalski. 1970. The intracellular origin of flagellar hairs in the Chrysophyceae and Xanthophyceae. J. Cell Sci. 6: 701-719.

Leadbeater, B. S. C. 1969. A fine structural study of *Olisthodiscus luteus* Carter. Br. Phycol. J. 4: 3-17.

Manton, I. and K. Harris. 1966. Observations on the microanatomy of the brown flagellate *Sphaleromantis tetragona* Skuja with special reference to the flagellar apparatus and scales. J. Linn. Soc. (Bot.) 59: 397-403.

Manton, I. and G. F. Leedale. 1961. Observations on the fine structure of *Paraphysomonas vestita* with special reference to the Golgi apparatus and the origin of scales. Phycologia 1: 37-57.

Meeks, J. C. 1974. Chlorophylls. In Stewart, W. D. P. [Ed.] *Algal Physiology and Biochemistry*. Botanical Monographs, No. 10. Blackwell Scientific.

Moore, R. C. *et al.* 1968. Developments, trends and outlooks in paleontology. J. of Paleontology 42: 1327-1377.

Ostroff, C. R. and S. D. Van Valkenburg. 1978. The fine structure of *Pseudopedinella pyriforme* Carter (Chrysophyceae). Br. Phycol. J. 13: 35-49.

Pascher, A. 1932. Über die Verbreitung endogener bzw. endoplasmatisch gebildeter Sporen bei den Algen. Bot. Zentralbl. Beih. 49: 293-308.

Peterfi, L. S. 1969. The fine structure of *Poterioochromonas malhamensis* (Pringsheim) comb. nov. with special reference to the lorica. Nova Hedwigia 17: 93-103.

Peterfi, L. S. and L. Momeu. 1977. Remarks on the taxonomy of some *Synura* species based on the fine structure of scales. Stduii si Comubicari-St. nt. 21: 15-23.

Pienaar, R. N. 1976. The microanatomy of *Sphaleromantis marina* sp. nov. (Chrysophyceae). Br. Phycol. J. 11: 83-92.

Pienaar, R. N. 1977. The microanatomy of an unusual marine species of *Chromulina* (Chrysophyceae). Proc. Electron Microscopy Society of Southern Africa, 7: 101-102.

Prescott, G. W. 1969. *The Algae.* A Review. Riverside Studies in Biology. Houghton Mifflin Company. Boston. 436 pp.

Reimann, B. E. F. 1964. Deposition of silica inside a diatom cell. Expl. Cell Res. 34: 605-608.

Ricketts, T. R. 1965. Chlorophyll c in some members of the Chrysophyceae. Phytochem. 4: 725-730.

Rouiller, C. and E. Faure-Fremiet. 1958. Structure fine d'un flagelle chrysomonadien: *Chromulina psammobia*. Expl. Cell Res. 14: 47-67.

Sandgren, C. D. 1978. Resting cysts of the Chrysophyceae: their induction, development and strategic significance in the life history of planktonic species. Ph.D. Thesis, University of Washington. 176 pp.

Schnepf, E. and G. Deichraber. 1969. Über die fein struktur von *Synura petersenii* unter besonderer Berucksichtigung der Morphogenese ihrer Kieseksehuppen. Protoplasma 68: 85-106.

Sheath, R. E., J. A. Hellebust and T. Sau. 1975. The statospore of *Dinobryon divergens* Imhof: formation and germination in a subarctic lake. J. Phycol. 11: 131-138.

Starr, R. C. 1964. The culture collection of algae at Indiana University. Am. J. Bot. 51: 1013-1044.

Stein, J. R. [Ed.]. 1973. *Handbook of Phycological Methods*. Culture methods and growth measurements. Cambridge University Press. 448 pp.

Swale, E. M. F. 1969. A study of the nanoplankton flagellate *Pedinella hexacostata* Vysotskii by light and electron microscopy. Br. Phycol. J. 4: 65-86.

Takahashi, E. 1972. Studies on the genera *Mallomonas* and *Synura*, and other plankton in freshwater with the electron microscope. III. On three new species of Chrysophyceae. Bot. Mag. Tokyo 85: 293-302.

Takahashi, E. 1975. The fine structure of the scales and flagella of the Chrysophyta. In Tokida, J. and H. Hirose [Eds.] *Advances of Phycology in Japan.* G. Fischer Verlag, Jena. pp. 67-97.

Takahashi, E. 1978. Electron Microscopical Studies of the Synuraceae (Chrysophyceae) in Japan. Tokai University Press.

Throndsen, J. 1969. Flagellates of Norwegian coastal waters. Nytt. Mag. Bot. 16: 161-216.

Throndsen, J. 1971. *Apedinella* gen. nov. and the fine structure of *A. spinifera* (Throndsen) comb. nov. Norw. J. Bot. 18: 47-64.

Throndsen, J. 1973. Motility in some marine nanoplankton flagellates. Norw. J. Zool. 21: 193-200.

Tomas, C. R. 1978a. *Olisthodiscus luteus* (Chrysophyceae). I. Effects of salinity and temperature on growth, motility and survival. J. Phycol. 14: 309-313.

Tomas, C. R. 1978b. *Olisthodiscus luteus* (Chrysophyceae). II. Formation and survival of a benthic phase. J. Phycol. 14: 314-319.

Vesk, M. and S. W. Jeffrey. 1977. Effect of blue-green light on photosynthetic pigments and chloroplast structure in unicellular marine algae from six classes. J. Phycol. 13: 280-288.

Wujek, D. E. 1969. Ultrastructure of flagellated Chrysophytes. I. *Dinobryon.* Cytologia 34: 71-79.

Wujek, D. E. and J. Kristiansen. 1978. Observations on bristle and scale production in *Mallomonas caudata* (Chrysophyceae). Arch. Protistenk. 120: 213-221.

Copyright 1980 by Elsevier North Holland, Inc.

XANTHOPHYTES

DAVID J. HIBBERD
Culture Centre of Algae and Protozoa, Institute of Terrestrial Ecology, 36, Storey's Way, Cambridge, CB3 ODT, England.

ABSTRACT

The majority of xanthophytes are freshwater coccoid unicells with parietal yellow-green chloroplasts. Photosynthetic pigments include chlorophylls *a*, c_1 and c_2, β-carotene, diatoxanthin, diadinoxanthin, heteroxanthin and vaucheriaxanthin ester. The cell wall in many species is bipartite or delicately sculptured and shows a variety of characteristic chemical reactions. Cellulose has been identified with certainty only in one species. Xanthophytes do not produce starch and a β-1,3-linked glucan has not yet been identified as a storage metabolite even though structures resembling "leucosin" vesicles in other groups are formed.

The most common method of reproduction is by the formation of autospores; zoospores are produced less abundantly. Sexual reproduction appears to be confined to the advanced forms with siphoneous organization; isogamy, anisogamy and oogamy are known. Zoospores possess two unequal anteriorly-directed flagella, the longer bearing tubular mastigonemes. There are usually two chloroplasts per cell and one, terminating below the position of flagellar insertion, typically bears a red eyespot at its anterior end. The short flagellum has a swelling which overlies the eyespot at its proximal end. The flagellar transition region contains a transitional helix. Both microtubular and cross-banded fibrillar flagellar roots are present. One to several Golgi bodies lie with their forming faces against the nuclear envelope with one edge near the flagellar basal bodies. Chloroplasts in both zoospores and vegetative cells possess regularly spaced three-thylakoid lamellae and a girdle lamella is present in all but three of the species investigated so far. Pyrenoids, in species in which they occur, are present in both vegetative cells and zoospores and are penetrated by normal chloroplast lamellae. The chloroplasts are surrounded by a layer of chloroplast ER and a periplastidial network occurs between the latter and the chloroplast envelope.

GENERAL CHARACTERISTICS

The Xanthophyceae or yellow-green algae are not strictly phytoflagellates since about two-thirds of the 600 or so species so far described (Pascher, 1937-

1939; Ettl, 1978) exist vegetatively as coccoid unicells, about 20% are filamentous and 10%, mostly species of *Vaucheria*, are siphoneous in organization. The proportion of monadoid, rhizopodial and palmelloid forms is, therefore, very small, and doubt must be cast even on the assignation of many of these to this class pending their reinvestigation using modern methods. Nevertheless, the xanthophytes are dealt with here because flagellated cells, mainly asexual zoospores, are produced by the majority of species.

The Xanthophyceae[1] has been monographed by Pascher (1937-1939), by far the greatest single contributor to our knowledge of the group. This account has recently been revised by Ettl (1978) and outstanding nomenclatural problems have been dealt with by Silva (1979). The systematics of the Xanthophyceae are therefore well served and the present account will deal mainly with cytological features, particularly those of the motile cells on which the majority of ultrastructure research has so far been concentrated.

The color of most xanthophytes is predominantly greenish, and while this is usually of a more yellow shade than is typical for the Chlorophyceae confusion with members of this class can be a problem particularly when the structure of the zoospores is not known. Indeed, the distinguishing characters of the group were not recognized until the latter part of the nineteenth century (Borzi, 1895; Bohlin, 1897a; Bohlin, 1897b) and the earliest known species were regarded as Chlorophyceae. The features differentiating Xanthophyceae from the Chlorophyceae which led to its establishment (as the Heterokontae) by Luther in 1899 (Luther, 1899) are principally those of chloroplasts and cell wall and storage metabolites; these are still the main characters by which vegetative cells of the class may be recognized.

Chloroplasts and pigment composition. In addition to their yellow-green rather than grass-green color, there is generally a larger number of chloroplasts in species of Xanthophyceae than Chlorophyceae and they are mostly discoid in shape, parietal in position (Figs. 1, 6, 9) (though becoming deep-seated during zoospore or autospore formation) and show little variation in form (Pascher, 1937-1939; Ettl, 1978). Absence of pyrenoids from xanthophycean chloroplasts has also been considered as characteristic but detailed light microscope investigations (Pascher, 1937-1939; Korshikov, 1930) and more recently electron microscopy (Falk, 1967; Massalski and Leedale, 1969; Hibberd and Leedale, 1971) has shown that many species do, in fact, possess pyrenoids (Fig. 7).

---

[1] See Ettl (1977) for discussion on use of this name.

The pigment composition of xanthophytes is the only area of their biochemistry in which there is any reliable information on a reasonably comprehensive range of species. Of the chlorophylls, only chlorophylls *a* (Strain, 1958) and *c* (both c_1 and c_2) (Guillard and Lorenzen, 1972; Jeffrey, 1976) are found in true xanthophytes and the presence of chlorophyll *b* in any presumed xanthophyte is an *a priori* reason for its transference to one of the classes of green algae (Whittle and Casselton, 1969; Bates and Craigie, 1978). Presence or absence of chlorophyll *b* is the most reliable taxonomic criterion when the gross assignment of very small planktonic species is in question; when chlorophyll *b* is lacking xanthophytes may be distinguished from eustigmatophytes by their distinctive xanthophyll composition (Whittle and Casselton, 1975; Whittle, 1976). Unlike the several other algal classes lacking chlorophyll *b* (the Cryptophyceae and Dinophyceae also lack chlorophyll c_1) the *c* chlorophylls are present in only extremely low concentrations in the Xanthophyceae and are likely to be missed unless specifically looked for. Weight ratios of chlorophylls *a*:*c* range from 55:1 to 116:1 in the 9 species so far investigated (Jeffrey, 1976). Chlorophyll *e* has also been reported in early investigations (Strain, 1958) but it has never been found again and so its existence seems unlikely.

The yellow-green color of xanthophytes is partly due to the masking of the green color of the chlorophyll *a* by carotenoid pigments but probably primarily to the absence of fucoxanthin, the pigment responsible for the overall brown color of chrysophytes, prymnesiophytes, diatoms and the brown seaweeds (Liaan-Jensen, 1977). Work during the last 10 years has established that the carotenoids typically present in the Xanthophyceae include β-carotene and the xanthophylls diatoxanthin, which is dominant, diadinoxanthin, heteroxanthin and vaucheriaxanthin ester (Stransky and Hager, 1970; Whittle and Casselton, 1975; Whittle, 1976; Liaan-Jensen, 1977). Minor xanthophylls include neoxanthin and cryptoxanthin monoepoxide and several others have been identified less commonly. A blue-green color, produced by some species when treated with concentrated hydrochloric acid, thought to be caused by reaction with xanthophyll pigments, is sometimes said to be a test for the group but it is unreliable as some species produce no color change, some turn a shade other than blue-green and others give a variety of colors (Pascher, 1937-1939).

Cell wall structure and composition. The cell wall of many xanthophytes is delicately sculptured, and in others, particularly the filamentous and larger coccoid species, the wall consists of two overlapping halves (Bohlin, 1897a Pascher, 1937-1939; Ettl, 1978). When present, both these characters have diagnostic value. In most cases the bipartite nature is not obvious, becoming apparent only after staining or swelling with alkalis or when the two halves separate for some reason such as zoospore release (Pascher, 1937-1939) (Fig. 35). It must be stressed, however, that a bipartite wall is present in by no means all, perhaps not even a majority of species. In species with bipartite walls the parts may be more or less equal in size or very unequal, but in all cases studied in detail (Falk and Kleinig, 1968; Deason, 1971a; Hibberd and Leedale, 1971) the gradually tapering edges overlap widely (Figs. 10, 12, 13). In the case of filamentous species the walls are formed of interlocking H-pieces (Bohlin, 1897a; Pascher, 1937-1939; Ettl, 1978). Treatment with dilute KOH or NaOH characteristically swells the wall and reveals a complex lamellate construction (Bohlin, 1897a), the walls appearing to be composed of closely stacked deeply U-shaped elements. They also stain strongly with dilute basic dyes such as methylene blue and congo red and are very resistant to strong acids. For all these reasons the walls have often been described as being composed predominantly of an acid pectic compound. However, cellulose has now been identified by a variety of techniques as a major constituent of the continuous tubular wall of the coenocyte *Vaucheria* (Parker *et al.*, 1963; Maeda *et al.*, 1966; Cleare and Percival, 1973). Although *Vaucheria* may not typify the whole class in this respect, its cell wall resembles that of most species in giving a strong positive reaction with Schiff's reagent without prior acid hydrolysis. This neglected and so far little understood reaction is largely confined to

---

Abbreviations used in the legends. C, chloroplast; CER, chloroplast ER; CEN, centrioles; CV, contractile vacuole; ES, eyespot; F^1, long, mastigoneme-bearing flagellum; F^2, short smooth flagellum; FB, flagellar basal body; FS, flagellar swelling; GL, girdle lamella; L, "leucosin" vesicle; M, mitochondrion; ml, m2, m3, microtubular flagellar roots; N, nucleus; O, lipoidal globule; P, pyrenoid; PL, plastoglobuli; PN, periplastidial network; R, rhizoplast; V, "empty" vacuole; VF, vesicle with radiate fibrillar contents; VO, vesicle containing a dense osmiophilic globule; X, vesicle with dense fibrillar contents.

Figs. 1-4. Vegetative cells of *Botrydiopsis* spp. Fig. 1. *B. alpina* (x1500). Fig. 2. Autospores of *B. arhiza* (x1000). Fig. 3. Zoosporangium of *B. arhiza* (x 1000). Fig. 4. Section of *B. alpina* (x10,000).

Figs. 5-7. Vegetative cells of *Mischococcus sphaerocephalus*. Fig. 5. A dichotomously branched colony with the cells borne on the ends of mucilage stalks (x 750). Fig. 6. Vegetative cells showing the parietal chloroplasts (x1500). Fig. 7. Median section showing the parietal chloroplasts with semi-immersed pyrenoids (x10,000).

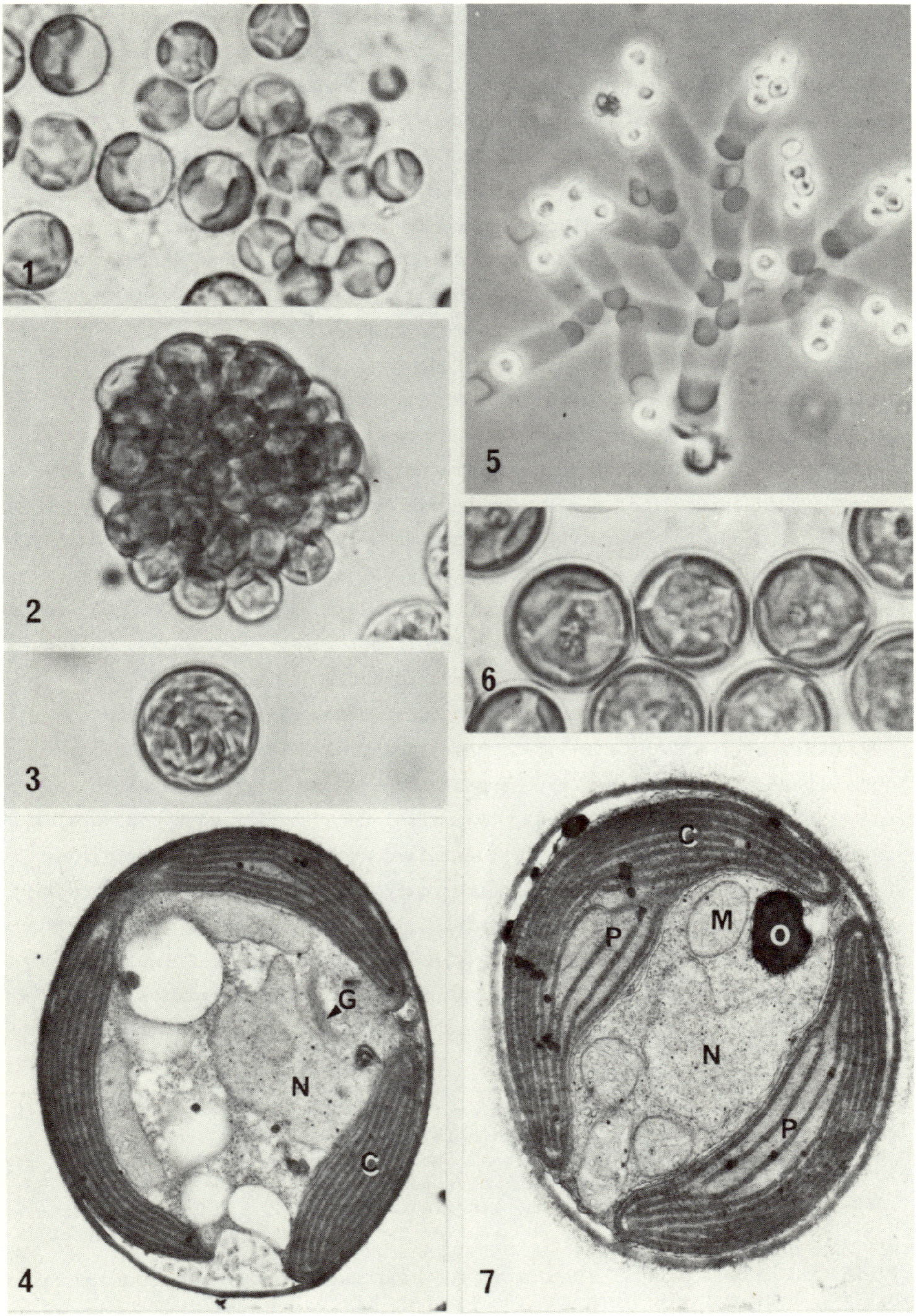
1
5
2
6
3
G
N
C
4
C
M
O
P
N
P
7

the Xanthophyceae (Prat, 1947) and forms a further useful diagnostic character, though since all species do not react a negative reaction is inconclusive. Electron microscopy of sectioned material shows that the cell walls of the species investigated have a distinctly layered structure (Falk, 1967; Falk and Kleinig, 1968; Hibberd and Leedale, 1971) and that the tapering edges of the overlapping walls are held together by a separate layer of cementing material (Falk and Kleinig, 1968; Hibberd and Leedale, 1971) (Fig. 13).

Storage metabolites. Xanthophytes do not store starch and a negative reaction with dilute iodine in potassium iodide is still a useful means of distinguishing them from members of the Chlorophyceae, though the test is often inconclusive on very small cells. Fats, oils and leucosin (chrysolaminaran) have been described as reserves (Pascher, 1937-1939), but little is in fact known of the identity of the storage metabolite in the Xanthophyceae.

Although it appears to be generally accepted that fat accumulation is particularly characteristic of the Xanthophyceae and diatoms, a comparison of the capacity to accumulate fat and determination of the fatty acid, unsaponifiable lipid, total cell nitrogen and hydrolysable polysaccharide contents of representatives of several algal classes (Collyer and Fogg, 1954) showed no important differences in the physiological relations of fat accumulation in the Chlorophyceae, Euglenophyceae, Xanthophyceae and Bacillariophyceae. Fat accumulation in all these groups was associated with low cell nitrogen contents.

Vesicles of leucosin, a β-1,3-linked glucan, have been identified in the Xanthophyceae by their glistening appearance and solubility in a variety of reagents including water (Pascher, 1937-1939) and vesicles with contents similar in appearance to those usually identified as leucosin vesicles in the Chrysophyceae are conspicuous in electron micrographs of the vegetative cells of several species (Hibberd and Leedale, 1971) (Figs. 11, 12). Studies of the carbohydrates of *Tribonema*, however (Cleare and Percival, 1972; Cleare and Percival, 1973), have shown the presence of only a small proportion of β-1,3-linked glucose units and reaction with a specific β-1,3, glucanase failed to

Figs. 8-13. Vegetative cells of *Ophiocytium majus*. Figs. 8-9. Mid- and high-focus levels showing the band-shaped chloroplasts and the positions of the nuclei (x1500). Fig. 10. Empty cell wall showing the tapering mouth (x1500). Fig. 11. Oblique section showing the relative positions of the chloroplasts, a nucleus, Golgi body and a "leucosin" vesicle (x10,000). Fig. 12. Longitudinal section through the distal end showing the two overlapping parts of the cell wall and the position, relative to a nucleus and Golgi body, of a pair of centrioles (x15,000). Fig. 13. Detail of the cell wall showing the intercalary material between the cap and the tubular part of the wall (x25,000).

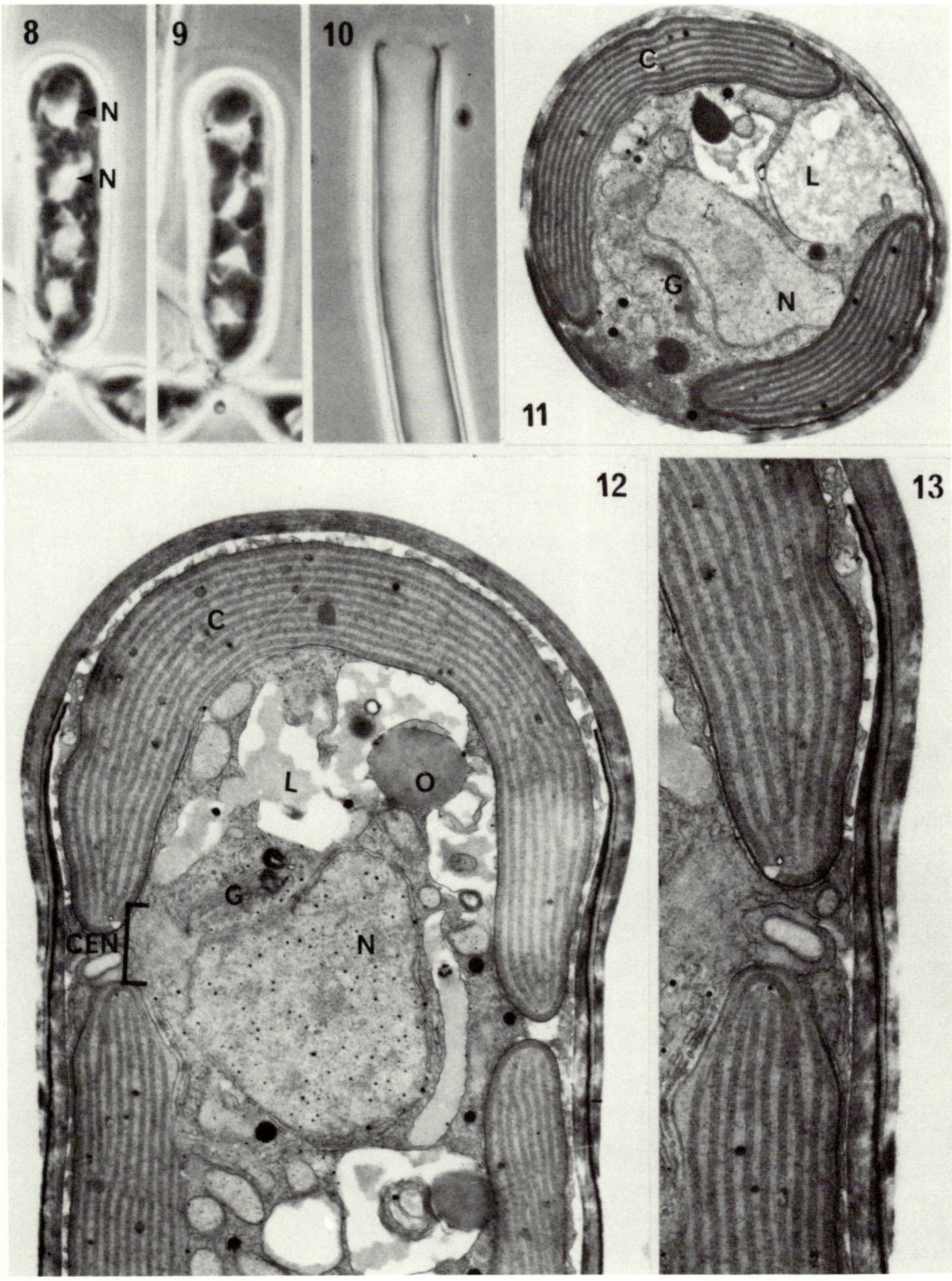
8
N
N
9
10
11
C
L
G
N
12
C
L
O
G
CEN
N
13

confirm their existence. The major alkali-soluble polysaccharides are, in fact, a 1,6-linked glucan and 1,4- and 1,3-linked xylans, and glycerol, erythrytol, xylitol, glucose and mannitol have been characterised as low molecular weight carbohydrates.

Motile cell structure. While the vegetative cells of many coccoid xanthophytes may be difficult to distinguish from green algae or eustigmatophytes, their motile stages (mainly zoospores) are characteristic (Pascher, 1937-1939; Massalski and Leedale, 1969; Hibberd and Leedale, 1971; Ettl, 1978). Although they have a similar basic organization to members of the Chrysophyceae (q.v.) they have yellow-green and not golden-brown chloroplasts and are transitory stages in the life history unlike chrysophytes most of which are permanently motile. Xanthophycean motile cells, with the exception of the spermatozoids and extremely large compound zoospores of *Vaucheria*, are remarkably constant in structure, and in this respect again contrasts with the Chrysophyceae which show a considerable amount of variation on the same basic organization.

Xanthophycean zoospores are mostly ovoid or pyriform in shape, vary from about 5-20 μm in length and are completely naked, being separated from their environment by only a plasmalemma. A degree of plasticity of shape and amoeboid movement are therefore also common characteristics of these cells. The zoospores bear two unequal, anteriorly directed flagella and are bilaterally symmetrical with an obliquely truncate anterior end into which the flagella are inserted in the median plane of symmetry (Figs. 14-16). The long flagellum is mostly approximately the same length as the body of the zoospore while the second is from half to one quarter this length and emerges laterally. There are one to several but most commonly two chloroplasts per cell depending on species. One, usually designated as occupying a ventral position, typically terminates immediately below the truncate face of the cell while the dorsal chloroplast extends into the anterior end (Figs. 15, 16). When there is only a single chloroplast this occupies the ventral position. Zoospores of most species possess a reddish refractile eyespot at the anterior end of the ventral chloroplast (Fig. 16). One or two contractile vacuoles occur in the anterior end of the zoospore and the nucleus, difficult to distinguish in living cells, lies centrally towards the anterior end. The posterior end of the zoospore usually appears vesiculate and dense granules often occur at the periphery.

Early studies using staining techniques developed for bacterial flagella (Vlk, 1931; Vlk, 1938) demonstrated that the long flagellum of xanthophycean zoospores typically was a "Flimmergeissel" bearing lateral "Flimmer" (= tinsel) and that the short flagellum was a "Peitschengeissel" (= whiplash flagellum) not bearing Flimmer but with a thick basal part and a thinner distal part. These observations have been confirmed and extended using electron microscopy as described in detail below.

DETAILS OF ULTRASTRUCTURE

Flagella. Electron microscopy shows (Massalski and Leedale, 1969; Hibberd and Leedale, 1971) that the long flagellum of xanthophycean zoospores bears lateral hairs (Fig. 17) as does the shorter front flagellum of the two relatively long flagella of the spermatozoids of *Vaucheria* (Koch, 1951; Møestrup, 1970). The flagellar hairs are straight and stiff, vary in length between 1 and 2 μm depending on species and are about 20 nm thick with a short tapering basal part attached to the flagellar membrane at the narrow end, a tubular shaft and two long fine terminal filaments of equal or unequal length (Fig. 18). Appendages of this kind are now commonly referred to as tubular mastigonemes (Bouck, 1972). Lateral filaments of the type attached to the shaft of the tubular mastigonemes of some Chrysophyceae and Cryptophyceae (Bouck, 1972) have not yet been found in the Xanthophyceae, and may be absent from this group. The short flagellum of the zoospores and the long posterior flagellum of *Vaucheria* spermatozoids are always free of appendages but may terminate in a hair point (Massalski and Leedale, 1969; Hibberd and Leedale, 1971) (Fig. 17) probably produced by a reduction in the number of axonemal microtubules. The synzoospores of *Vaucheria* are exceptional in that both members of the numerous pairs of subequal flagella are smooth (Koch, 1951; Greenwood *et al.*, 1956; Ott and Brown, 1974a).

The two flagella on a typical zoospore are inserted at an obtuse angle to each other into a raised dome of cytoplasm on the anterior truncate face of the cell (Massalski and Leedale, 1969; Hibberd and Leedale, 1971) (Fig. 20). The long flagellum emerges freely but about the proximal 1 μm of the short flagellum bears an asymmetrically wedge-shaped swelling with electron-dense contents at its proximal end (Figs. 20-22). The swelling lies in a shallow depression in the cell immediately over the eyespot (Figs. 21, 22).

The basal bodies are constructed of the usual 9 triplet-microtubules but the transition region between the basal body and flagellar axis has a charac-

Figs. 14-18. Light microscopy of zoospores of *Ophiocytium majus* (Fig. 14), *Mischococcus sphaerocephalus* (Fig. 15) and *Botrydiopsis alpina* (Fig. 16) (x 2000). Fig. 17. Shadowcast zoospore of *Ophiocytium majus* (x5000). Fig. 18. Tubular mastigonemes on the long flagellum of the zoospore of *Ophiocytium majus* (x20,000).

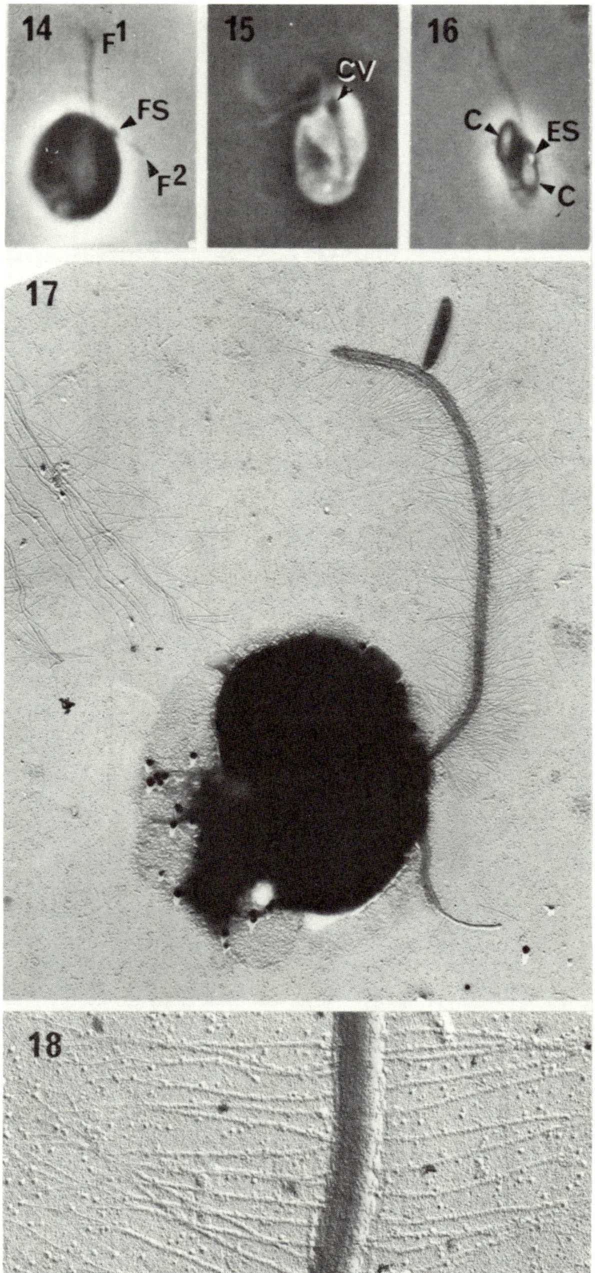
14
F1
FS
F2
15
CV
16
C
ES
C
17
18

teristic structure shared with and best known in the Chrysophyceae (q.v.) (Hibberd, 1980). Thus the lumen of the flagellar axoneme is traversed by a partition with a central axosomal thickening which always occurs exactly at the level at which the flagellum enters the cell body. Close to this and surrounding the proximal few nm of the central pair of tubules of the flagellar axis is a dense helix with mostly 3-4 gyres (Hibberd, 1979) (Fig. 23). This structure, the transitional helix (Hibberd, 1976; Hibberd, 1978a; Hibberd, 1980) has also been called a coiled fiber (Bouck, 1971) and a "*Spiralkörper*" (Casper, 1972) in members of the Chrysophyceae.

There is as yet no complete analysis of the flagellar root system in any xanthophycean motile cell, and only a very limited range of species has been examined. It is clear, nevertheless, that two types of root may originate near the basal bodies: a single narrow cross-banded fibrous root (rhizoplast) which runs closely against the inner surface of the anterior prolongation of the nucleus (Fig. 24); and at least three tubular roots extending in various directions through the cytoplasm (Figs. 25, 26). The number of component microtubules in each of these roots is three or four near the basal bodies. The very small size and inconspicuous nature of the rhizoplast in the Xanthophyceae contrasts with the massiveness of this structure in some Chrysophyceae. The spermatozoids of *Vaucheria* are unusual in having an anterior "proboscis", similar to that in the spermatozoids of the brown alga *Fucus*, which contains a band of mostly 8 or 9 microtubules originating near the basal bodies (Moestrup, 1970).

Chloroplast, pyrenoid and eyespot. The chloroplast lamellae are usually composed of 3 thylakoids which show varying degrees of coherence depending on species. In the majority of investigated forms a single lamella, the girdle lamella or girdle band, encircles the others round the edge of the chloroplast (Fig. 27), enclosing between itself and the ends of the parallel lamellae a region of less dense matrix (Falk and Kleinig, 1968; Massalski and Leedale, 1969; Hibberd and Leedale, 1971), almost certainly the location of the chloroplast DNA as in the Chrysophyceae and Phaeophyceae (Bisalputra, 1974). The absence of a girdle lamella in *Bumilleria sicula* (Massalski and Leedale, 1969),

Figs. 19-23. Structure of the zoospores of *Ophiocytium majus*. Figs. 19, 20. Longitudinal sections (x10,000). Figs. 21, 22. Longitudinal and transverse sections of the flagellar swelling and eyespot. Fig. 21 (x 25,000); Fig. 22 (x 30,000). Fig. 23. Longitudinal section of the flagellar basal body and transition region with a transitional helix (arrowhead) (x50,000).

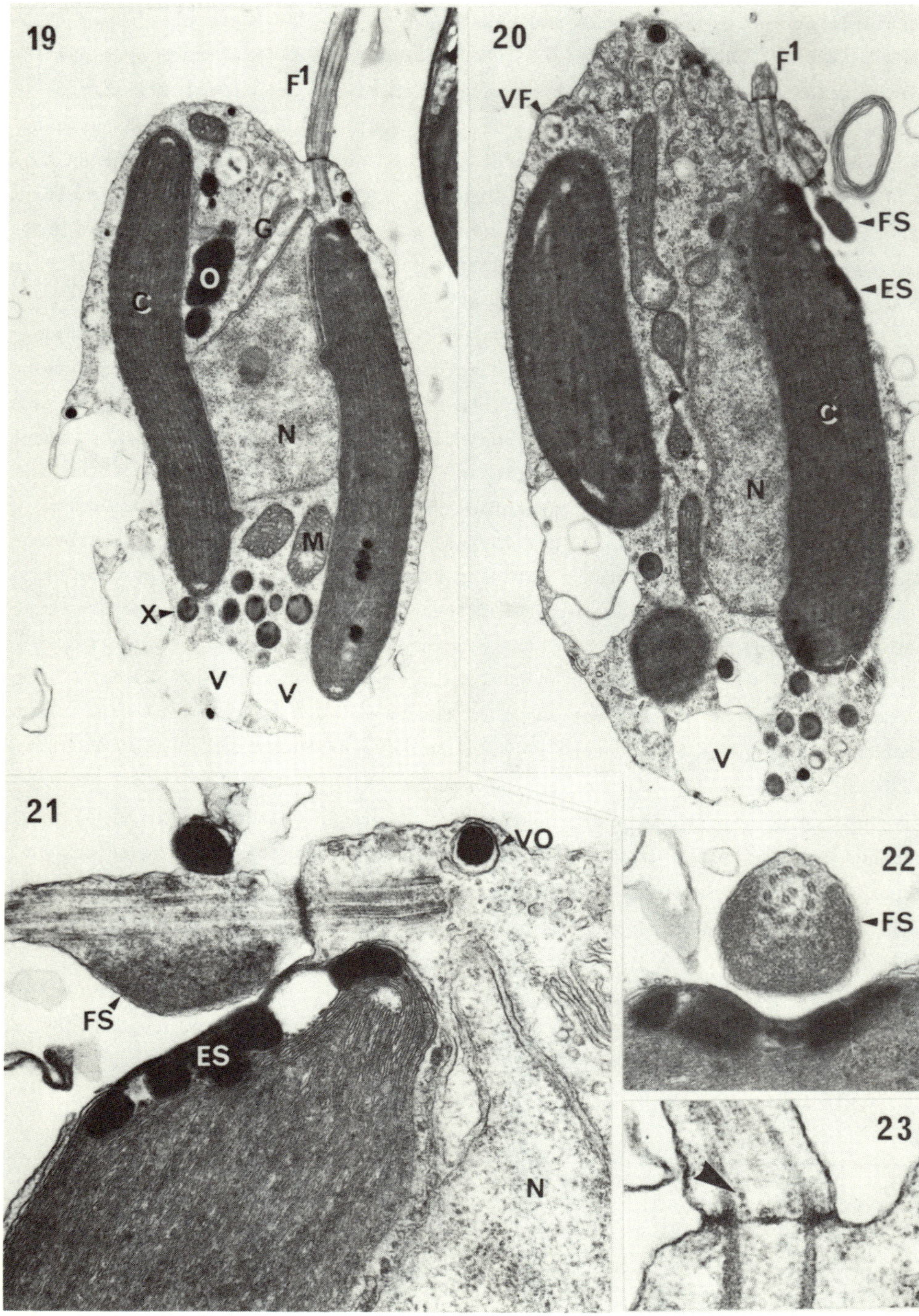
19
F1
G
O
C
N
M
X
V
V
20
F1
VF
FS
ES
C
N
V
21
VO
FS
ES
N
22
FS
23

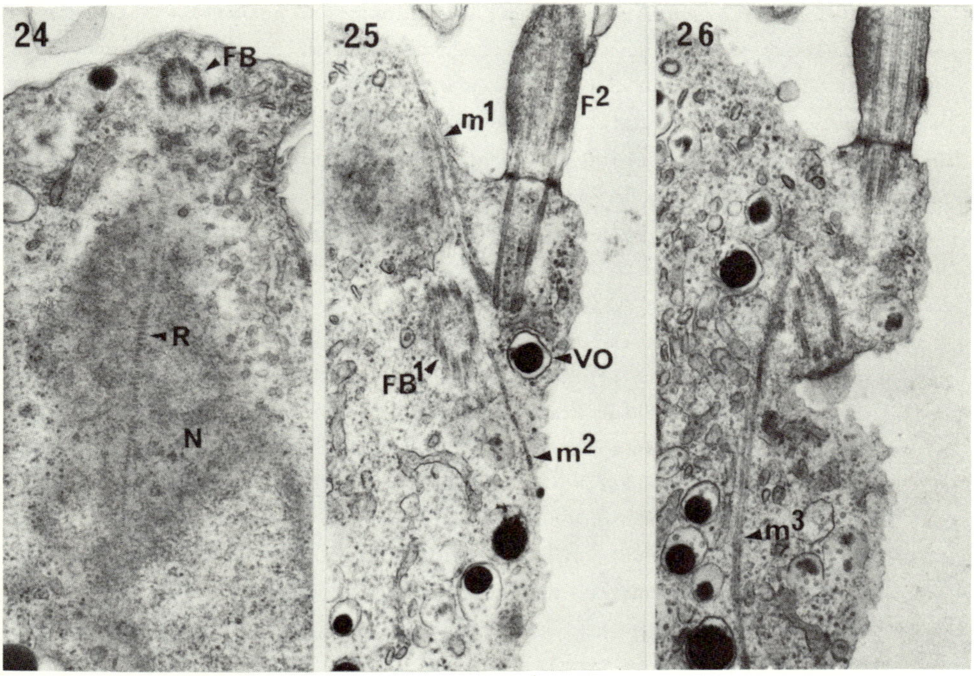

Figs. 24-26. Flagellar roots in zoospores of *Ophiocytium majus*. Fig. 24. Rhizoplast and nucleus (x25,000). Figs. 25, 26. Microtubular roots (x25,000).

Bumilleriopsis filiformis (Hibberd and Leedale, 1971; Böger and Kiermayer, 1974) and *Pseudobumilleriopsis pyrenoidosa* (Deason, 1971a) is probably due to secondary loss, since these three species are relatively advanced and seem closely related.

The chloroplasts are bounded by the usual double-membraned chloroplast envelope and also by a layer of chloroplast ER, the outermost membrane of which is continuous with the outermost membrane of the nuclear envelope (Figs. 21, 29, 30). The region between the chloroplast envelope and chloroplast ER, the perichloroplastic space (Bisalputra, 1974) or chloroplast cell compartment (Coombs and Greenwood, 1976), contains a membranous reticulum, best developed where the nuclear envelope and CER are confluent, which appears either tubular or vesicular in transverse section (Massalski and Leedale, 1969; Hibberd and Leedale, 1971) (Fig. 29). It was first described from a xanthophyte by Falk and Kleinig (1968) as the "Periplastidär-Reticulum" (periplastidial network). This network shows occasional connections with the inner membrane of the CER (Falk and Kleinig, 1968), and clusters of 3-6 spherules with dense walls can commonly be seen in tangential sections (Massalski and Leedale, 1969; Hibberd and Leedale, 1971). These structures are particularly evident in xanthophycean zoospores (Fig. 29) though they are also found in the vegetative cells. They also occur in the Chrysophyceae (Hibberd, 1971; 1978b). The composition of the spherules and their exact relationship with the periplastidial network has not yet been established.

Pyrenoids occur in both the zoospore and vegetative cell in species in which they are present. They are typically of a semi-immersed type and form a bulge on the inner face of the chloroplast (Descomps, 1963; Falk, 1967; Massalski and Leedale, 1969; Hibberd and Leedale, 1971; Deason, 1971a) (Fig. 28). They are traversed by normal three-thylakoid lamellae but these are more widely spaced than in the remainder of the chloroplast and may also show discontinuities (Massalski and Leedale, 1969). Thylakoid-free outpushings in *Tribonema viride* (Falk and Kleinig, 1968) and the germlings of *Vaucheria woroniniana* (Marchant, 1972) have been described as projecting pyrenoids but these attributions must remain in doubt pending more thorough investigation. No storage material or capping vesicle has been found to be associated with pyrenoids in the Xanthophyceae but lipid droplets (plastoglobuli), which are normally distributed randomly in the chloroplast matrix (Fig. 27) are often seen to be concentrated at the periphery of the pyrenoid (Falk, 1967; Massalski and Leedale, 1969; Hibberd and Leedale, 1971; Deason, 1971a).

Xanthophycean eyespots consist of a single layer of densely osmiophilic globules which lies at the anterior end of the ventral chloroplast of the zoospore immediately within the chloroplast envelope and directly beneath the swelling on the short flagellum (Fig. 21) (Massalski and Leedale, 1969; Hibberd and Leedale, 1971). Eyespots appear to be absent from the zoospores of many xanthophycean species (Pascher, 1937-1939; Ettl, 1978), but this may simply be a result of their small size. In *Vaucheria*, however, their absence is genuine; the spermatozoids lack a chloroplast (Moestrup, 1970) and the zoospores have chloroplasts without eyespots (Ott and Brown, 1974a).

Nucleus, Golgi body and mitochondria. The nucleus in xanthophycean zoospores is pyriform and occupies a characteristic position lying against the inner face of the ventral chloroplast with its tip close to the flagellar basal bodies (Falk, 1967; Massalski and Leedale, 1969; Deason, 1971a; Hibberd and Leedale, 1971) (Figs. 19, 20, 30). In the multiflagellate zoospores of *Vaucheria* one nucleus is attached to each pair of basal bodies but there is no association with the chloroplasts (Greenwood *et al.*, 1956; Ott and Brown, 1974a). The nucleus has a normal double-membraned envelope but in *Bumilleria sicula* patches of a delicate network with 15 nm openings occur in the perinuclear space (Massalski and Leedale, 1969). A further unusual feature in this species is that endoplasmic reticulum has been found organized parallel to areas of the nuclear envelope with circular fenestrations exactly opposed to the nuclear pores. The nucleus in vegetative cells is more regular in shape than that in the zoospores (Figs. 4, 7, 11, 12) but one end is still extended, in this case towards a centriole pair (Deason, 1971a; Hibberd and Leedale, 1971) (Fig. 12). While connections between the nuclear envelope and CER can be found in vegetative cells (Figs. 4, 11) they are much less conspicuous than in the zoospore.

Golgi bodies in the Xanthophyceae are typically small, consisting of only 3-7 cisternae (Falk, 1967; Falk and Kleinig, 1968; Massalski and Leedale, 1969; Hibberd and Leedale, 1971) (Fig. 30). Although these are not usually particularly inflated at their edges they may be considerably fenestrated (Massalski and Leedale, 1969). In zoospores there is generally only a single Golgi body which lies with its forming face close against the nuclear envelope and with one edge near the flagellar basal bodies (Massalski and Leedale, 1969; Hibberd and Leedale, 1971) (Figs. 19, 30). When more than one Golgi body is present all of them lie against the anterior prolongation of the nucleus (Massalski and Leedale, 1969). The same positional relationship between the nucleus, Golgi body and centrioles exists in the vegetative cells and the Golgi body lies against a flattened or concave face of the nuclear envelope (Falk and Kleinig, 1968; Deason, 1971a; Hibberd and Leedale, 1971) (Figs. 4, 11, 12). Species of *Vaucheria* are unusual in that each of the numerous Golgi bodies is associated with an ER cisterna and a mitochondrion (Greenwood, 1959; Ott and Brown, 1974a). Mitochondria in the Xanthophyceae are of the microvillar type,

---

Figs. 27-30. Fig. 27. The edge of a chloroplast in a zoospore of *Ophiocytium majus* (x35,000). Fig. 28. Pyrenoid of *Mischococcus sphaerocephalus* (x30,000). Fig. 29. Detail of nuclear envelope, chloroplast ER and periplastidial network in a zoospore of *Ophiocytium majus* (x50,000). Fig. 30. Structure and position of the Golgi body in a zoospore of *Ophiocytium majus* (x50,000).

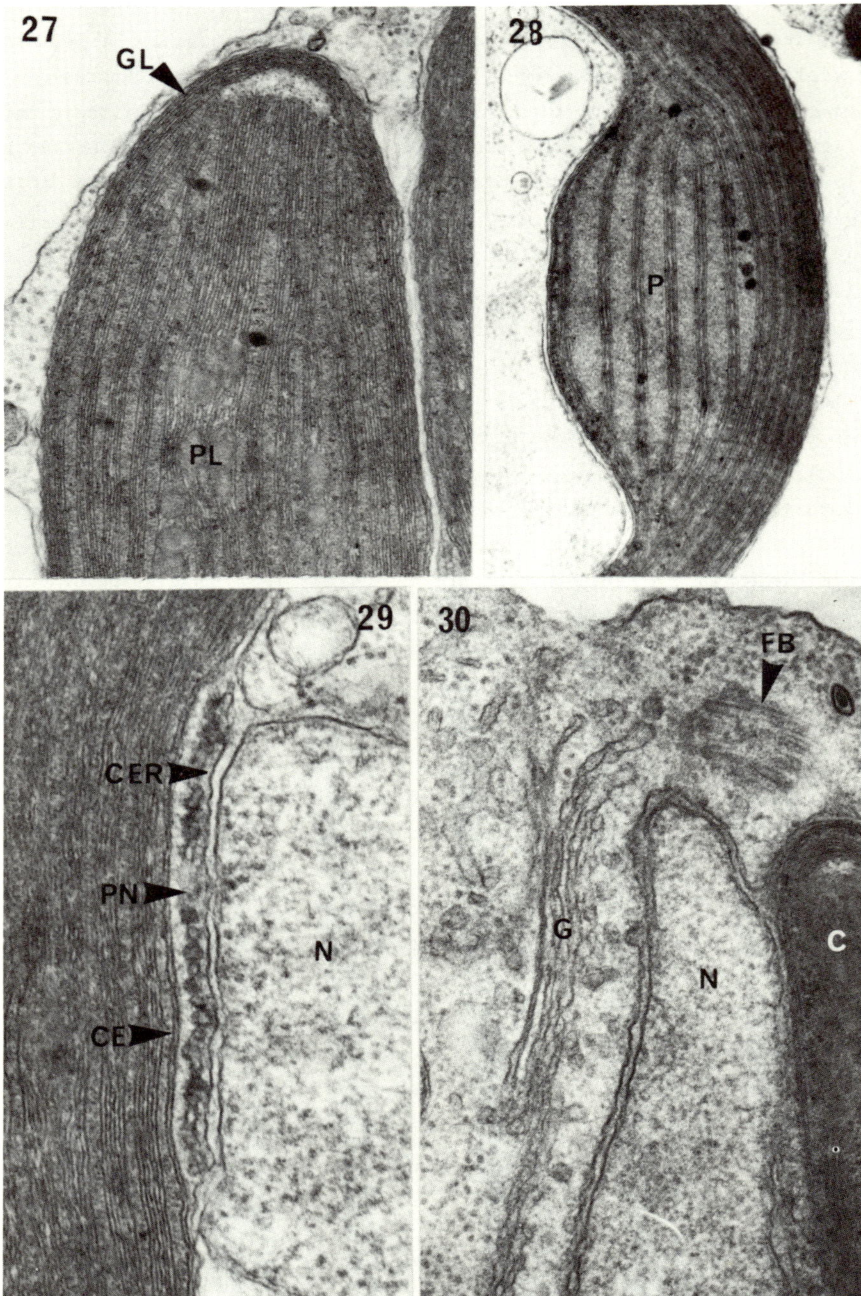
27
GL
PL
28
P
29
CER
PN
CE
N
30
FB
G
N
C

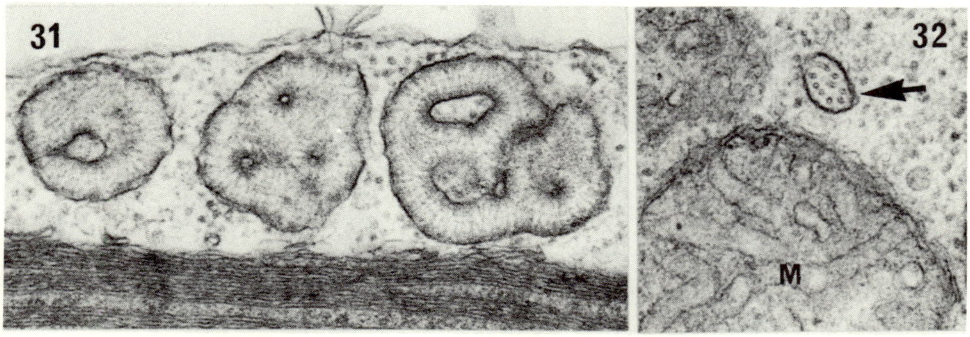

Figs. 31, 32. Fig. 31. Perforated disc-shaped vesicles with radiate fibrous contents at the periphery of a zoospore of *Ophiocytium majus* (x50,000). Fig. 32. Part of a mitochondrion in a zoospore of *O. majus* showing the tubular cristae and a transverse section of an ER cisterna containing tubular mastigonemes (arrow) (x50,000).

that is, the invaginations of the inner mitochondrial membrane into the matrix are tubular in shape (Fig. 32).

Contractile vacuoles, vesicles and droplets. The contractile vacuoles occur in the anterior region of the cell adjacent to the Golgi body and flagellar bases (Massalski and Leedale, 1969; Hibberd and Leedale, 1971) (Fig. 15). They show no particularly noteworthy structural features, but coated vesicles may often be found in close proximity. Zoospores of some species contain several characteristic types of vesicles and inclusions (Massalski and Leedale, 1969; Hibberd and Leedale, 1971). Most commonly seen are small peripheral vesicles containing a single spherical densely osmiophilic body (Massalski and Leedale, 1969; Deason, 1971a; Hibberd and Leedale, 1971) (Figs. 21, 25, 26), possibly produced in the Golgi body (Deason, 1971b). Others with less dense spirally wound fibrous contents occur at the posterior end of the zoospores of *Ophiocytium majus* (Hibberd and Leedale, 1971) (Figs. 19, 20) and appear to contain the adhesive material by which the zoospore attaches itself during settling (Fig. 39). Peculiar fibrous disc-like vesicles apparently produced in the perinuclear space have been found in the zoospore of more than one species (Massalski and Leedale, 1969; Deason, 1971a) (Figs. 20,31) and possibly give rise to the cell wall material as the cells settle. Irregularly shaped droplets, probably of lipoidal material, are generally distributed in the cell (Figs. 7, 19) and large, apparently empty vacuoles regularly occur in the posterior ends of zoospores (Massalski and Leedale, 1969; Hibberd and Leedale, 1971) (Figs. 19, 20).

MOVEMENT PATTERNS

The long flagellum of biflagellate xanthophycean zoospores beats in an approximate sine wave propagated at the tip of the flagellum which, as in other flagella bearing tubular mastigonemes, "pulls" the cell through the water (Jahn *et al.*, 1964). The cells rotate about their axis and describe a somewhat helical path. This may be due to the action of the short flagellum, which is difficult to see during life but appears to beat laterally. The smooth flagellum in the spermatozoids of *Vaucheria*, which is longer than the mastigoneme-bearing flagellum, shows little or no movement (Møestrup, 1970).

REPRODUCTION AND LIFE HISTORIES

Sexual reproduction in the Xanthophyceae appears to be rare and virtually confined to the phylogenetically advanced siphoneous forms. The majority of the group reproduce asexually in a variety of ways, the method being dependent on the basic cell form (Pascher, 1937-1939; Ettl, 1978). The few monadial or rhizopodial forms uncertainly attributed to this class divide bilaterally and also appear to produce endogenous cysts similar but not identical to those characteristic of the Chrysophyceae. The palmelloid forms can reproduce by division but also produce zoospores or their non-flagellate equivalent "hemiautospores" (Ettl, 1978) or "aplanospores". The majority of species, the coccoid forms, reproduce primarily by the production of autospores, small replicas of the parent cells. Zoospore formation is also common among the coccoid species though in culture large numbers are normally produced only after a change to fresh medium or some other stimulation. The most usual number of both aplanospores and zoospores produced by each mother cell is two or four, though many more may be found in multinucleate species showing a particularly large increase in vegetative cell size (Fig. 2). Multiple zoospores or synzoospores, probably resulting from incomplete cleavage, occur in siphoneous, filamentous and some multinucleate coccoid forms; species producing them usually also have normal biflagellate zoospores (Pascher, 1939). Filamentous forms reproduce either by cell division leading to filament growth or by zoospores. As in the coccoid forms zoospores are common but are produced in large numbers in culture only following some environmental shock (Massalski and Leedale, 1969; Pecora and Rhodes, 1973).

In the siphoneous genus *Botrydium*, asexual reproduction is by means of either biflagellate zoospores or aplanospores. Sexual reproduction involving biflagellate gametes, which may be either isogamous or anisogamous, has also been described (Moewus, 1940). Asexual reproduction in *Vaucheria* is exclusively by the production of synzoospores in zoosporangia formed by septation of the tips of filaments. Sexual reproduction in this genus is oogamous. Antheridia and oogonia are formed either on special lateral branches of the main filament or are sessile. The antheridium produces a large number of colorless spermatozoids which are different structurally in a number of ways from the xanthophycean zoospores already described (Moestrup, 1970). The mature oogonium is uninucleate and fertilization through a special pore produces a thick-walled resting zygote. Meiosis occurs at germination which yields a new siphoneous filament.

With the exception of the oospores of *Vaucheria*, resting stages are seen only rarely in the Xanthophyceae. Endogenously produced cysts with a bipartite wall appear to be confined to the few monadial or rhizopodial forms. The most common resting stage in coccoid and filamentous forms is the akinete, a single-celled spore in which the spore wall is formed by a thickening of the parent cell wall. A particular type of aplanospore with a bipartite cell wall of quite a different shape from that of the parent cell has also been described for some coccoid and filamentous forms, including *Tribonema* (Pascher, 1937-1939; Ettl, 1978). Both akinetes and this type of aplanospore produce zoospores or "hemiautospores" on germination.

MITOSIS

Mitosis has been studied at the ultrastructural level in only one species of *Vaucheria* (Ott and Brown, 1972). The most noteworthy feature of nuclear division is that the nuclear envelope remains completely intact at all stages, and is not penetrated at the poles by gaps or fenestrae as commonly found in other examples of "closed" division. Following migration to opposite poles of each of the pair of centrioles normally associated with each nucleus, a spindle consisting of both continuous and non-continuous microtubules forms within the nuclear envelope. The nucleolus fragments by metaphase, and the centrioles are found to have duplicated by early anaphase. Anaphase is characterized by the formation of an extremely long interzonal spindle, probably by extension of the continuous spindle microtubules. By early telophase the spindle is very narrow and consists only of about 24 microtubules. Each nucleus is then cut off from the spindle by invagination of the nuclear envelope and the nucleolus reforms. Even if this type of closed division is found to be typical of the Xanthophyceae as a whole it is likely that the very long interzonal spindle is confined to *Vaucheria* and perhaps the other coenocytic forms.

Conventional chromosome cytology of the Xanthophyceae is extremely difficult owing to the generally small cell and nuclear size, but counts of 17 chromosomes have been published for four species of *Tribonema* (Iorya, 1974).

ULTRASTRUCTURAL ALTERATIONS DURING ZOOSPOROGENESIS, GAMETOGENESIS AND ZOOSPORE SETTLING

Zoosporogenesis. Fine structure details are available for only a few species (Falk, 1967; Hibberd, 1969; Deason, 1971b) but a similar pattern of events apparently occurs. *Vaucheria* once again diverges from the basic pattern and will be dealt with separately later. In the light microscope zoospore formation is indicated by the chloroplasts moving away from the walls of the cell to take up a more internal position and by the appearance of eyespots in the chloroplasts (Pascher, 1937-1939; Ettl, 1978) (Figs. 3, 33). Electron microscopy shows that one or two chloroplasts come to lie against each nucleus, the degree of association usually being greater than in the normal vegetative cell (Fig. 34). The number of chloroplasts surrounding the nuclei is the same as the number in the fully-formed zoospore which, as previously indicated, is species specific. An eyespot appears in the chloroplast or in one of the chloroplast pair at a relatively late stage in zoospore formation (Fig. 33). The pair of centrioles normally associated with each nucleus comes to lie at the end of one of the chloroplasts and each pair produces two flagella in a vesicle (Deason, 1971b). The Golgi body maintains its position relative to the centrioles and nuclear envelope.

Flagellar hair formation is one of the better-documented aspects of zoospore development (Leedale *et al.*, 1970; Deason, 1971). The hairs appear to be formed in the perinuclear space and are then transported to the cell surface in vesicles of the endoplasmic reticulum (Fig. 32). The manner of transfer and attachment of the hairs to the flagellar membrane has not been determined or why they become attached to one flagellum and not the other. The Golgi body does not appear to be involved in flagellar hair formation as it is in the Chrysophyceae where it adds the lateral filaments to the hairs (Bouck, 1971).

Zoospore cleavage occurs by the formation and coalescence of vesicles which

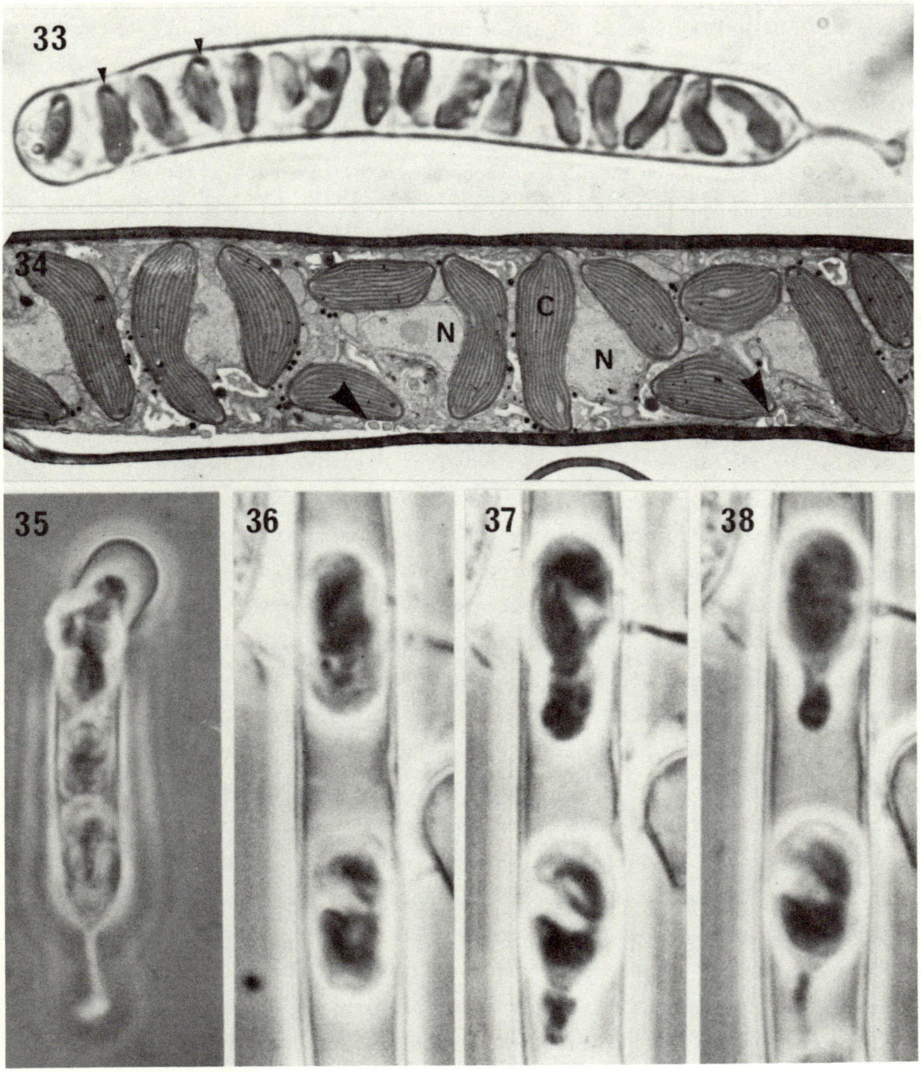

Figs. 33-38. Fig. 33. Zoosporangium of *Ophiocytium majus*; some of the chloroplasts contain eyespots (arrowheads) (x2000). Fig. 34. Longitudinal section of a zoosporangium of *O. majus*; arrowheads, flagellar profiles (x5000). Fig. 35. Release of an amoeboid zoospore of *O. majus* by displacement of the cap (x1500). Figs. 36-38. Stages in zoospore settling in *O. majus*; the cells are trapped within the zoosporangium wall (x2000).

separates the nuclei and associated chloroplasts (Deason, 1971b). Zoospore release is by rupture or gelatinization of part of the mother cell wall or, in species with bipartite walls, by separation of the two sections of the wall (Fig. 35). Zoospores are amoeboid immediately after release but swim rapidly when completely free.

The compound zoospores of *Vaucheria* are formed in the apical region of a vegetative filament where organelles accumulate before separation of the zoosporangium by a septum (Ott and Brown, 1974b). The normal association between the mitochondria and Golgi bodies disappears and flagella form in internal vesicles from the pair of centrioles accompanying each nucleus. These vesicles eventually coalesce to produce large internal flagellar pools which then migrate to the surface of the maturing zoospore and evaginate so that the flagella become situated on the zoospore surface. Septation of the zoospore from the vegetative filament occurs during this stage. The compound nature of the zoospore is revealed by the production of an incipient cleavage furrow which "attempts" to cut off the chloroplasts from the remainder of the zoospore.

Gametogenesis. Information on gametogenesis is virtually restricted to spermatogenesis in *Vaucheria* (Ott and Brown, 1978). The process is very similar to that which takes place during zoosporogenesis until flagellar pool formation. Flagellar hairs originate in the ER during later stages of flagellum formation and mitochondria become closely adpressed to each nuclear envelope. The Golgi bodies at first lie external to the mitochondria but later themselves come to lie against the nucleus. Nuclear division continues to take place after the flagella are formed. During the final stage of spermatogenesis the nuclei elongate and the proboscis develops, extending into the flagellar vacuole formed from the coalescence of the flagellar vesicles. Cleavage furrows, typically having ER but only rarely microtubules associated with them, gradually delimit the individual spermatozoids which come to lie in a large internal vacuole. The chloroplasts and mitochondria remain in the peripheral cytoplasm. There are no residual nuclei, all being used in spermatozoid formation.

Zoospore settling and germination. Published information on zoospore settling is so far only available for *Vaucheria* (Ott and Brown, 1975). The flagella are retracted into the body of the zoospore and the peripheral layer of cytoplasm, which contains only nuclei and vesicles, alters in appearance. The nuclei sink into the cytoplasm and the vesicles disappear, probably giving rise to the cell wall, and are replaced by chloroplasts and other organelles. The withdrawn flagellar axonemes disintegrate and a bulge forms at one end of the settled cell. The various organelles and vesicles then become arranged as in

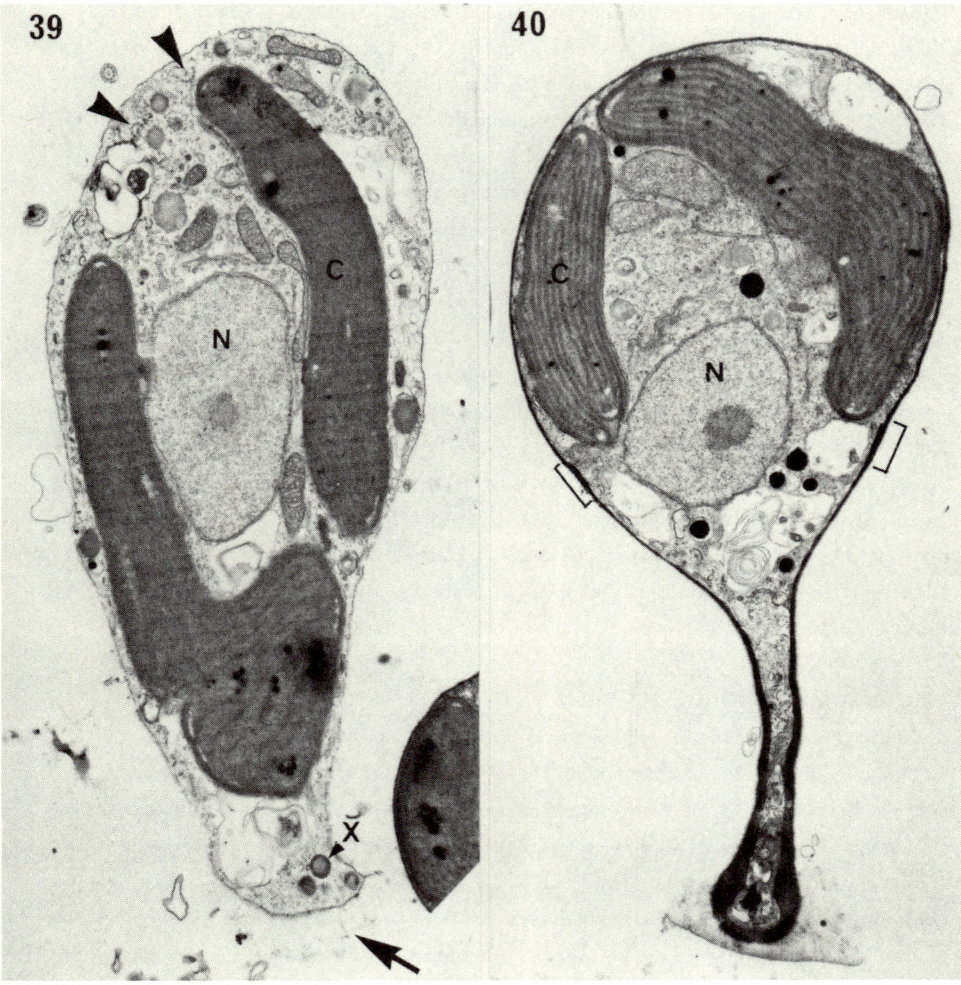

Figs. 39, 40. Fig. 39. Longitudinal section through a settling zoospore of *Ophiocytium majus*. The flagellar axonemes have been withdrawn and are disintegrating (arrowheads) and the stipe is forming finely fibrillar adhesive material (arrow) (x10,000). Fig. 40. Germling of *O. majus*. Brackets indicate the region of overlap of the two parts of the first formed very dense wall. A fibrillar pad attaches the cell to the substratum (x7500).

the vegetative cells and a large central vacuole forms.

In typical zoospores, settling takes place by gradual rounding off of the cell body, withdrawal of the flagella, re-organization of organelles and wall formation. In the case of species with bipartite walls their bipartite nature is established very early in germination. In species in which the vegetative cells are normally attached, settling involves the rapid elongation and attachment of the posterior end of the zoospore. Details are available only for *Ophiocytium majus*. After withdrawal of the flagella the posterior end of the cell, which contains dense granular material, elongates and then narrows within seconds forming the stipe of the new vegetative cell (Figs. 36-38). The distal end of the forming stipe becomes surrounded by finely fibrillar adhesive material (Fig. 39) which is apparently released from the dense vesicles with fibrillar contents since these are not present in later stages of germination. The flagellar axonemes disintegrate and the eyespot disappears. The first-formed wall is very dense and two overlapping parts can be seen even at this early stage (Fig. 40). This initial wall is probably formed from the fibrous material contained within the discoid vesicles illustrated in Fig. 31 since these also disappear during settling. The chloroplasts come to lie closely against the wall, the nucleus becomes more regular in shape and a normal, lamellate and less dense wall is laid down, thus re-establishing the typical vegetative cell.

BIOCHEMISTRY AND PHYSIOLOGY

With the exception of pigment composition, so little is known of the biochemistry and physiology of xanthophytes that comparisons with any other groups are not yet possible. The most detailed investigations have been those testing the ability of a number of species to grow in the dark using organic substances as sources of carbon (Belcher and Miller, 1960; Casselton, 1966). Most species behave as facultative chemo-organotrophs but it is interesting that all but one of the species found to be obligate phototrophs are now placed in the Eustigmatophyceae. In addition, "*Monodus subterraneus*", a eustigmatophyte, was found to utilize glutamine as a nitrogen source by deamidation to glutamic acid whereas the xanthophytes *Botrydiopsis intercedens* and *Tribonema minus* did not (Belmont and Miller, 1965).

Isolated chloroplasts of *Bumilleriopsis filiformis* have been found to be particularly suitable for study of enzymological problems of photosynthetic electron transport (Böger, 1969) and interaction of isolated carriers (Boger, 1971). In addition, 95-98% of division synchrony has been induced in this species by changes of daylength, light intensity and temperature (Hesse, 1974).

HABITAT

Xanthophytes occur predominantly in freshwater habitats (Pascher, 1937-1939; Ettl, 1978). Most taxa are seen only rarely and many have not been found again since they were first described. This is probably because they mostly occur in small numbers, and they are possibly not as rare as is generally supposed, 56 species and varieties having been described in one recent survey (Tarapchak, 1972). Most unattached coccoid forms occur in still waters among other algae or submerged vegetation, particularly in low pH environments which are also often rich in iron. Some species are clearly calcicolous, however, and others are truely planktonic. Coccoid Xanthophyceae are relatively well represented in the algal flora of soils (Vischer, 1945). However, several soil algae attributed to the Xanthophyceae have now been shown to be eustigmatophytes, and this may also be true of others. Species of the filamentous *Tribonema* occur relatively common among or forming floating mats in still water, particularly in late winter. The most conspicuous forms in nature are the siphoneous genera *Botrydium* and *Vaucheria*. The former, though relatively uncommon, occurs as large green vesicles sometimes in vast numbers on the surface of drying mud, produced, for example, after the draining of a pond or reservoir. The well-known *Vaucheria* is widespread in freshwater, brackish and marine habitats, growing either submerged or on damp or splashed surfaces.

CULTURE MEDIA

Xanthophytes are relatively easy to cultivate in the laboratory. They grow well, though not as rapidly as most green algae, on most defined mineral media. Both dilute (Chu No. 10) and rich (Bold's Basal Medium) media have been used successfully (Massalski and Leedale, 1969; Hibberd and Leedale, 1971; Guillard and Lorenzen, 1972). These and other suitable mineral media are listed by Nicholls (1973). Although the majority of species investigated so far are auxotrophic (Provosoli and Carlucci, 1974) growth can be improved by supplementing the mineral medium with either soil extract or vitamins. The standard methods of purification suffice for most species and a simple but effective method for producing axenic cultures of *Vaucheria* has recently been devised (Åberg and Fries, 1976). Strains may be routinely maintained on nutrient agar slopes or in soil-water tubes.

NEEDED RESEARCH

With the exception of *Vaucheria* the Xanthophyceae is a relatively neglected group and virtually any aspect of its biology would repay further investigation. Biochemical and physiological studies are particularly lacking and not even the nature of the main storage metabolite is known with certainty. Although the zoospore structure of several species has been examined in some detail, comparative observations on a greater range of species would be useful. The processes of zoospore formation and settling have not yet been investigated ultrastructurally and an attached species such as *Ophiocytium majus* may provide a useful subject for studying the biochemical aspects of wall formation. Probably the main gap in knowledge of the structure of the Xanthophyceae concerns the nucleus and cell division.

Although the Xanthophyceae have been thoroughly monographed, knowledge of their occurrence and distribution is poor and a large number of forms originally described by Pascher (1937-1939) have not been found again since. In addition, relatively few species are available in culture, most of these having been isolated over 30 years ago by Vischer. Several of these have recently been transferred to the new algal class Eustigmatophyceae but until more forms are isolated it will not be possible to say whether any of the described species of coccoid Xanthophyceae should be included in the Eustigmatophyceae as the descriptions mostly do not include reference to the characters diagnostic for this group.

ACKNOWLEDGEMENTS

I am grateful to Dr. F. K. Kupicha for help in the preparation of this chapter.

REFERENCES

Åberg, H. and L. Fries. 1976. On cultivation of the alga *Vaucheria dichotoma* (Xanthophyceae) in axenic culture. Phycologia 15: 133-141.

Bates, S. S. and J. S. Craigie. 1978. Chloroplast pigments of a green phytoplankter from the Hudson Estuary, U.S.A. Phycologia 17: 79-84.

Belcher, J. H. and J. D. A. Miller. 1960. Studies on the growth of the Xanthophyceae in pure culture. IV. Nutritional types amongst the Xanthophyceae. Arch. Mikrobiol. 36: 219-228.

Belmont, L. and J. D. A. Miller. 1965. The utilization of glutamine by algae. J. Exp. Bot. 16: 318-324.

Bisalputra, T. 1974. Plastids. In Stewart, W. D. P. [Ed.], *Algal Physiology and Biochemistry*. Blackwell, Oxford.

Böger, P. 1969. Photophosphorylierung mit Chloroplasten aus *Bumilleriopsis filiformis* Vischer. Z. Pflanzenphysiol. 61: 85-97.

Böger, P. 1971. Zusammenhang zwischen Transhydrogenase- und Diaphorase-Aktivität der Ferredoxin-NADP-Reduktase und der photosynthetischen NADP-Reduktion. Z. Naturforsch. 26b: 807-815.

Böger, P. and O. Kiermayer. 1974. Electron microscopy of plastids of *Bumilleriopsis filiformis*. Arch. Mikrobiol. 98: 207-214.

Bohlin, K. 1897a. Studier öfver några slägten af alggruppen Confervales Borzi. Bih. K. Svenska VetenskAkad. Handl. 23, Afd. III, No. 3: 1-56.

Bohlin, K. 1897b. Zur morphologie und biologie einzelliger algen. Öfvers. K. VetenskAkad. Förh. 54: 507-529.

Borzi, A. 1895. Studi Algologici. Fasc. II. A. Reber, Palermo.

Bouck, G. B. 1971. The structure, origin and composition of the tubular mastigonemes of the *Ochromonas* flagellum. J. Cell Biol. 50: 362-384.

Bouck, G. B. 1972. Architecture and assembly of mastigonemes. Adv. Cell Molec. Biol. 2: 237-271.

Casper, S. J. 1972. Zum Feinbau der Geisseln der Chrysomonaden. I. *Uroglena americana* Calkins. Arch. Protistenk. 114: 65-82.

Casselton, P. J. 1966. Chemo-organotrophic growth of xanthophycean algae. New Phytol. 65: 134-140.

Cleare, M. and E. Percival. 1972. Carbohydrates of the fresh water alga *Tribonema aequale*. I. Low molecular weight and polysaccharides. Br. Phycol. J. 7: 185-193.

Cleare, M. and E. Percival. 1973. Carbohydrates of the freshwater alga *Tribonema aequale*. II. Preliminary photosynthetic studies with ^{14}C. Br. Phycol. J. 8: 181-184.

Collyer, D. M. and G. E. Fogg. 1954. Studies on fat accumulation by algae. J. Exp. Bot. 6: 256-275.

Coombs, J. and A. D. Greenwood. 1976. Compartmentation of the photosynthetic apparatus. In Barber, J. [Ed.] *The Intact Chloroplast*. Elsevier, Amsterdam.

Deason, T. R. 1971a. The origin of flagellar hairs in the xanthophycean alga *Pseudobumilleriopsis pyrenoidosa*. Trans. Amer. Micros. Soc. 90: 441-448.

Deason, T. R. 1971b. The fine structure of sporogenesis in the xanthophycean alga *Pseudobumilleriopsis pyrenoidosa*. J. Phycol. 7: 101-107.

Descomps, S. 1963. Contribution à l'étude infrastructurale des Vauchérie (Xanthophycées, Chromophytes). C. R. Hebd. Séanc. Acad. Sci., Paris 256: 1333-1335.

Ettl, H. 1977. Taxonomische Bemerkungen zu den Xanthophyceen. Nova Hedwigia 28: 555-568.

Ettl, H. 1978. Xanthophyceae. In Ettl, H. J. Gerloff and H. Heynig [Eds.] *Süsswasserflora von Mitteleuropa*, Bd. 3. 1. Teil. Gustav Fischer, Stuttgart.

Falk, H. 1967. Zum Feinbau von *Botrydium granulatum* Grev. (Xanthophyceae). Arch. Mikrobiol. 58: 212-227.

Falk, H. and H. Kleinig. 1968. Feinbau und Carotinoide von *Tribonema* (Xanthophyceae). Arch. Mikrobiol. 61: 347-362.

Greenwood, A. D. 1959. Observations on the structure of the zoospore of *Vaucheria*. II. J. Exp. Bot. 10: 55-68.

Greenwood, A. D., I. Manton and B. Clarke. 1956. Observations on the structure of the zoospores of *Vaucheria*. J. Exp. Bot. 8: 71-86.

Guillard, R. R. L. and C. J. Lorenzen. 1972. Yellow-green algae with chlorophyllide c. J. Phycol. 8: 10-14.

Hesse, M. 1974. Wachstum und Synchronisierung der alge *Bumilleriopsis filiformis* Vischer (Xanthophyceae). Planta (Berl.) 120: 135-146.

Hibberd, D. J. 1969. Cytological studies on the coccoid Xanthophyceae. Ph.D. Thesis, University of Leeds.

Hibberd, D. J. 1971. Observations on the cytology and ultrastructure of *Chrysamoeba radians* Klebs. (Chrysophyceae). Br. Phycol. J. 6: 207-223.

Hibberd, D. J. 1976. The ultrastructure and taxonomy of the Chrysophyceae and Prymnesiophyceae (Haptophyceae); a survey with some new observations on the ultrastructure of the Chrysophyceae. Bot. J. Linn. Soc. 72: 55-80.

Hibberd, D. J. 1978a. Possible phylogenetic value of the transitional helix in some chromophyte algal classes and some colourless protists. BioSystems 10: 115-116.

Hibberd, D. J. 1978b. The fine structure of *Synura sphagnicola* (Korsh.) Korsh. (Chrysophyceae). Br. Phycol. J. 13: 403-412.

Hibberd, D. J. 1980. The structure and phylogenetic significance of the flagellar transition region in the chlorophyll c-containing algae. BioSystems (in press).

Hibberd, D. J. and G. F. Leedale. 1971. Cytology and ultrastructure of the Xanthophyceae. II. The zoospore and vegetative cell of coccoid forms, with special reference to *Ophiocytium majus* Naegeli. Br. Phycol. J. 6: 1-23.

Iorya, T. 1974. Chromosome numbers in four species of *Tribonema* (Xanthophyceae). J. Jap. Bot. 49: 29-31.

Jahn, T. L., N. D. Landman and J. R. Fonseca. 1964. The mechanism of locomotion of flagellates. II. Function of the mastigonemes of *Ochromonas*. J. Protozool. 11: 291-296.

Jeffrey, S. W. 1976. The occurrence of chlorophyll c_1 and c_2 in algae. J. Phycol. 12: 349-354.

Koch, W. J. 1951. A study of the motile cells of *Vaucheria*. J. Elisha Mitchell Scient. Soc. 67: 123-131.

Korshikov, A. A. 1930. On the occurrence of pyrenoids in Heterocontae. Beih. Bot. Zbl. 46: 470-478.

Leedale, G. F., B. S. C. Leadbeater and A. Massalski. 1970. The intracellular origin of flagellar hairs in the Chrysophyceae and Xanthophyceae. J. Cell Sci. 6: 701-719.

Liaan-Jensen, S. 1977. Algal carotenoids and chemosystematics. In Faulkener, D. J. and W. H. Fenical [Eds.] *Marine Natural Products Chemistry.* Plenum, New York.

Luther, A. 1899. Ueber *Chlorosaccus* eine neue Gattung der Süsswasseralgen, nebst Bemerkungen zur Systematik verwandter algen. Bih. K. Svenska VetenskAkad. Handl. 24, Afd. III, No. 13: 1-22.

Maeda, M., K. Kuroda, Y. Iriki, M. Chihara, K. Nisizawa and T. Miwa. 1966. Chemical nature of major cell wall constituents of *Vaucheria* and *Dichotomosiphon* with special reference to their phylogenetic positions. Bot. Mag. Tokyo 79: 422-426.

Marchant, H. 1972. Pyrenoids of *Vaucheria woroniniana* Heering. Br. Phycol. J. 7: 81-84.

Massalski, A. and G. F. Leedale. 1969. Cytology and ultrastructure of the Xanthophyceae. I. Comparative morphology of the zoospores of *Bumilleria sicula* Borzi and *Tribonema vulgare* Pascher. Br. Phycol. J. 4: 159-180.

Moestrup, Ø. 1970. On the fine structure of the spermatozoids of *Vaucheria sescuplicaria* and on the later stages in spermatogenesis. J. Mar. Biol. Assn. U. K. 50: 513-523.

Moewus, F. 1940. Über die Sexualität von *Botrydium granulatum*. Biol. Zentralbl. 60: 484-498.

Nicholls, H. W. 1973. Growth media - freshwater. In Stein, J. R. [Ed.] *Handbook of Phycological Methods, Culture Methods and Growth Measurements*. Cambridge University Press.

Ott, D. W. and R. M. Brown. 1972. Light and electron microscopical observations on mitosis in *Vaucheria litorea* Hofman ex C. Agardh. Br. Phycol. J. 7: 361-374.

Ott, D. W. and R. M. Brown. 1974a. Developmental cytology of the genus *Vaucheria*. I. Organisation of the vegetative filament. Br. Phycol. J. 9: 111-126.

Ott, D. W. and R. M. Brown. 1974b. Developmental cytology of the genus *Vaucheria*. II. Sporogenesis in *V. fontinalis* (L.) Christensen. Br. Phycol. J. 9: 333-351.

Ott, D. W. and R. M. Brown. 1975. Developmental cytology of the genus *Vaucheria*. III. Emergence, settlement and germination of the mature zoospore of *V. fontinalis* (L.) Christensen. Br. Phycol. J. 10: 49-56.

Ott, D. W. and R. M. Brown. 1978. Developmental cytology of the genus *Vaucheria*. IV. Spermatogenesis. Br. Phycol. J. 13: 69-85.

Parker, B. C., R. E. Preston and G. E. Fogg. 1963. Studies on the structure and chemical composition of the cell walls of Vaucheriaceae and Saprolegniaceae. Proc. R. Soc. B. 158: 435-445.

Pascher, A. 1937-1939. Heterokonten. In *L. Rabenhorst's Kryptogamen-Flora von Duetschland, Osterreich und der Schweiz*. 2nd. Ed. Bd. XI. Akad. Verlagsges., Leipzig.

Pascher, A. 1939. Über gelegentliche Synzoosporenbildungen bei Algen und über die Verbreitung synzoosporer Organisationen. Beih. Bot. Zbl. 59: 389-408.

Pecora, R. A. and R. G. Rhodes. 1973. Zoospore production in selected xanthophycean algae. Br. Phycol. J. 8: 321-324.

Prat, S. 1947. The reaction of algal cells with Schiff's reagent. Spisy Vyadáv. Pvír. Fak. Karl. Univ. 177: 1-16.

Provasoli, L. and A. F. Carlucci. 1974. Vitamins and growth regulators. In Stewart, W. D. P. [Ed.] *Algal Physiology and Biochemistry*. Blackwell, Oxford.

Silva, P. C. 1979. Review of the taxonomic history and nomenclature of the yellow-green algae. Arch. Protistenk. 121: 20-63.

Strain, H. H. 1958. Chloroplast pigments and chromatographic analysis. Pennsylvania State University Press.

Stransky, H. and A. Hager. 1970. Das Carotinoidmuster und die Verbreitung des lichtinduzierten Xanthophyllcyclus in verschiedenen Algenklassen. II. Xanthophyceae. Arch. Mikrobiol. 71: 164-190.

Tarapchak, S. J. 1972. Studies on the Xanthophyceae of the Red Lake Wetlands, Minnesota. Nova Hedwigia 23: 1-43.

Vischer, W. 1945. Heterokonten aus Alpinen Böden, Speziell dem Schweizerischen NationalPark. Ergbn. wiss. Unters schweiz. NatnParks 1: 481-511.

Vlk, W. 1931. Über die Struktur der Heterokontengeisseln. Beih. Bot. Zbl. 48: 214-220.

Vlk, W. 1938. Über den Bau der Geissel. Arch. Protistenk. 90: 448-488.

Whittle, S. J. 1976. The major chloroplast pigments of *Chlorobotrys regularis* (West) Bohlin (Eustigmatophyceae) and *Ophiocytium majus* Naegeli (Xanthophyceae). Br. Phycol. J. 11: 111-114.

Whittle, S. J. and P. J. Casselton. 1969. The chloroplast pigments of some green and yellow-green algae. Br. Phycol. J. 4: 55-64.

Whittle, S. J. and P. J. Casselton. 1975. The chloroplast pigments of the algal classes Eustigmatophyceae and Xanthophyceae. II. Xanthophyceae. Br. Phycol. J. 10: 192-204.

Copyright 1980 by Elsevier North Holland, Inc.

PRYMNESIOPHYTES (= HAPTOPHYTES)

DAVID J. HIBBERD

Culture Centre of Algae and Protozoa, Institute of Terrestrial Ecology, 36, Storey's Way, Cambridge, CB3 ODT, England.

ABSTRACT

The majority of prymnesiophytes are planktonic marine unicells and most are either flagellate or produce flagellate stages. Prymnesiophycean zoids bear two equal, subequal or unequal flagella which characteristically lack flagellar appendages. Only in species of the order Pavlovales does the long flagellum or both flagella of the unequal pair bear fine hairs. The flagellar transition region has a characteristic structure with two relatively widely spaced cross partitions, and a stellate pattern occurs over a very short distance. Flagellar roots are of the microtubular type, either simple or compound and cross-banded fibrillar roots are absent. Between the flagella is inserted a third filiform organelle, the haptonema. This may be long and capable of coiling, short and stiff or reduced in various ways. Haptonemata are basically composed of a core of cytoplasm with a ring of 6 or 7 microtubules, a surrounding tube of ER continuous with the peripheral ER of the cell, and an outer layer of cytoplasm bounded by the haptonema membrane which is continuous with the cell membrane. Haptonemata have a characteristic transition region and basal structure. Cells are covered by one to several layers of delicate unmineralised scales usually visible only by electron microscopy. These are simple to extremely complex in shape and are composed of a cellulose matrix and mainly pectic polysaccharides. Coccolithophorids have, in addition to scales, a continuous peripheral covering of coccoliths. Scales are produced within the cisternae of the Golgi body. Species in the Pavlovales do not form typical scales but most bear dense bodies on the long flagellum, these also occurring on the haptonema and cell body in some forms. Cells possess mostly 1-2 golden-brown chloroplasts, surrounded by chloroplast ER and containing three-thylakoid lamellae; a girdle lamella is absent. Each chloroplast contains usually a single pyrenoid, its matrix penetrated by 2-thylakoid lamellae. Photosynthetic pigments include chlorophylls a, c_1 and c_2, β-carotene, fucoxanthin, diadinoxanthin and diatoxanthin. The storage product in the group is generally thought to be chrysolaminaran. Eyespots are known only in the order Pavlovales. There is a single nucleus which lies immediately posterior to the large Golgi body. The latter lies close to the flagellar basal bodies and has a structure highly characteristic for the class, being both polarised towards the basal bodies and strongly heterogeneous from face to face. The forming face lies against cisternae of rough ER and scales are produced in the cisternae towards the opposite face, distal to a central group of adjacent cisternae which characteristically show median distensions. Mitochondria contain tubular cristae. Many species possess a peripheral layer of ER underlying the plasmalemma. Knowledge of life histories is still fragmentary but may involve alternation between various flagellate forms or flagellate and benthic stages.

GENERAL CHARACTERISTICS

Most prymnesiophytes exist as biflagellate unicells for the whole or greater part of their life cycle (Figs. 1, 6, 7, 45). However, the range of form also includes amoeboid and palmelloid stages, coccoid unicells, non-motile and motile colonies and variously constructed branched colonies or filamentous forms (Figs. 11-14). Until relatively recently all known forms now contained within this class, including the coccolithophorids, were classified as chrysophytes, although within that group they had been segregated in the orders Isochrysidales (Pascher, 1910) and Prymnesiales (Papenfuss, 1955) on the basis of the number and relative lengths of the flagella. Accumulation of data on cell structure and life histories led Parke (1961) to recognise within the Chrysophyceae *sensu lato* the existence of a completely separate assemblage which was formally established as the new class Haptophyceae by Christensen in 1962. Subsequent research, particularly that involving electron microscopy, has fully substantiated this decision though in some classifications (e.g., Bourrelly, 1968) prymnesiophytes are still included within the Chrysophyceae and separated at the next lower taxonomic rank of subclass. The alternative name for the group, Prymnesiophyceae, based on the genus *Prymnesium*, was proposed (Hibberd, 1976) following changes in the rules of botanical nomenclature concerning typification and priority of names of taxa above the rank of family.

The class is presently divided into four orders: Isochrysidales, Coccosphaerales, Prymnesiales and Pavlovales (Parke and Green, 1976). With the exception of the Pavlovales, a natural group including species, mainly in the genus *Pavlova*, differing in several distinct ways from the remainder of the Prymnesiophyceae, the classification is to some extent artificial and is likely to change with advancing knowledge of cell structure and life histories (see Parke and Green, 1976). The nomenclature used in this account follows that of Parke and Green (1976).

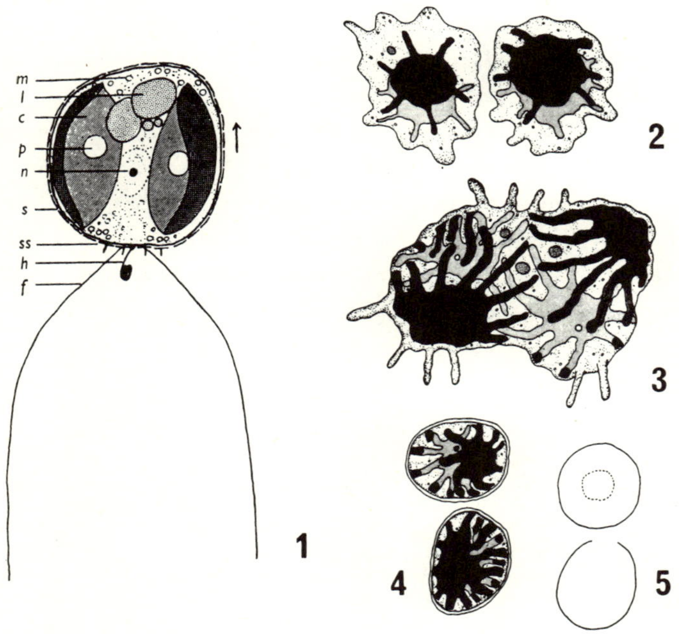

Figs. 1-5. Fig. 1. Drawing of a swimming cell of *Chrysochromulina kappa*. The flagella and haptonema are held behind the body in the position characteristic for this species during rapid swimming; the arrow indicates the direction of motion. c, chloroplast; f, flagellum; h, haptonema; l, leucosin vesicle; m, muciferous body; n, nucleus; p, pyrenoid; s, scales; ss, spined scales (x4000). Figs. 2-5. Non-flagellate cells of *C. kappa*. Fig. 2. Two amoeboid cells with slightly lobed chloroplasts. Fig. 3. A large amoeboid cell with four deeply lobed chloroplasts. Fig. 4. Small walled cells produced from two divisions of a large walled cell, each with two lobed chloroplasts. Fig. 5. Empty wall of the small walled cells showing the pore through which the motile stage may have been liberated (x2500). (Reproduced with the permission of Dr. M. Parke and Cambridge University Press, from Parke, Manton and Clarke, 1955.)

Prymnesiophycean zoids have a unique and distinctive structure which resembles that of chrysophytes only superficially. The cells are spherical, oval, flattened or saddle-shaped and show varying degrees of metaboly. Each bears two flagella which are equal, subequal or distinctly unequal in length (Figs. 7, 45). Relative flagellar lengths may vary within a genus and sometimes also within a species. Most species can swim in several different ways. In the great majority of species both flagella are completely smooth without any kind of lateral appendage (Fig. 7); tubular mastigonemes (*sensu* Bouck, 1971) are characteristically absent from this class.

In the majority of species a third filiform organelle, the haptonema, is inserted into the cell between the two flagella (Figs. 1, 6, 7, 45). Initially the haptonema was referred to as a third flagellum (Massart, 1920; Lackey, 1939), but while it is usually only slightly smaller in diameter than the flagella it differs from them in both its structure and behaviour. The haptonema may be long and capable of coiling (Fig. 56), short and stiff (Figs. 6, 7), short and narrow (Fig. 45) or vestigial (Figs. 48, 50); it is completely absent in only a few cases.

Figs. 6-14. Fig. 6. A living cell of *Prymnesium parvum* flattened by coverslip pressure. F, "fat" body; Fl, flagellum; H, haptonema; L, "leucosin" vesicle; P, plastid; Py, pyrenoid (x2000). Fig. 7. Shadowcast cell of *Prymnesium* sp. showing the two equal flagella and short haptonema; the field is littered with detached scales (arrowheads) (x6000). Fig. 8. Living cell of *Hymenomonas carterae* showing the protoplast squeezed out of the coccolith shell (x2000). Fig. 9. Dried cell of *Chrysochromulina chiton* showing the two flagella and extended haptonema. Light microscopy (x1000). Fig. 10. Liquid preparation of a cell of *Chrysochromulina kappa* killed with osmium vapour showing the effects of graphite feeding. A pellet of ingested graphite lies between the two plastids at the pole away from the flagella (x2000). Figs. 11, 12. Vegetative phase of *Ruttnera spectabilis*. Fig. 11. Cells arranged in "blocks" (the "*Sarcinochrysis*" condition) (x5000); Fig. 12. Cells from an old culture showing strongly asymmetric development of the surrounding mucilage (x1200). Fig. 13. Cells from an old culture of *Chrysotila stipitata* showing cells at the ends of mucilage stalks stained with toluidine blue to show the lamellations (arrows) (x800). Fig. 14. Lateral view of an annuloid colony of *Corymbellus aureus*. The arrowhead indicates the sub-acute appearance of the anterior end in lateral view (x625). (Fig. 6 reproduced with the permission of Prof. I. Manton and Springer-Verlag, from Manton and Leedale, 1963; Fig. 8 with the permission of Prof. I. Manton and Cambridge University Press, from Manton and Leedale, 1969; Fig. 9 with the permission of Dr. M. Parke and Cambridge University Press, from Parke, Manton and Clarke, 1958; Fig. 10 with the permission of Dr. M. Parke and Cambridge University Press, from Parke, Manton and Clarke, 1955; Figs. 11 and 12 with the permission of Dr. J. C. Green and Cambridge University Press, from Green and Parke, 1975; and Fig. 14 with the permission of Dr. J. C. Green and Cambridge University Press, from Green, 1975b.)

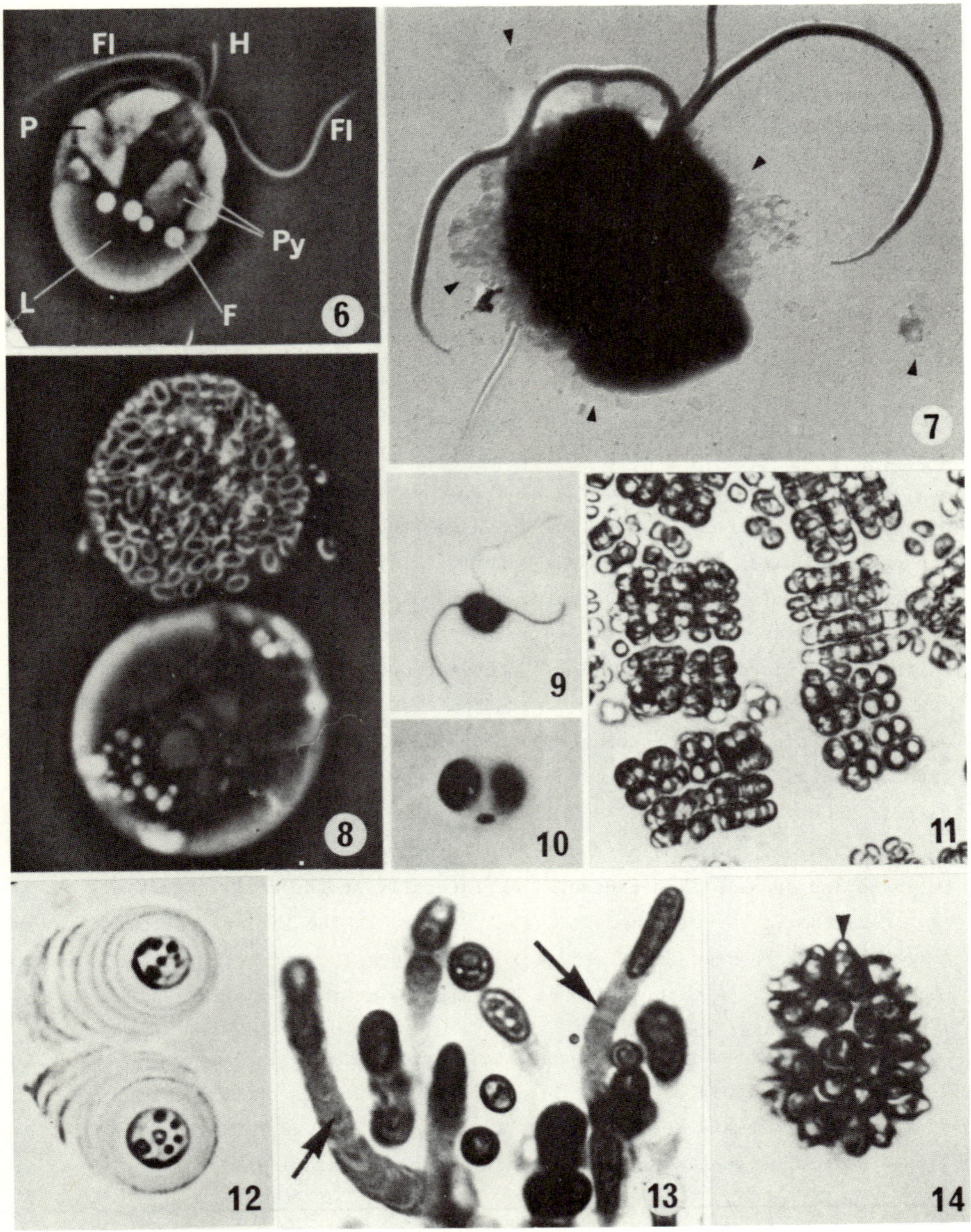
Fl
H
P
Fl
Py
L
F
6
7
8
9
10
11
12
13
14

Prymnesiophytes usually possess one or two golden-brown saucer-shaped parietal chloroplasts per cell, each with a single pyrenoid (Figs. 1, 6). Pigments include chlorophylls *a* and *c*; both chlorophylls c_1 and c_2 have been found in all species examined (Jeffrey, 1976). Modern investigations (Hager and Stransky, 1970; Berger *et al.*, 1977; Liaan-Jensen, 1977) show that fucoxanthin is the major carotenoid except in *Emiliania huxleyi* where it is replaced by the fucoxanthin derivative 19'-hexanoyloxyfucoxanthin; β-carotene, diadinoxanthin and diatoxanthin are also typically present. Ten other minor carotenoids have also been identified. It is not yet clear to what extent the carotenoid composition of prymnesiophytes differs from that of chrysophytes though data available at present suggest that the latter lack acetylenic carotenoids such as diatoxanthin and possess instead violaxanthin, antheraxanthin and zeaxanthin.

Cells are covered by one to several layers of delicate unmineralised scales, though taxa in the order Pavlovales have cells without typical scales. Prymnesiophycean scales are usually visible only by electron microscopy (Figs. 15, 16), though the whole scale covering may be detected in the light microscope by staining with cresyl blue. The scales of the most proximal layer or layers are oval in shape but those of the outermost layer, when more than one is present, are usually more elaborate and sometimes extremely complex (Fig. 83). In a few species the very large spine-shaped scales of this layer can be seen in the light microscope (Fig. 82). Coccolithophorids have, in addition to scales, a continuous peripheral covering of coccoliths (Fig. 8). The cell wall of benthic species or stages consists of numerous closely adpressed layers of scales of a similar type to one of the layers on the swimming stage, where this is known.

Prymnesiophycean zoids possess a single nucleus which is roughly spherical in shape and usually lies centrally or laterally in the cell immediately posterior to the large Golgi body (Figs. 1, 15, 16). The latter lies close to the flagellar basal bodies (Figs. 15, 16) and is usually not clearly visible in the light microscope. A contractile vacuole occurs at the flagellar end of the cells of freshwater species.

The nature of the storage metabolite in prymnesiophytes is not yet clear. The refringent liquid occurring in conspicuous vesicles occupying a large part of the volume of many species (Figs. 1, 6) is usually identified as the β-1,3-linked glucan chrysolaminaran (leucosin) on the basis of the rose-red staining

---

Figs. 15, 16. Longitudinal sections of *Prymnesium* sp. C, chloroplast; ER, peripheral endoplasmic reticulum; F, flagellum; G, Golgi body; M, mitochondrion; N, nucleus; S, scales (x10,000).

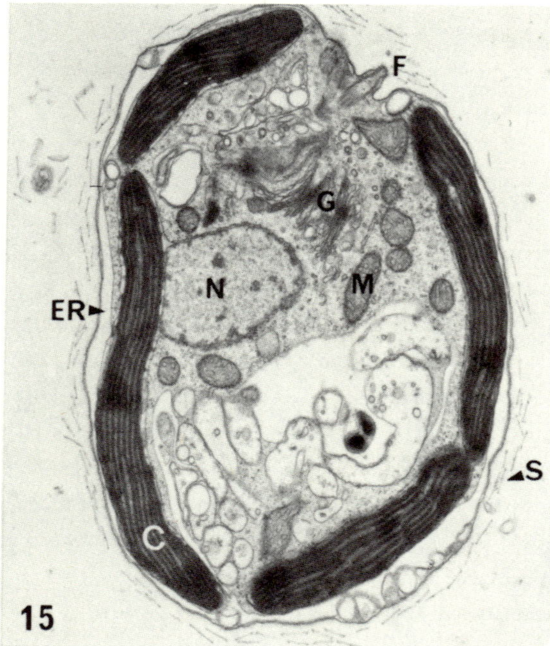
F
G
N
M
ER
S
C
15

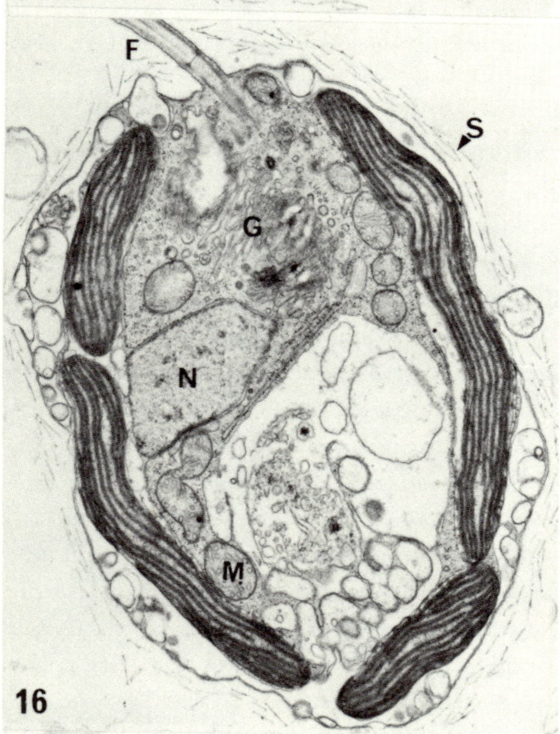
F
S
G
N
M
16

reaction produced by very dilute cresyl blue. This reaction is not specific for leucosin, however, and the identification has not been confirmed by chemical analysis. Species of *Pavlova* are unusual in the class in mostly possessing, in addition to "leucosin", two large refringent storage granules per cell which appear to be composed of a substance having the same X-ray powder pattern as the paramylon in *Euglena* (Kreger and Van der Veer, 1970). Lipid globules are also frequently present in prymnesiophytes (Fig. 6), and these often increase in number with ageing.

Several species in the genus *Chrysochromulina* possess the ability to ingest suspended particles, bacteria and small plant forms up to 2.5 μm in diameter (e.g. Parke, Manton and Clarke, 1955). Ingestion is rapid and occurs at the non-flagellar pole of the cell, but once taken in the ingested material may be transferred to the flagellar pole. A colourless, slightly granular substance seems to flow out of the cell to surround the particles being ingested which are then quickly drawn within the body. Phagotrophy is most easily demonstrated by incorporating fine graphite particles into the culture medium since these become ingested and gradually compacted into a pellet clearly visible in the light microscope (Fig. 10). This body is then ejected from the cell. The whole process from initial uptake to ejection can take place in about 5 minutes.

DETAILS OF ULTRASTRUCTURE

Flagella

The two smooth flagella are mostly inserted at an acute angle to each other, often in a depression or trough in the cell surface. Both flagella emerge freely (Figs. 15-17), neither (with one exception, see later) bearing a swelling or other paraxial structure. In *Prymnesium parvum* there is a pit up to 1 μm deep with a "hairy" coating on the cytoplasmic side of the membrane, at the side of each flagellar base (Manton, 1966b). The basal bodies are constructed of 9 triplet microtubules in the usual way (Figs. 23, 24). The structure of the transition region between the basal body and flagellar axoneme is known in detail only in *Prymnesium parvum* (Figs. 18-25; Manton, 1964a), though there is evidence that a similar situation exists in other species (Manton, 1964b, 1968; Parke, Green and Manton, 1971; Green, 1973; Green and Hibberd, 1977). In *P. parvum* two relatively widely spaced partitions cross the flagellar axoneme, both of these lying above the level of entry of the flagellum into the cell body (Fig. 25). Between the partitions the flagellar membrane is inflated and transverse sections of this region show that a stellate pattern of fibres sim-

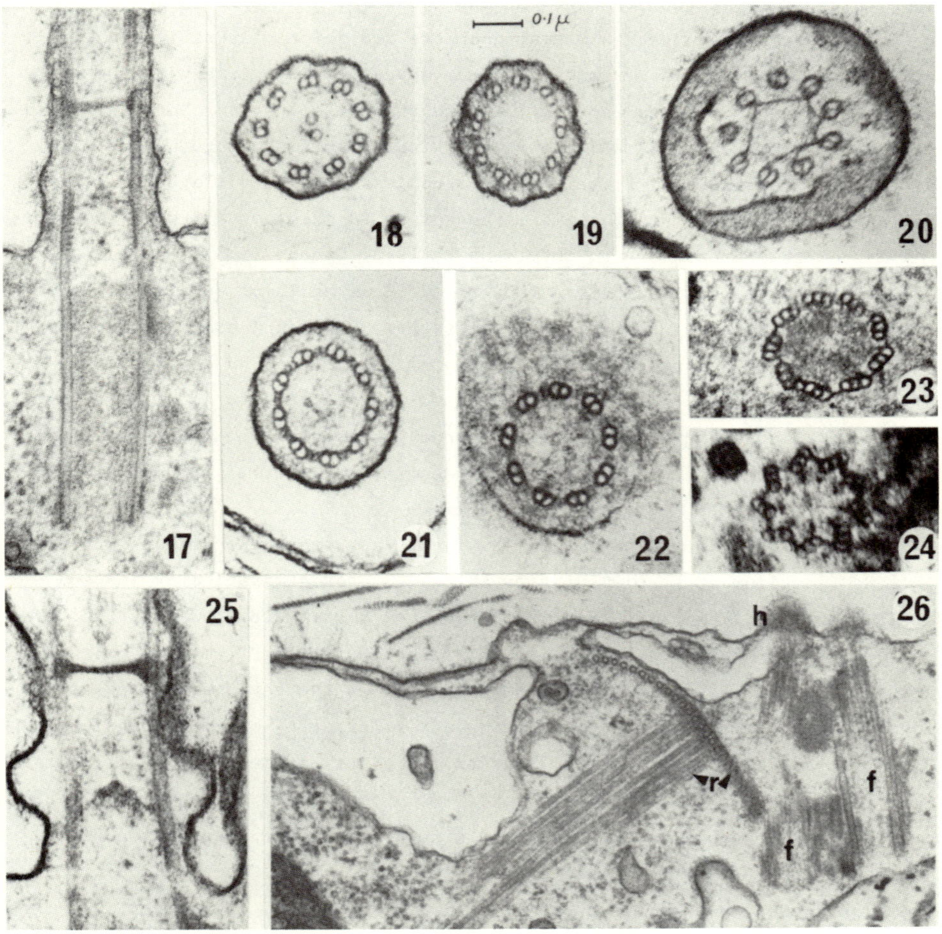

Figs. 17-26. Fig. 17. Longitudinal section of the transition region and flagellar base of *Prymnesium* sp. (x50,000). Figs. 18-24. Transverse sections at successively deeper levels through the flagellum, transition region and flagellar base of *Prymnesium parvum*. Fig. 18. Normal flagellar axoneme. Fig. 19. At the level of the distal partition. Fig. 20. Through the swelling and stellate pattern. Fig. 21. At the level of the proximal partition. Fig. 22. At the level of the cell surface. Fig. 23. The flagellar basal body. Fig. 24. The extreme proximal end of the basal body (x66,667). Fig. 25. Longitudinal section of the transition region of *P. parvum* (x66,667). Fig. 26. Compound root (r) in *Hymenomonas carterae*. f, flagellar bases; h, haptonema (x50,000). Figs. 18-25 reproduced with the permission of Prof. I. Manton and the Royal Microscopical Society, from Manton, 1964b; Fig. 26 with the permission of Prof. I. Manton and the Royal Society, from Manton and Peterfi, 1969).

ilar to that in chlorophyll *b*-containing plants occurs over a very short distance (Fig. 20). At the level of the most proximal and less distinct of the two cross partitions, the 9 axonemal doublets are joined laterally (Fig. 22).

Flagellar roots in the Prymnesiophyceae appear to be entirely of the microtubular type, and cross-banded fibrillar roots have not so far been described. The tubular roots may consist simply of a single sheet of microtubules (Fig. 43) running from near the flagellar basal bodies around the flagellar depression and into the cell body, commonly appearing to make contact with the inner face of the chloroplasts (e.g. Chretiennot, 1973; Green, 1973). In the genus *Hymenomonas* the situation is more complex since in addition to a simple root composed of a small number of microtubules there is also a compound root (Fig. 26) consisting of a broad sheet of closely aligned microtubules from which arises almost at right angles a close-packed bundle of up to 100 microtubules which pass towards and along the inner concave face of a plastid to terminate finally at the opposite end of the cell (Manton and Peterfi, 1969; Gayral and Fresnel, 1976).

Although flagellar appendages are typically absent in the Prymnesiophyceae, the long flagellum in species of the Pavlovales characteristically bears a tomentum of long fine hairs (Fig. 27; Green and Manton, 1970; Green, 1976; Van der Veer, 1976; Green and Hibberd, 1977). The short flagellum also often bears a garniture of finer hairs. Most species in the Pavlovales are also unusual in that the long flagellum bears a regular array of small dense bodies (Figs. 28, 29) which vary in form between different species (Green and Manton, 1970; Van der Veer, 1972, 1976; Green, 1973, 1975a, 1976; Billard, 1976). In some species these bodies also occur in a less regular arrangement and in smaller numbers on the haptonema and cell body (Fig. 30; Green, 1975a). They have never been found on the short flagellum of any species, however, and they are completely absent in *Diacronema vlkianum* (Green and Hibberd, 1977).

Haptonema

Structure. The haptonema in its typical form, as in the genera *Chrysochromulina* and *Prymnesium*, has been the subject of several detailed studies (Manton, 1964b, 1967a, 1968; Leadbeater and Manton, 1969a). In transverse section it appears as a ring of mostly 7 microtubules surrounded by three concentric unit membranes (Fig. 31). The outermost membrane is the plasmalemma, which is continuous with the cell membrane at the proximal end of the organelle, while the remaining pair represent a tube of endoplasmic reticulum continuous with the peripheral ER of the cell body (Fig. 43). The ring of microtubules is embedded in the cytoplasmic core of the organelle and a layer of cytoplasm occurs

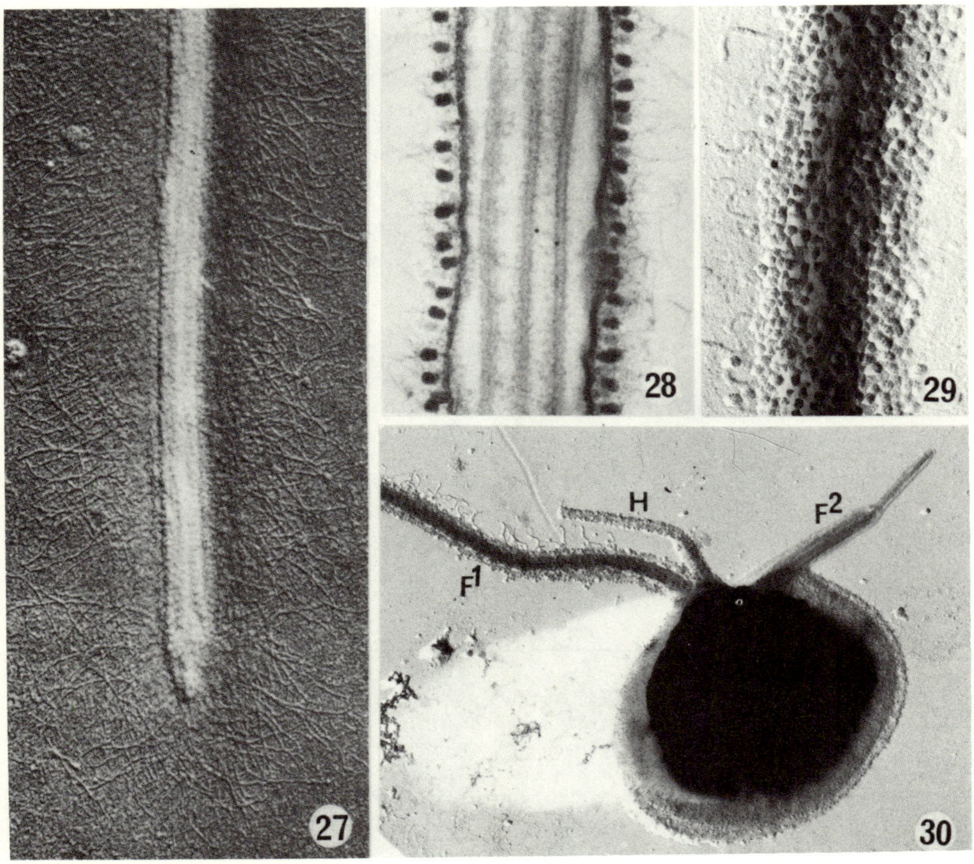

Figs. 27-30. Fig. 27. Tip of the front flagellum of *Pavlova gyrans* showing the attached long fine hairs. Longitudinal rows of dense bodies are also faintly discernable (x *ca*. 20,000). Fig. 28. Longitudinal section through a front flagellum of *P. gyrans* showing the dense bodies and tomentum on the surface (x60,000). Fig. 29. Shadowcast long flagellum of *P. lutheri* showing the dense bodies and appendages (x40,000). Fig. 30. Shadowcast whole cell of *P. lutheri* showing the dense bodies on the long flagellum (F_1), haptonema (H) and cell body. F_2, short flagellum. (x10,000). (Figs. 27, 28 reproduced with the permission of Dr. J. C. Green and Cambridge University Press, from Green and Manton, 1970.)

between the outermost ER membrane and the plasmalemma (Fig. 32). The two cytoplasmic zones are locally confluent through lacunae in the ER tube, and at the tip of the organelle the haptonema cavity continues smoothly over the tip of the core; externally, the tip appears either rounded, tapered or slightly spatulate (Fig. 32) depending on species.

While the structure of the major part of the organelle is thus relatively straight forward, a number of complex changes take place at the proximal end. The first change from normal structure which indicates the distal end of the transition region is that the ring of microtubules becomes organised into a C and then into a compact arc, and the ER space forms a crescentic cavity on the curved side of the tubules (Figs. 33-35). Immediately before the microtubular core enters the cell, a flat-ended, finger-like process protrudes from one side of the crescentic ER cavity and enters the central cytoplasm bounded by the C (Figs. 36, 43, 44, 46). Bands of amorphous dense material join the membrane bounding the tip of this process to that of the cavity on the concave side of the C (Figs. 44, 46).

Further rearrangements in the microtubular component of the haptonema take place as the cell body is entered. Just within the cell the arc of microtubules becomes arranged into a curved zig-zag and the number is increased by one, usually from 7 to 8 (Fig. 37). These microtubules then close up still

---

Figs. 31-44. Typical haptonemal structure in *Chrysochromulina chiton* and *Prymnesium* sp. Fig. 31. Section of one side of a coiled haptonema of *C. chiton* showing three gyres in transverse view (x85,000). Fig. 32. Longitudinal section of the haptonemal tip in *Prymnesium* sp. (x50,000). Figs. 33-36. Transverse sections showing successive changes in the arrangement of the ER and microtubules at the proximal end of the haptonema. Fig. 33. *C. chiton*. The ring of tubules begins to open out into a C (x85,000). Fig. 34. *C. chiton*. The tubules close up to form a compact arc and the haptonemal cavity starts to become discontinuous (x85,000). Fig. 35. *C. chiton*. The base is swollen and the ER forms a relatively small arc (x85,000). Fig. 36. *Prymnesium* sp. The haptonemal cavity enters the concave side of the arc of microtubules immediately before they enter the cell. F, flagellum (x50,000). Figs. 37-42. Transverse sections of the haptonemal base of *C. chiton* showing changes in the number and arrangement of microtubules at successively deeper levels (x66,667). Fig. 43. Longitudinal section of the haptonema in *Prymnesium* sp. passing transversely through the finger-like process of ER (arrowhead). Note the continuity of the haptonemal cavity with the superficial ER of the cell body, and the microtubular root (x50,000). Fig. 44. Longitudinal section of the base of a haptonema of *C. chiton* showing the bulbous swelling and the flat-tipped finger-like process of ER in side view. Note the band of dense material (arrow) joining the tip or side of the finger to the inner membrane bounding the cavity on the opposite side (x28,650). (Figs. 31, 33-35, 37-42 and 44 reproduced with the permission of Prof. I. Manton and Springer-Verlag, from Manton, 1963.)

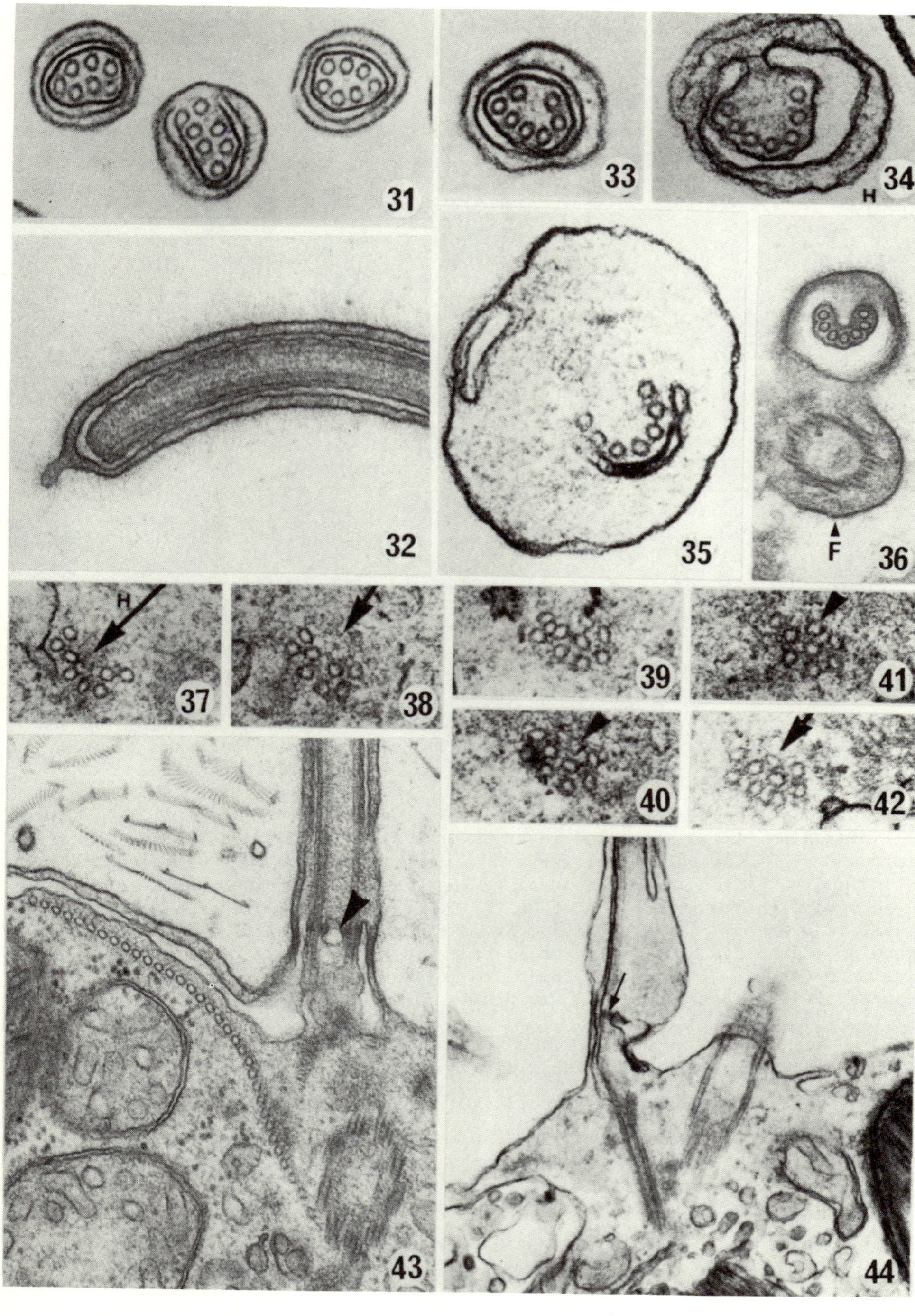
31
33
H
34
32
35
F
36
H
37
38
39
41
40
42
43
44

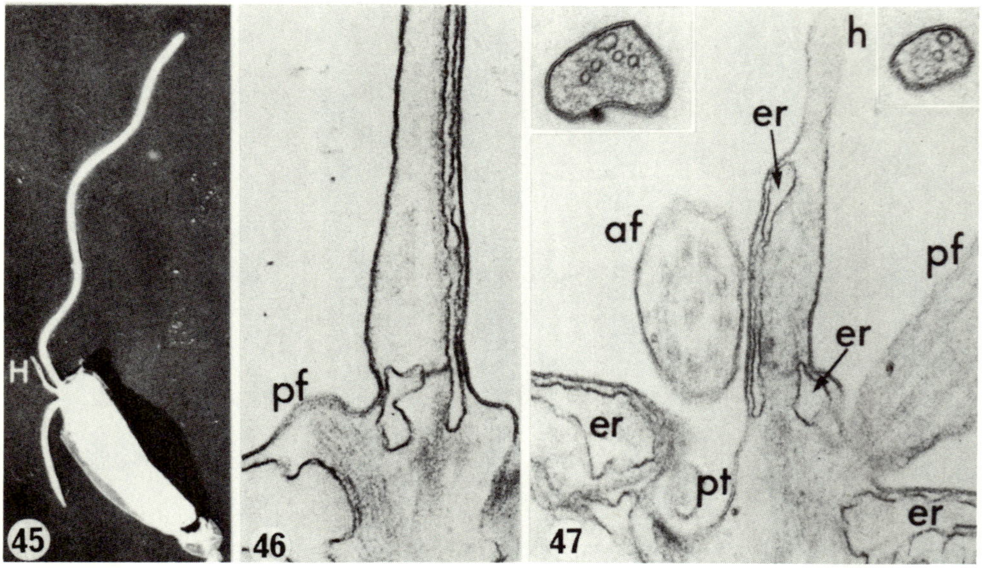

Figs. 45-47. Aspects of haptonemal structure in the Pavlovales. Fig. 45. Shadowcast whole cell of *Pavlova gyrans* showing the two unequal flagella and the narrow haptonema (H) (x3000). Fig. 46. Median longitudinal section of the proximal part of the haptonema of *Pavlova salina* passing through the single elongated vesicle of ER on the side of the haptonema towards the anterior flagellum, linked by a dense strand at the base to a finger-like ER projection. pf, posterior flagellum (x55,000). Fig. 47. Haptonema of *Diacronema vlkianum* in longitudinal section (h) and in transverse section towards the base and tip (left and right insets, respectively). af, anterior flagellum; er, endoplasmic reticulum; h, haptonema; pf, posterior flagellum; pt, pit (x60,000). (Fig. 45 reproduced with the permission of Dr. J. C. Green and Cambridge University Press, from Green and Manton, 1970; Fig. 46 with the permission of Dr. J. C. Green and Cambridge University Press, from Green, 1976; Fig. 47 with the permission of Dr. J. C. Green and Cambridge University Press, from Green and Hibberd, 1977; insets to Fig. 47 originals, also used in Green and Hibberd, 1977.)

further to form two superposed arcs with 4 microtubules in each, after which a ninth tubule is added (arrow, Figs. 38-40) converting the arrangement into a hexagonal close packing (Figs. 41, 42) before the whole group terminates together.

Reduced haptonemata. In some prymnesiophycean species both the microtubular and ER components are reduced from the typical condition described above, and in some cases the organelle is so small as to be detectable only in sectioned material. Such conditions are usually interpreted as derived by evolutionary

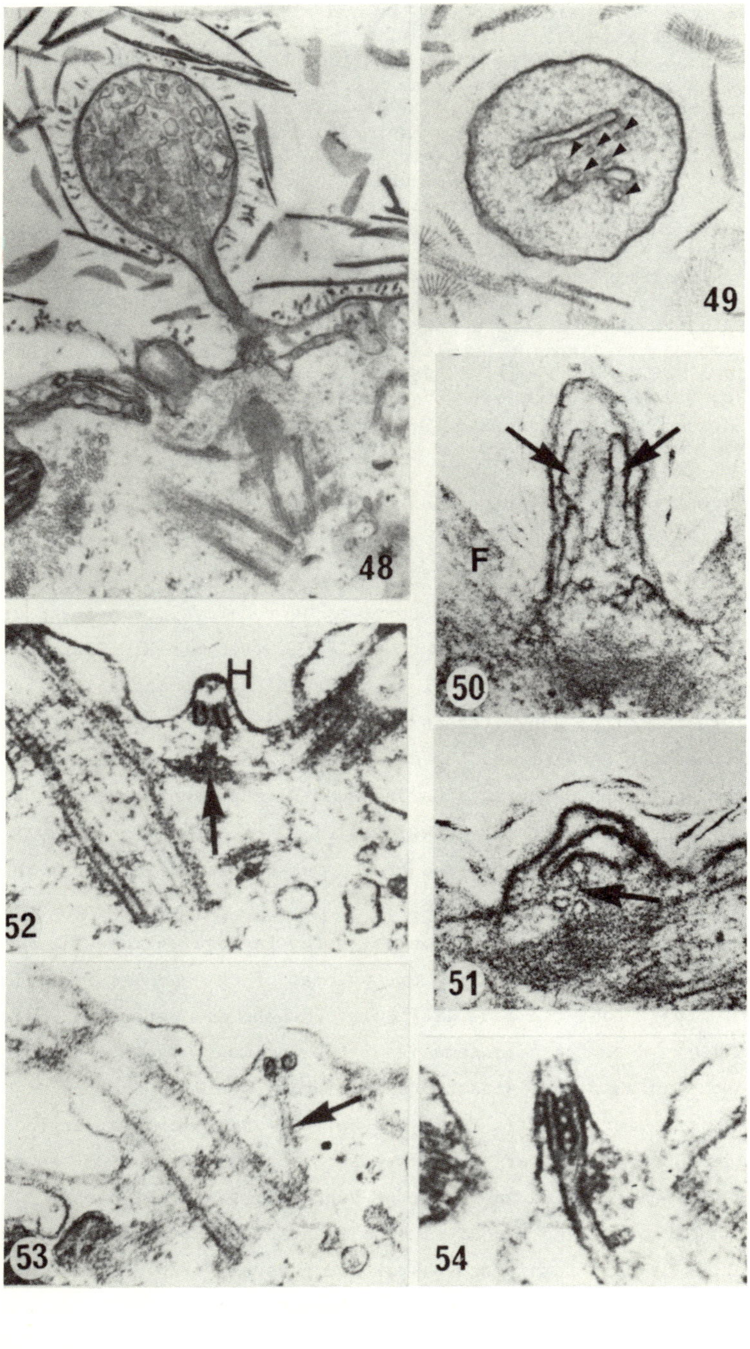
48
49
F
50
H
52
51
53
54

Figs. 48-54. Fig. 48. Longitudinal section through the bulbous haptonema of *Hymenomonas roseola* showing microtubules and vesicles inside and a covering of columnar material and scales (x30,000). Fig. 49. Transverse section through the bulbous haptonema of *Hymenomonas carterae* showing microtubules (arrowheads) and ER (x26,667). Figs. 50, 51. The haptonema in *Isochrysis galbana*. Fig. 50. Longitudinal section. F, flagella; arrows, ER profiles (x55,500). Fig. 51. Transverse section at the base of a haptonema showing an arc of 4 microtubules within the cell (arrow). The free part of the haptonema in this species contains three microtubules. (x85,000). Figs. 52-54. The haptonema in *Ruttnera spectabilis*. Figs. 52, 53. Longitudinal sections passing through both flagella showing profiles of ER, the band of material joining the flagellar bases (Fig. 52, arrow) and the microtubules (Fig. 53, arrow) passing between the ER profiles and terminating against the base of the flagellum. (Fig. 52, x65,000 ; Fig. 53, x57,000.) Fig. 54. Transverse section of the haptonema showing the entering ER bifurcating around a row of three microtubules (x80,000). (Figs. 48 and 49 reproduced with the permission of Prof. I. Manton and the Royal Society, from Manton and Peterfi, 1969; Figs. 50 and 51 with the permission of Dr. J. C. Green and Cambridge University Press, from Green and Pienaar, 1977; Figs. 52-54 with the permission of Dr. J. C. Green and Cambridge University Press, from Green and Parke, 1974.)

reduction from the typical state and are thus generally described as reduced or vestigial, though this cannot be certain until more is known about haptonematal function.

In genera of the Pavlovales the haptonema is relatively short and attenuated distally (Figs. 30, 45; Green and Manton, 1970; Green, 1973, 1975a, 1976; Green and Hibberd, 1977). While the basal parts are closely comparable with the typical state described above, transverse sections in the free part above the level of transition usually show between 4 and 1 microtubules depending on the level of the section and the species and the ER is present either as separate vesicles or as a single elongated vesicle (Figs. 46, 47). In *Pavlova lutheri* (Green, 1975a) both microtubules and ER are missing completely from the very narrow distal region of the haptonema.

While the narrowness of the haptonema in the Pavlovales makes it difficult to see in the light microscope, in some species of *Hymenomonas* it is its shortness which brings about the same difficulty (Manton and Peterfi, 1969; Leadbeater, 1971a). Here the haptonema is of the "Stummel" type (*sensu* von Stosch, 1958), bulbous in shape and attached by a slender stalk to the cell (Fig. 48). It contains irregular profiles of ER and up to 6 irregularly arranged microtubules (Fig. 49). The basal part appears normal in longitudinal sections but details of the number and arrangement of tubules are not yet clear. In some species of *Hymenomonas* and *Ochrosphaera* the haptonema appears to be completely reduced, only the base remaining (Gayral and Fresnel, 1976; Pienaar, 1976a).

The most extreme cases of haptonematal reduction occur in *Isochrysis galbana* and *Ruttnera spectabilis*. In the former a short proboscis contains ER profiles and three microtubules, these increasing to four in the base (Figs. 50, 51; Green and Pienaar, 1977), while in the latter a small papilla contains three microtubules and barely detectable ER profiles (Figs. 52-53; Green and Parke, 1974). In *Imantonia rotunda*, although no recognisable haptonema appears between the flagella, many cells bear a protuberance in which ER profiles are developed (Green and Pienaar, 1977). However, in *Dicrateria inornata* (Green and Pienaar, 1977), the flagellate stage of the coccolithophorid *Emiliania huxleyi* (Klaveness, 1972b), and one swarmer type of *Hymenomonas carterae* (Leadbeater, 1971a), both the external and basal parts of the haptonema are completely missing.

Certain species in the Pavlovales are so far unique in possessing a haptonematal root (Green, 1976; Green and Hibberd, 1977). This originates in a dense disc to which the proximal end of the haptonematal microtubules are attached

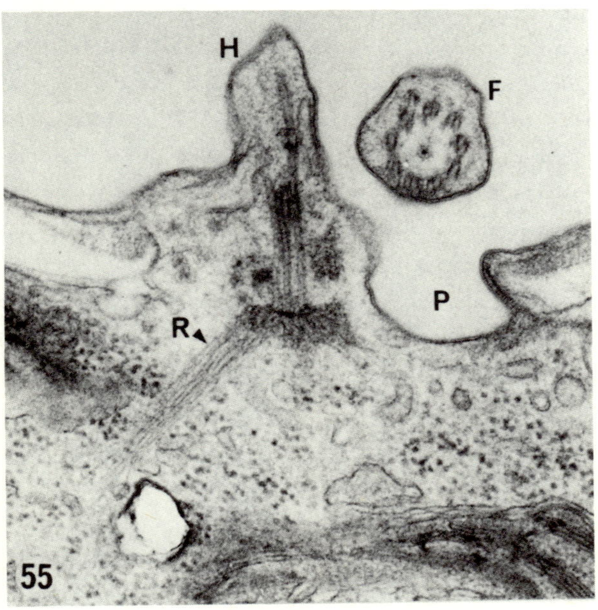

Fig. 55. Section passing longitudinally through the haptonemal base of *Diacronema vlkianum* showing the fibrous haptonematal root (R) and the electron-dense disc at the point of junction with the haptonematal microtubules. F, flagellum; H, haptonema; P, pit (x50,000). (Original, also used in Green and Hibberd, 1977.)

and appears to be composed of non-striated microfibrils which diverge from their point of origin widely and deeply into the cell for a distance of at least 1.25 µm (Fig. 55).

Behaviour. The movements of the haptonema have been described in the early studies of Parke, Manton and Clarke (1955, 1956, 1958, 1959) and in more detail recently by Leadbeater (1971b) who used high speed ciné photography. Most prymnesiophytes can attach to a microscope slide or coverslip by means of their haptonema and it was this ability, together with the thread-like form, which suggested the name selected for this organelle (Parke, Manton and Clarke, 1955). When the cell is attached in this way, the flagella usually continue to beat, often rotating it around the point of attachment apparently without the haptonema becoming twisted. Attachment is usually by the "adhesive" tip but may be at other points along the length in these species with a relatively long haptonema. Haptonemata can also undergo gliding movements once attached, sometimes leaving a faint trace. Haptonemata in general show bending and flicking movements, but in most species in which the haptonema is longer than the cell body it can coil reversibly (Figs. 57-62). Coiling is very rapid, taking only 1/60th - 1/100th sec. Uncoiling follows immediately and is very much slower, taking between 2 and 10 sec. (Figs. 57, 58). Although the coiling process is too rapid to be followed in detail, the slower coiling which occurs in dying cells appears to begin at the distal end and proceed proximally. The helix is always formed on the side towards the cell and to the left of the haptonema when the coil is viewed from the ventral surface (Figs. 59-62). Haptonematal coiling may be induced in some species by slight mechanical stimuli such as tapping the microscope or coverslip, but repeated shock-induced coiling leads to a gradual loss of sensitivity so that after 20 successive responses the haptonema may cease to coil. In other species coiling occurs less readily and in some occurs only as a precursor of death. During haptonematal coiling the flagella cease to beat and also appear to coil up, uncoiling again at the same time as the haptonema.

Haptonemata appear to have no obvious mechanical function and the mechanism of coiling in those species possessing this ability is unknown. However, Manton (1964b, 1968) has hypothesised a mechanism involving changes in hydrostatic pressure (turgor) in the haptonema cavity controlled in some way at the base of the organelle, where the haptonema cavity is greatly reduced and where the base is swollen in species possessing coiling haptonemata. Temporary closure of the haptonema cavity by some change in this region could possibly lead to uncoiling by a build-up of internal turgor pressure and the re-opening of this

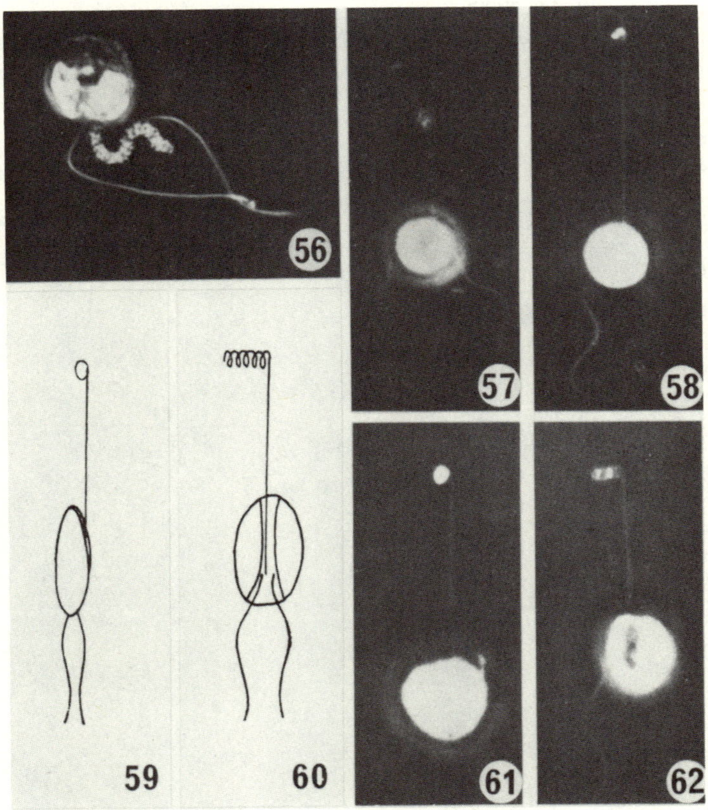

Figs. 56-62. Fig. 56. Cell of *Chrysochromulina strobilus* killed with osmic vapour to show the coiled haptonema (x1330). Figs. 57, 58. Living cells of *Chrysochromulina* sp. Two stages in haptonematal uncoiling, with the haptonema two-thirds and three-quarters uncoiled, respectively (x1330). Figs. 59-60. Drawings showing the relationship between the direction of coiling of the helix and the orientation of the cell. Fig. 59 shows the coil directed dorsally in lateral view and Fig. 60 shows the coil forming towards the left side when the cell is seen from the ventral surface. Figs. 61, 62. Living cells of *Chrysochromulina* sp. showing the haptonemata coiling as in Figs. 59, 60, respectively (x1330). (Figs. 56-62 reproduced with the permission of Dr. B. S. C. Leadbeater and Cambridge University Press, from Leadbeater, 1971b.)

connection, with consequent rapid dissipation of pressure, might explain the extremely rapid coiling movement. The asymmetry required for such changes to effect coiling could result from a particular pattern of connections between the core and wall, the cytoplasmic regions of the haptonema.

Chloroplast, pyrenoid and eyespot

The chloroplast lamellae are normally composed of three thylakoids (Figs. 63, 64). A girdle lamella is absent without exception. The chloroplasts are bound-

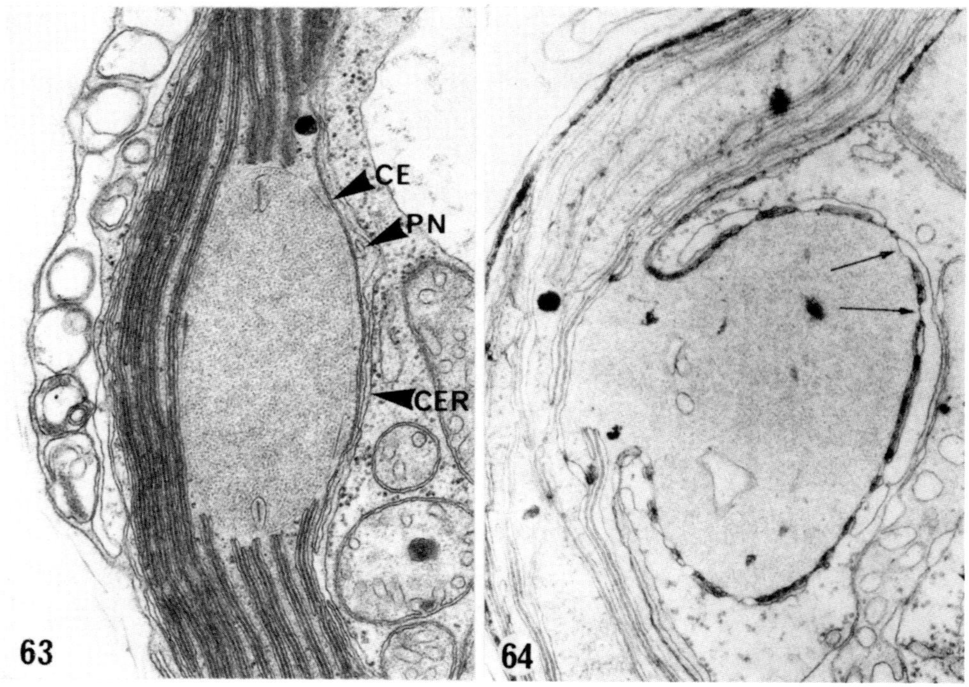

Figs. 63, 64. Fig. 63. Semi-immersed pyrenoid of *Prymnesium* sp. The matrix is penetrated by narrow two-thylakoid lamellae and traces of a pyrenoid "membrane" are also visible. CE, chloroplast envelope; CER, chloroplast ER; PN, periplastidial network (x30,000). Fig. 64. Bulging pyrenoid of *Chrysochromulina chiton* with a capping vesicle with transparent contents. The arrows indicate places at which the contents of the vesicle are separated from the pyrenoid core by only one unit membrane (for further explanation see text) (x33,330). (Reproduced with the permission of Prof. I. Manton and Cambridge University Press, from Manton, 1966a.)

ed by a double-membraned chloroplast envelope and a layer of chloroplast ER (Fig. 63), the outer membrane of which is typically continuous with the outer membrane of the nuclear envelope. The region between these two pairs of membranes, the chloroplast cell compartment (*sensu* Coombs and Greenwood, 1976) can occasionally be seen to contain vesicles which possibly represent a periplastidial network similar to but rather less well developed than that in the Chrysophyceae (Gibbs, 1979) and Xanthophyceae (Falk and Kleinig, 1968).

Pyrenoids are present in the great majority of species investigated. They are traversed by two-thylakoid lamellae and may be either immersed (Klaveness, 1972a), semi-immersed (Fig. 63), bulging into the cytoplasm (Manton and Peterfi, 1969; Leadbeater, 1971a) or projecting (Fig. 64), being attached to the inner face of the chloroplast by only a narrow stalk (Manton, 1966a). The pyrenoids in *Phaeocystis* zoids are unusual in being fusiform in shape and bounded by a delicate "membrane" (Parke, Green and Manton, 1971). A pyrenoid "membrane" may also be present in other species (Fig. 63). A crystalline matrix has been described for the pyrenoids of *Emiliania huxleyi* (Klaveness, 1972a) and *Chrysochromulina acantha* (Leadbeater and Manton, 1971).

A pyrenoid-associated storage metabolite or capping vesicle is not usually apparent but the projecting pyrenoids of *Chrysochromulina chiton* have a capping vesicle containing electron-translucent material (Fig. 64) which possibly contains a liquid carbohydrate reserve (Manton, 1966a). This vesicle appears to be formed by dilation of the outer membrane of the chloroplast envelope which pushes outwards between the meshes of the overlying chloroplast ER and then spreads laterally. In another species of this genus the projecting pyrenoid is enveloped by the nucleus (Manton and Leedale, 1961).

The lack of an eyespot was at first thought to be one of the distinguishing features of the class, but this rule is broken by some species now classified in the order Pavlovales (Green, 1976). In these the eyespot is intraplastidial and consists of a sheet of osmiophilic droplets located beneath the proximal end of the posterior flagellum (Figs. 65, 66; Green and Manton, 1970; Green, 1973, 1975a; Van der Veer, 1976). Only in a single species, *Diacronema vlkianum* (Green and Hibberd, 1977), does the short flagellum bear a swelling where it passes over the eyespot (Figs. 67, 68), but this swelling is unique in structure and completely different from the eyespot-associated swellings in the Chrysophyceae, Xanthophyceae and Phaeophyceae. In this species the short flagellum consists of a proximal part expanded on the side adjacent to the flagellum and a long thin distal part into which only the central pair of microtubules of the axoneme extend. Within the expanded part a series of teeth is attached to the axonemal doublet nearest the cell at intervals of 50 nm. Faintly fibrillar material which appears to be arranged radially with respect to the doublet bearing the teeth occurs between the teeth and fills the remainder of the swelling (Figs. 67, 68). A more common eyespot association in prymnesiophytes is with a characteristic pit or canal which, in most species possessing it, enters

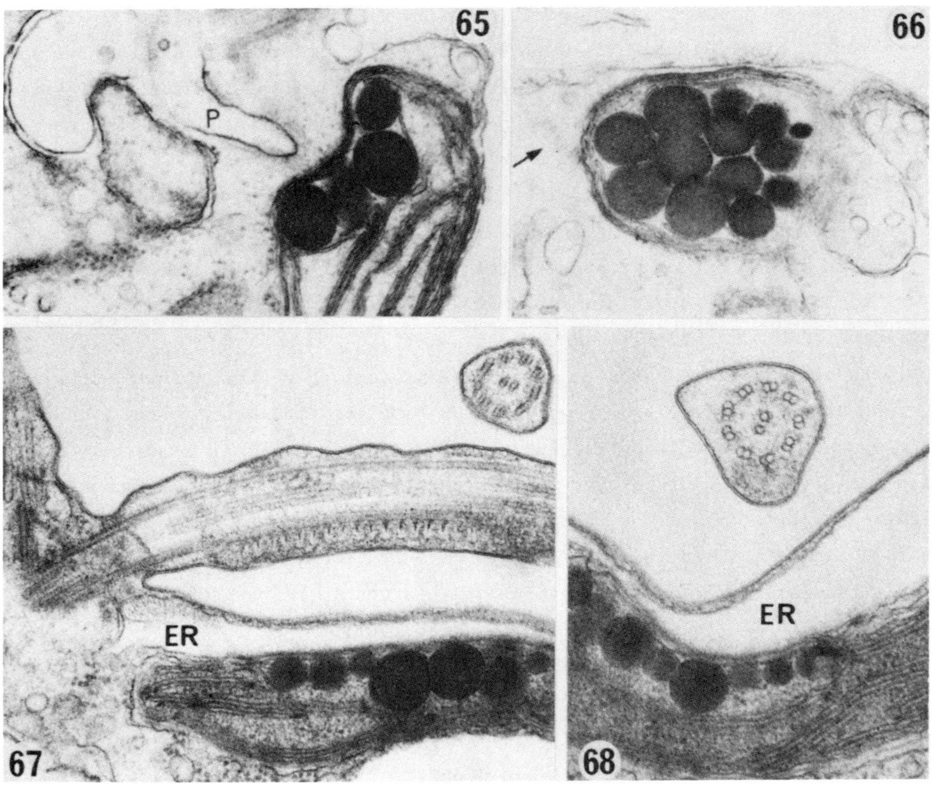

Figs. 65-68. Figs. 65, 66. Vertical and tangential sections of the eyespot in *Pavlova gyrans*. Fig. 65. The tip of the pit (P) is covered with dense material and ends opposite the concave face of the eyespot after passing below the bases of the appendages (x30,000). Fig. 66. A microtubular root (arrow) approaches the eyespot (x30,000). (Reproduced with the permission of Dr. J. C. Green and Cambridge University Press, from Green and Manton, 1970). Figs. 67, 68. Longitudinal (Fig. 67) and transverse (Fig. 68) sections of the posterior flagellum of *Diacronema vlkianum* showing the teeth attached to the axoneme and the diffuse material in the flagellar swelling and between the teeth. ER, endoplasmic reticulum. (Fig. 67, x40,000; Fig. 68, x50,000.) (Originals, also used in Green and Hibberd, 1977.)

the cell beneath the long anterior flagellum (Figs. 47, 55), curves beneath the flagella and ends against the eyespot (Fig. 65; Green and Manton, 1970; Green, 1973).

Nucleus and Mitochondria

Though it is often flattened against the internal face of one of the chloroplasts, the nucleus is more or less regular in shape and is never pulled out towards the flagellar basal bodies as in several other groups. Indeed, such a relationship would not be possible in the majority of forms since the large Golgi body occupies a characteristic position between the nucleus and the basal bodies (Figs. 15, 16, 69, 70).

Prymnesiophycean mitochondria have tubular cristae (Figs. 63, 73). The only unusual feature so far described is a marked orientation of the microvillae towards the flagellar or haptonematal bases in a large mitochondrion which always lies near the flagellar bases in species of *Prymnesium* (Manton, 1964b, 1966b).

Golgi body

Structure. The form of the Golgi body is one of the most characteristic features of the Prymnesiophyceae, including members of the Pavlovales, and it appears that it can now be regarded as important a diagnostic character as the possession of a haptonema, as was originally suggested by Manton (1967a). Prymnesiophycean Golgi bodies are widely illustrated but the highly characteristic structure can be seen particularly well in a series of papers on *Prymnesium parvum* and *Chrysochromulina chiton* (Manton, 1966b, 1967a, 1967b).

The Golgi body always lies close to the flagellar basal bodies and haptonematal base and this seems to be also true for non-motile stages where the flagellar bases are still present but carry no flagella (Manton and Leedale, 1969). The Golgi body has a distinct positional relationship with the flagellar bases, the edges of the central cisternae being bunched towards the entering flagella and haptonema (Figs. 69-72). These edges have a narrow smooth appearance, while the opposite edges are dilated and often fenestrated and bear coated (hairy) diverticula or vesicles (Fig. 69). For these reasons, the Golgi body as a whole presents a strongly polarised appearance. It is not only polarised from edge to edge, however, but is also heterogeneous from face to face. The forming face lies against cisternae of rough ER which show connections with the nuclear envelope (Figs. 69, 70) and here the Golgi cisternae are relatively narrow. Coated vesicles often occur between the ER and the most proximal cisterna (Fig. 70). Towards the opposite face, the cisternae show a generally swollen appearance, and scales in the process of formation can often be seen within them (Figs. 71, 73, 75) (see below). Large cavities containing many contorted tubes or whorls of membranes and sometimes scales occur between the most distal Golgi cisternae and the cell surface (Fig. 70).

A further characteristic feature of this type of Golgi body is that the central cisternae possess median dilations of a particular type (Figs. 71, 72, 75). These dilations, together with the particular asymmetry of the organelle, are the features which so clearly differentiate prymnesiophycean Golgi bodies from those of other groups.

The metabolite causing the cisternal distensions shows a variety of appearances after fixation (Figs. 69, 71, 72), apparently depending on time of fixation in relation to the diurnal cycle. The distensions appear virtually empty after a long period of darkness but under normal conditions dense material usually appears distributed unevenly on the cisternal wall with less dense material internally (Fig. 71). Intense staining of the contents appears to take place only in material fixed at the end of a long exposure to light (Fig. 72). The cisternal membranes in the distended regions appear to be more delicate than in the remainder of the Golgi body since they normally break down after fixation in osmium tetroxide alone, producing the central cavity which originally led to the epithet "peculiar" now generally applied to this particular condition (Parke, Manton and Clarke, 1959).

Scale formation and composition. As already mentioned, scales identical to those covering the cell body are commonly found within the cisternae of the Golgi body and this is undoubtedly the site of their production. Detailed studies of scale formation (Manton, 1966b, 1967a, 1967b; Leadbeater and Manton, 1969a, 1969b; Manton and Leedale, 1969) have established that recognisable scales are found, usually singly, only in those cisternae distal to the central group possessing the "peculiar" dilations; that scale production takes place progressively as the cisternae move from the Golgi centre to the distal (maturing) face of the Golgi body; and that scale production appears to be cyclic, probably following a diurnal rhythm, production being most active in the late

Figs. 69-72. Golgi structure in the Prymnesiophyceae. Note the various appearances of the dilations of the central cisternae, and the unusual orientation of the forming cup-shaped scale of *Chrysochromulina strobilus* in Fig. 71 (large arrow), with the morphologically outer (concave) face towards the Golgi centre. ER, endoplasmic reticulum; F, flagellum; N, nucleus; arrows, nascent scales. (Figs. 69-70, x20,000; Figs. 71-72, x33,330.) (Fig. 71 reproduced with the permission of Dr. B. S. C. Leadbeater and Springer-Verlag, from Leadbeater and Manton, 1969a; Fig. 72 with the permission of Prof. I. Manton and Cambridge University Press, from Manton, 1967a.)

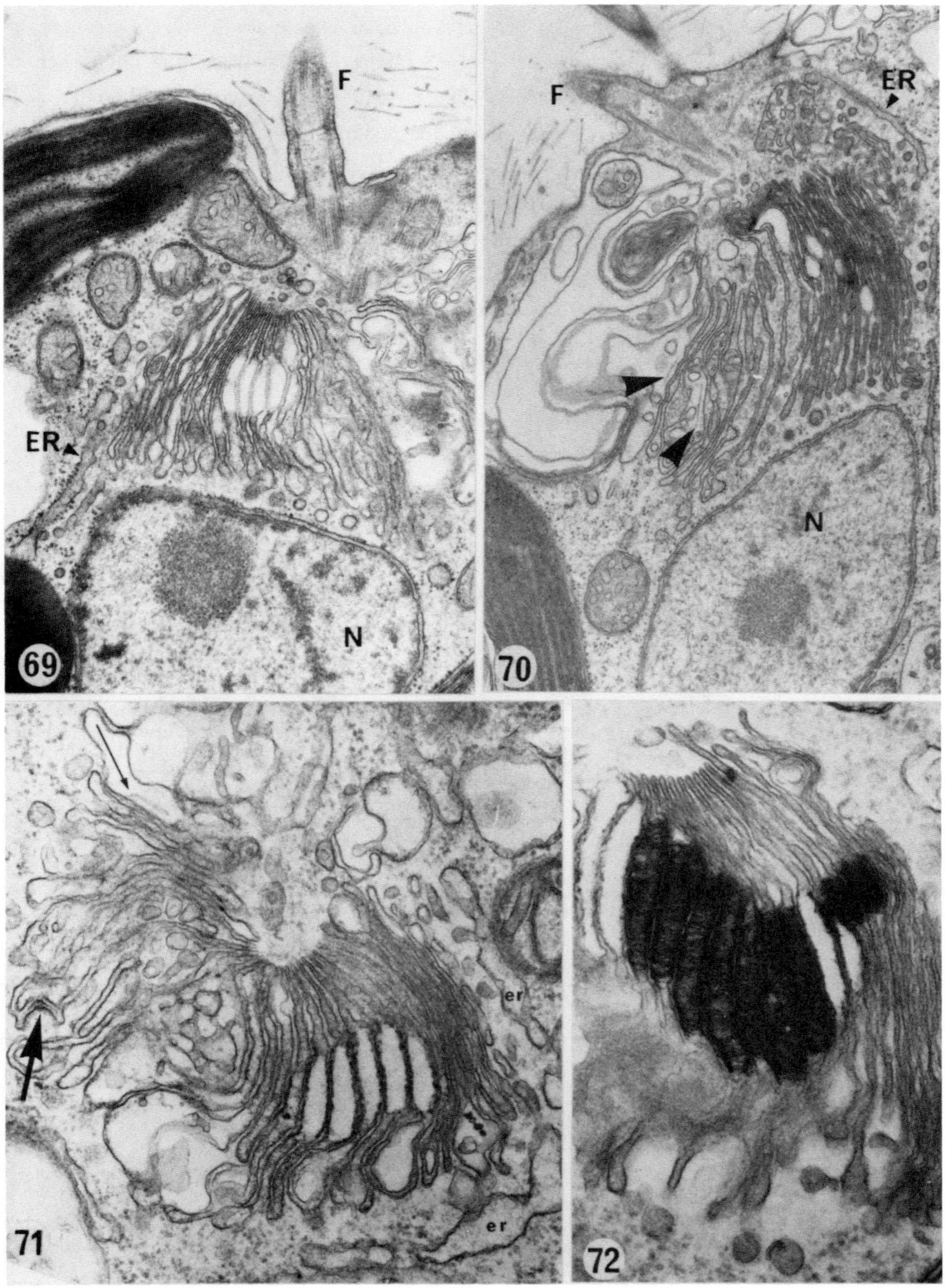
F
ER
N
69
F
ER
N
70
er
er
71
72

afternoon or early evening (Manton, 1967a). In many species there is also a regular orientation of the developing scales such that they always face one way, with the morphologically outer surface directed away from the Golgi centre and the morphologically under surface towards it. Exceptions to this rule occur in the production of very large scales which during the course of development become displaced from the Golgi stack into T-shaped cisternae, the stem of the T remaining in the stack, the scale occupying the cross-piece beyond and at right angles to the stack (Fig. 75; Manton, 1967b). This arrangement is lost during the final phase in development. Also exceptional are certain mutually related species of *Chrysochromulina* possessing unusual cup-shaped scales since in these the orientation of the scale relative to the Golgi centre is at random, with the morphologically outer surface either away from or towards the centre of the stack (Fig. 71; Leadbeater and Manton, 1969a, 1969b).

Immature scales fill the cisternal cavity in which they are formed but the most distal scale-containing vesicles become larger and more complicated internally as the cell surface is approached. Branched tubes appear to invaginate into the cisternae from their bounding membranes, greatly enlarging the lumen. As a last development before liberation the branched tubes become replaced by spirally wound membrane systems. Scale liberation takes place by fusion between the membranes of the mature scale-containing cisternae and the plasmalemma (Fig. 74), though the process is difficult to follow and may be complicated by the presence of bulbous inclusions and possible involvement of the superficial layer of ER.

Some further information on scale composition and formation has been provided by Brown and co-workers in a study of cell wall formation in *Hymenomonas pringsheimii* (Brown, 1969; Brown *et al.*, 1970). The concentric microfibrils are laid down first followed by the radial ones, the entire process of microfibrillar formation taking place in 3-4 adjacent cisternae. These fibrils then become coated with polysaccharides of a mainly pectic character, rich in galactose and

---

Figs. 73-75. Fig. 73. Transverse section of the Golgi body of *Prymnesium* sp. showing the formation of scales (arrowheads) in the cisternae. The morphologically outer surfaces are directed away from the Golgi centre. C, chloroplast; HD, hairy diverticulum; M, mitochondrion; N, nucleus (x40,000). Fig. 74. Part of the cell exterior near the maturing face of the Golgi body in *Prymnesium* sp. showing the possible manner of release of a scale (arrowhead) through a vesicle opening to the surface (x20,000). Fig. 75. Vertical section of a Golgi body of *Chrysochromulina chiton* showing the position of a large rimmed scale (arrow) within a T-shaped vesicle at this stage of formation (x20,000). (Reproduced with the permission of Prof. I. Manton and Cambridge University Press, from Manton, 1967b.)

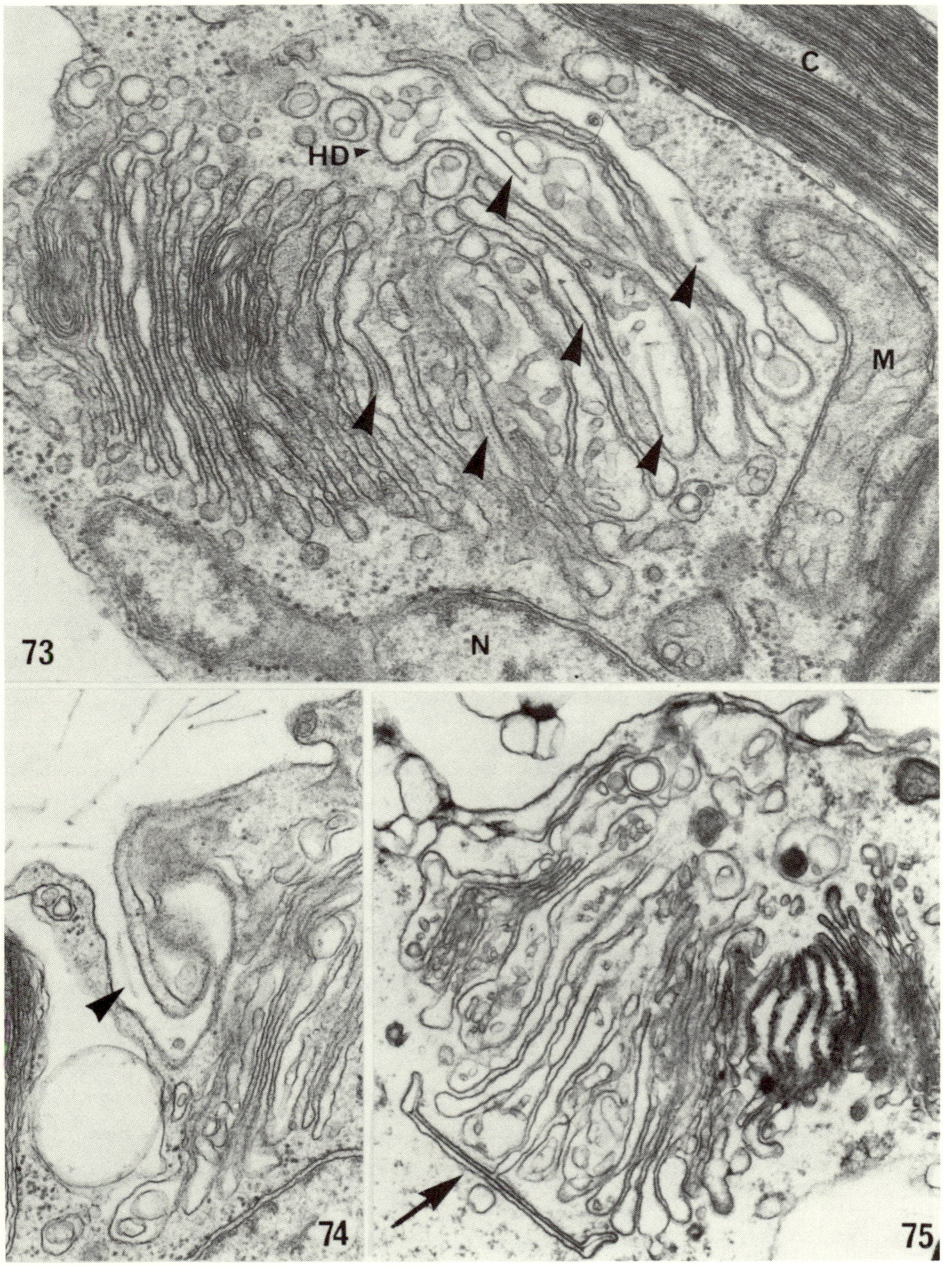
C
HD
M
N
73
74
75

ribose. Isolation and purification of the scales shows that the microfibrils forming the scale matrix are ribbon-shaped, with cross-sections of 12-25 x 25-40 Å and they have been identified as cellulose by a variety of chemical and physical methods. The central Golgi dilations have been interpreted to be the sites of polymerisation of glucose monomers into the crystalline network of the cellulose chains.

In a study using time-lapse ciné photography of cell formation in *Hymenomonas pringsheimii*, it has been shown that the whole protoplast rotates within the mother cell wall every 15-20 minutes, thus explaining how the scale layers comprising the wall are deposited evenly over the cell surface. It has been estimated from this study that during active wall synthesis one Golgi cisterna must mature approximately every two minutes (Brown, 1969).

Coccolith production may also take place in the Golgi cisternae. The process has been most intensively investigated in *Hymenomonas carterae* (see Pienaar, 1976b for bibliography) in which it has been shown that coccoliths are assembled in the Golgi cisternae from a single scale baseplate and multiple granular elements called coccolithosomes (Outka and Williams, 1971). Both elements are formed within the same cisterna but the coccolithosomes are segregated in diverticulae until the baseplate is complete, when they then progressively associate with each other to form the complex elements of the coccolith rim in which calcium carbonate then becomes crystallised.

Scales and other external structures

The most primitive scale type in the Prymnesiophyceae is generally considered to be a two-layered plate without a rim (Manton and Leedale, 1969). The most proximal (inner) layer has a radiating pattern of fibrils while the distal (outer) layer has a concentric fibril pattern (Fig. 76). Scales of this type compose the innermost layer in many species and the only type in others. Variations on this basic structure include the addition of a patternless rim, normally directed outwards; elaboration of the outer of the two scale layers leading to a wide range of morphological diversity (e.g. Leadbeater, 1972) culminating in the spectacular spine scales of such species as *Chrysochromulina ericina* (Parke, Manton and Clarke, 1956), *C. parkeae* (Green and Leadbeater, 1972) and *C. pringsheimii* (Figs. 83, 86; Parke and Manton, 1962); and loss of the outer scale layer as in the large plate scales of *C. pringsheimii* (Fig. 85) and both scale types of a deviant (probably mutant) strain of *C. chiton* (Manton, 1967a, 1967b). Apart from these generally recognisable patterns, three species of *Chrysochromulina*, *C. strobilus*, *C. camella* and *C. cymbium*, each with two layers of scales, are unusual in that the scales of the underlayer are very thin and rimless and appear to be only single-layered, and that the more complex scales of the outer layer, possibly also composed of only a single layer of material, are cup-shaped, a form otherwise known in the class (Figs. 80, 81; Leadbeater and Manton, 1969a, 1969b). It is in these species that the nascent scales are randomly orientated within the Golgi body as already described and they also differ from other species in the genus in having 6 rather than 7 tubules in their haptonemata. No evaluation of the taxonomic or functional significance of these observations has yet been proposed but it has been suggested that the three species in question may eventually have to be separated from *Chrysochromulina* into a genus of their own (Leadbeater and Manton, 1969b).

The scales in some species are embedded (Parke, Green and Manton, 1971; Green and Parke, 1974, 1975) or covered (Figs. 80, 81; Leadbeater and Manton, 1969a, 1969b) in a dilute mucilaginous material which may obscure them completely in shadowcast preparations and they have sometimes been demonstrated only with difficulty after previously being thought absent (Parke, 1972). The only well-documented case where scales genuinely appear to be absent is the coccolith-bearing stage of *Emiliania huxleyi* where there seems to be neither an underlayer of scales nor a coccolith baseplate (Klaveness, 1972a). In place of the scales this cell type has an unusual three-layered covering separated by a space from the plasmalemma. This material stains with alcyan blue and might therefore be pectic in composition.

In addition to scales, some prymnesiophytes, most notably species of *Hymenomonas*, have a layer of fibrillar material of unknown composition sometimes termed a columnar deposit (Manton and Peterfi, 1969; Pienaar, 1976a) between the scales and the plasmalemma.

The various scale layers are mostly distributed evenly over the cell body but in *Corymbellus aureus* the underlayer of scales is confined to the flagellar depression (Green, 1975b) and in *Hymenomonas pringsheimii* the vestigial haptonema is covered by a small scale type not found elsewhere on the cell (Leadbeater, 1971a). In addition, the spined scales of some species of *Chrysochromulina* are restricted to one or both ends of the cell (Fig. 83; Parke, Manton and Clarke, 1955, 1956; Parke and Manton, 1962).

The dense bodies which occur on the long flagellum and sometimes also the cell body and haptonema in most species in the Pavlovales have already been described. The term "scale" is not usually applied to them because of lack of information on their chemical nature but there is now evidence that they are produced within the cisternae of the Golgi body (Van der Veer, 1972, 1976).

The final extracellular structures to be described are small calcified bodies produced in old cultures of the benthic stage of some species. These apparently arise extracellularly and usually first appear as small rods or cruciform structures. Subsequent growth then takes place by accretion which may eventually lead to the formation of complex structures up to 50 μm across. These are of interest because of their similarity to certain enigmatic Upper Cretaceous/Early Tertiary micro-fossils such as *Tetralithus* and *Marthasterites* which may thus have been produced as a result of extracellular deposition of calcium carbonate by prymnesiophytes (Green and Parke, 1975).

Peripheral ER

A common but generally little-recognised feature of a number of prymnesiophycean zoids is an apparently empty membrane-bounded space underlying the plasmalemma over almost the whole cell except in the region adjacent to the flagella and haptonema (Figs. 15, 16, 43, 67, 68; Manton, 1966b; Green and Parke, 1974, 1975; Green, 1975a, 1975b, 1976; Gayral and Fresnel, 1976; Pienaar, 1976a). This has been termed the subcutaneous space (Manton and Leedale, 1963) but is now identifiable as endoplasmic reticulum. When the lumen is more or less uninterrupted the cells appear to have a three-layered periplast composed of the plasmalemma, the outer membrane of the ER and the narrow zone of cytoplasm between them. The ER usually appears empty but faint fibrillar material can be seen in some cases, especially in species in the Pavlovales in the region near the flagellar apparatus (Green, 1975a; Green and Hibberd, 1977).

Figs. 76-81. Scale types in the Prymnesiophyceae. Fig. 76. Two oval plate scales of *Prymnesium parvum*, the uppermost showing a concentric pattern of fibres and a narrow raised rim, the lower the radiating ridges on the proximal (under) face (x7500). Figs. 77, 78. Large scales from the outer scale layer of *Chrysochromulina chiton*. Fig. 77. Shadowcast scales showing the proximal face (left) with a radiating pattern of fibrils and the patternless distal face (right) (x20,000). Fig. 78. Section of a scale showing the projecting rim and "hairy" lining which gives the amorphous appearance seen in shadowcast scales (x40,000). Fig. 79. A coccolith of *Hymenomonas carterae* showing the radiating ridges of the proximal face of the subtending baseplate scale (x26,667). Figs. 80, 81. Scales of *Chrysochromulina camella* in vertical (Fig. 81) and tangential (Fig. 80) section. A layer of oval plate scales with a thickened margin, a cruciform centre and regular radiating ridges underlies cup-shaped scales with a short basal stalk, a bowl with four superposed rows of rectangular perforations and a patternless straight-sided rim. The latter are covered by a mucilage deposit appearing fibrillar after fixation (x40,000). (Figs. 77, 78 reproduced with the permission of Prof. I. Manton and Cambridge University Press, from Manton, 1976b; Fig. 79 with the permission of Prof. I. Manton and Cambridge University Press, from Manton and Leedale, 1969; Figs. 80, 81 with the permission of Dr. B. S. C. Leadbeater and Springer-Verlag, from Leadbeater and Manton, 1969b.)

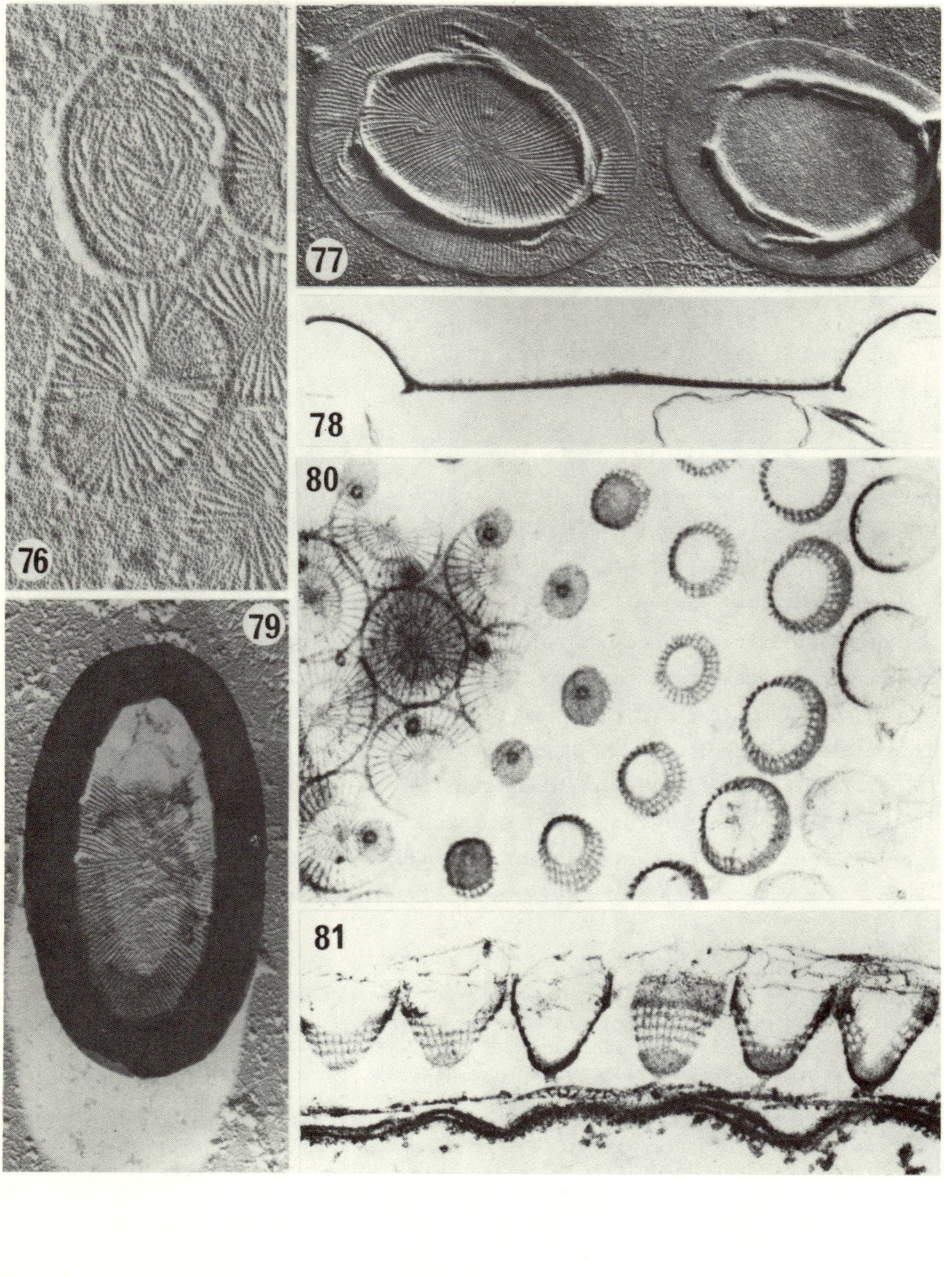
77
78
80
76
79
81

Other vesicles and vacuoles

The contractile vacuole present in freshwater species appears to show no features peculiar to the group except that in the freshwater Pavlovales it lies close to the pit and probably discharges into it (Green, 1973; Green and Hibberd, 1977). The chrysolaminaran vesicles are bounded by a single membrane in the usual way. More noteworthy are the muciferous bodies which occur peripherally in some species. These stain blue with very dilute cresyl blue which can also discharge them, producing long mucilage streamers (Parke, Manton and Clarke, 1955; Parke and Manton, 1962; Manton and Leedale, 1963). Release may also be caused by bright light. Slower discharge may also occur, this producing a large balloon-like vesicle externally. In *Platychrysis pigra* the muciferous bodies are so numerous as to be the main characteristic feature of the species at the light microscope level (Chretiennot, 1973).

MOVEMENT PATTERNS

In general prymnesiophytes exhibit varied patterns of swimming and any one species may swim in several different ways. As with many of the features mentioned already, description may be divided between the Pavlovales and the remainder of the class. Taking the latter and larger group first, the flagellar beat may be homo- or heterodynamic though forms with equal length flagella do not necessarily beat homodynamically. The most detailed and careful studies of swimming patterns are those of Parke and co-workers on various species of *Chrysochromulina* (Parke, Manton and Clarke, 1955, 1956, 1958, 1959; Parke and Manton, 1962; Parke, Lund and Manton, 1962). Although these show variations in detail, when the cell is swimming rapidly the flagella and haptonema are at the posterior end of the cell with respect to the direction of motion and the haptonema is usually tightly coiled (Figs. 1, 88). Though their overall course is straight, the cells swim in a helical path and also rotate about

---

Figs. 82-87. Scales of *Chrysochromulina pringsheimii*. Fig. 82. Empty scale case. Dark ground light microscopy (x667). Fig. 83. A dried empty scale case showing the massive terminal spines and smaller spined scales covering the remainder of the cell body (x1330). Fig. 84. Two smaller spined scales, the upper exposing the proximal face, the lower the distal face (x20,000). Fig. 85. Two sizes of plate scale, all three exposing the proximal face (x20,000). Fig. 86. Part of a large spine with the baseplate folded, exposing the lower surface (x20,000). Fig. 87. Tip of a large spine (x20,000). (Figs. 82-87 reproduced with the permission of Dr. M. Parke and Cambridge University Press, from Parke and Manton, 1962.)

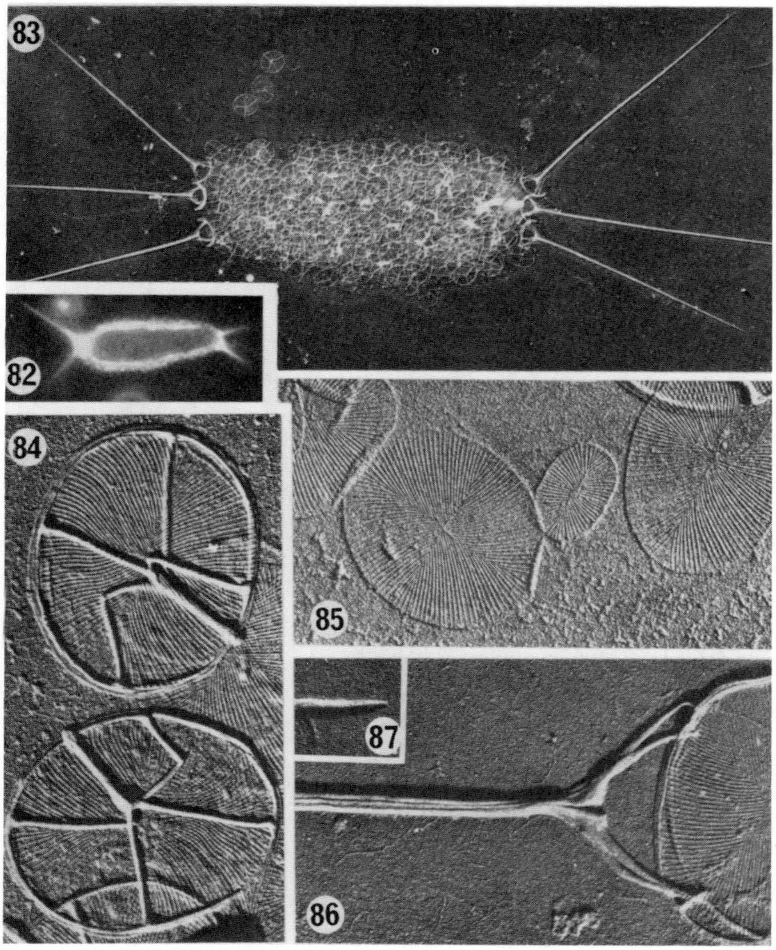
83
82
84
85
87
86

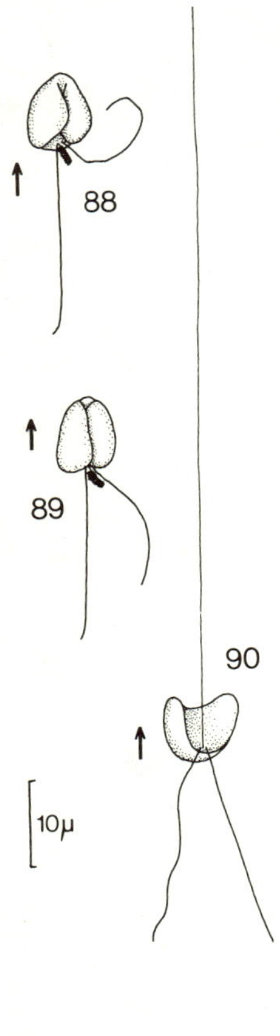

Figs. 88-90. Cells of *Chrysochromulina ephippium* showing the shape of the body and position of the flagella and haptonemata when moving in different ways. Fig. 88. A rapidly swimming cell, haptonema coiled, one flagellum trailing, one beating laterally. Fig. 89. A cell swimming more slowly, one flagellum trailing, the second beating posteriorly. Fig. 90. A slowly gliding cell, the haptonema fully extended, the flagella posterior and heterodynamic. (Redrawn after Parke, Manton and Clarke, 1956.)

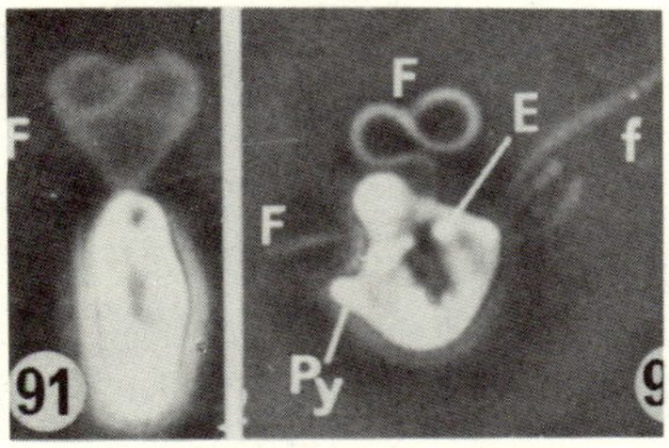

Figs. 91-92. Fig. 91. A living cell of *Pavlova gyrans* showing the characteristic S form of the beating long flagellum (x2000). Fig. 92. A living flattened cell of *P. gyrans* showing the S-shaped resting position of the long flagellum. E, eyespot; F, flagella; f, filopodium; Py, pyrenoid (x2000). Figs. 91, 92 reproduced with the permission of Dr. J. C. Green and Cambridge University Press, from Green and Manton, 1970.)

their axis. When swimming more slowly, the cell may be orientated in the same way but with the flagella held differently (Fig. 89) and the haptonema may also be extended. Alternatively, the cell may glide slowly with the haptonema extended forwards (Fig. 90) and the flagella attached either anteriorly or posteriorly. When homodynamic, both flagella beat with an undulating wave motion and in species where equal or subequal flagella beat heterodynamically, one flagellum propels the cell while the other trails behind the body during rapid swimming (Fig. 88).

One of the most characteristic features of members of the Pavlovales, where the flagella are always prominently unequal in length (Fig. 45), is the wave action of the long flagellum (Green and Manton, 1970; Van der Veer, 1972, 1976; Green, 1975a). This is always markedly S-shaped, the S being orientated sideways with respect to the direction of motion and at the forwardly directed end of the cell (Fig. 91). This flagellum still assumes this characteristic shape even when beating slowly and at rest (Fig. 92). This action produces a characteristic heart-shaped blur in rapidly beating flagella (Fig. 91). The posterior flagellum on the other hand is directed backwards and beats with a stiff jerky action.

LIFE HISTORIES

Knowledge of prymnesiophycean life histories is still fragmentary. Thus, while several different patterns have been recognised, information on all is incomplete. Culture studies have, however, gone some way towards clarifying the situation with regard to some coccolithophorids, many of which, perhaps all, have more than one distinct cell type in their life cycle.

Possibly the simplest life cycle is the flagellate/palmelloid alternation found in several species of *Chrysochromulina* (e.g. Parke, Manton and Clarke, 1955). In these, non-flagellate amoeboid cells apparently lacking the normal scale covering of the monads and possessing stellate or finely lobed chloroplasts form a layer on the bottom of ageing cultures (Figs. 2, 3). Very large amoeboid cells are possibly produced by the fusion of smaller ones. Regularly shaped cells with a smooth wall are also produced, probably from the amoeboid stage, and each divides to produce four walled daughter cells (Fig. 4). The motile phase appears to be reconstituted from the products of this division since empty cell walls with a circular pore are also found (Fig. 5) but this and other stages have not been followed in detail and nuclear cytology remains unknown.

A somewhat better-documented alternation, this time between a diploid free-living motile coccolithophorid and a haploid benthic pseudofilamentous stage, occurs in *Hymenomonas carterae* (Rayns, 1962; von Stosch, 1967; Leadbeater, 1971a; Gayral, Haas and Lepailleur, 1972). The coccolith-bearing "*Hymenomonas*-stage" reproduces mitotically but under certain conditions meiotic divisions produce four small non-coccolith-bearing swarmers which grow into the benthic pseudofilamentous "*Apistonema*-stage". This appears to produce non-coccolith-bearing asexual zoids which form new *Apistonema* thalli or motile gametes, similar to the swarmers but apparently smaller in size and with a reduced number or complete lack of chloroplasts (von Stosch, 1967). Fusion of two gametes produces a zygote which develops coccoliths and thus re-establishes the *Hymenomonas*-stage. Although it is establised with reasonable centainty that the *Hymenomonas*-stage is diploid and the *Apistonema*-stage haploid (Rayns, 1962), the function of the various non-coccolith-bearing cells needs further investigation in both this and other species in which fusion has not so far been detected. The whole life cycle may, in fact, be far more complicated than described here since it has been shown that the non-motile coccolithophorids *Ochrospheara neopolitana* (Gayral and Fresnel-Morange, 1971) and *O. verrucosa* both can produce motile cells with coccoliths corresponding to those of *H. carterae*, and a number of other *Hymenomonas* strains have been found to contain cells corresponding to *O. verrucosa*. Futhermore *O neopolitana* is known to reproduce by means of biflagellate asexual swarmers, as well as by division into two or four within the coccosphere, and also sexually by the fusion of haploid isogametes, producing a diploid zygote with a similar structure to the vegetative cells. Meiotic division of the latter produces four vegetative cells or swarmers (Schussnig, 1930, 1940; Schwartz, 1932). Many of these cell types may thus constitute different phases of a common life cycle.

Alternation between two completely different free-living coccolithophorids has been clearly established in the case of *Coccolithus pelagicus* (Parke and Adams, 1960). This species, which has a well known non-motile "*Coccolithus*-phase" bearing coccoliths of the placolith type, has been shown to produce in cultured isolates, motile cells bearing two equal or subequal flagella and a haptonema, and coccoliths of a completely different structural type (crystall-oliths) from the non-motile phase. These were originally described separately as the coccolithophorid *Crystallolithus hyalinus* (Gaarder and Markali, 1956) but this must now be regarded as a synonym of *Coccolithus pelagicus*. Transformation from the non-motile to the motile phase has not been seen and the nuclear relations are unknown. The finding of two distinct structural types of coccoliths in the life history of one coccolithophorid shows a serious danger in basing the taxonomy of recent-as opposed to fossil-species wholly or largely on coccolith morphology. Parke and Green (1976) consider it probable that recent flagellate taxa presently provisionally classified in the order Coccosphaerales are not distinct entities but only alternate motile phases of non-flagellate taxa. As in the example of *Coccolithus*/*Crystallolithus*, the solution to these problems will necessitate fine structure and culture studies.

A further type of life history is demonstrated by *Emiliania huxleyi* (Klaveness, 1972b). The dominant phase here is a non-motile coccolithophorid reproducing vegetatively by fission. In culture this species can produce cells bearing two equal length flagella and scales, but not haptonema or coccoliths, and also completely naked non-motile cells. As noted previously the non-motile coccolith-bearing stage is unusual in apparently not possessing normal inorganic scales either as an underlayer to the coccoliths or as coccolith baseplates and the complete absence of a haptonema from the motile scale cells is also unusual. Little is known concerning the environmental conditions triggering the transition between these various cell types but ageing of cultures favours the formation of motile scaly cells and naked cells from the typical form with coccoliths and also the formation of scaly cells in cultures of naked cells. Nuclear staining shows that the coccolith-bearing and naked cells have the same amount of DNA per cell and the scaly flagellate cell may have a smaller chromosome number than either of these two cell types (Paasche and Klaveness, 1970). The difference in the anatomy of the motile cells of this species from that in *Coccolithus pelagicus* supports its generic separation as *E. huxleyi* based on coccolith structure (Hay *et al.*, 1967).

Prymnesiophycean cysts are not well documented and are probably not produced by species which have an alternate benthic phase. They are best known in species of *Prymnesium* (Carter, 1937; Conrad, 1941) were they are ovate or elliptical and have a wide mouth which mostly appears not to be closed by a separate plug. No ultrastructural observations are yet available, but preliminary results (Prof. R. E. Pienaar, personal communication) show that they are constructed in a completely different way from those in the Chrysophyceae which have a silica wall with a pore which is closed by a plug at maturity (Hibberd, 1977).

MITOSIS

Mitosis has been studied so far at the fine-structural level only in *Prymnesium parvum* (Manton, 1964c), and even here only preliminary data are available for some stages. The first indication of cell division is flagellar replication and the chloroplasts and Golgi body also divide at an early stage. Microtubules emanating from near the flagellar bases proliferate and are directed towards various parts of the cell. Some of these make contact with the outer membrane of the nuclear envelope in prophase. Separation of the two new flagellar poles is presumably rapid since it has not been followed, though one pair of basal bodies comes to lie near but not at the poles of the spindle and a Golgi body occurs near to each. The nuclear envelope then breaks up into tubular cisternae and views of telophase show that all the spindle fibres pass between the chromosomes, none apparently making contact with them. The cell and spindle then elongate and a nuclear envelope reforms round each of the two groups of daughter chromosomes, beginning on the polar side of each group, the spindle side remaining open for much longer. The fragments of the old nuclear envelope appear to be the main source of the new ones. Cleavage starts at the cell surface and the spindle fibres become bunched in a compact strand. The nucleus at telophase is deeply indented and a scar which remains detectable in interphase cells appears to be formed at the presumed region of closure of the nuclear envelope.

BIOCHEMISTRY AND PHYSIOLOGY

The Prymnesiophyceae cannot yet be seen to differ appreciably in their physiology or biochemistry from other flagellate groups. Since these aspects are adequately treated elsewhere they are not dealt with further here with the exception of the fish poisoning properties of *Prymnesium parvum*. Reviews of the literature may be found in Allen (1969), which also includes the Chrysophyceae; Paasche (1968), which deals particularly with the coccolithophorids and the experimental work on coccolith formation; and Boney (1970), which includes the scaly flagellates in general. More recent aspects of the biochemistry of the group are contained in Stewart (1974), Ragan & Chapman (1978) and Aaronson (in press). In consulting this literature the reader should be aware of the fact that a number of organisms still listed as chrysophytes in physiological and biochemical investigations have now been demonstrated to be members of the Prymnesiophyceae. Thus the well-known "chrysophyte" *Monochrysis lutheri* Droop, used in numerous laboratory studies on phytoplankton growth kinetics, nutrition, biochemistry and as a food organism in animal feeding experiments, is now classified as the prymnesiophyte *Pavlova lutheri* (Droop) Green on the basis of its ultrastructure (Green, 1975a).

The biochemistry and physiology of *Prymnesium parvum* have been intensively studied because this flagellate secretes potent toxins which are the cause of extensive fish mortalities in brackish water localities. Disastrous consequences followed its establishment in all the brackish water fish ponds in Israel, for example, and under bloom conditions cell concentrations of 10^6 - 10^7 cells per ml may be reached. Some of the features of the toxigenic reaction are that the toxicity may vary with the physiological conditions of the cells, so that high cell numbers are commonly but not necessarily highly toxic; toxin formation occurs only in the light; stability of the toxins is influenced by a variety of factors including pH, phosphate levels, ultraviolet radiation and suspended colloids; and the toxin depends for its ichthyotoxic activity on a variety of co-factors, the dialized toxin being inactive to fish. The initial target of poisoning appears to be the gill, producing reversible specific damage to its tissues resulting in loss of their selective permeability; mortality results from a second stage response of the sensitised fish to non-specific toxicants normally present in the water in concentrations sublethal to normal fish. Control measures for fishponds have been devised and involve the utilisation of ammonium sulphate at high pH, *P. parvum* being unusually sensitive to lysis by ammonia. Detailed reviews of the extensive literature on this subject are given by Shilo (1967) and Boney (1970).

HABITAT

The very great majority of prymnesiophytes are marine and planktonic and they are often important components of the phytoplankton in both coastal and oceanic waters. Some coccolithophorids may reach very high concentrations, forming one of the main plankton constituents. *Phaeocystis pouchetii*, a colonial species with the cells embedded in mucilage, may occur in such vast quantities as to render the sea oily and yellow-brown in colour. Large blooms of this species have been thought to act as barriers bringing about deviations in the migratory paths of herring shoals in the North Sea, resulting in low catches (see Boney, 1970 for a fuller account).

Only a few species seem to occur exclusively in freshwater, the most notable of which are *Chrysochromulina parva*, the type species of this important and otherwise marine genus, and the coccolithophorid *Hymenomonas roseola*. Even though more will probably be discovered as knowledge of the group increases, it is already clear that the group is mainly represented in the sea.

Some benthic forms, probably stages in the life history of planktonic monads, often form conspicuous supralittoral "chrysophyte" bands, particularly on chalk cliffs. These may be important as a source of food for smaller inshore animals, especially the economically significant shellfish larvae, since they may release considerable numbers of swarmers (Green and Parke, 1974).

In some cases the distinction between marine and freshwater species is not clearcut. Probably the best example of this is *Diacronema vlkianum*, a member of the Pavlovales, which was first discovered in freshwater (Vlk, 1938; Prauser, 1958) but which has also been isolated from a marine locality. A recent freshwater isolate has been gradually transferred to full-strength seawater and the marine strain has been grown in dilute freshwater media (Green and Hibberd, 1977). The growth conditions in nature may not, however, be so different since the freshwater locality was highly eutrophic, but the species undoubtedly is remarkably euryhaline. Strains of other species such as *Prymnesium parvum*, *Pavlova lutheri* and *Hymenomonas carterae* show almost as great a tolerance as *D. vlkianum* but usually demand a salinity of 3% or above for survival in culture even though some may also tolerate greater than full strength seawater.

CULTURE MEDIA

Marine media are dealt with comprehensively by McLaughlin (1973) and the compositions of 10 of the most widely used are given. One enriched natural seawater medium such as Erdschreiber and one of the artificial seawater media to which trace elements and vitamins have been added, e.g. ASP-2 (Provasoli, McLaughlin and Droop, 1957; Provasoli, 1963) should suffice for the great majority of studies, the media being used either alone or mixed in various proportions. The reader is referred to both of the latter articles for a treatment of marine media in genera.

NEEDED RESEARCH

Fine structure studies are relatively well-advanced though there is as yet little information on nuclear and cell division. By far the most neglected area in the structural biology of prymnesiophytes is the study of life histories, even though it is already clear that they may have a profound significance for the systematics of the group. Studies on the geographical distribution and systematics of recent coccolithophorids are yielding results of the greatest interest (e.g. Manton, Sutherland and Oates, 1976) and can be expected to do so for some time to come. It is clear from a recent review (Ragan and Chapman, 1978) that virtually every aspect of the biochemistry of the group would repay further study.

ACKNOWLEDGEMENT

I am grateful to Dr. F. K. Kupicha for help in the preparation of this chapter.

REFERENCES

Allen, M. B. 1969. Structure, physiology and biochemistry of the Chrysophyceae. Ann. Rev. Microbiol. 23: 29-46.

Berger, R., S. Liaan-Jensen, V. McAlister and R. R. L. Guillard. 1977. Carotenoids of Prymnesiophyceae (Haptophyceae). Biochem. Syst. Ecol. 5: 71-75.

Billard, C. 1976. Sur une nouvelle espèce de *Pavlova*, *P. virescens* nov. sp., (Haptophycées). Bull. Soc. Phycol. Fr. 21: 18-27.

Boney, A. D. 1970. Scale-bearing phytoflagellates: an interim review. Oceanogr. Mar. Biol. Ann. Rev. 8: 251-305.

Bouck, G. B. 1971. The structure, origin, isolation and composition of the tubular mastigonemes of the *Ochromonas* flagellum. J. Cell Biol. 50: 362-384.

Bourrelly, P. 1968. *Les algues d'eau douce.* 2: Algues Jaunes et Brunes. Boubée et Cie, Paris.

Brown, R. M. 1969. Observations on the relationship of the Golgi apparatus to wall formation in the marine chrysophycean alga, *Pleurochrysis scherffelii* Pringsheim. J. Cell Biol. 41: 109-123.

Brown, R. M., W. W. Franke, H. Kleinig, H. Falk and P. Sitte. 1970. Scale formation in chrysophycean algae. I. Cellulose and noncellulosic wall components made by the Golgi apparatus. J. Cell Biol. 45: 246-271.

Carter, N. 1937. New or interesting algae from brackish water. Arch. Protistenk. 90: 1-68.

Chretiennot, M. J. 1973. The fine structure and taxonomy of *Platychrysis pigra* Geitler (Haptophyceae). J. Mar. Biol. Assn. U.K. 53: 905-914.

Christensen, T. 1962. *Alger* In Böcher, T. W., M. Lange and T. Sørensen [Eds.] Botanik, II, No. 2., Systematic Botanik. Munksgaard, Copenhagen.

Conrad, W. 1941. Notes protistologiques. XXI. Sur les Chrysomonadines à trois fouets. Aperçu synoptique. Bull. Mus. R. Hist. Nat. Belg. 17 (45): 1-16.

Coombs, J. and A. D. Greenwood. 1976. Compartmentation of the photosynthetic apparatus. In Barber, J. [Ed.] *The Intact Chloroplast.* Elsevier, Amsterdam.

Falk, H. and H. Kleinig. 1968. Feinbau und Carotinoide von *Tribonema* (Xanthophyceae). Arch. Mikrobiol. 61: 347-362.

Gaarder, K. R. and J. Markali. 1956. On the coccolithophorid *Crystallolithus hyalinus* n. gen., n. sp. Nytt Mag. Bot. 5: 1-5.

Gayral, P. and J. Fresnel. 1976. Nouvelles observations sur deux Coccolithophoracées marines: *Cricosphaera roscoffensis* (P. Dangeard) comb. nov. et *Hymenomonas globosa* (F. Magne) comb. nov. Phycologia 15: 339-355.

Gayral, P. and J. Fresnel-Morange. 1971. Résultats préliminaires sur la structure et la biologie de la coccolithacée *Ochrosphaera neopolitana* Schussnig. C.R. Hebd. Séanc. Acad. Sci., Paris 273: 1683-1686.

Gayral, P., C. Haas and H. Lepailleur. 1972. Alternance morphologique de générations et alternance de phases chez les Chrysophycées. Mem. Soc. Bot. Fr. 1972: 215-230.

Gibbs, S. P. 1979. The route of entry of cytoplasmically synthesized proteins into chloroplasts of algae possessing chloroplast ER. J. Cell Sci. 35: 253-266.

Green, J. C. 1973. Studies in the fine structure and taxonomy of flagellates in the genus *Pavlova*. II. A freshwater representative, *Pavlova granifera* (Mack) comb. nov. Br. Phycol. J. 8: 1-12.

Green, J. C. 1975a. The fine-structure and taxonomy of the haptophycean flagellate *Pavlova lutheri* (Droop) comb. nov. (=*Monochrysis lutheri* Droop). J. Mar. Biol. Assn. U.K. 55: 785-793.

Green, J. C. 1975b. *Corymbellus aureus* gen. et sp. nov., a new colonial member of the Haptophyceae. J. Mar. Biol. Assn. U.K. 5: 31-38.

Green, J. C. 1976. Notes on the flagellar apparatus and taxonomy of *Pavlova mesolychnon* Van der Veer, and on the status of *Pavlova* Butcher and related genera within the Haptophyceae. J. Mar. Biol. Assn. U.K. 56: 595-602.

Green, J. C. and D. J. Hibberd. 1977. The ultrastructure and taxonomy of *Diacronema vlkianum* (Prymnesiophyceae) with special reference to the haptonema and flagellar apparatus. J. Mar. Biol. Assn. U.K. 57: 1125-1136.

Green, J. C. and B. S. C. Leadbeater. 1972. *Chrysochromulina parkeae* sp. nov. (Haptophyceae) a new species recorded from S.W. England and Norway. J. Mar. Biol. Assn. U.K. 52: 469-474.

Green, J. C. and I. Manton. 1970. Studies in the fine structure and taxonomy of flagellates in the genus *Pavlova*. I. A revision of *Pavlova gyrans*, the type species. J. Mar. Biol. Assn. U.K. 50: 1113-1130.

Green, J. C. and M. Parke. 1974. A re-investigation by light and electron microscopy of *Ruttnera spectabilis* Geitler (Haptophyceae) with special reference to the fine structure of the zoids. J. Mar. Biol. Assn. U.K. 54: 539-550.

Green, J. C. and M. Parke. 1975. New observations upon members of the genus *Chrysotila* Anand, with remarks upon their relationships within the Haptophyceae. J. Mar. Biol. Assn. U.K. 55: 109-121.

Green, J. C. and R. N. Pienaar. 1977. The taxonomy of the order Isochrysidales (Prymnesiophyceae) with special reference to the genera *Isochrysis* Parke, *Dicrateria* Parke and *Imantonia* Reynolds. J. Mar. Biol. Assn. U.K. 57: 7-17.

Hager, A. and H. Stransky. 1970. Das Carotinoidmuster und die Verbreitung des lichtinduzierten Xanthophyllcyclus in verschiedenen Algenklassen. V. Einzelne Vertreter der Cryptophyceae, Euglenophyceae, Bacillariophyceae, Chrysophyceae und Phaeophyceae. Arch. Mikrobiol. 73: 77-89.

Hay, W. W., H. P. Mohler, P. H. Roth, R. R. Schmidt and J. R. Boudreaux. 1967. Calcareous nannoplankton zonation of the Cenozoic of the Gulf Coast and Caribbean-Antillean area and transoceanic correlation. Trans. Gulf-Cst. Assn. Geol. Socs 17: 426-480.

Hibberd, D. J. 1976. The ultrastructure and taxonomy of the Chrysophyceae and Prymnesiophyceae (Haptophyceae): a survey with some new observations on the ultrastructure of the Chrysophyceae. Bot. J. Linn. Soc. 72: 55-80.

Hibberd, D. J. 1977. Ultrastructure of cyst formation in *Ochromonas tuberculata* (Chrysophyceae). J. Phycol. 13: 309-320.

Jeffery, S. W. 1976. The occurrence of chlorophyll c_1 and c_2 in algae. J. Phycol. 12: 349-354.

Klaveness, D. 1972a. *Coccolithus huxleyi* (Lohmann) Kamptner I. Morphological investigations on the vegetative cell and the process of coccolith formation. Protistologica 8: 335-346.

Klaveness, D. 1972b. *Coccolithus huxleyi* (Lohm.) Kamptn. II. The flagellate cell, aberrant cell types, vegetative propagation and life cycles. Br. Phycol. J. 7: 309-318.

Kreger, D. R. and J. Van der Veer. 1970. Paramylon in a chrysophyte. Acta Bot. Neerl. 19: 401-402.

Lackey, J. B. 1939. Notes on plankton flagellates from the Scioto River. Lloydia 2: 128-143.

Leadbeater, B. S. C. 1971a. Observations on the life history of the haptophycean alga *Pleurochrysis scherffelii* with special reference to the microanatomy of the different types of motile cells. Ann. Bot. 35: 429-439.

Leadbeater, B. S. C. 1971b. Observations by means of ciné photography on the behaviour of the haptonema in plankton flagellates of the class Haptophyceae. J. Mar. Biol. Assn. U.K. 51: 207-217.

Leadbeater, B. S. C. 1972. Fine structural observations on six species of *Chrysochromulina* (Haptophyceae) from Norway. Sarsia 49: 65-80.

Leadbeater, B. S. C. and I. Manton. 1969a. New observations on the fine structure of *Chrysochromulina strobilus* Parke and Manton with special reference to some unusual features of the haptonema and scales. Arch. Mikrobiol. 66: 105-120.

Leadbeater, B. S. C. and I. Manton. 1969b. *Chrysochromulina camella* sp. nov. and *C. cymbium* sp. nov., two new relatives of *C. strobilus* Parke and Manton. Arch. Mikrobiol. 68: 116-132.

Leadbeater, B. S. C. and I. Manton. 1971. Fine structure and light microscopy of a new species of *Chrysochromulina* (*C. acantha*). Arch. Mikrobiol. 78: 58-69.

Liaaen-Jensen, S. 1977. Algal carotenoids and chemosystematics. In Faulkener, D. J. and W. H. Fenical [Eds.] *Marine Natural Products Chemistry*. Plenum, New York. pp. 239-259.

Manton, I. 1964a. The possible significance of some details of flagellar bases in plants. J. Roy. Microsc. Soc. 82: 279-285.

Manton, I. 1964b. Further observations on the fine structure of the haptonema in *Prymnesium parvum*. Arch. Mikrobiol. 49: 315-330.

Manton, I. 1964c. Observations with the electron microscope on the division cycle in the flagellate *Prymnesium parvum* Carter. J. Roy. Microsc. Soc. 83: 317-325.

Manton, I. 1966a. Further observations on the fine structure of *Chrysochromulina chiton*, with special reference to the pyrenoid. J. Cell Sci. 1: 187-192.

Manton, I. 1966b. Observations on scale production in *Prymnesium parvum*. J. Cell Sci. 1: 375-380.

Manton, I. 1967a. Further observations on the fine structure of *Chrysochromulina chiton* with special reference to the haptonema, "peculiar" Golgi structure and scale production. J. Cell Sci. 2: 265-272.

Manton, I. 1967b. Further observations on scale formation in *Chrysochromulina chiton*. J. Cell Sci. 2: 411-418.

Manton, I. 1968. Further observations on the microanatomy of the haptonema in *Chrysochromulina chiton* and *Prymnesium parvum*. Protoplasma 66: 35-53.

Manton, I. and G. F. Leedale. 1961. Further observations on the fine structure of *Chrysochromulina minor* and *C. kappa* with special reference to the pyrenoids. J. Mar. Biol. Assn. U.K. 41: 519-526.

Manton, I. and G. F. Leedale. 1963. Observations on the fine structure of *Prymnesium parvum* Carter. Arch. Mikrobiol. 45: 285-303.

Manton, I. and G. F. Leedale. 1969. Observations on the microanatomy of *Coccolithus pelagicus* and *Cricosphaera carterae* with special reference to the origin and nature of coccoliths and scales. J. Mar. Biol. Assn. U.K. 49: 1-16.

Manton, I. and L. S. Peterfi. 1969. Observations on the fine structure of coccoliths, scales and the protoplast of a freshwater coccolithophorid, *Hymenomonas roseola* Stein, with supplementary observations on the protoplast of *Cricosphaera carterae*. Proc. Roy. Soc. B 172: 1-15.

Manton, I., J. Sutherland and K. Oates. 1976. Arctic coccolithophorids: two species of *Turrisphaera* gen. nov. from West Greenland, Alaska, and the Northwest Passage. Proc. Roy. Soc. B 194: 179-194.

Massart, J. 1920. Recherches sur les organismes inferieurs. VIII. Sur la motilité des flagellates. Bull. Acad. R. Belg. Cl. Sci., Ser. 5, 6: 116-141.

McLaughlin, J. 1973. Growth media - marine. In Stein, J. R. [Ed.] *Handbook of Phycological Methods, Culture Methods and Growth Requirements*. Cambridge University Press. pp. 25-51.

Outka, D. E. and D. C. Williams. 1971. Sequential coccolith morphogenesis in *Hymenomonas carterae*. J. Protozool. 18: 285-297.

Paasche, E. 1968. Biology and physiology of coccolithophorids. Ann. Rev. Microbiol. 22: 71-86.

Paasche, E. and D. Klaveness. 1970. A physiological comparison of coccolith-forming and naked cells of *Coccolithus huxleyi*. Arch. Mikrobiol. 73: 143-152.

Papenfuss, G. F. 1955. Classification of the algae. In *A Century of Progress in the Natural Sciences*, 1853-1953. California Academy of Sciences, San Francisco. pp. 115-224.

Parke, M. 1961. Some remarks concerning the class Chrysophyceae. Br. Phycol. Bull. 2: 47-55.

Parke, M. 1972. The production of calcareous elements by benthic algae belonging to the class Haptophyceae (Chrysophyta).

Parke, M. and I. Adams. 1960. The motile (*Crystallolithus hyalinus* Gaarder and Markali) and non-motile phases in the life history of *Coccolithus pelagicus* (Wallich) Schiller. J. Mar. Biol. Assn. U.K. 39: 263-274.

Parke, M. and J. C. Green. 1976. Haptophyta. In Parke, M. and P. S. Dixon, *Check list of British marine algae - third revision*. J. Mar. Biol. Assn. U.K. 56: 527-594, pp. 551-555.

Parke, M. and I. Manton. 1962. Studies on marine flagellates VI. *Chrysochromulina pringsheimii* sp. nov. J. Mar. Biol. Assn. U.K. 42: 391-404.

Parke, M., J. C. Green and I. Manton. 1971. Observations on the fine structure of zoids of the genus *Phaeocystis* (Haptophyceae). J. Mar. Biol. Assn. U.K. 51: 927-941.

Parke, M., J. W. G. Lund and I. Manton. 1962. Observations on the biology and fine structure of the type species of *Chrysochromulina* (*C. parva* Lackey) in the English Lake District. Arch. Mikrobiol. 42: 333-352.

Parke, M., I. Manton and B. Clarke. 1955. Studies on marine flagellates. II. Three new species of *Chrysochromulina*. J. Mar. Biol. Assn. U.K. 34: 579-609.

Parke, M., I. Manton and B. Clarke. 1956. Studies on marine flagellates. III. Three further species of *Chrysochromulina*. J. Mar. Biol. Assn. U.K. 35: 387-414.

Parke, M., I. Manton and B. Clarke. 1958. Studies on marine flagellates. IV. Morphology and microanatomy of a new species of *Chrysochromulina*. J. Mar. Biol. Assn. U.K. 37: 209-228.

Parke, M., I. Manton and B. Clarke. 1959. Studies on marine flagellates. V. Morphology and microanatomy of *Chrysochromulina strobilus* sp. nov. J. Mar. Biol. Assn. U.K. 38: 169-188.

Pascher, A. 1910. Chrysomonaden aus dem Herschberger Grossteiche. Monogr. Abh. Intern. Rev. Gesammt. Hydrobiol. Hydrogr. 1: 1-66.

Pienaar, R. N. 1976a. The microanatomy of *Hymenomonas lacuna* sp. nov. (Haptophyceae). J. Mar. Biol. Assn. U.K. 56: 1-11.

Pienaar, R. N. 1976b. The rhythmic production of body covering components in the haptophycean flagellate *Hymenomonas carterae*. In Watabe, N. and K. M. Wilbur [Eds.] *The Mechanisms of mineralization in the invertebrates and plants*. Columbia, University of South Carolina Press. pp. 203-229.

Prauser, H. 1958. *Diacronema vlkianum*, eine neue Chrysomonade. Arch. Protistenk. 103: 117-128.

Provasoli, L. 1963. Growing marine seaweeds. In De Virville, D. and J. Feldman [Eds.] *Proceedings of the International Seaweed Symposium, 4*. Oxford, Pergamon. pp. 9-17.

Provasoli, L., J. J. A. McLaughlin and M. R. Droop. 1957. The development of artificial media for marine algae. Arch. Mikrobiol. 25: 392-428.

Ragan, M. A. and D. J. Chapman. 1978. A biochemical phylogeny of the protists. New York, Academic Press.

Rayns, D. G. 1962. Alternation of generation in a coccolithophorid, *Cricosphaera carterae* (Braarud and Fagerland) Braarud. J. Mar. Biol. Assn. U.K. 42: 481-484.

Schussnig, B. 1930. *Ochrosphaera neopolitana* n. gen. et sp., eine neue Chrysomonade mit Kalkhülle. Öst. Bot. Z. 79: 164-170.

Schussnig, B. 1940. Über einige neue Protophyten aus der Adria. Arch. Protistenk. 93: 317-330.

Schwartz, E. 1932. Der Formwechsen von "*Ochrosphaera neopolitana*". Arch. Protistenk. 77: 434-462.

Shilo, M. 1967. Formation and mode of algal toxins. Bact. Rev. 31: 180-193.

Stewart, W. D. P. [Ed.]. 1974. *Algal Physiology and Biochemistry*. Oxford, Blackwell.

Stosch, H. A. von. 1958. Der Geiselapparat einer Coccolithophoride. Naturwissenschaften 45: 140-141.

Stosch, H. A. von. 1967. Haptophyceae. In Ruhland, W. [Ed.] *Vegetative Fortpflanzung, Parthenogenese und Apogamie bei Algen*. Encyclopaedia of Plant Physiology 18: pp. 646-656.

Van der Veer, J. 1972. *Pavlova helicata* (Haptophyceae), a new species from the Frisian Island Schiermonnikoog, the Netherlands. Nova Hedwigia 23: 131-143.

Van der Veer, J. 1976. *Pavlova calceolata* (Haptophyceae), a new species from the Tamar Estuary, Cornwall, England. J. Mar. Biol. Asso. U.K. 56: 21-30.

Vlk, W. 1938. Über den Bau der Geissel. Arch. Protistenk. 90: 448-488.

Copyright 1980 by Elsevier North Holland, Inc.

EUSTIGMATOPHYTES

DAVID J. HIBBERD
Culture Centre of Algae and Protozoa, Institute of Terrestrial Ecology, 36, Storey's Way, Cambridge, CB3 ODT, England

ABSTRACT

The Eustigmatophyceae is a very small class of algae all members of which are yellow-green coccoid unicells, freshwater or marine. Photosynthetic pigments include chlorophyll *a*, β-carotene, violaxanthin and vaucheriaxanthin ester. Eustigmatophytes do not produce starch and the storage metabolite seems to be a material of unknown composition which appears finely lamellate in sections. Vegetative cells characteristically contain a single lobed chloroplast with a conspicuous pyrenoid, one to several nuclei, a large bright red vesicle and a vacuole with vibrating granular contents. Eustigmatophytes reproduce primarily by the formation of autospores and this is the only method of reproduction in some species. Others also produce zoospores, though always in relatively small numbers.

Zoospores have an organization unique in the plant kingdom. They are elongate and lageniform with, in most species, only a single anteriorly-directed flagellum, a very large extraplastidial red eyespot at the extreme anterior end of the cell and a single long flattened chloroplast. The nucleus lies in the narrow end of the cell between the anterior end of the chloroplast and the eyespot. The flagellum bears tubular mastigonemes and at its proximal end the expanded flagellar membrane forms a T-shaped swelling which is flattened against the cell over the eyespot. In species with biflagellate zoospores the long flagellum has exactly the same structure while the short flagellum is smooth and does not bear a swelling at its proximal end. The flagellar transition region contains a transitional helix. Both microtubular and cross-banded fibrillar roots are present. A single Golgi body occurs only in biflagellate zoospores where it lies against the anterior end of the nucleus, close to the flagellar basal bodies. Chloroplasts in both zoospores and vegetative cells possess three-thylakoid lamellae but a girdle lamella is absent in all cases. Pyrenoids occur in all species but are found in the vegetative cells only, never in zoospores. They are mostly large, polyhedral or spherical, are attached to the face of the chloroplast by only a narrow stalk and are not penetrated by chloroplast lamellae. Flat vesicles with contents appearing finely lamellate after fixation lie against the face of the pyrenoid. The chloroplasts are surrounded by a layer of chloroplast ER and a periplastidial network occurs between the latter and the chloroplast envelope.

GENERAL CHARACTERISTICS

This very small class of algae, so far containing only 11 species, was set up in 1971 (Hibberd and Leedale, 1971) as a result of the discovery that a number of forms originally classified as unicellular coccoid xanthophytes had motile cells with an organization completely different from that characteristic of the Xanthophyceae and also from all the other classes of algae (Hibberd and Leedale, 1970, 1972). The vegetative cells also have some unique features. All the known species exist vegetatively as coccoid unicells. They are either spherical (Figs. 1-5), polyhedral, stellate (Figs. 6, 7) or ovoid (Figs. 8-11) and are mostly free-floating though one species is normally attached.

The separation of eustigmatophytes from the Xanthophyceae has produced a number of taxonomic and nomenclatural problems (Hibberd and Leedale, 1972; Silva, 1979). Only the previously xanthophycean genera *Chlorobotrys* and *Vischeria* can be referred to the Eustigmatophyceae since the strains studied have been identified with the type species of these genera. A taxonomic and nomenclatural revision of several other taxa is currently being prepared and so to avoid confusion names of species not in either of these two genera will be placed in quotation marks. Culture collection strain numbers are appended. These names may thus be linked with a nomenclaturally correct name at a later date.

Because of their yellow-green color, eustigmatophytes may be confused with coccoid members of both the Xanthophyceae and Chlorophyceae and many forms still classified as xanthophytes may yet be assigned to the Eustigmatophyceae with increased knowledge of their cell structure and pigment composition. The form of the zoospore is by far the best diagnostic character, though the vegetative cells of most species contain one or more of the highly distinctive pyrenoids described below. However, some species lack these pyrenoids and others do not produce zoospores so examination of the xanthophyll pigments or electron microscopy of sections may be necessary to assign an unknown organism to the correct class (Antia *et al.*, 1975).

Chloroplasts and pigment composition. Eustigmatophytes all seem to contain a single lobed parietal chloroplast (Hibberd and Leedale, 1972; Hibberd, 1974; Antia *et al.*, 1975) and thus differ from most xanthophytes which contain more than one discoid chloroplast. Pyrenoids are typically present but in the vegetative cells only, never in the zoospores. In the larger species with spherical cells the pyrenoids have a highly characteristic form, appearing in the light microscope as large polyhedral crystals, rhombohedral or pentagonal in optical section (Figs. 1-4, 12). One species with ellipsoidal cells differs in having spherical pyrenoids (Lee and Bold, 1973). Electron microscopy is necessary to demonstrate positively that the pyrenoids are attached to the inner face of the chloroplast by only a narrow stalk (Fig. 3) and they are usually referred to in the older literature as albumen crystals (Petersen, 1932; Pascher, 1937-1939; Vischer, 1945). Plates of refractile material lie against and usually overlap the faces of the pyrenoid (Figs. 3, 4, 12). In some small ellipsoid azoosporic forms the pyrenoids do not have this characteristic structure and the region of attachment covers a greater area of the chloroplast surface (Antia *et al.*, 1975) (Fig. 11), and it has been suggested for two species that the pyrenoids may be only transitory structures (Antia *et al.*, 1975).

The pigment composition of eustigmatophytes has been investigated to a similar degree to that of Xanthophytes (Whittle and Casselton, 1969; Guillard and Lorenzen, 1972; Norgard *et al.*, 1974; Whittle and Casselton, 1975; Whittle, 1976). Of the chlorophylls, only chlorophyll *a* has so far been found. Since all the remaining groups of algae possess either chlorophyll *b* or one or both of the *c* chlorophylls in addition to chlorophyll *a*, it seems possible that further investigation may reveal the existence of low concentrations of chlorophyll *c* as in the Xanthophyceae. Fucoxanthin is absent from eustigmatophyes as in the Xanthophyceae and their yellow-green color is due to this lack and to the presence of a range of other carotenoids. Those typically present include β-carotene and the xanthophylls violaxanthin, which is dominant, and vaucheriaxanthin ester. Several minor xanthophylls have also been identified. Although the pigment composition of eustigmatophytes is similar to that of xanthophytes the replacement of diatoxanthin in the latter with violaxanthin in the former seems to be a consistent and highly characteristic difference which is significant enough to be detected even in the total spectrum of extracted pigments (Whittle and Casselton, 1975).

Cell wall structure and composition. Unlike many xanthophytes, the cell walls of eustigmatophytes appear always to be in one piece. They also differ from those in many Xanthophyceae in showing no swelling in dilute alkalis, no intense staining with basic dyes and in never giving a strong positive reaction with Schiff's reagent without prior acid hydrolysis. There have so far been no

Abbreviations used in the legends. C, chloroplast; CE, chloroplast envelope; CW, cell wall; ES, eyespot; F^1, mastigoneme-bearing flagellum; F^2, short smooth flagellum; FB^2, second basal body; FS, flagellar swelling; LV, vesicle with finely lamellate contents; M, mitochondrion; m, microtubular root; N, nucleus; P, pyrenoid; PN, periplastidial network; R, rhizoplast; V, vacuole; VR, vesicle with red refractile contents.

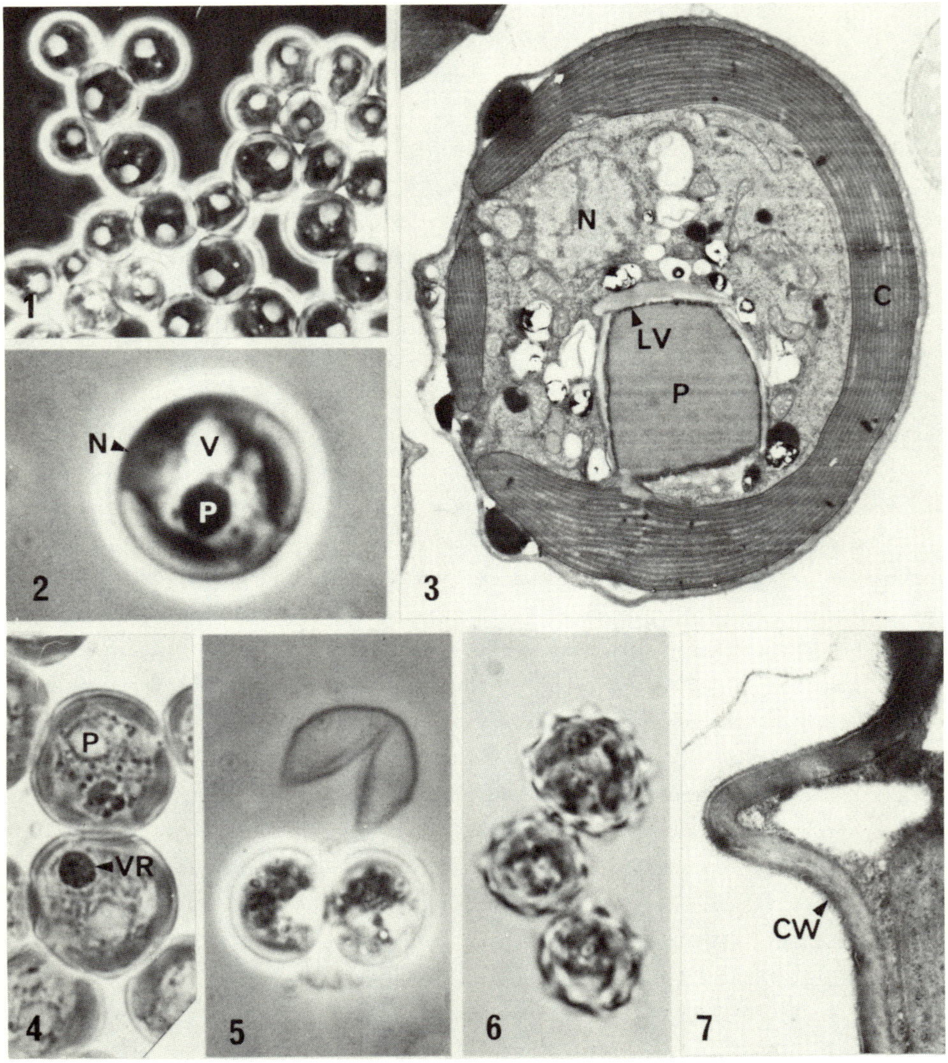

Figs. 1-7. Eustigmatophycean vegetative cells. Fig. 1. "*Polyedriella helvetica*" (x 1000). Fig. 2. *Vischeria punctata* (x 3000). Fig. 3. Section of "*P. helvetica*" (x 7500). Fig. 4. "*Pleurochloris*" sp. (x2000). Fig. 5. Autospores and mother cell wall of "*Pleurochloris commutata*" (x 2000). Fig. 6. *V. stellata* (x 2000). Fig. 7. Cell wall of *V. stellata* (x 20,000).

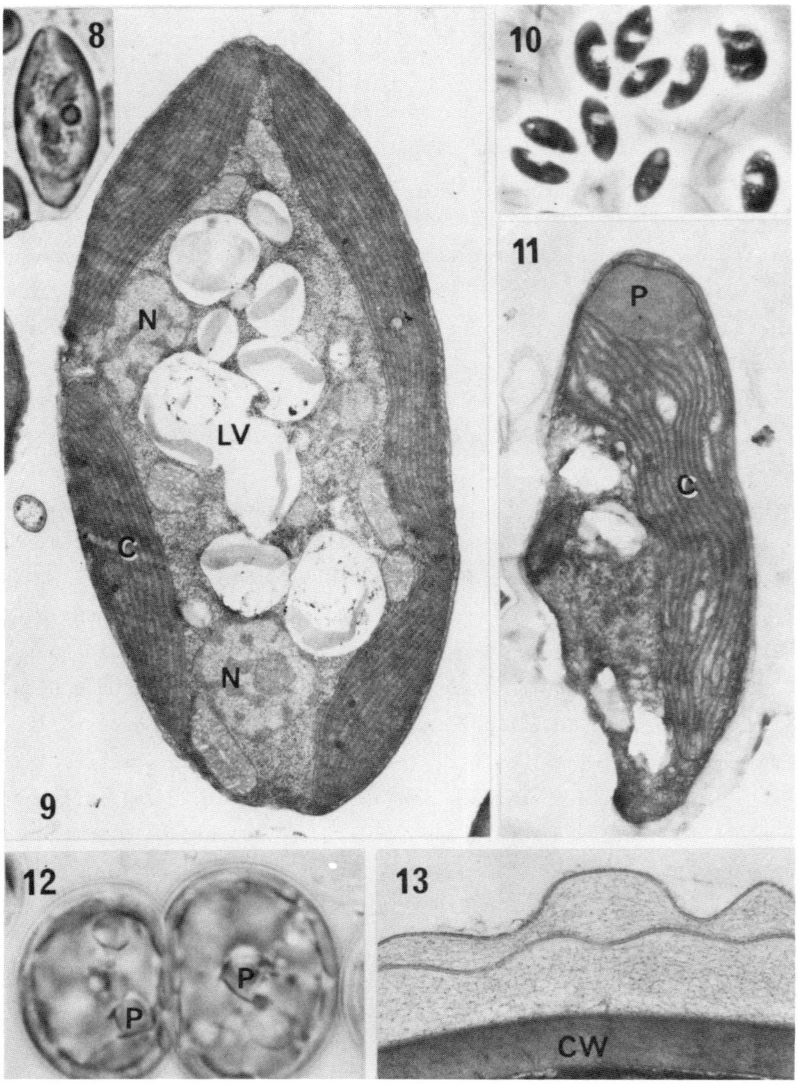

Figs. 8-13. Eustigmatophycean vegetative cells. Fig. 8. "*Ellipsoidion acuminatum*" (x1000). Fig. 9. Section of "*E. acuminatum*" (x10,000). Fig. 10. "*Monodus subterraneus*" (x1500). Fig. 11. "*M. subterraneus*" (x15,000). Fig. 12. *Chlorobotrys regularis* (x2000). Fig. 13. Section of cell wall and mucilage layers of *C. regularis* (x30,000).

observations of the chemical composition of the cell wall of eustigmatophytes, though that of *Chlorobotrys regularis* gives negative reactions in chemical tests for cellulose (Hibberd, 1974). The wall of this species is unusual, however, in its refractile appearance and high degree of flexibility, and in being surrounded by concentric layers of mucilaginous material separated by tripartite membrane-like structures (Hibberd, 1974) (Figs. 12, 13). In this latter feature *C. regularis* is so far unique in the group.

Storage metabolites. Eustigmatophytes do not store starch and give a negative reaction with dilute iodine in potassium iodide. Unlike the Xanthophyceae, a recognizable material in the form of refractile plates is associated with the pyrenoids of eustigmatophytes and almost certainly represents the storage metabolite in this group (Figs. 3, 12). Notwithstanding the unusual appearance of this material in sections (see below) and the inherent taxonomic importance of the nature of storage metabolites in general, its chemical composition has not yet been investigated.

Other features of vegetative cell structure. Species with relatively large, more or less spherical cells characteristically contain a conspicuous vacuole with vibrating granular contents (Fig. 2) and a pigmented body 3-5 μm in diameter (Fig. 4). This varies in color from pale yellow-brown to red-brown, becoming larger and darker in color as the cultures age. In the zoospore-producing species, e.g. *Vischeria stellata*, it appears to be composed of a number of smaller droplets but in *Chlorobotrys regularis*, which is azoosporic, it is always homogeneous (Hibberd, 1974). The composition of this material is not known but its refractile appearance and osmiophilic nature indicate that it contains a high proportion of lipoidal material. In comparison with the other cell contents, the nucleus is inconspicuous and is best seen in younger cells (Fig. 2). It is about 4 μm in diameter and lies between the edges of the chloroplast lobes.

Motile cell structure. Eustigmatophycean zoospores have an organization unique in the plant kingdom and are therefore highly characteristic for the class (Hibberd and Leedale, 1972). In most species they bear only one emergent flagellum though two species have a pair of unequal flagella (Hibberd and Leedale, 1972; Lee and Bold, 1973). The uniflagellate zoospores are elongate and lageniform (flask-shaped), with the posterior two-thirds of the body flattened (Figs. 14, 15). They are usually 10-15 μm long, about 5 μm at maximum width in face view and only 2-3 μm wide in side view. The zoospores are completely naked and always show a degree of plasticity of shape. The flagellum is approximately as long as the cell body and appears to be attached termin-

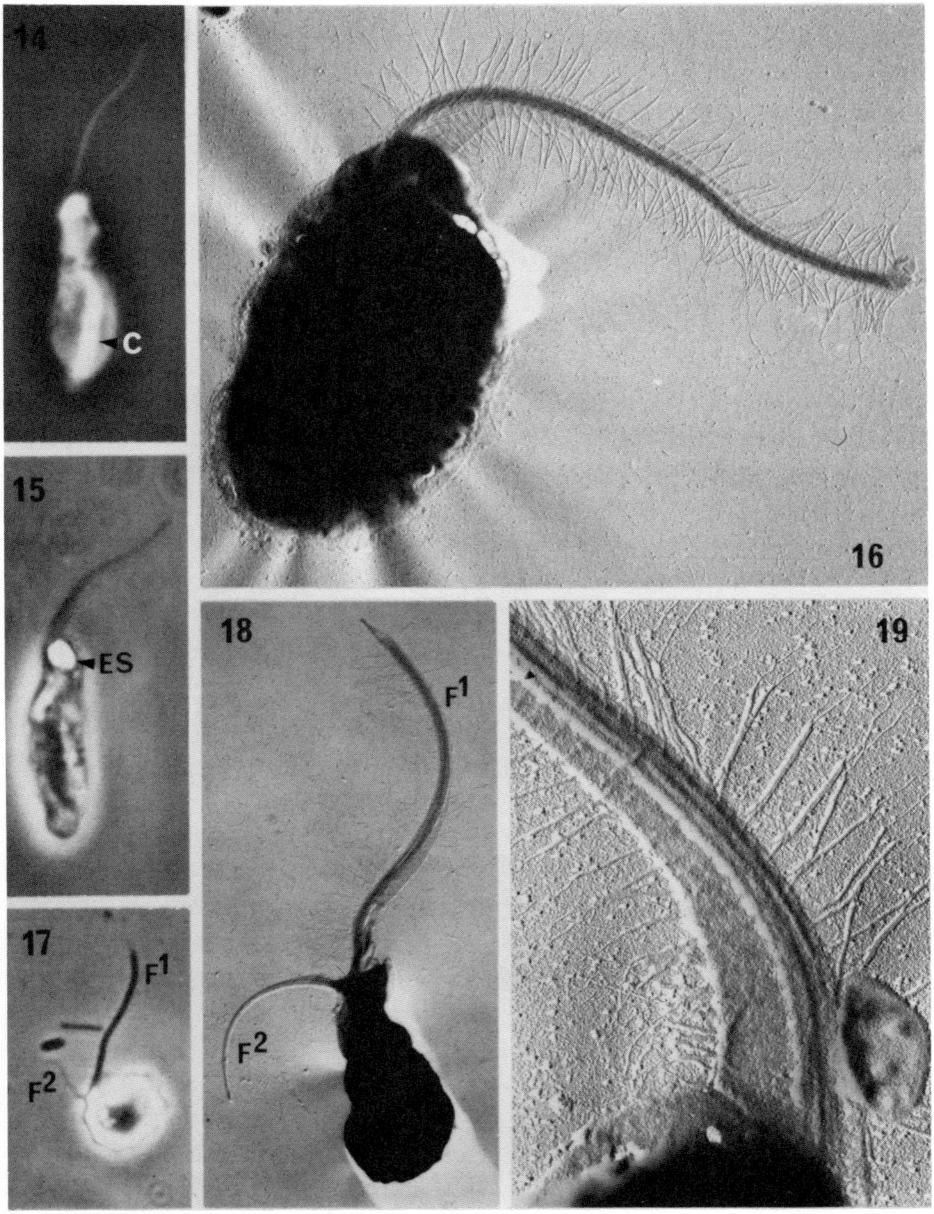

Figs. 14-19. Eustigmatophycean zoospores. Fig. 14. *Vischeria stellata* (x 2000). Fig. 15. "*Pleurochloris commutata*" (x 2000). Fig. 16. Shadowcast cell of *V. punctata* (x 7500). Fig. 17. Fixed and dried cell of "*Ellipsoidion acuminatum*" (x 2000). Fig. 18. Shadowcast cell of "*E. acuminatum*" (x 4300). Fig. 19. Shadowcast flagellum of *V. punctata* (x 15,000).

ally at the narrow end of the cell which is directed anteriorly during swimming (Figs. 14, 15). However, favorably orientated cells show that the flagellum is inserted just below the anterior end of the cell and that it widens at its proximal end for a length of about 3 μm, reaching a maximum width of about 1 μm near to the cell (Figs. 14, 15). Each zoospore contains only one long flattened chloroplast which usually lies with its broadest face against the broad face of the zoospore (Fig. 14). By far the most conspicuous feature of eustigmatophycean zoospores, and that on which the name of the group is based, is the eyespot. This is very large, filling almost the whole of the extreme anterior end of the cell (Figs. 14, 15, 20), and is quite separate from the chloroplast. It is red-orange in color and can often be seen to be made of smaller globules of variable size. Because of this feature and their long thin shape, eustigmatophycean zoospores bear a general resemblance to small euglenoid flagellates, though they are, of course, quite unrelated to that group. The nucleus lies in the narrow end of the cell directly behind the eyespot and separating it from the chloroplast (Fig. 20). Contractile vacuoles have not been identified with certainty in these zoospores. Two types of characteristic vesicles occurring only in the zoospore are described in detail below. One early attempt to demonstrate that the flagellum of one species was a "Flimmergeissel" was unsuccessful (Petersen, 1932), but lateral hairs can be detected on fixed and dried material in the light microscope, and this is confirmed by electron microscopy (Fig. 17).

Zoospores with two emergent flagella (Hibberd and Leedale, 1972; Lee and Bold, 1973) are very similar to those described above, but do not appear to be as generally long and thin. In liquid preparations of living or fixed zoospores only a single terminally inserted flagellum can be seen. A much narrower short flagellum becomes apparent only when the cells are vapour-fixed and dried for light microscopy (Fig. 16) or in shadowcast preparations examined in the electron microscope (Fig. 18). The long flagellum bears a bilateral array of hairs but the second flagellum is smooth and very narrow for the greater part of its length, only a short proximal part having the same diameter as the long flagellum (Hibberd and Leedale, 1972; Lee and Bold, 1973) (Figs. 17, 18).

DETAILS OF ULTRASTRUCTURE

Flagella. Electron microscopy shows that eustigmatophycean flagellar appendages are tubular mastigonemes of the same type as found in the Xanthophyceae, Chrysophyceae and Cryptophyceae (Figs. 17, 19). As in the Xanthophyceae, however, lateral filaments have not yet been found attached to the tubular

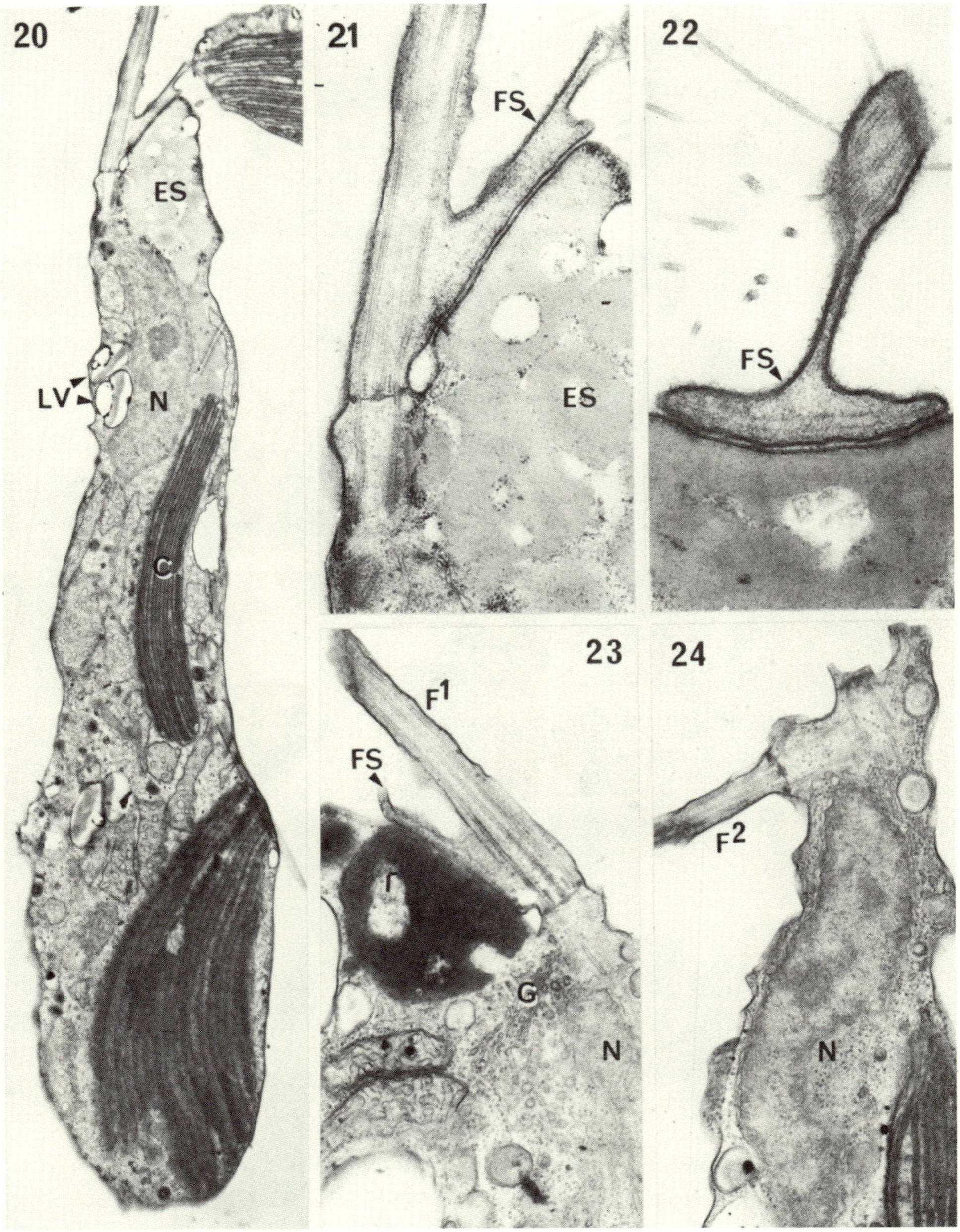

Figs. 20-24. Sections of eustigmatophycean zoospores. Fig. 20. "*Polyedriella helvetica*" (x7500). Fig. 21. LS flagellar swelling of "*P. helvetica*" (x25,000). Fig. 22. TS flagellar swelling of "*P. helvetica*" (x40,000). Fig. 23. LS flagellar swelling of "*Ellipsoidion acuminatum*" (x20,000). Fig. 24. LS through short flagellum of "*E. acuminatum*" (x20,000).

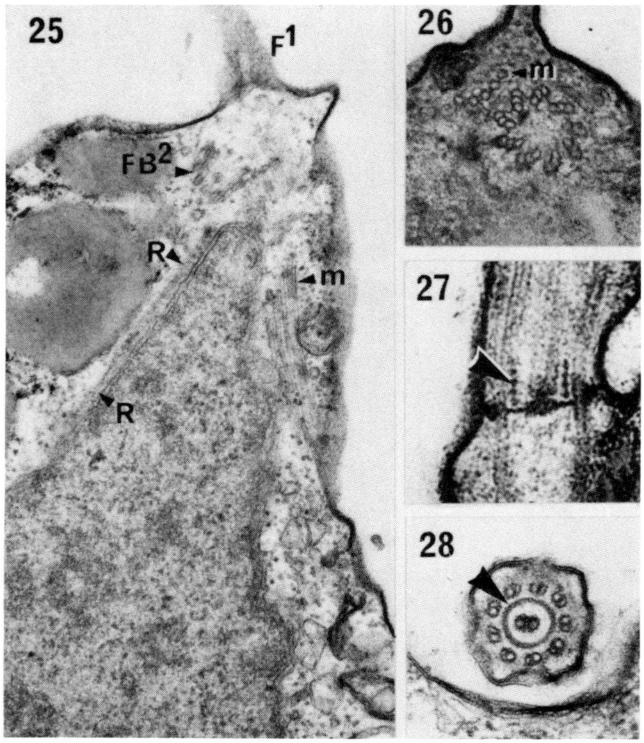

Figs. 25-28. Fig. 25. Flagellar roots of "*Polyedriella helvetica*" (x30,000). Fig. 26. TS basal body and microtubular root of *Vischeria punctata* (x50,000). Fig. 27. LS transition region of "*P. helvetica*" (x50,000). Fig. 28. TS transition region of "*Ellipsoidion acuminatum*" (x50,000).

shaft of the hairs.

Shadowcast preparations of the flagellum of uniflagellate zoospores and the long flagellum of biflagellate zoospores (Hibberd and Leedale, 1972) show that the flagellar membrane becomes progressively wider and looser as it nears the cell, closely surrounding the flagellar axis for only the distal one third of its length (Figs. 17, 18, 19).

In uniflagellate zoospores the flagellum is inserted into the cell subapically, being pressed against the obliquely truncate anterior face of the cell before entering the body about 2 μm below the anterior end (Fig. 20). The expanded flagellar membrane is elaborated into a T-shaped swelling over this region and the part flattened against the cell contains layered electron-dense material (Figs. 21, 22). The tubular mastigonemes are always attached to the part of the flagellar membrane surrounding the axoneme and never to the flagellar swelling (Fig. 22).

In biflagellate zoospores the long flagellum has exactly the same structure as described above with the swelling at the base of the mastigoneme-bearing flagellum (Fig. 23). The short flagellum has no swelling at its proximal end and leaves the cell freely (Fig. 24). It is only one-third the width of the long one for the greater part of its length, the distal part of the axoneme consisting of only two separate microtubules (Hibberd and Leedale, 1972), probably a continuation of the central pair of the usual 9 + 2 tubule pattern which occurs in the wider proximal region.

In uniflagellate zoospores a second basal body lies close to the proximal end of the flagellar base and approximately at right angles to it (Fig. 25). It is only one-third its length, however, though it is similarly constructed of nine triplet-fibers. In biflagellate zoospores the two flagella are inserted laterally with the basal bodies inclined to each other at an angle of about 90° (Fig. 24).

The flagellar transition region has an identical structure to that described for the Xanthophyceae, with a transitional helix surrounding the central pair of tubules of the flagellar axoneme (Figs. 21, 27, 28) immediately distal to a transverse partition across the axoneme at the level at which the flagellum enters the cell body (Fig. 27).

Flagellar roots have not yet been investigated in detail but both microtubular and cross-banded fibrillar types (rhizoplasts) are present. The former, consisting of 2-5 microtubules, arise close to the basal bodies (Figs. 25, 26) and pass anteriorly and posteriorly in the cell. The cross-banded roots are narrow and pass from close to the basal bodies along the anterior flattened face of the nucleus (Fig. 25).

The eyespot is composed of a number of osmiophilic globules of variable size and neither the eyespot as a whole nor the component globules are membrane bounded (Fig. 21). One large D-shaped droplet lies closely against the cell membrane opposite the flagellar swelling.

<u>Chloroplast and pyrenoid</u>. The chloroplast lamellae (Fig. 29) are evenly spaced and are composed of three non-cohering thylakoids. Additional thylakoids commonly appear between some of the lamellae, particularly at the edge of

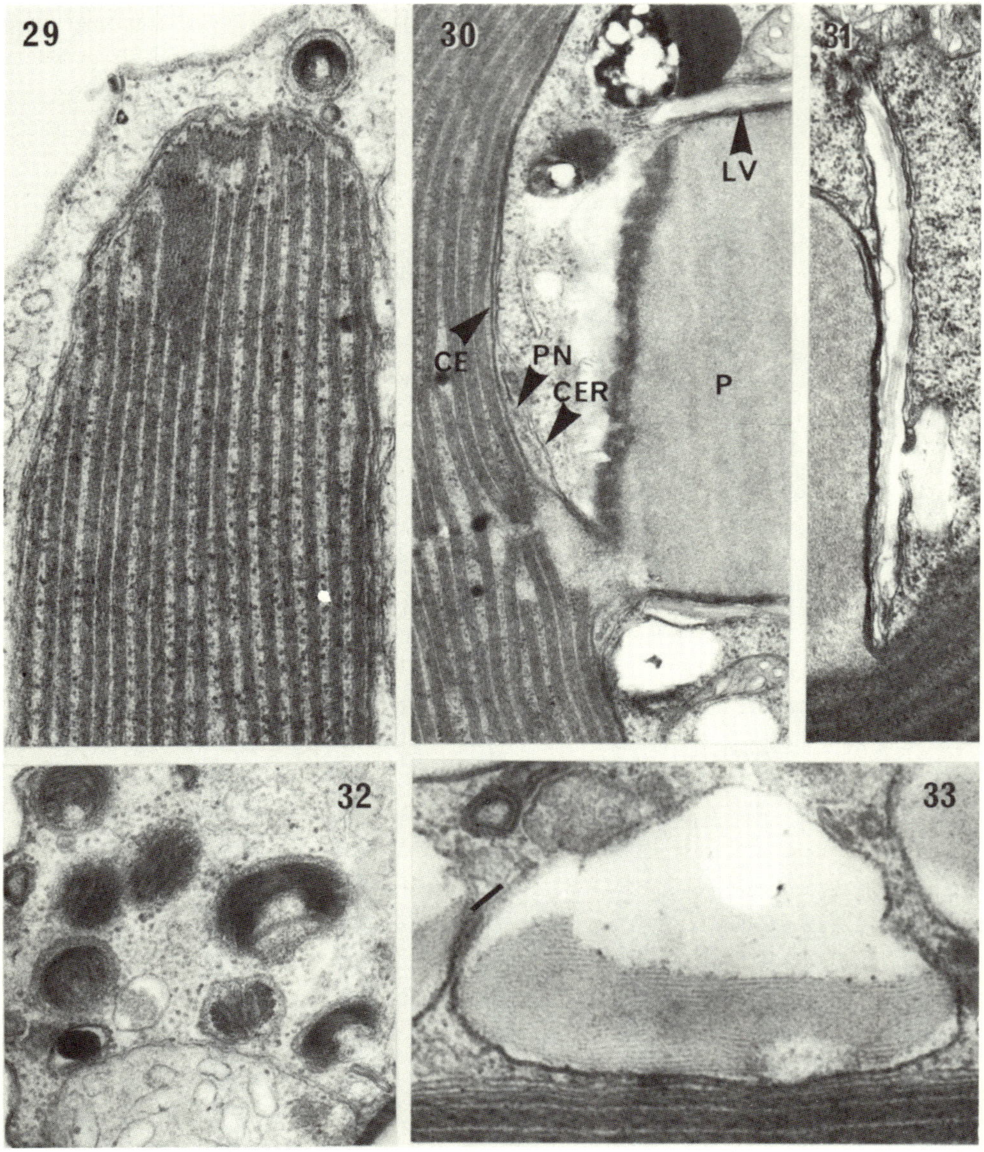

Figs. 29-33. Fig. 29. LS chloroplast of *Vischeria stellata* (x40,000). Fig. 30. Chloroplast-pyrenoid connection in "*Polyedriella helvetica*" (x25,000). Fig. 31. Flattened vesicle lying against pyrenoid in "*P. helvetica*" (x25,000). Fig. 32. Dense vesicle with fibrous contents in a zoospore of *V. stellata* (x40,000). Fig. 33. Vesicle with lamellate contents in a zoospore of *V. punctata* (x50,000).

the chloroplast, forming granum-like stacks. The chloroplasts do not contain a girdle lamella and the longitudinal lamellae terminate close to the chloroplast envelope so there is no peripheral region of less dense matrix; the chloroplast genophores are thus probably scattered throughout the chloroplast matrix.

The chloroplasts are bounded by a double-membraned envelope and by a layer of ER (Fig. 30). The chloroplast ER is not continuous with the nuclear envelope in the zoospores, however, and in the vegetative cells connections have been found only in two very small azoosporic species (Antia *et al.*, 1975). A periplastidial network occurs in the perichloroplastic space, particularly near the pyrenoid (Fig. 30) (see xanthophyte chapter for terminology).

The pyrenoid is closely surrounded by the chloroplast envelope and the chloroplast ER which together form a single dense layer in which the individual membranes cannot be clearly recognized (Fig. 30). The matrix appears finely granular or homogeneous and is not penetrated by either chloroplast lamellae or individual thylakoids (Fig. 30). The plates of refractive material which can be seen lying against the faces of the pyrenoid in the light microscope are represented in sections by flattened vesicles containing material which appears very finely lamellate after fixation (Fig. 31). Smaller vesicles with identical contents also lie freely in the cytoplasm of both the vegetative cells and zoospores (Fig. 33) and their appearance in some sections suggests that they originate from the pyrenoid vesicles by a process of budding (Fig. 33). Lamellate material has been described within the chloroplast matrix of two small marine azoosporic species (Antia *et al.*, 1975) but since lamellate vesicles also surround the pyrenoid the two substances may be completely different.

<u>Nucleus, Golgi body, mitochondria and vesicles</u>. In eustigmatophycean zoospores the nucleus is pyriform with the anterior end elongated towards the flagellar basal body (Fig. 20), while in vegetative cells it is more or less spherical (Figs. 2, 3, 9). In neither case is the outer membrane of the nuclear envelope confluent with the chloroplast ER or rough ER in the cytoplasm.

An unusual feature of the species producing uniflagellate zoospores is that a Golgi body is present only in the vegetative cells, never in the zoospores. In biflagellate zoospores, on the other hand, a Golgi body lies close against the anterior end of the nucleus with one edge near the flagellar basal bodies (Hibberd and Leedale, 1972; Lee and Bold, 1973) (Fig. 23). There are only a few observations on vegetative cells but here the Golgi body appears to lie close to and parallel with the nuclear envelope (Hibberd, 1974). Eustigmatophycean Golgi bodies are always relatively small, consisting of up to only five cisternae. Mitochondria in the Eustigmatophyceae have tubular cristae (Fig. 32).

In zoospores, but apparently not vegetative cells, spherical vesicles 200-500 nm in diameter with fibrous, spirally wound contents are found, mostly in a peripheral position. They are most likely involved in some way in the settling process, though there is as yet no information on this. Small vesicles containing a single very dense globule also occur peripherally in the zoospores and large, apparently empty vacuoles are present in the posterior end of the cell. The vesicles with lamellate contents have been described above.

MOVEMENT PATTERNS

The mastigoneme-bearing flagellum of zoospores beats in a tip-generated approximate sine wave which pulls the cell rapidly and smoothly in a more or less straight path, with little rotation of the cell body. Movement becomes slower and less regular as the cells gradually round up prior to settling, a process greatly accelerated by exposure to light.

REPRODUCTION AND LIFE HISTORIES

Eustigmatophytes reproduce primarily by the formation of autospores (Fig. 5). The larger spherical species produce either two D-shaped or four tetrahedral autospores while in the more elongate species the autospores are arranged more or less longitudinally within the mother cell wall. Zoospores are produced in relatively small quantities by the majority of forms, though they appear to be the only means of reproduction of one species (Lee and Bold, 1973) and completely absent in others (Hibberd, 1974; Antia *et al.*, 1975). Sexual reproduction has not yet been observed in any species.

HABITAT

Eustigmatophytes occur in both freshwater and marine habitats. The species with spherical vegetative cells have all been isolated from soils (Petersen, 1932; Vischer, 1945) and may be confined to that habitat, as also might "*Monodus subterraneus*", a smaller species with narrowly ovoid cells. Some species have been isolated from streams (Lee and Bold, 1973) however, and other small forms belong to the marine plankton (Antia *et al.*, 1975). It seems likely that eustigmatophytes will be found to be more prevalent in nature once the group is more widely recognized.

CULTURE MEDIA

Similar methods of cultivation may be used for eustigmatophytes as for xanthophytes and they grow well though relatively slowly in a variety of defined mineral media. Marine species grow in enriched seawater media (Antia *et al.*, 1975). A study has been made of the growth and mineral nutrition of "*Monodus subterraneus*" and an optimal medium developed for this species (Miller and Fogg, 1957).

NEEDED RESEARCH

It is clear from the foregoing account that with the exception of motile cell structure and pigment composition, basic knowledge of the biology of the Eustigmatophyceae is very limited. Attention will therefore be drawn only to the more obvious gaps. From a taxonomic point of view many species may still remain hidden in the Xanthophyceae, though none can presently be transferred on the basis of existing descriptions. The isolation into culture of yellow-green algae in general is therefore a basic requirement for further taxonomic investigation.

Fixation difficulties are responsible for the relative paucity of information on vegetative cell structure, though this is likely to be valuable in the systematics of the group, particularly for the azoosporic species. The processes of zoospore formation and settling remain to be studied at both the light and electron microscope level though this will depend on the development of a reliable method of zoospore production. There are no life history studies and sexuality has not been discovered. The type of nuclear division remains unknown. With the exception of "*Monodus subterraneus*" which has been studied as a xanthophyte, biochemical and physiological studies are virtually non-existent and notwithstanding the obvious nature of the storage metabolite it has not yet been characterized biochemically.

ACKNOWLEDGEMENTS

I am grateful to Dr. F. K. Kupicha for help in the preparation of this chapter.

APPENDIX

Strain numbers of the species names referred to, in the collections of the Culture Centre of Algae and Protozoa, Cambridge (CCAP) and the University of Texas, Austin (UTEX). Names in quotation marks are either nomenclaturally incorrect or are to be changed for taxonomic reasons.

	CCAP	UTEX
Chlorobotrys regularis (W. West) Bohlin	810/1	-
"*Ellipsoidion acuminatum* Pascher"	822/1	-
"*Monodus subterraneus* Petersen"	848/1	151
"*Pleurochloris commutata* Pascher"	860/1a	310
"*Polyedriella helvetica* Vischer & Pascher"	861/1	49
Vischeria punctata Vischer	887/1	153
Vischeria stellata (Poulton) Pascher	887/2b	312

REFERENCES

Antia, N. J., T. Bisalputra, J. Y. Cheng and J. P. Kalley. 1975. Pigment and cytological evidence for reclassification of *Nannochloris oculata* and *Monallantus salina* in the Eustigmatophyceae. J. Phycol., 11: 339-343.

Guillard, R. R. L. and C. J. Lorenzen. 1972. Yellow-green algae with chlorophyllide c. J. Phycol., 8: 10-14.

Hibberd, D. J. 1974. Observations on the cytology and ultrastructure of *Chlorobotrys regularis* (West) Bohlin with special reference to its position in the Eustigmatophyceae. Br. Phycol. J., 9: 37-46.

Hibberd, D. J. and G. F. Leedale. 1970. Eustigmatophyceae- a new algal class with unique organization of the motile cell. Nature (Lond.) 225: 758-760.

Hibberd, D. J. and G. F. Leedale. 1971. A new algal class - the Eustigmatophyceae. Taxon, 20: 523-525.

Hibberd, D. J. and G. F. Leedale. 1972. Observations on the cytology and ultrastructure of the new algal class, Eustigmatophyceae. Ann. Bot., 36: 49-71.

Lee, K. W. and H. C. Bold. 1973. *Pseudocharaciopsis texensis* gen. et sp. nov., a new member of the Eustigmatophyceae. Br. Phycol. J., 8: 31-37.

Miller, J. D. A. and G. E. Fogg. 1957. Studies on the growth of Xanthophyceae in pure culture. I. The mineral nutrition of *Monodus subterraneus* Petersen. Arch. Mikrobiol., 28: 1-17.

Norgard, S., W. A. Svec, S. Liaaen-Jensen, A. Jensen and R. R. L. Guillard. 1974. Chloroplast pigments and algal systematics. Biochem. Syst. Ecol., 2: 3-6.

Pascher, A. 1937-1939. Heterokonten. In L. Rabenhorst's *Kryptogamen-Flora von Deutschland, Osterreich und der Schweiz.* 2nd Edn., Bd. XI. Akad. Verlagsges. Leipzig.

Petersen, J. B. 1932. Einige neue Erdalgen. Arch. Protistenk. 76: 395-408.

Silva, P. C. 1979. Review of the taxonomic history and nomenclature of the yellow-green algae. Arch. Protistenk., 121: 20-63.

Vischer, W. 1945. Heterokonten aus Alpinen Böden, Speziell dem Schweizerischen NationalPark. Ergebn. wiss. Unters Schweiz. NatnParks, 1: 481-511.

Whittle, S. J. 1976. The major chloroplast pigments of *Chlorobotrys regularis* (West) Bohlin (Eustigmatophyceae) and *Ophiocytium majus* Naegeli (Xanthophyceae). Br. Phycol. J., 11: 111-114.

Whittle, S. J. and P. J. Casselton. 1969. The chloroplast pigments of some green and yellow-green algae. Br. Phycol. J., 4: 55-64.

Whittle, S. J. and P. J. Casselton. 1975. The chloroplast pigments of the algal classes Eustigmatophyceae and Xanthophyceae. II. Eustigmatophyceae. Br. Phycol. J., 10: 179-191.

Copyright 1980 by Elsevier North Holland, Inc.
Cox, ed. Phytoflagellates

SILICOFLAGELLATES

SHIRLEY D. VAN VALKENBURG
Department of Botany, University of Maryland, College Park, Maryland 20742, U.S.A

ABSTRACT

Silicoflagellates display a number of characters common among the Chrysophyceae -- in pigmentation, in mitochondrial and chloroplast structure, and in the deposition of biological silica -- although the number of dictyosomes is unusually high (∿72) for members of this class. Their uniqueness lies in their morphology. Four distinct forms of the cultured *Dictyocha fibula* have been examined by transmission electron microscopy. Postulates are offered concerning the possible function of the variations in form observed. In addition, the fossil record, cultural requirements, and systematic problems of silicoflagellates are briefly summarized. Directions for future research are suggested.

INTRODUCTION

Silicoflagellates are planktonic, microscopic algae with numerous discoid, golden-brown chloroplasts, a single anterior flagellum and a delicate, exterior siliceous skeleton composed of tubular elements. They occur in all oceans, although usually in relatively low numbers. Preliminary pigment analysis gives indication of the presence of chlorophylls *a* and *c*, carotenes, diadinoxanthin, fucoxanthin, diatoxanthin, and lutein (Van Valkenburg, Ph.D. thesis 1970). Although the storage products have not been analyzed, ultrastructural evidence does not refute the expectation of chrysolaminarin and lipids. A few early workers evidently mistook cells in late division stages as evidence of sexual union. However, there are no modern accounts of sexuality in silicoflagellates from either field or laboratory. It is generally agreed that silicoflagellates are highly temperature-sensitive, a given strain tolerating only a few degrees of temperature variation from its normal range. Silicoflagellates have been variously assigned class, subclass, or ordinal rank under the division of golden algae (termed Chrysophyta, Chrysophycophyta, or Chromophyta depending upon the system used). This flexibility of taxonomic treatment reflects the uniqueness of this group. Silicoflagellates are obviously Chrysophytes, but beyond that, they have no near neighbors in that the siliceous skeletons of the non-photosynthetic Ebridians are both internal and solid rather than tubular.

Still greater confusion exists at the generic and specific levels concerning the biological validity of the various taxa. The taxonomy of silicoflagellates has always been based on the configuration of the skeleton, i.e., the number of sides of the basal ring, the presence of a central arch (or alternatively of one or more apical rings), and the number of spines on the basal ring. Although mature skeletons are external to the plasmalemma (Van Valkenburg, 1971b), they are largely hidden by the myriad pseudopods which entwine among them, such that it is only possible to determine the genus and species of cleaned skeletons, i.e., skeletons unencumbered by protoplasts. This is an important cause of the difficulty experienced in resolving the biosystematic problems of this group, because although a clone can be cultured, it is not possible to know the skeletal configuration of the initial cell from which the clone was established.

Silicoflagellates evolved nearly 120 million years ago, in the early Cretaceous, with wide fluctuations in abundance and diversity over geologic time. They reached their greatest development during the Tertiary, at which time they included larger, more elaborate forms than exist today. Early work with fossil silicoflagellate material resulted in an ever-expanding number of genera and species (Loeblich *et al.*, 1968). A number of paleontologists (Mandra and Brigger, 1972; Mandra and Mandra, 1971; Perch-Nielsen, 1975) defend the biological validity (i.e., the genetic distinctness) of the traditional taxa because they say (a) relatively little statistical variation in skeletal form exists within fossil populations, and/or (b) the *Dictyocha/Distephanus* ratio is a reliable indicator of seawater temperature. Greta Fryxell finds a high degree of uniformity of skeletal form in extant Antarctic populations from the Weddell Sea (pers. comm.). However, other workers with modern forms (Hovasse, 1946; Frenguelli, 1935) stressed the variations within populations. Our own studies with the cultured *Dictyocha fibula* (Van Valkenburg and Norris, 1970) revealed three genera, established by traditional taxonomy, within a single clonal culture. Lipps (1970) postulated that all biologically valid taxa of silicoflagellates that have ever existed could probably be reduced to 8 genera, and that the initial form from which others evolved was probably *Dictyocha*. According to Lipps (1970), the prevailing opinion is that there are currently 2 living species, *Dictyocha fibula* and *Dictyocha octonaria*. *Distephanus speculum*, later changed to *Dictyocha speculum*, is now thought to intergrade with *Dictyocha fibula*, an opinion with which I agree. However, reports continue to refute, as well as support this assessment. Thus it appears this issue is not likely to be settled by additional survey data on either fossil or extant forms. What is needed fundamentally, I believe, is a thorough understanding of how the skeleton is formed, determined through transmission electron microscopy of the sequence of stages of skeleton formation in synchronized cultures.

Dictyocha was first cultured in 1967 (Van Valkenburg and Norris, 1970) and grows in Provasoli enriched seawater medium (Provasoli, 1968) at reduced enrichment concentration. Nearly all published works on the ultrastructure of silicoflagellate skeletons have been principally concerned with the presence or absence of a surface reticulum on fossil forms (Jerković, 1966, 1969; Mandra, 1969; Lipps, 1970; Mandra and Mandra, 1971, 1972; Perch-Nielsen, 1975). Although some fossil skeletons do have a reticular surface, most specimens are smooth. I know of no convincing evidence of spinose collars on spine tips of fossil material. Only one paper on the ultrastructure of the silicoflagellate protoplast has been published (Van Valkenburg, 1971b); only one of the extant species, *Dictyocha fibula*, was investigated.

ULTRASTRUCTURE

The Skeleton. Conical projections, collectively termed a spinose collar, surround the tip of each spine except where the tips have obviously been broken (Fig. 1). The collars correspond to those reported by Hovasse (1946) except that the component conical projections are often more numerous in the cultured material (8-10 projections) than those reported from nature (4-5 projections). The apparent partial fusion of a spine with a rod observed in the cultured *D. fibula* (Van Valkenburg, 1971a) corresponds with an observation by Frenguelli (1935) of some excessively long spines on natural populations of *D. fibula* which acquire contact with other parts of the skeleton and fuse with them.

Transmission electron micrographs of carbon replicas made from skeletons of cultured *Dictyocha* demonstrate that junctures of the skeletal elements such as rings, bars, and spines are of two kinds: smooth dichotomies and irregular junctures that clearly result from fusion (Fig. 1). This indicates a sequence of skeleton formation analogous to linear growth, as compared to formation of the entire structure concurrently (as indicated in line drawing, Fig. 1). Accordingly, it was postulated that the extent and complexity, and therefore the "genus" and "species" of a skeleton could be the result of the length of time the formative process proceeds. Variations in formation time could thus account for clonal cultures composed of several "taxa" (Van Valkenburg, 1971a).

The Protoplast. The ultrastructure of the protoplast summarized diagrammatically in Fig. 2 is of the sunburst form, which describes about 90% of the cells sectioned. A centrally-positioned nucleus is surrounded by a dense band of cytoplasm, the perikaryon, which contains all of the dictyosomes but none of the chloroplasts. Extending outward from the perikaryon are numerous thin cytoplasmic strands connected distally to globular cytoplasmic masses. These globules, or pseudopodia, contain all of the chloroplasts but none of the dictyosomes. Numerous assorted vesicles and mitochondria are found both in the perikaryon and in the pseudopodia. This configuration gives the protoplast a frothy appearance and accounts for the long-standing debate among light microscopists as to whether *Dictyocha* possessed a cell membrane. This network of thin strands was reported by Hovasse (1932) from light microscopy. The strands, approximately 0.5 μm in width, are visible with light optics on a crushed live cell when the condenser aperture is strongly reduced. The dotted line in Fig. 2 marks the exterior boundary of a mucilagenous matrix that was demonstrated through use of Ruthenium Red. Note the positioning of small, dense leaflets within the matrix and close to the perikaryon in Fig. 2. The leaflets, of about 0.3 X 0.03 μm, are not removed by hydrofluoric acid. Although never observed to be attached to any other structure, they appear to be firmly retained in the matrix. In all cases observed, the formed skeletal elements of sunburst cells are exterior to the convoluted cell membrane, rather than interior to it. Chloroplast and mitochondrial structure are typical for members of the Chrysophyceae. Chloroplasts contain 3-disk lamellar bands which do not enter the pyrenoids. Both a chloroplast envelope and chloroplast endoplasmic reticulum are present, but whether the chloroplast ER is continuous with the nuclear envelope has not been determined. The nuclear envelope does not remain intact during division. The number of dictyosomes within each individual is unusually high (estimated at 72) for members of this class.

Skeleton Formation. Evidence indicates that the silicoflagellate skeleton, a structure of greater diameter than the protoplast at rest, is nevertheless constructed internally and then extruded by the cell. The unusually high numbers of dictyosomes in this organism may indicate some type of involvement of dictyosomes in this process.

Because the sunburst form is by far the most common, it is reasonable to postulate that the sunburst is the normal form of the single cell, between the periodic and successive events of skeleton formation, skeleton extrusion and cell division. Of the remaining 10% of cells sectioned (discounting cells

---

Fig. 1. A carbon replica of a skeleton of *Dictyocha fibula*. BR, basal ring; CA, central arch; RS, radial spine; SB, supporting bar; s, spinose collar; S_1, smoothly dichotomous junctures; S_2, junctures which appear to be the result of fusion. Note the spinose structure at an angle of the open basal ring (arrow). The diagram below indicates the suggested manner of formation of this skeleton (after Van Valkenburg, 1971a; permission of J. Phycol.)

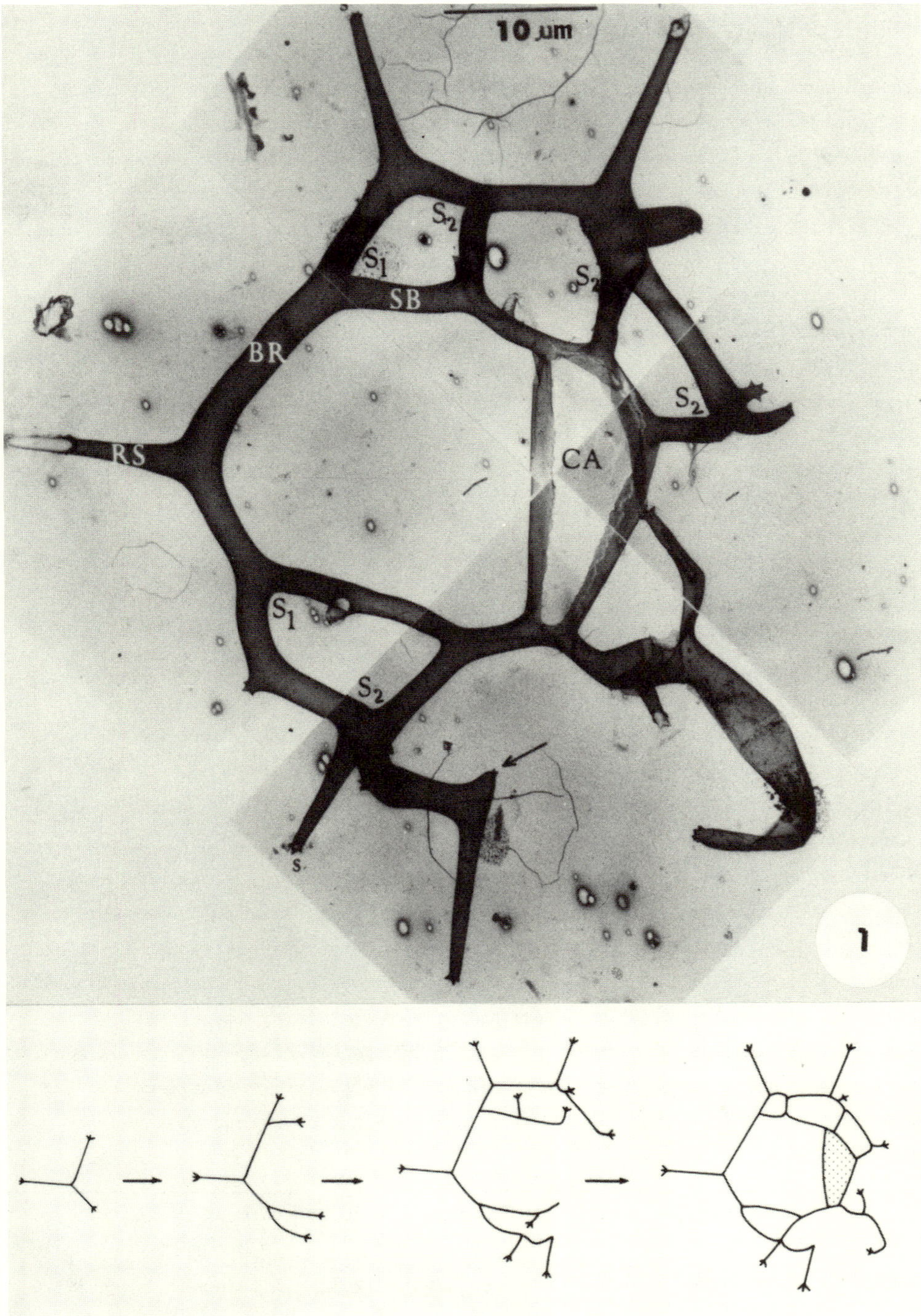
10 µm
S2
S1
SB
S2
BR
S2
RS
CA
S1
S2
s
1

judged to be scenescent) only a few were observed of a ring form, of a spherical form, and of a spherical-with-embayments form (Van Valkenburg, 1970). Although some of these observations appear to correspond with those by light microscopists of what they thought to be skeleton formation, they are too few in number to constitute proof. Rather they are being presented here as indications of the possible existence of some unusual phenomena. When considered together with other unexplained structures, they invite speculation about function from the static evidence of form in micrographs of *Dictyocha*.

The following account of an observation of *Dictyocha fibula* by light microscopy, written by Marshall (1934), is quoted here because of its importance to this discussion:

> "In one specimen observed....the cytoplasm outside the skeleton was much more extensive than usual and showed a definite external layer, which is not commonly seen. It formed broad lobes round the spines, from the tips of which extended very mobile pseudopodia....From about the middle of one side projected a long oval vacuole with a thin wall, and this remained constant in form and position for a considerable time, and even after fixation. The chromatophores were mostly massed in the center, but the condition of the pseudopodia shows that this was a healthy specimen. Gemeinhardt states that before division the amount of protoplasm in *Distephanus* is increased, and it seems possible that this *Dictyocha* was about to divide and that the vacuole indicated where a future spine would be laid down. If this were so, the resulting skeleton would be abnormal; but dividing forms are frequently seen, in which one member of the pair is normal and the other abnormal."

The ring form, which appears to correspond to Marshall's observation, was examined only once by transmission electron microscopy (Fig. 3). Note the broad cytoplasmic layer external to the perikaryon, and the presence of chloroplasts in the perikaryon as well as in the outer ring. The characteristic densities of the mature skeleton are located exterior to the cell membrane, hollow, with the characteristic chipping of sectioned silica. Hovasse (1932) reported from light microscopy that the new skeleton was positioned close to and facing the "old" one during the formative process, which he thought took place externally (pers. comm.), and that they were mirror images of each other. The chipped silica structure at the bottom of the micrograph in Fig. 3 is a part of the basal ring, which indicates that the two cross-sectioned tubular elements exterior to the cytoplasm are of support bars rather than radial spines.

Fig. 2. A diagrammatic representation of the sunburst form of *Dictyocha fibula*, demonstrating the system of fine cytoplasmic strands that extend into the matrix, connecting the perikaryon with the peripheral cytoplasmic components (skeleton incomplete) (after Van Valkenburg, 1971b; permission of J. Phycol.)

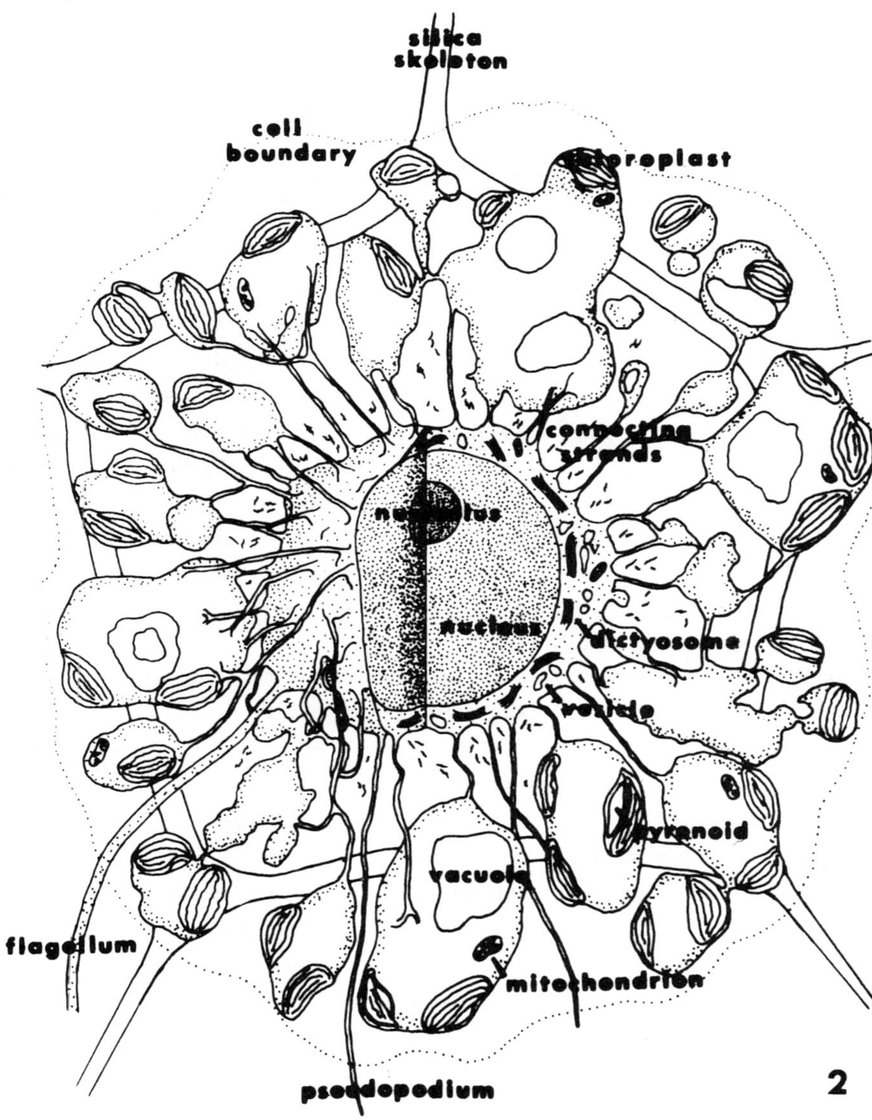
silica
skeleton
cell
boundary
chloroplast
connecting
strands
nucleolus
nucleus
dictyosome
vesicle
pyrenoid
vacuole
flagellum
mitochondrion
pseudopodium

2

In the vicinity of these elements, but within the protoplast are four large vacuoles with dense, relatively thick walls (Fig. 2, at arrows) which appear more amorphous than the profiles of the formed skeleton.

Two electron-opaque materials of differing appearance occur, one interior and one exterior to the membranes of these vacuoles in Fig. 3, and in Figs. 4-10. A variety of dense particles including short microtubule-like structures about 350 Å in diameter are often close to, or in contact with, these thick-walled vacuoles (Figs. 12-13). Similar particles occasionally occur in vesicles near the vacuoles at all levels sectioned. Note the vacuole at the double arrows in Fig. 3. This vacuole was sectioned at 9 approximately equidistant levels through 1/2 the diameter of the perikaryon. Level 1 (Fig. 4) represents the peripheral surface of the perikaryon and level 9 (Fig. 10) represents its equator. Fig. 3 is of level 5. This vacuole, which is about 8-10 μm long, is reconstructed in Fig. 11. Note the connecting bridge between the perikaryon and outer ring. Similar vacuoles were not found in the sunburst form, nor in either of the other forms observed. It can be postulated that this ring form is involved in skeleton formation and that this vacuole represents a forming skeletal element. However, hydrofluoric acid does not remove these densities. This would seem to indicate either that this material is not silica, or else that the silica is in some form, or some combination, that hydrofluoric acid does not remove.

The spherical form (Fig. 14) lacks pseudopodia but has approximately the same diameter as that of the perikaryon plus pseudopodia of the sunburst form. The spherical form thus can be viewed to have a greatly enlarged perikaryon, which differs from the perikaryon of the sunburst form in that it contains scattered chloroplasts, as well as large vesicles containing leaflets similar to those found free in the matrix of the sunburst form.

Examination of the "spherical-with-embayments" form (Fig. 15) suggests a possible transition between the spherical and sunburst forms. Vesicles containing leaflets are further enlarged, and some, but not all of them, give the appearance of having broken open to the outside. Chloroplasts are located within

---

Fig. 3. Transmission electron micrograph. Cacodylate buffered glutaraldehyde and non-distilled acrolein fixation, osmium tetroxide postfixation. A tangential section of a rarely-seen form of *Dictyocha fibula* (the ring form). A cytoplasmic bridge at another level connects the ring to the perikaryon. Thick-walled vacuoles at arrows are postulated to be new skeletal elements in the process of formation. Note chipped elements of "old" skeleton exterior to the protoplast. Vacuole at double arrows is examined in greater detail in Figs. 4-11 (after Van Valkenburg, 1970).

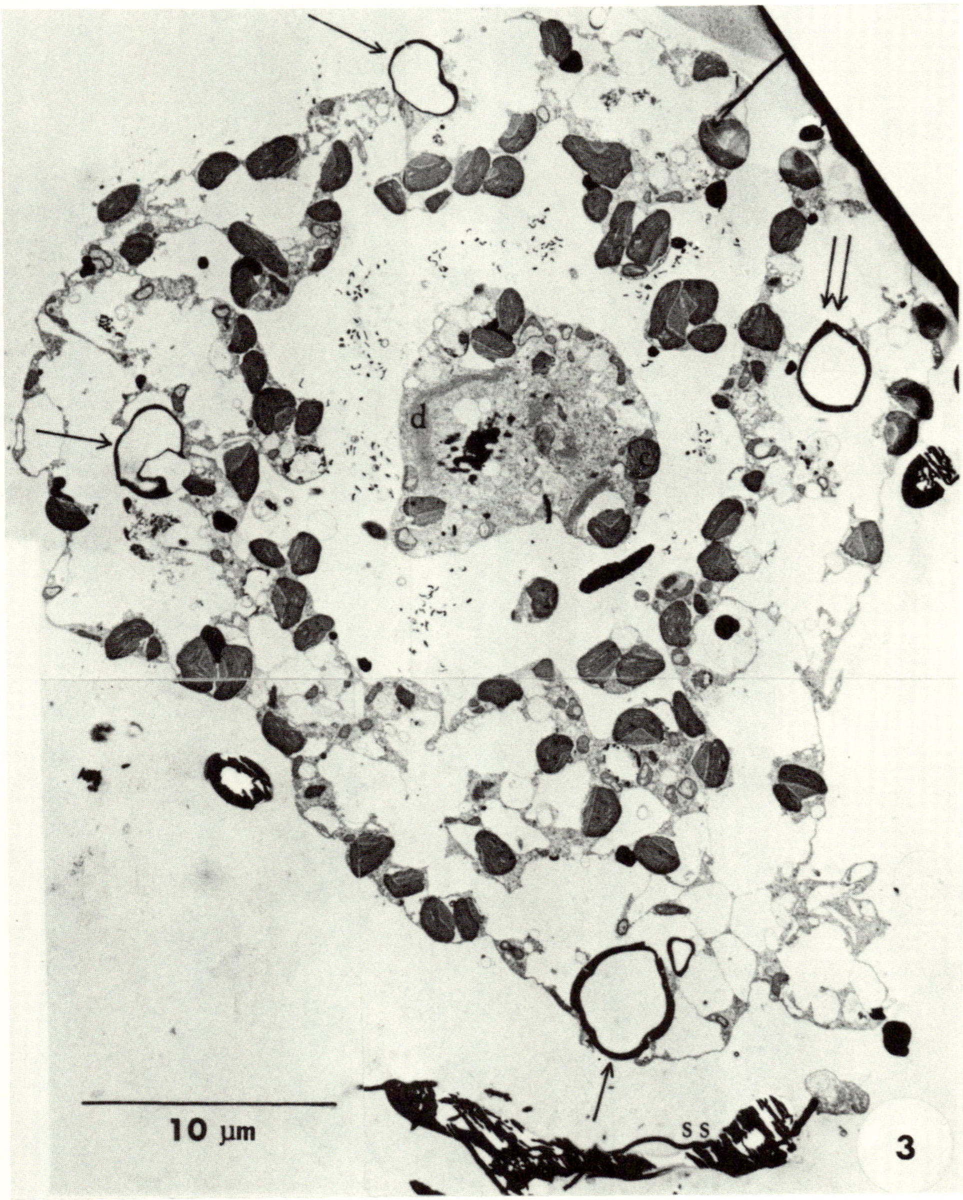
d
c
10 µm
ss
3

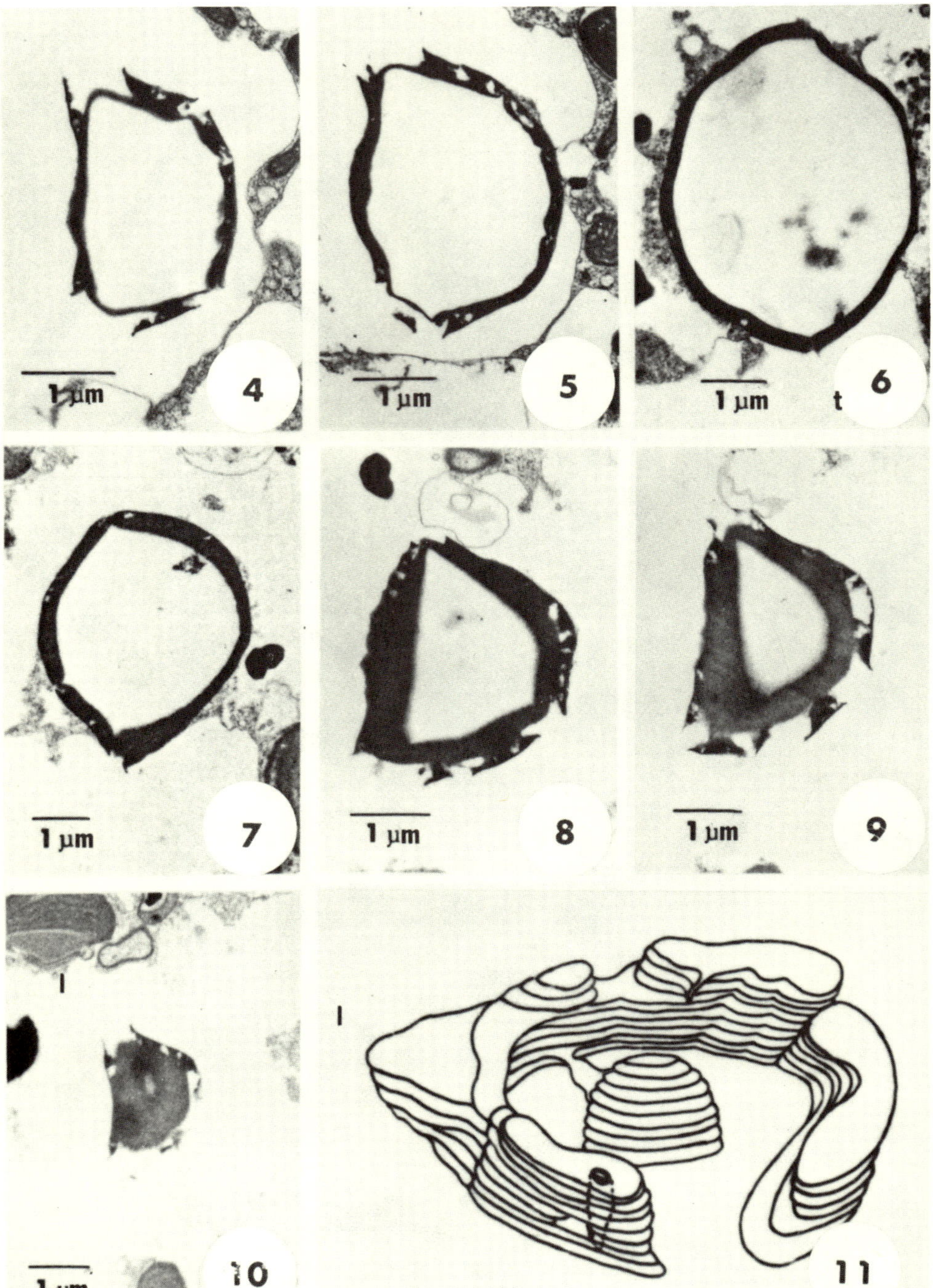
1 µm
4
1 µm
5
1 µm
t
6
1 µm
7
1 µm
8
1 µm
9
l
1 µm
10
l
11

Figs. 4-10. Transmission electron micrographs of the thick-walled vacuole at double arrows in Fig. 3, shown at 7 of 9 approximately equidistant sampling levels through one-half of the diameter of the perikaryon. The section between Figs. 5 and 6 was interrupted at the location of this vacuole; the section beyond Fig. 10 contained no evidence of this vacuole. An outer concretion of more dense particles appears to be added to an inner ring of amorphous material of somewhat lesser density. A membrane, if present, is not evident in these micrographs (after Van Valkenburg, 1970). Fig. 11. A diagrammatic reconstruction of the thick-walled vacuole shown in Figs. 4-10, and at double arrows in Fig. 3, showing only that part of the cell which was sectioned (after Van Valkenburg, 1970).

peripheral strands of cytoplasm where this has occurred, but are still within the perikaryon on the side where vesicles containing leaflets remain intact. Thus one can postulate that if the leaflet-containing vesicles observed in the spherical form all expanded and burst, a transition would be made, through the spherical-with-embayments form, to the sunburst form. This postulate requires a mechanism for the rapid expansion and bursting of these leaflet carrying vesicles. I suggest that the matrix material may be either rapidly formed or rapidly expanded within these vesicles, such as to enlarge and break them. This

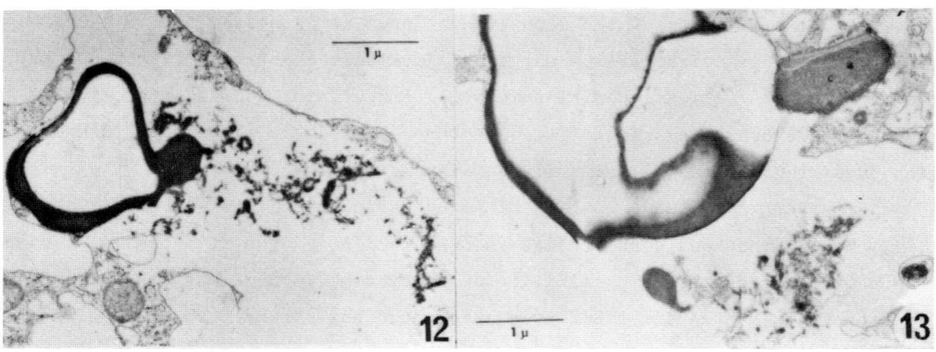

Figs. 12-13. Cacodylate-buffered glutaraldehyde and non-distilled acrolein fixation, osmium tetroxide postfixation. Fig. 12. A membrane is visible around part of the circumference of this thick-walled vacuole, which is more dense at its periphery. Short, microtubule-like profiles scattered among the dense particles to the right of the vacuole may be involved in the formation of the skeleton (after Van Valkenburg, 1970). Fig. 13. In this micrograph, the dense material is within a membranous vacuole which appears to determine its contours (after Van Valkenburg, 1970).

postulate would account for the production and delivery of large quantities of matrix material, for the formation of the "pseudopodia", and for the change in placement of the chloroplasts and the leaflets from what is found in the spherical forms, to what is found in the sunburst forms.

This theorizing does not account for a transition to or from the ring form, because intermediate stages, if indeed they have been observed, have not yet been recognized as such. Synchronized cultures are needed in order to study the sequence and the stages of this proposed morphological cycle of the protoplast, with its accompanying phenomena of skeleton formation and extrusion.

An extensive, electron-opaque material is found both free in the cytoplasm and associated primarily with the exterior boundary of the lamellar side of the chloroplasts, in all fixations which are post-fixed with osmium. This material does not fit published descriptions of lipid of the type represented by osmiophilic globules, which are absent in material fixed with distilled acrolein (Luft, 1959). Note that this density appears either within compartmentalized vesicles attached to chloroplasts, or else within the chloroplast ER (Fig. 3).

Other densities, whether of the same or of differing material, sometimes assume bizarre geometric forms which have bilateral symmetry (Figs. 16-20). None of these densities was removed by hydrofluoric acid (Van Valkenburg, 1970). The striking similarity of the depositional product in nearly all of the groups which deposit biological silica suggests a common mechanism of formation, but the mechanism of silica deposition is still largely an unsolved mystery. A number of workers have studied it, some through electron microscopy, others with physiological tools, primarily on diatoms (Darley, 1969).

Stoermer *et al.* (1965), working with the naviculoid diatom *Amphipleura pellucida*, found amorphous dense bodies in elaborations of the dictyosomes, a phenomenon which suggests the possibility of dictyosome involvement in silica deposition in this organism. But this phenomenon has not been found in silica deposition studies on other organisms. In *A. pellucida*, vesicles both with and

---

Figs. 14-15. Transmission electron micrographs: phosphate-buffered glutaraldehyde fixation, osmium tetroxide postfixation. Fig. 14. A tangential section of the spherical form of *Dictyocha fibula*, showing a small portion of the nucleus. Large vesicles in perikaryon contain leaflets (after Van Valkenburg, 1970). Fig. 15. A tangential section of *D. fibula* in which nucleus is not seen. Leaflets appear both in large vesicles and in embayments within the cell, which may have been formed by expansion and eruption of the vesicles (after Van valkenburg, 1970).

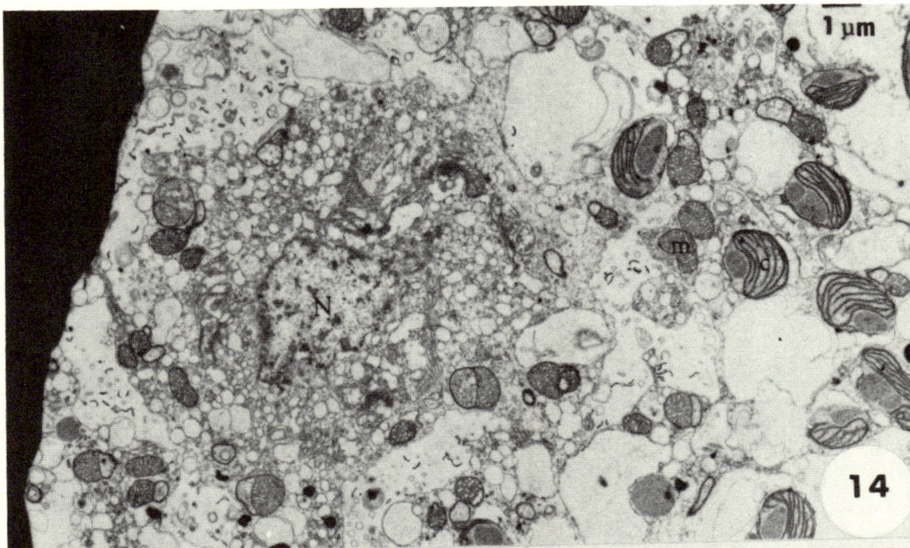
1 µm
N
m
c
14

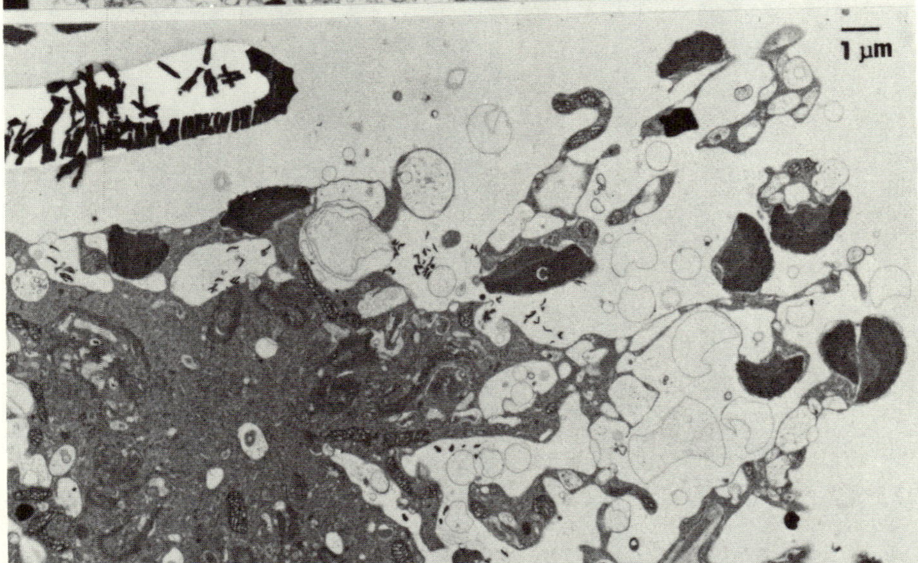
1 µm
c
15

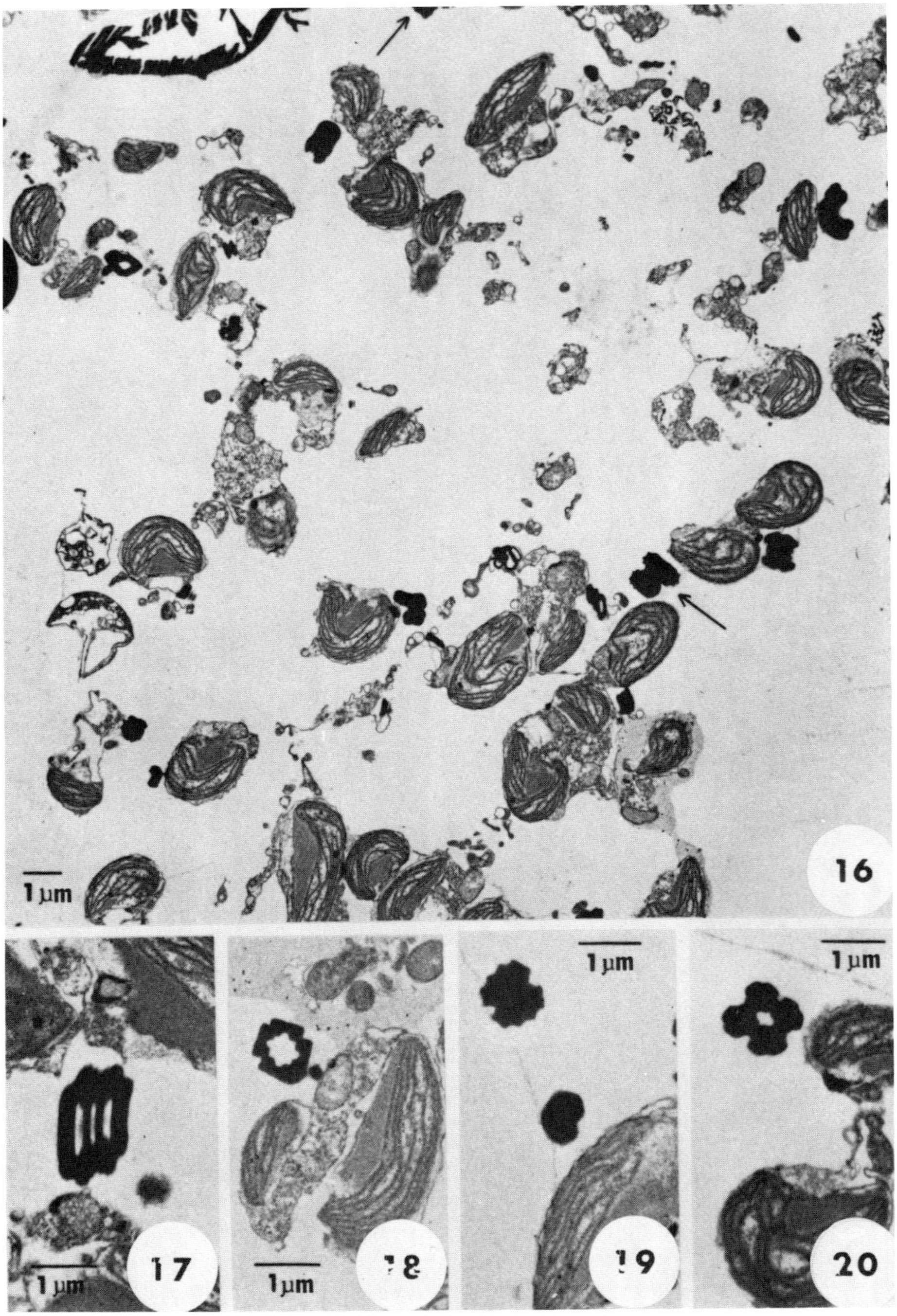
1 µm
16
1 µm
17
1 µm
18
1 µm
19
1 µm
20

Fig. 16-20. Transmission electron micrographs. Cacodylate-buffered, distilled acrolein fixation, osmium postfixation. A dense material is both associated with the chloroplasts and found free in the cytoplasmic processes and the matrix of *Dictyocha fibula*. This material often assumes configurations with bilateral symmetry and hollow centers, and with scallops, lobes, or ridges at their periphery (after Van Valkenburg, 1970).

without dense inclusions are formed by the dictyosomes during cell division. The first evidence of frustule formation is the deposition of the siliceous median rib, within a membrane sac. This sac expands laterally to form the silica deposition vesicle (the silicalemma) which appears to serve as a mold for the formation of the valve. After the valve is formed, the membranes and the small amount of cytoplasm external to it are sloughed off. The only densities removed by hydrofluoric acid are those within the silicalemma. Densities associated with the Golgi, and densities in vesicles found between the Golgi and the new wall are not removed by hydrofluoric acid. The form in which silicic acid is present in the cells before it ends up in the silica shell has not been determined (Lewin and Reimann, 1969).

Because of the several recurring patterns of as-yet-unexplained densities found within the silicoflagellate protoplast, *Dictyocha* may be a worthwhile alternative to diatom material with which to investigate the morphological pathway of the deposition of biological silica. Synchronized cultures are also essential for this type of work. Thus it appears that further progress on silicoflagellate fine structure is dependent upon the attainment of synchronized cultures, for which the experience of the diatom workers (Lewin *et al.*, 1966) will probably be the best guide.

REFERENCES

Darley, W. M. 1969. Silicon and the division cycle of the diatoms *Navicula pelliculosa* and *Cylindrotheca fusiformis*. Proc. N. A. Paleontol. Conv. September 1969. Part G. 994-1009.

Frenguelli, J. 1935. Variaciones de *Dictyocha fibula* en el Golfo de San Matias. An. Mus. Argent. Cienc. Nat. 38: 263-381.

Fryxell, Greta. (pers. comm.)

Hovasse, R. 1932. Contribution á l'étude des silicoflagellés; multiplication, variabilité, hérédité, affinités. Bull. Biol. Fr. Belg. 66: 448-501.

Hovasse, R. 1946. Flagellés á squélette siliceux. Silicoflagellés et Ebriidés provenant du plancton recueilli au cours des campagnes scientifiques du Prince Albert I de Monaco (1885-1912). Resultats Campagnes Scientifiques 107: 21 pp.

Hovasse, R. (pers. comm., 1967).

Jerković, L. 1966. The surface structure of the skeleton of some fossil silicoflagellides of Yugoslavia studied with the electron microscope. Proc. 6th Intern. Congr. E. M., Kyoto, 1966. 2: 335-336.

Jerković, L. 1969. Les nouvelles recherches de la superficie du squelette des Silicoflagellides. Univerziteta u Sarajevu. Godisnjaka Bioloskog Instituta 22: 129-176.

Lewin, J. C., B. E. Reimann, W. F. Busby and B. E. Volcani. 1966. Silica shell formation in synchronously dividing diatoms. In Cameron, I. L. and B. M. Padillo [Eds.]. *Cell Synchrony Studies in Biosynthetic Regulation.* Academic Press, New York, pp. 169-188.

Lewin, J. C. and B. E. Reimann. 1969. Silicon and plant growth. Ann. Rev. Plant Physiol. 20: 289-304.

Lipps, J. H. 1970. Ecology and evolution of silicoflagellates. Proc. N. A. Paleontol. Conv. Part G: 965-993.

Loeblich, A. R. III, L. A. Loeblich, H. Tappan and A. R. Loeblich, Jr. 1968. Annotated Index of Fossil and Recent Silicoflagellates and Ebridians with Descriptions and Illustrations of Validly Proposed Taxa. Memoir 106, Geol. Soc. Am. Boulder, Colo., 319 pp.

Luft, J. H. 1959. The use of acrolein as a fixative for light and electron microscopy. Anat. Rec. 133: 305 (abstract).

Mandra, Y. T. 1969. Silicoflagellates: A new tool for the study of Antarctic tertiary climates. Antarctic J. of the U.S. 4: 172-174.

Mandra, Y. T. and H. Mandra. 1971. Upper Eocene silicoflagellates from New Zealand. Antarctic J. of the U.S. 6: 177-178.

Mandra, Y. T. and H. Mandra. 1972. Paleoecology and taxonomy of silicoflagellates from an Upper Miocene diatomite near San Felipe, Baja, California, Mexico. California Acad. Sci. no. 99: 35 pp.

Mandra, Y. T. and A. L. Brigger. 1972. Plate tectonics, paleonmagnetism, tropical climate, and Upper Eocene silicoflagellates. Antarctic J. of the U.S. 7: 191-193.

Marshall, S. M. 1934. The Silicoflagellata and Tintinnoinea. British Mus. Nat. Hist. Great Barrier Reef Expedition 1928-1929, Sci. Rept. 4: 623-664.

Perch-Nielsen, K. 1975. Late Cretaceous to Pleistocene silicoflagellates from the southern southwest Pacific, DSDP, Leg 29. Initial Rept. Deep Sea Drilling Project, Vol. 24, Wash., D.C.

Provasoli, L. 1968. Media and prospects for the cultivation of marine algae. In Watanabe, A. and A. Hattori [Eds.]. *Cultures and Collections of Algae.* Proc. U.S.-Japan Conf. Hakone, September 1966. Jap. Soc. Plant Physiol. 63-75.

Stoermer, E. F., H. S. Pankratz and C. C. Bowen. 1965. Fine structure of the diatom *Amphipleura pellicida.* II. Cytoplasmic fine structure and frustule formation. Am. J. Bot. 52: 1067-1078.

Van Valkenburg, S. D. 1970. The ultrastructure of the silicoflagellate *Dictyocha fibula* Ehrenberg. Ph.D. thesis, University of Washington.

Van Valkenburg, S. D. and R. E. Norris. 1970. The growth and morphology of the silicoflagellate *Dictyocha fibula* Ehrenberg in culture. J. Phycol. 6: 48-54.

Van Valkenburg, S. D. 1971a. Observations on the fine structure of *Dictyocha fibula* Ehrenberg. I. The skelton. J. Phycol. 7: 113-118.

Van Valkenburg, S. D. 1971b. Observations on the fine structure of *Dictyocha fibula* Ehrenberg. II. The protoplast. J. Phycol. 7: 118-132.

Copyright 1980 by Elsevier North Holland, Inc.

CHLOROMONADS*

PETER HEYWOOD
Division of Biology and Medicine, Brown University, Providence, Rhode Island, 02912, U.S.A.

ABSTRACT

Chloromonads are biflagellate unicells that occur as either motile or palmelloid individuals. These organisms usually range in length from approximately 30-80 μm. Their shapes vary from almost spherical to ovoid, and they are frequently flattened dorsoventrally. Chloromonads lack cell walls and eyespots. Trichocysts or mucocysts are frequently present. A complex flagellar root system connects the anterior surface of the nucleus to the basal bodies. The two flagella arise near the apex of the cell: one extends anteriorly and bears tubular flagellar hairs that originate in the endoplasmic reticulum, while the other lacks appendages and trails posteriorly. Chloroplasts are usually numerous, discoid and parietal; each possesses a girdle band, chloroplast ER, a ring-shaped nucleoid, chlorophylls *a* and *c*, and several xanthophylls (but not fucoxanthin). The reserve food material appears to be oil. Pyrenoids are not present in freshwater chloromonads. The Golgi apparatus is found anterior to the nucleus at interphase. During mitosis the nuclear envelope remains intact (aside from polar gaps) until telophase and a Golgi apparatus is associated with each of the poles. The marine genus *Chattonella* lacks a contractile vacuole but this structure is present in freshwater chloromonads where it occurs anterior to the Golgi apparatus. Chloromonads usually occur in habitats rich in vegetation where the water is acidic or neutral. Many details of their biochemistry and physiology remain uninvestigated.

INTRODUCTION

The term chloromonads will be used in this chapter because it is brief, and taxonomically noncommittal. Other designations for these algae include Chloromonadales (Smith, 1950), Chloromonadineae (Fritsch, 1935), Chloromonadophyceae (Klein and Cronquist, 1967; Fott, 1968), Chloromonadophyta (Graffius, 1966; Loeblich and Fine, 1977) and Raphidophyceae (Christensen, 1964).

The genera *Chattonella, Gonyostomum, Merotricha* and *Vacuolaria* are recognized as being chloromonads by most authorities. It is generally agreed that

*This chapter is dedicated to the memory of Professor Dr. Bohuslav Fott (1908-1976) of Charles University, Prague, Czechoslovakia.

Hornellia marina (Subrahmanyan, 1954) is synonymous with *Chattonella subsalsa* (Hollande and Enjumet, 1956; Mignot, 1976). Similarly, *Trentonia* and *Vacuolaria* are probably congeneric, and the organism originally described as *Trentonia flagellata* (Stokes, 1886, 1888) was probably *Vacuolaria viridis* (Fott, 1968). There is no strong evidence to favor the inclusion in the chloromonads of the chloroplast-less genera *Colpomena, Hyaloselene, Reckertia, Monomastix* and *Thaumatomastix*. Loeblich and Fine (1977) have suggested that the marine genera *Fibrocapsa, Heterosigma* and *Olisthodiscus* are chloromonads, and that they are congeneric with *Chattonella*. However, this has been disputed (Gibbs, Chu and Magnussen, in press). Until further information is available for these organisms it seems appropriate to adopt a conservative view of the chloromonads and to limit this account to a discussion of the genera *Chattonella, Gonyostomum, Merotricha* and *Vacuolaria*. The species that are reçognized are as follows: *Chattonella subsalsa* Biecheler, *Gonyostomum intermedium* Skuja, *G. latum* Ivanov, *G. ovatum* Fott, *G. semen* (Ehrenberg) Diesing, *Merotricha bacillata* Mereschowsky, *Vacuolaria penardii* Fott, *V. tropicalis* Bicudo and Cardoso, *V. virescens* Cienkowsky and *V. viridis* (Dangeard) Senn.

METHODS

Cultures of *Vacuolaria virescens* are available from the Cambridge Culture Collection (Number LB 1195/1). Collections of algae containing *Gonyostomum semen* are obtainable from the Freshwater Biological Association at Ferry House, Ambleside, England. Chloromonads can be concentrated by phototaxis (Chapman and Haxo, 1966; Heywood, 1968; Spencer, 1971), and unialgal isolates can be obtained by carefully pipetting the chloromonad cells through a series of watch glasses containing culture medium before inoculating them into culture vessels. Media and conditions for culturing chloromonads have been described by Drouet and Cohen (1935), Kirby (1950), Provasoli (1958), Chapman and Haxo (1966) and Spencer (1971). *Gonyostomum semen* and *Vacuolaria virescens* have been successfully grown on Guillard's freshwater medium WC (Guillard and Lorenzen, 1972) and on a defined medium (Heywood, 1973b).

For light microscopy of nuclear structure and chromosome morphology cells were fixed in a mixture of 3 parts 95% ethanol to 1 part of glacial acetic acid, mordanted with aqueous ferric alum, and stained with acetocarmine. Slides were made permanent by removing the coverslip in an acetic acid solution, dehydrating in ethanol, and mounting in euparol (Heywood and Godward, 1974).

Methods of preparation for ultrastructural examination have been described on a number of occasions (for example, Schnepf and Koch, 1966b; Heywood, 1972,

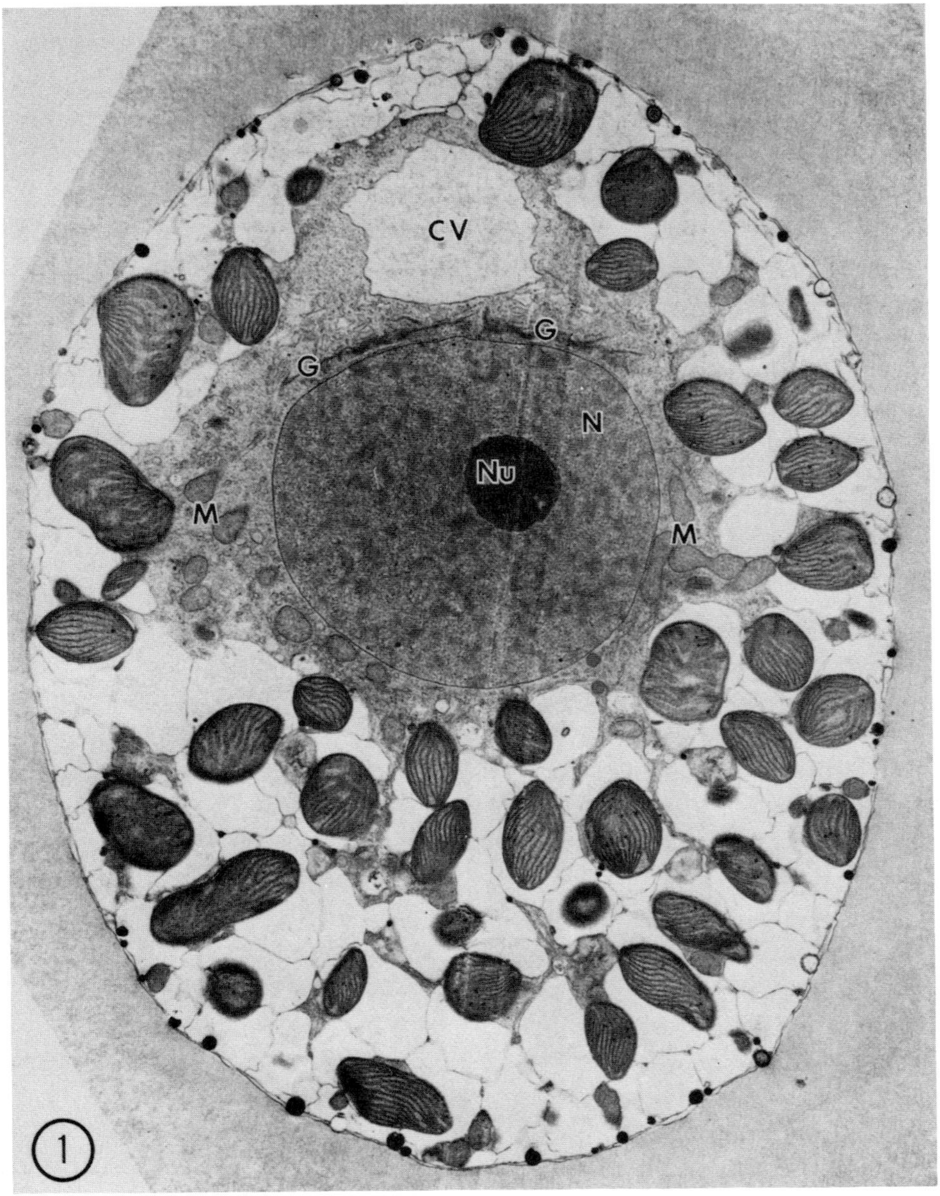

Fig. 1. Longitudinal section through a cell of *Vacuolaria virescens* showing the nucleus (N), nucleolus (Nu), Golgi apparatus (G) and contractile vacuole (CV). Mitochondria (M) are present in the layer of cytoplasm surrounding the nucleus; chloroplasts occur between this layer and the cell membrane (x4,300).

1977b; Mignot, 1976). Most ultrastructural studies have employed fixation in glutaraldehyde solutions (usually of concentration 2-3%) suspended in cacodylate or phosphate buffer (usually of pH 6.0-7.0, and molarity 0.07-0.1 M). Specimens fixed in this manner have been postfixed in osmium tetroxide solution, dehydrated, embedded in Epon, sectioned and stained sequentially with uranyl acetate and lead citrate. Other fixative solutions used include potassium permanganate fixative (Schnepf and Koch, 1966b), osmium tetroxide fixative (Schnepf and Koch, 1966b; Heywood, 1973a) and Dalton's osmium tetroxide-potassium dichromate fixative (Heywood, 1978a). The electron micrographs illustrating this chapter were obtained from cells which had been fixed in glutaraldehyde followed by postfixation in osmium tetroxide solution (Figs. 2-10, 12) or from cells which had been fixed directly in Dalton's chrome-osmium fixative (Figs. 1, 11, 13, 14).

CHLOROPLASTS

Chloromonad cells contain hundreds of discoid chloroplasts of approximate size 2-3 μm to 3.5-5 μm. They occupy the region between the cell membrane and the layer of cytoplasm surrounding the nucleus (Fig. 1). Some chloroplasts extend into this layer of cytoplasm whereas others are in contact with the cell membrane, chloroplasts or other cell organelles. Individual chloroplasts are bounded by a chloroplast envelope external to which is endoplasmic reticulum; connections between the latter and the nuclear envelope are not evident. In *Gonyostomum semen* chloroplasts are present as a single layer below the cell membrane; in some preparations the longitudinal axes of the chloroplasts are oriented parallel to the surface of the cell (Hovasse, 1945; Heywood, 1968) but this is not always the case (Mignot, 1967). This may reflect differences in intensity of illumination and position in the diurnal cycle at the time of fixation. Subrahmanyan (1954) has observed changes in chloroplast orientation in *Chattonella subsalsa* which he believed were in response to illumination.

Chloroplast ultrastructure has been studied in *Chattonella subsalsa* (Mignot, 1976), *Gonyostomum semen* and *Vacuolaria virescens* (Koch and Schnepf, 1967; Mignot, 1967; Heywood, 1968, 1977a). Many features of their ultrastructure are

Fig. 2. Part of a cell of *Vacuolaria virescens* containing endoplasmic reticulum (er), abundant ribosomes, several mitochondria (M) and chloroplasts. Note the arrangement of chloroplast bands, the occurrence of small lipid droplets (L) in the chloroplast stroma, and the presence of girdle bands (g) which are bounded by chloroplast envelope and endoplasmic reticulum (ce/er) (x15,000) (Reproduced from Heywood, 1977a by kind permission of the Journal of Phycology.)

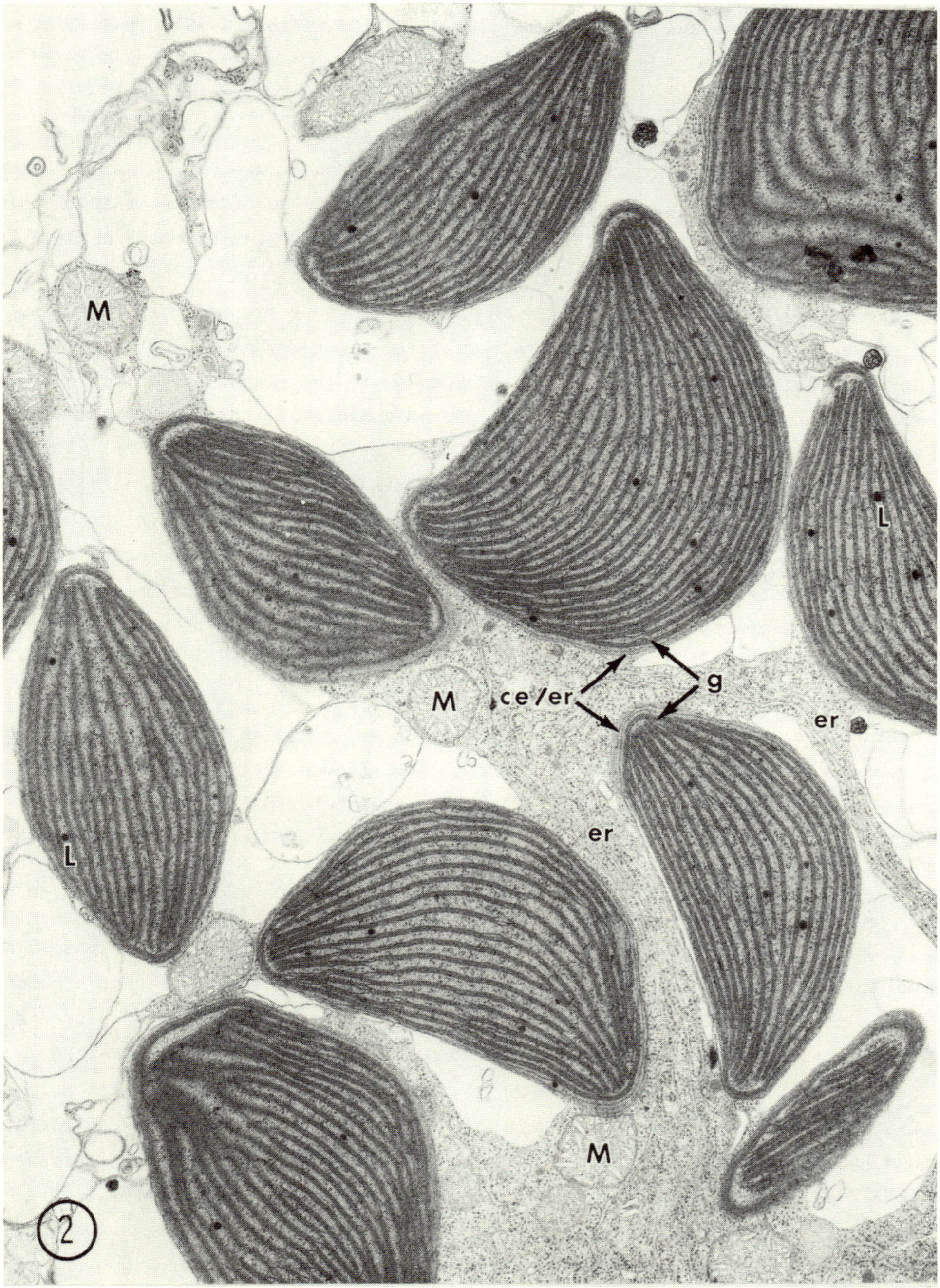
M
M
ce/er
g
er
er
L
L
M
2

shared with algae belonging to the Bacillariophyceae, Chrysophyceae, Haptophyceae, Phaeophyceae and Xanthophyceae (Dodge, 1973; Taylor, 1976). In median longitudinal sections through chloroplasts between 10 and 25 bands are apparent which are arranged approximately parallel to the longitudinal axis of the chloroplast (Fig. 2). Chloroplast bands consist of three thylakoids; occasional thylakoids extend across the stroma from one band to another. A girdle band covers the entire organelle in *Gonyostomum semen* and *Vacuolaria virescens* but appears to be absent in *Chattonella subsalsa*. Spherical osmiophilic structures which occur both in the cytoplasm and in the chloroplast stroma presumably serve as a lipid food reserve. Pyrenoids appear to be present only in the marine genus *Chattonella* where they occur at the pole of the chloroplast. Fibrillar material resembling chloroplast DNA occurs in the ring-shaped electron-translucent nucleoid region which is located immediately below the girdle lamella around the periphery of the chloroplast, and electron-opaque spherical structures resembling chloroplast ribosomes are present in the stroma.

Chloromonad chloroplasts usually appear as bright green in color which has sometimes been described as "maigrün" or "vert-pré" in the German and French descriptions (Poisson and Hollande, 1943; Fott, 1968). A yellowish-green coloration of the chloroplasts has been reported in *Chattonella subsalsa* (Subrahmanyan, 1954), *Merotricha bacillata* (Skuja, 1934) and *Vacuolaria virescens* (Spencer, 1971). In the former case Subrahmanyan noted that in the laboratory an overcrowded sample of *Chattonella subsalsa* showed a change in chloroplast coloration from green to yellowish-green to yellow which he believed accompanied a deterioration in the condition of the organism. A similar change in chloroplast color from green to brown occurred in natural conditions when this species reached a high density (Hollande and Enjumet, 1956). Presumably the change in color indicated an accumulation of carotenoid pigments.

Chapman and Haxo (1966) demonstrated the presence of chlorophyll *a* and four carotenoid pigments in *Vacuolaria virescens*, and showed that there was a chlorophyll to carotenoid ratio of 0.29:1. They identified the carotenoid pigments as β-carotene, antheraxanthin, lutein monoepoxide and a trollixanthin-like xanthophyll. Chapman and Haxo noted the similarity between these carotenoids and those of the xanthophycean alga *Tribonema aequale*. Subsequently Guillard and Lorenzen (1972) demonstrated the presence of chlorophyll *c* in the chloromonads and reported minor differences between the xanthophylls of these organisms and those of the xanthophycean algae that they examined. These two classes of algae are similar in their lack of a fucoxanthin pathway (Ragan and Chapman, 1978).

MITOCHONDRIA AND MICROBODIES

Mitochondria of chloromonads occur in large numbers in the layer of cytoplasm surrounding the nucleus; they are often present in extensions of this cytoplasm which may reach outwards as far as the cell membrane (Figs. 1, 2). Mitochondrial ultrastructure has been determined in *Chattonella subsalsa* (Mignot, 1976), *Gonyostomum semen* and *Vacuolaria virescens* (Mignot, 1967; Heywood, 1977b). Serial sections through mitochondria of *Gonyostomum semen* revealed that most were ellipsoidal structures in the size range 0.9-2.0 µm by 0.5-1.3 µm (although occasional ones were larger); mitochondria of *Vacuolaria virescens* had similar dimensions (Heywood, 1977b). This agrees with Mignot's (1967) observation on the morphology of mitochondria from these species where he reported that most mitochondria were approximately spherical and of a diameter 1-2 µm. It is therefore apparent that these species do not have the complex mitochondrial reticulum which occurs in some algae (Atkinson, John and Gunning, 1974; Grobe and Arnold, 1975). Mitochondria of chloromonads have abundant tubular cristae which occupy much of the volume of the organelle. In some electron-translucent areas lacking cristae there occur fine fibrils resembling the mitochondrial DNA of other organisms (Fig. 3).

Gonyostomum semen and *Vacuolaria virescens* possess organelles which are morphologically similar to microbodies (Heywood, 1974c). These are ellipsoidal or circular in section and usually of diameter 0.3-0.7 µm, although occasionally some are larger. They are limited by a single membrane and possess a homogeneous finely granular matrix which lacks crystalline inclusions (Fig. 10). Microbodies frequently occur adjacent to elements of the endoplasmic reticulum and are often located close to both chloroplasts and mitochondria. Microbodies appear to be few in number in chloromonad cells since most sections of whole cells do not contain them. However, it is certain that there is more than one microbody per cell since occasional sections contain two of these organelles sufficiently far apart that they could not be part of a single elongated microbody.

FLAGELLAR APPARATUS

Chloromonads are biflagellate: one flagellum bears tubular flagellar hairs on its surface and extends in front of the cell, while the other lacks appendages and trails posteriorly. During swimming movement the anterior flagellum projects directly forwards and beats rapidly, whereas the trailing flagellum usually lies close to the ventral surface of the cell and moves little if at all. Observations of particle movement in the surrounding medium show that particles in front of the swimming cell are swept into rapid currents around the anterior flagellum and then pass over the surface of the cell. The organism then moves forward to occupy the displaced fluid (Heywood, 1968). Accompanying the forward movement there is a spiral rotation which may be quite pronounced in some cells. Occasionally the anterior flagellum stops its rapid beating and twists back on itself before recommencing its normal motion.

The anterior flagellum is usually approximately as long as the cell while the trailing flagellum is the same length or longer. Although there is some

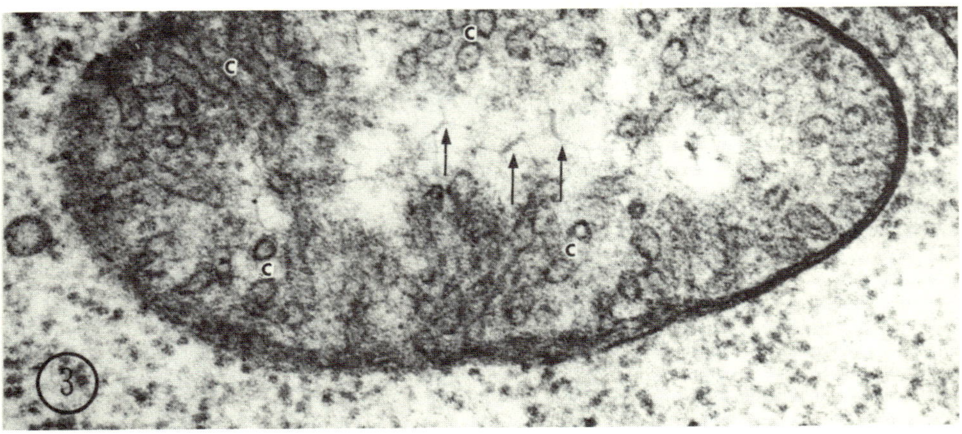

Fig. 3. A mitochondrion of *Gonyostomum semen*. Tubular cristae (c) are abundant around a central electron-translucent region that contains fibrillar material (arrows) which resembles the mitochondrial DNA of other organisms (x70,000).

Figs. 4-7. The flagellar apparatus of *Vacuolaria virescens*. Fig. 4. Longitudinal section through the two basal bodies (b) and the basal portion of one flagellum (f). Part of the descending microtubule root (dmr) system is present in oblique section (x35,000). Fig. 5. Transverse section across the descending microtubule root system (x70,000). Fig. 6. The flagellar groove (fg) containing two flagella and flagellar hairs (x35,000). Fig. 7. Higher magnification view of Fig. 6. The tubular flagellar hairs (fh) which are present in longitudinal and oblique section are narrower and less densely stained than the axonemal microtubules of the flagella. Asterisks indicate the points of attachment of the flagellar hair bases (fhb) on the flagellar membrane (x70,000).

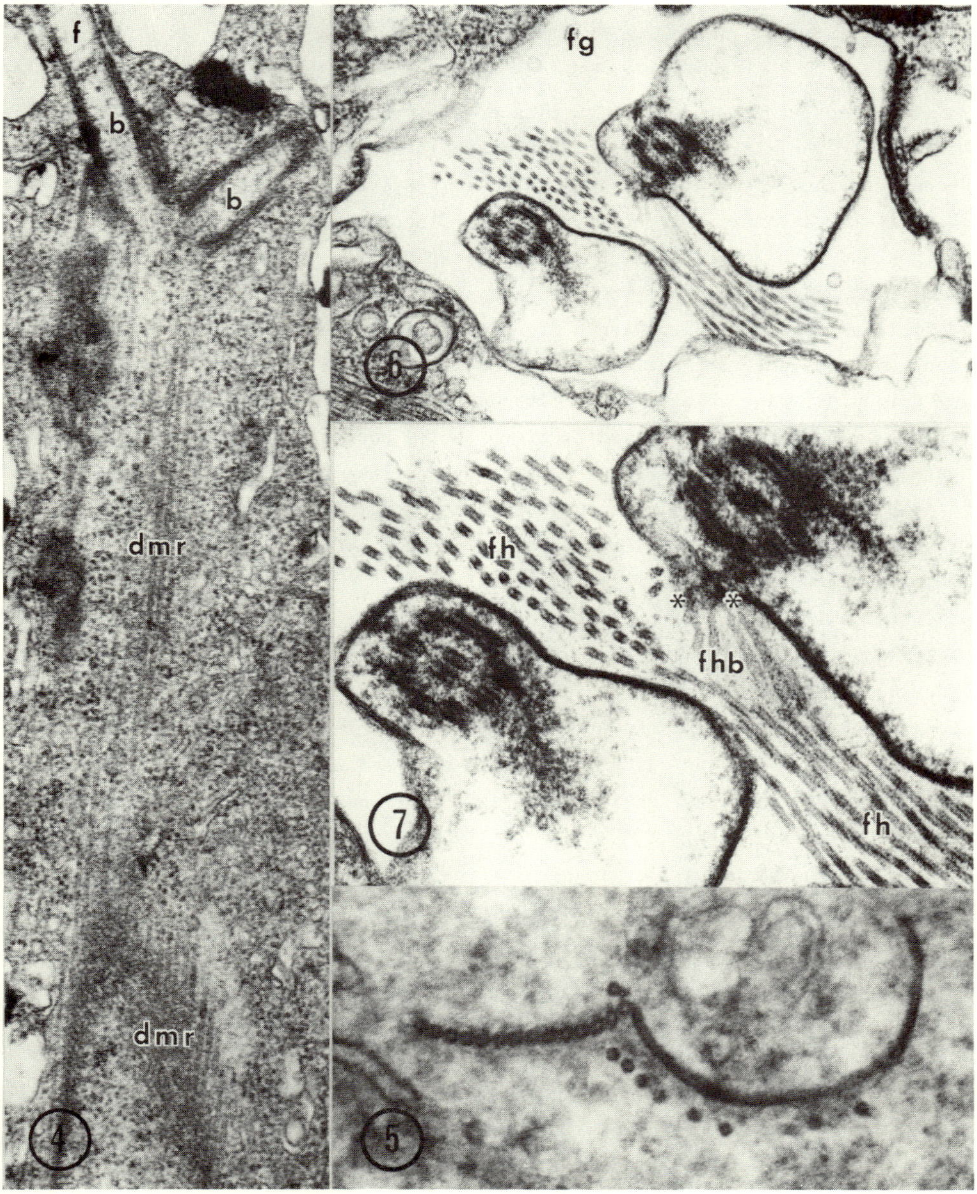
f
b
b
dmr
dmr
4
fg
6
fh
fhb
fh
7
5

variability in flagellar length within a species, this is even more pronounced between species. For example, the anterior flagellum in *Gonyostomum ovatum* and both flagella in *Gonystomum semen* are approximately as long as the cell but the trailing flagellum in the former is approximately twice this length (Fott, 1968).

The flagella arise in shallow pits on the ventral surface near the apex of the cell. The ultrastructure of the flagella and flagellar root system has been described for *Chattonella subsalsa* (Mignot, 1976), *Gonyostomum semen* and *Vacuolaria virescens* (Mignot, 1967). The flagellar basal bodies lie close together and form an angle of approximately 90° or more to each other in *Gonyostomum semen* and approximately 80° to each other in *Vacuolaria virescens* (Fig. 4). A fibrous flagellar root extends from the flagellar basal bodies and makes contact with the anterior surface of the nucleus (Fig. 8). A linear array of microtubules arises close to the fibrous flagellar root and extends posteriorly (Figs. 4, 5, 8). This descending microtubule root system consists of approximately 10-20 microtubules (Fig. 5) and may continue for almost the whole length of the nucleus. Other microtubules extend anteriorly from the region of the flagellar basal bodies. Microtubules also occur immediately below the cell membrane in the region of flagellar insertion.

The tubular flagellar hairs on the anterior flagellum are similar in structure and origin in *Chattonella subsalsa* (Mignot, 1976) and *Vacuolaria virescens* (Heywood, 1972). Each flagellar hair consists of a tapered basal region of length 200-250 nm whose narrower proximal end terminates on the flagellar membrane and whose broader distal portion is attached to a tubular shaft. In *Vacuolaria virescens* the shaft is of diameter 17 nm (Figs. 6, 7) and is at least 1.3 μm in length. In *Chattonella subsalsa* the shaft is of diameter 20 nm and length 2.5 μm; it terminates in fine filamentous structures. The orientation of flagellar hairs is the same in both species: the basal region makes an angle of 80-90° with the longitudinal axis of the axoneme, and the shaft extends anteriorly at an angle to the basal region.

The flagellar hairs of *Chattonella subsalsa* and *Vacuolaria virescens* resemble those of other "heterokont" algal groups both in their structure and in their origin within the cell (Bouck, 1969; Heath, Greenwood and Griffiths, 1970; Leedale, Leadbeater and Massalski, 1970; Heath and Darley, 1972). Some cisternae of the endoplasmic reticulum in *Chattonella subsalsa* and *Vacuolaria virescens* are enlarged and contain parallel arrays of tubular structures whose morphology is similar to that of the flagellar hairs (Koch and Schnepf, 1967; Heywood, 1972; Mignot, 1976). These presumptive flagellar hairs are thought

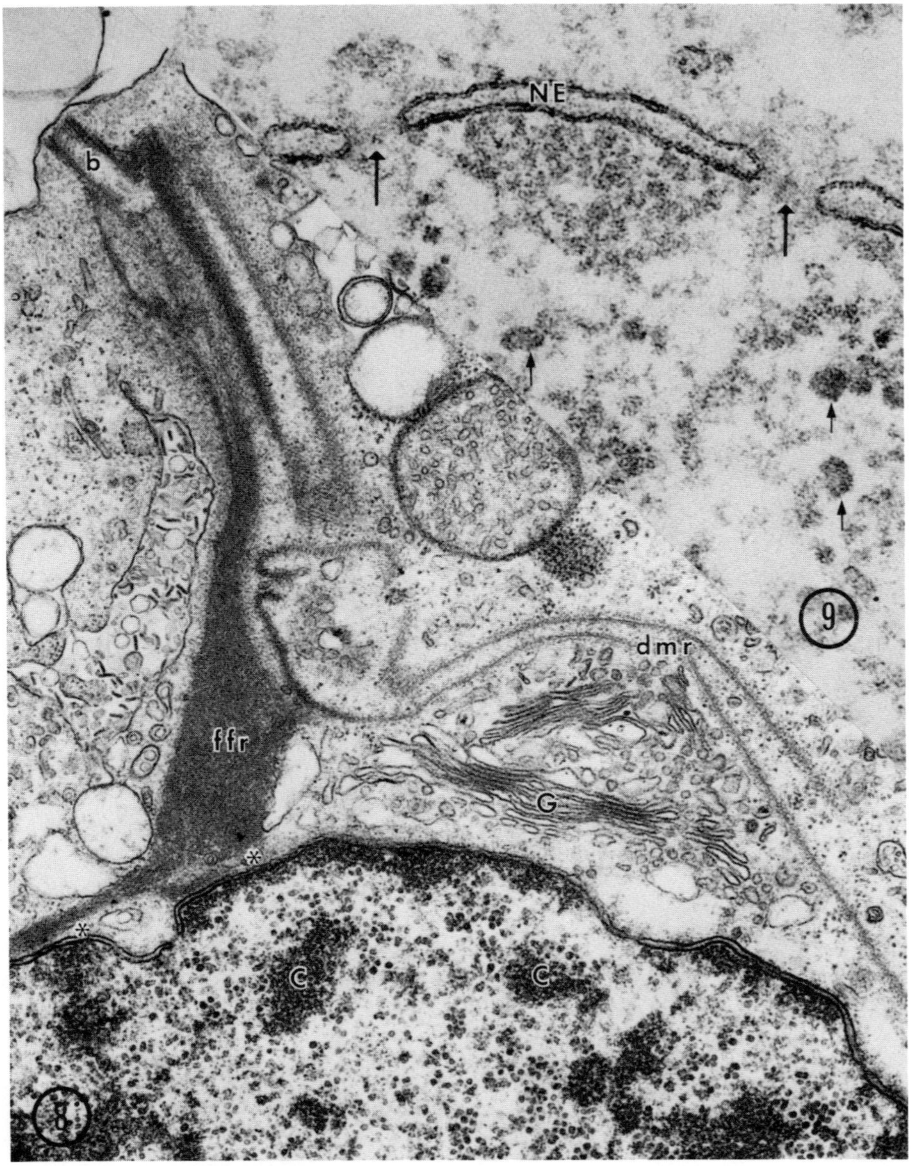

Figs. 8 and 9. Flagellar apparatus, nucleus and Golgi apparatus of *Gonyostomum semen*. Fig. 8. Longitudinal section through the flagellar apparatus including the basal body (b), the descending microtubule root (dmr) and the fibrous flagellar root (ffr) which makes contact with the nuclear envelope (asterisks). A Golgi apparatus (G) occurs anteriorly to the nucleus. Nuclear granules are present in the nucleoplasm between the chromatin (C) (x24,000). Fig. 9. Part of the nucleus including the nuclear envelope (NE) and nuclear pores (large arrows). Small arrows indicate some of the nuclear granules present within the nucleoplasm (x120,000).

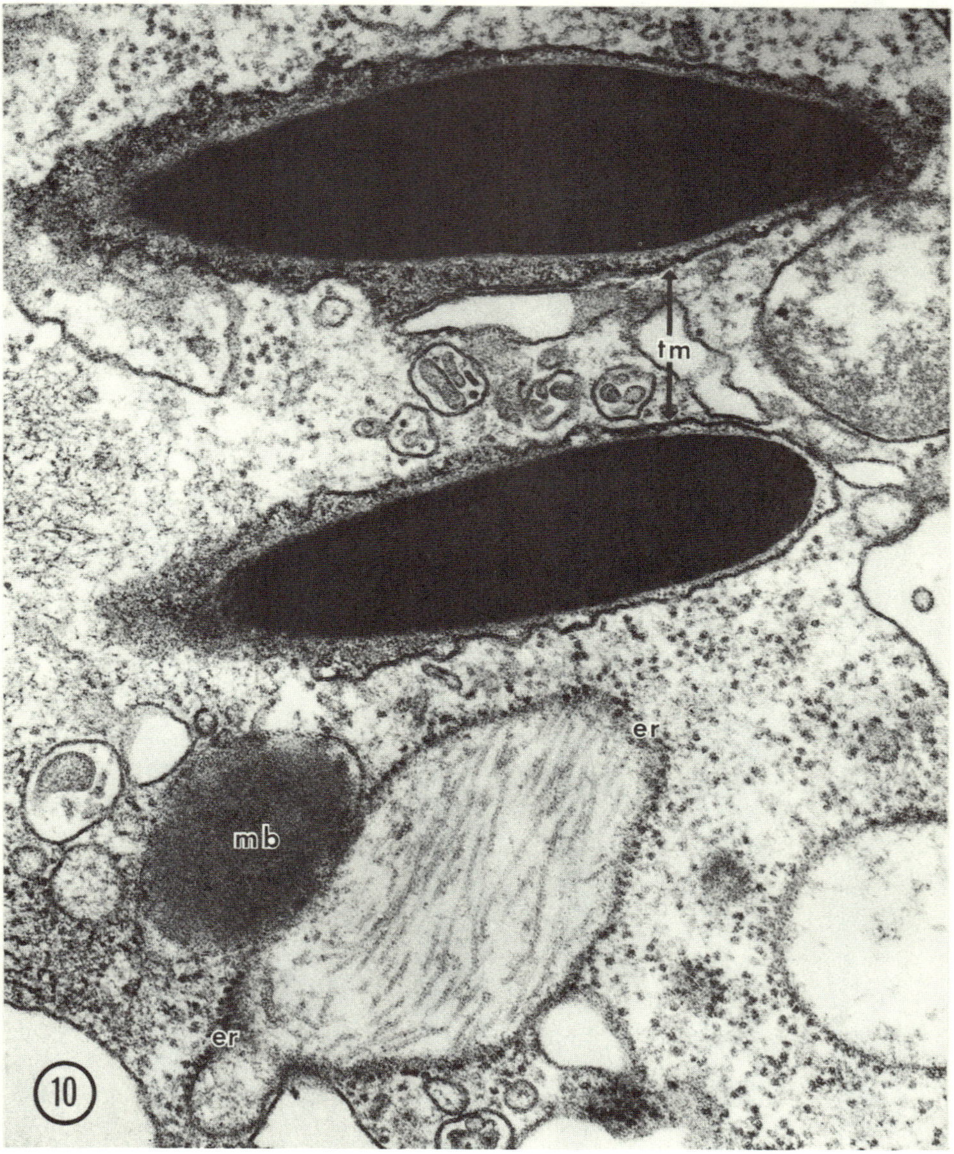

Fig. 10. Oblique section through two trichocysts of *Gonyostomum semen*. Each trichocyst consists of an electron-dense structure which is separated from the cytoplasm by a membrane (tm). Also present is a microbody (mb) adjacent to a portion of the rough endoplasmic reticulum (er) whose cisterna contains presumtive flagellar hairs (x52,500).

to be assembled and stored within the endoplasmic reticulum before being transported to their definitive sites on the anterior flagellum. As Mignot (1976) has observed, vesicles containing flagellar hairs may be guided in this migration by the microtubules of the flagellar root system. A vesicle containing flagellar hairs and opening to the exterior close to the flagellar bases has been observed in a cell of *Vacuolaria virescens* which had been halted in metaphase by treatment with colchicine. This type of structure has not been seen in nondividing cells and it is thought to be an intermediate stage between the intracellular formation of flagellar hairs and their ultimate disposition on the anterior flagellum (Heywood, 1972). Frequently one end of the presumptive flagellar hair is attached to the inner surface of the cytoplasmic vesicle. It is possible that this connection is maintained when the flagellar hairs are externalized, and that the vesicle membrane and flagellar hairs together become part of the flagellar surface. Although the flagellar hairs of *Gonyostomum semen* have not been studied in the same detail, these are also tubular structures which are formed in the cisternae of the endoplasmic reticulum (Heywood, 1973a) as illustrated in Fig. 10.

Eyespots appear to be absent in the chloromonads examined so far. However, there are reports that *Chattonella subsalsa, Gonyostomum semen* and *Vacuolaria virescens* are positively phototactic (Poisson and Hollande, 1943; Chapman and Haxo, 1966; Heywood, 1968; Spencer, 1971; Mignot, 1976). A diurnal movement has been reported to occur in natural populations of *Gonyostomum semen:* cells migrated close to the surface of the water in the early hours of daylight, and in the afternoon began their downward descent (Cowles and Brambel, 1936). *Vacuolaria viridis* also exhibits vertical migration: in diffuse light cells occurred near the surface of the water but in bright sunlight they descended to a depth of 2 meters (Fott, 1935).

CONTRACTILE VACUOLE

Contractile vacuoles (sometimes termed "water expulsion vesicles") are present in freshwater chloromonads. It is thought that they serve in osmo-regulation by discharging a hypotonic solution from the cell but relatively little is known about their physiology; for example, the effects on the contractile vacuole cycle of the temperature, osmotic pressure, and pH of the culture medium have not been reported. The duration of the contractile vacuole cycle in *Vacuolaria virescens* is in the range of 36-80 seconds (Poisson and Hollande, 1943; Schnepf and Koch, 1966b; Spencer, 1971). This range is broad because of the variety of conditions in which the cells were maintained. But the contractile vacuole cycle is more constant under a particular set of environmental conditions; for example, its duration was 49-62 seconds at 23° C for cells in defined medium (Heywood, 1978b).

The contractile vacuole in chloromonads may reach 8-10 μm in diameter. Because it is a conspicuous structure it was noted in some of the earliest descriptions of these organisms (Cienkowsky, 1870), and has been examined on a number of occasions (for example, Poisson and Hollande, 1943; Hovasse, 1945; Schnepf and Koch, 1966b; Mignot, 1967; Heywood, 1978b). Light microscope observations of *Gonyostomum semen* and *Vacuolaria virescens* reveal a distinctive region between the anterior surface of the nucleus and the contractile vacuole (Poisson and Hollande, 1943; Hovasse, 1945). This was termed the "supranuclear cap" and it was recognized that this region produced vesicles which eventually formed the contractile vacuole (Poisson and Hollande, 1943). Electron microscope observations demonstrate that the supranuclear cap consists of a large Golgi apparatus and its products (Schnepf and Koch, 1966b; Mignot, 1967). Between the Golgi cisternae and the contractile vacuole there occurs a number of subsidiary vacuoles. With appropriate techniques of fixation it is apparent that the Golgi vesicles, subsidiary vacuoles and contractile vacuole contain electron-dense material (Fig. 11). It has been suggested (Heywood, 1978b) that the Golgi apparatus is producing a highly hydrophilic substance. This will attract water from the cytoplasm into the Golgi vesicles, subsidiary vacuoles and contractile vacuole whereupon the water and hydrophilic substance are discharged from the cell. For the moment this proposed mechanism remains a hypothesis, but it is worth noting that electron-dense material occurs in the Golgi apparatus of *Glaucocystis* which is also involved in secretion of water (Schnepf and Koch, 1966a), and that mucus has, on occasion, been observed in the water secreted by the contractile vacuole of *Vacuolaria virescens* (Tschermak-Woess, 1954).

The contractile vacuole of *Vacuolaria virescens* is a permanent structure

---

Figs. 11 and 12. Osmoregulation in *Vacuolaria virescens*. Fig. 11. The Golgi apparatus (G) is situated immediately adjacent to the nuclear envelope (NE). Golgi cisternae give rise to subsidiary vacuoles (asterisks) which empty into the contractile vacuoles (CV). Golgi cisternae, subsidiary vacuoles and the contractile vacuole contain electron-dense material which is thought to be hydrophilic. A flagellum (f) is situated in the flagellar groove (fg) immediately anterior to the contractile vacuole (x20,000). Fig. 12. Part of the contractile vacuole (CV) membrane. This differs from the other membranes present in this field in its possession of peg-like structures (arrowheads) which extend inward on the cytoplasmic side of the membrane (x112,000).

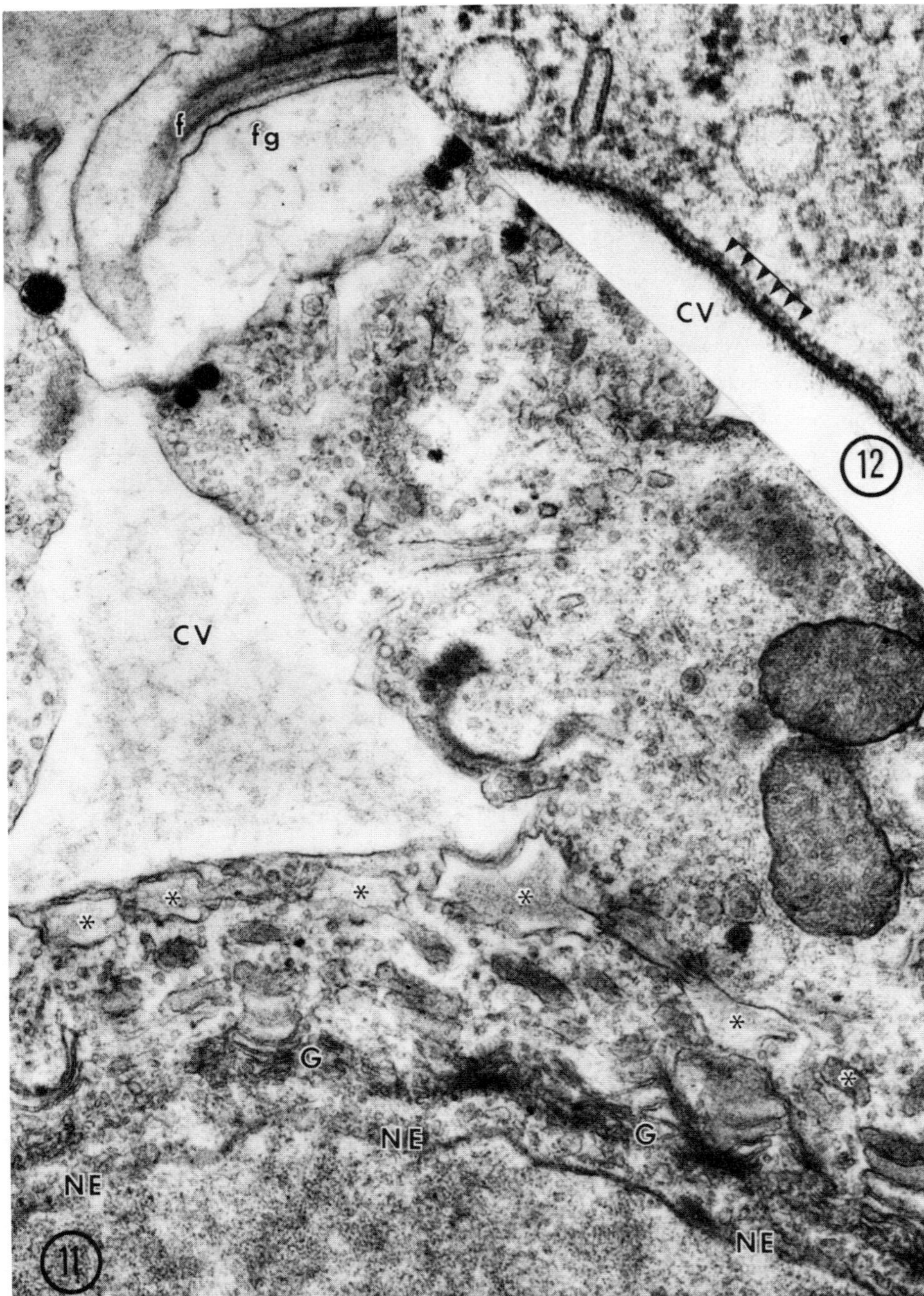
f
fg
CV
CV
12
*
G
NE
NE
NE
G
11

which is less conspicuous immediately after discharge of its contents. The membrane of the contractile vacuole consists of subunits of diameter 21-24 nm which are sometimes arrayed in a regular hexagonal pattern. The lateral walls of these subunits form peg-like structures which extend for a distance of 13-15 nm on the cytoplasmic side of the membrane (Fig. 12). Similar membranes (termed "coated membranes" or "bristle coat structures") have been reported from other systems where water and other substances are transported (Aaronson and Behrens, 1974; McKanna, 1976). The function of this specialized "coated membrane" in the contractile vacuole system of *Vacuolaria virescens* is not certain but it may be actively involved in the filling and emptying of the contractile vacuole, or it may serve to strengthen the membrane to withstand the pressure changes occurring during the contractile vacuole cycle.

A contractile vacuole is absent from *Chattonella subsalsa* which occurs in marine or brackish water habitats (Biecheler, 1936a; Subrahmanyan, 1954; Hollande and Enjumet, 1956). This observation is consistent with the proposed osmoregulatory function of the contractile vacuole, since this organelle would be superfluous in environments where the osmotic pressure of the external medium presumably approximates that of the cell. It is interesting to note that a Golgi apparatus is present near the anterior of the nucleus in *Chattonella subsalsa*: its function appears to be the production of material for the mucocysts (Mignot, 1976).

TRICHOCYSTS

Trichocysts or mucocysts occur in all genera of chloromonads. Trichocysts of *Gonyostomum* and *Merotricha* appear in the light microscope as rod-shaped structures which are often grouped together; for example, a battery of them occurs at the anterior end of the cell in *Gonyostomum semen* (Heywood, 1968) and *Merotricha bacillata* (Fott, 1968). In the former trichocysts are also distributed elsewhere in the cell (Drouet and Cohen, 1935) and may be grouped together in the posterior extremity of the cell (Heywood, 1968).

The structure and discharge of trichocysts has been examined in the most detail in *Gonyostomum semen* (Drouet and Cohen, 1935; Chadefaud, 1937; Hovasse, 1945; Mignot and Hovasse, 1975). They are membrane-limited structures approximately 2.0 to 2.5 μm in length and 0.5 to 0.6 μm in diameter (Fig. 10) consisting of osmiophilic fibrils originating in the Golgi apparatus (Heywood, in prep.) and packed so closely together it is frequently difficult to distinguish them in sections of undischarged trichocysts. In the laboratory trichocyst discharge may be induced by the addition of dyes to the medium (Drouet and Cohen, 1935).

But it is not known what triggers this release in natural circumstances, nor what function they may serve. Discharged trichocysts extend from the cell as thin mucilaginous threads which may reach 200 µm or more in length (Hovasse, 1945; Mignot and Hovasse, 1975). In electron microscope preparations it becomes apparent that each discharged trichocyst consisted of a complex bundle of interwoven fibrils (Mignot and Hovasse, 1975). Trichocysts bind with a variety of dyes (Drouet and Cohen, 1935) and heavy metal atoms (Mignot and Hovasse, 1975) but their exact chemical composition remains unknown.

Mucocysts of *Chattonella subsalsa* also consist of electron dense fibrils which probably originate in the Golgi apparatus (Mignot, 1976). Upon discharge the mucocysts extend for 30 µm as a bundle of fibrils which resembles those in the discharged trichocyst of *Gonyostomum semen*. Discharge of mucocysts in *Vacuolaria virescens* may create a layer of mucilage around a cell. Repeated cell divisions and continued production of mucilage leads to large numbers of palmelloid cells in an extensive gelatinous matrix.

NUCLEUS

The nuclei of chloromonads which may reach 20 µm or more in length occur either near the center of the cell or anteriorly to this. The fibrous root of the flagellar apparatus makes contact with the nuclear envelope at the anterior surface of the nucleus and an extensive Golgi apparatus also occurs in this region (Figs. 8, 11). The ultrastructure of the interphase nucleus has been described for *Chattonella subsalsa* (Mignot, 1976), *Gonyostomum semen* and *Vacuolaria virescens* (Mignot, 1967; Heywood, 1974a, 1976). The nuclear envelope (Fig. 9) resembles that of many other types of cells: there are inner and outer membranes and nuclear pores of diameter approximately 75 nm (Heywood, 1976). Two or more nucleoli are present; in *Gonyostomum semen* these are surrounded by peripheral heterochromatin but this is absent in *Chattonella subsalsa* (Mignot, 1976). Interphase chromatin is frequently visible as fine strands which may make contact with the nucleoli and the nuclear envelope. An investigation of some physical properties of chloromonad DNA revealed that the guanine plus cytosine (G+C) content is 35% and 34% in *Gonyostomum semen* and *Vacuolaria virescens* respectively (Rae, 1976). In *Gonyostomum semen* the nucleoplasm between the interphase chromatin contains structures of diameter 35-46 nm which are spherical to ellipsoidal in shape and which occur singly or in groups (Fig. 9). Their size is similar to that of perichromatin granules (Bouteille, Laval and Dupuy-Coin, 1974) but their cytochemistry and function remain unknown (Heywood, 1976).

Mitosis has been described for *Chattonella, Gonyostomum* and *Vacuolaria*. The most detailed accounts exist for *Vacuolaria virescens* (Dangeard, 1939; Poisson and Hollande, 1943; Mignot, 1967; Spencer, 1971; Heywood and Godward, 1972, 1973, 1974; Heywood, 1973c, 1973d, 1978a). It therefore seems appropriate to describe mitosis in this organism and then to compare this with information from other chloromonads.

Mitosis occurs in both motile and palmelloid cells of *Vacuolaria virescens*. Microtubules appear near the base of the flagella where they appear to be involved in the separation of the flagellar basal bodies. This stage is most appropriately referred to as preprophase since the chromosomes, nucleoli and nuclear envelope still possess their interphase morphology. At about this time a second contractile vacuole becomes apparent; the Golgi apparatus underlying the contractile vacuoles becomes extended and finally divides in two. A Golgi apparatus and contractile vacuole then become associated with each of the poles of the nucleus, and remain in this position throughout the subsequent stages of mitosis (Figs. 13, 14) and cytokinesis.

During prophase condensation of the chromosomes occurs and the nucleoli begin to disperse although fragments of these structures are frequently distinguishable as late as telophase. Spindle microtubules enter the nucleus at the poles through gaps in the nuclear envelope. Initially spindle microtubules lie close to the nuclear envelope but later in prophase they extend across the center of the nucleus towards the chromosomes. Aside from the polar gaps the nuclear envelope remains intact until telophase but its configuration changes throughout mitosis. For example, the nucleus is irregularly-shaped during prophase but it subsequently becomes elliptical with the metaphase plate of chromosomes extending across its longitudinal axis (Fig. 13). At this stage, frequently there are folds in the nuclear envelope near the equator and poles. As mitosis progresses into anaphase and telophase the nucleus becomes elongated and the two groups of progeny chromosomes move apart. This process continues

Figs. 13 and 14. Mitosis in *Vacuolaria virescens*. Fig. 13. The metaphase plate of chromosomes (CH) extends across the longitudinal axis of this elliptically shaped nucleus. At one pole there is a Golgi apparatus (G) and contractile vacuole (CV). At the other pole the contractile vacuole is not within the plane of section and a series of subsidiary vacuoles (asterisks) occur adjacent to the Golgi apparatus (G) (x5,500). Fig. 14. Longitudinal section through a telophase nucleus. New nuclear envelope is forming at the poleward surfaces of the progeny nuclei (short arrows), and the original nuclear envelope is beginning to disperse. A Golgi apparatus (G) occurs at each pole; a contractile vacuole (CV) is adjacent to one of the Golgi regions (x5,500). (Reproduced from Heywood, 1978a by kind permission of the Journal of Cell Science.)

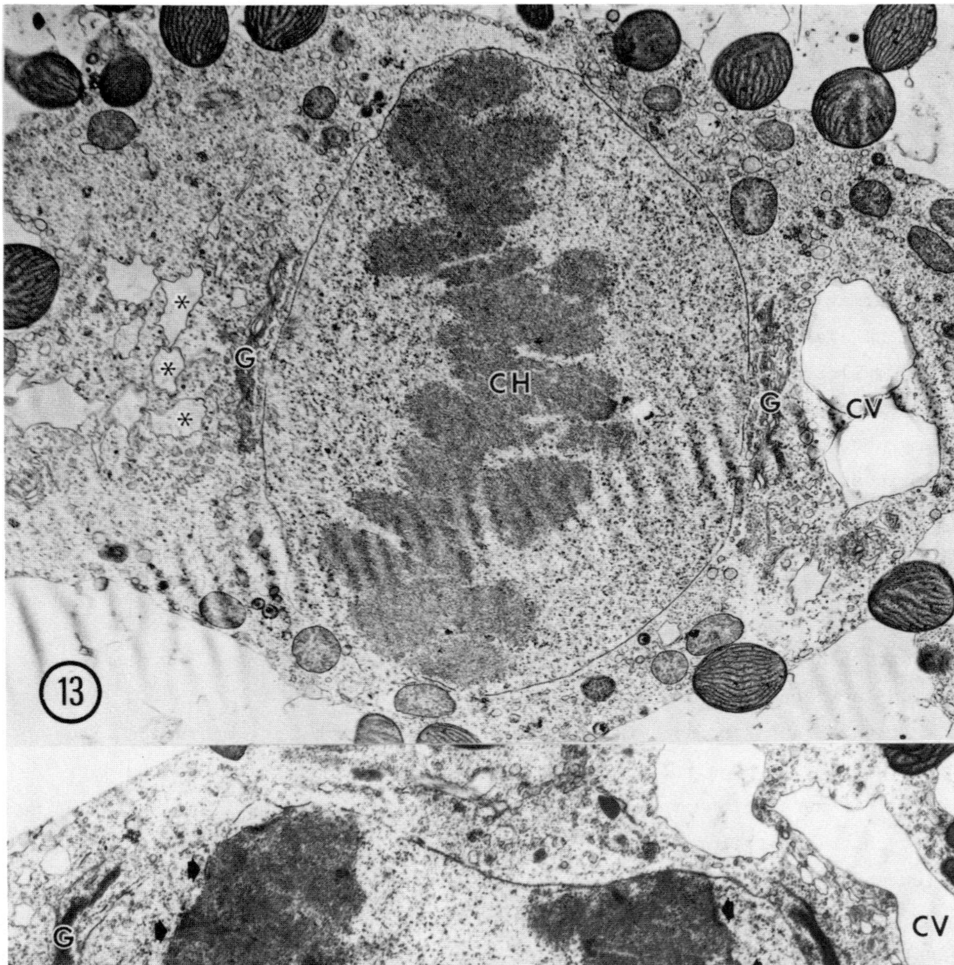
*
*
*
G
CH
G
CV
13

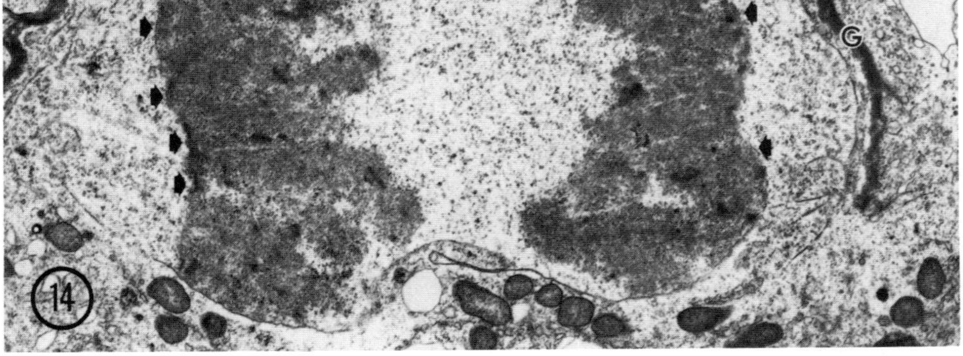
G
CV
G
14

until mid-telophase at which time the progeny chromosomes occur near opposite ends of a nucleus which is roughly cylindrical in shape (Fig. 14). At this stage the surface area of the nucleus is considerably greater than at the end of the preceding interphase. It has been suggested (Heywood, 1978a) that additions to the nuclear envelope early in mitosis result in the irregular surface of the nucleus at prophase and the folds in the nuclear envelope at metaphase. This additional nuclear envelope permits elongation of the nucleus at anaphase and telophase.

Contraction of the chromosomes occurs throughout prophase and becomes maximal during metaphase when the chromosomes vary in size from dot-like structures of 1 μm diameter to rod-shaped structures which reach up to 12 μm in length. The smaller chromosomes occur near the center of the metaphase plate and the longer chromosomes radiate outwards from this area. Centromeres are visible on many chromosomes as a constricted region which stains less densely than the remainder of the chromosome. The chromosome is frequently bent at the centromere, and this region is also recognizable at anaphase when it leads the chromosome in its poleward movement. In the electron microscope the kinetochore is distinguishable as a region of amorphous material which is approximately 0.3 μm in diameter (Heywood and Godward, 1972). Several chromosomal microtubules may be attached to each kinetochore. The nucleus also contains polar microtubules which lack connections with kinetochores.

At the transition from metaphase to anaphase chromatid separation occurs and the two groups of progeny chromosomes move apart parallel to the division axis. As separation proceeds the chromosome groups become tilted with the result that late telophase cells usually show one chromosome group in lateral view and the other in an approximately polar view; that is, the groups of progeny chromosomes form an angle of just less than 90° to each other. This change in the orientation of the chromosome groups during anaphase and telophase is illustrated by Figs. 7-10 in Poisson and Hollande (1943) and Figs. 13, 15 in Heywood and Godward (1974). During anaphase the poleward surfaces of the chromosome groups are flattened and the chromosomes are longer and less densely stained than at metaphase. Decondensation of the chromosomes continues during telophase, and at late telophase the nucleoli begin to reform.

During anaphase and early telophase membrane vesicles occur within the nucleus between the poles and the poleward surfaces of the chromosomes. These vesicles presumably originate in the cytoplasm and enter the nucleus by the polar gaps. Most vesicles are approximately spherical in shape but those in contact with the chromosomes are usually flattened. Fusion between vesicles on the surfaces of the chromosomes has been observed: presumably this process ultimately gives rise to the formation of new nuclear envelope on the poleward surfaces of the chromosomes (Fig. 14). As pointed out previously (Heywood, 1978a), membrane vesicles apparently contribute in the formation of nuclear envelopes in a variety of organisms. In *Vacuolaria virescens* the vesicles are probably cytoplasmic in origin rather than being derived from the original nuclear envelope since this is intact over most of its surface (except for the polar gaps) at the time of vesicle formation. However, some portions of this original nuclear envelope are incorporated into the nuclear envelopes of the progeny nuclei.

Several aspects of nuclear division in *Vacuolaria virescens* occur in many other organisms, for example, the extranuclear origin of spindle microtubules from the region of the flagellar bases, the presence of kinetochores and a metaphase plate of chromosomes, and the behavior of the chromosomes and nucleoli. But the persistence until telophase of the nuclear envelope in an almost intact condition and the formation of a new nuclear envelope within this original structure are highly unusual. It is thought that this may be a mechanism to guarantee that each of the cells receives a Golgi apparatus and contractile vacuole. That is, the duplication of these organelles early in mitosis and their close association with the nuclear envelope at the poles of the dividing nucleus (Figs. 13, 14) ensures that they occur at opposite ends of the cell. They remain there throughout cytokinesis and are partitioned to progeny cells with the nuclei when the division furrow divides the original cell in two. The involvement of mitosis in partitioning cytoplasmic components has been postulated on other occasions; for example, it is thought that the location of centrioles at the poles of the division spindle ensures their distribution to the progeny (Pickett-Heaps, 1971). A similar mechanism may exist for the segregation of microbodies to progeny cells since an association between these organelles and the nuclear envelope occurs during mitosis (Heywood, 1978a).

Fott's account of *Vacuolaria viridis* (Fott, 1935) indicates that mitosis in this species resembles that in *Vacuolaria virescens*; for example, the nuclear envelope is clearly present at metaphase, and the arrangement of chromosomes at metaphase, anaphase and telophase is similar in both species. Likewise, the light microscope appearance of mitotic stages in *Gonyostomum semen* (Hovasse, 1945; Heywood, 1974b) resembles those in *Vacuolaria virescens*. The nuclear envelope remains intact during mitosis in *Gonyostomum semen* and the contractile vacuoles occur on opposite sides of the dividing nucleus (Hovasse, 1945). The apparent dispersion of the nuclear envelope during mitosis in *Gonyostomum semen* and *Vacuolaria virescens* (Mignot, 1967) may have been an artifact of preparation. There is less information on mitosis in *Chattonella subsalsa* (Beicheler, 1936b) but from the diagrams it appears that mitosis is intranuclear; the "granular caps" occurring at opposite ends of the dividing nucleus are almost certainly the Golgi regions. There do not appear to be any accounts of nuclear division in the genus *Merotricha*.

The chromosome number of *Vacuolaria viridis* has been estimated to be approximately 30 (Fott, 1935) while in *Gonyostomum semen* there appear to be 65-75 chromosomes (Heywood, 1974b). The chromosome number of *Vacuolaria virescens* has been variously estimated as approximately 20 (Poisson and Hollande, 1943; Spencer, 1971) and 97±2 (Heywood and Godward, 1973). The latter determination was made on cells which had been in culture for several years and consequently it is possible that they were polyploid. Alternately, the lower estimates of chromosome number may reflect the difficulty in distinguishing all the chromosomes if the preparations were not sufficiently spread out. This issue should be resolved by detailed cytological investigation of natural populations.

In laboratory cultures of *Vacuolaria virescens* at 24° C the mitotic wave begins before the onset of the dark period and reaches a peak at approximately two hours after the beginning of the dark period. Mitosis lasts for approximately 2-2 1/2 hours and cytokinesis from 80 minutes to 4 hours (Heywood, 1968). Mitosis in *Chattonella subsalsa* also occurs at night and lasts for 35-45 minutes at 22-24° C (Biecheler, 1936b). Division stages have also been reported in cells of *Gonyostomum semen* examined during the dark period (Mignot, 1967; Heywood, 1974b); in this organism cytokinesis may last for several hours (Hovasse, 1945).

Meiosis has not yet been demonstrated in the chloromonads; that is, there are no reports of synaptonemal complexes, or observations of meiotic stages, or genetic evidence for meiosis (Heywood and Magee, 1976). A possible instance of sexual reproduction in *Chattonella subsalsa* was reported by Subrahmanyan (1954). He observed that two cells became attached anteriorly, stopped swimming and merged into a single binucleate individual which lost its flagella and became surrounded by a wall. He described this structure as a zygote but did not observe its subsequent fate, or fusion of the nuclei, or meiosis. This is insufficient evidence to draw any conclusions about sexual reproduction and the type of life cycle in the chloromonads. However, lack of evidence for meiosis should not be construed to indicate its absence from the chloromonads: it is more likely that this reflects the lack of detailed observations on natural populations and the absence of attempts to induce meiosis in laboratory conditions.

HABITAT

Chattonella subsalsa occurs in saline habitats where it may be periodically abundant. Collections made approximately 40 years apart from the same location near Sète in southern France revealed its presence during the months of August, September and early October (Biecheler, 1936a; Mignot, 1976). This habitat consists of saline canals which are rich in organic matter. During 1956 *Chattonella subsalsa* occurred in the water of the port of Algiers from late July to mid-August when the high cell density (between 1 million and 2 million cells per litre) imparted a rust color to the water (Hollande and Enjumet, 1956). The water was more saline (37 grams/litre) than in the French location (25 grams/litre), but it was similarly rich in organic matter. Hollande and Enjumet (1956) speculated that, since the water always contained high levels of nutrients, the bloom was probably caused by the high temperature (24° C) and the slight disturbance of the water. *Chattonella subsalsa* was also present in the port of Barcelona from July to September during the years 1963-1966 (Margalef, 1968). The usual density was 1,300-5,300 cells per millilitre. Although this species was rarely dominant, on one occasion it occurred at a density of 38,160 cells per millilitre and formed the major component of a red discoloration of the water. Subrahmanyan (1954) described a marine chloromonad occurring in the sea near Kozhikode, southern India. He named this organism *Hornellia marina* but it is widely presumed to have been *Chattonella subsalsa*. This organism occurred every year from 1949 to 1953: all records are from the months of August to December with the exception of one occurrence in April. The species was comparatively infrequent in some of the collections but at other times it occurred in such densities that the water was colored green or brownish-green. On these occasions *Chattonella subsalsa* was present to the exclusion of most of the other plankton, and when the organisms in 4 litres of sea water were allowed to settle out, it was found that they occupied a volume of 1 1/2 litres. Subrahmanyan pointed out that other algal blooms on this coast may have been due to this species and speculated that it may cause mortality amongst the marine life.

Gonyostomum semen is widely distributed throughout Europe and North America (Smith, 1933; Drouet and Cohen, 1935, 1937; Hovasse, 1945; Skuja, 1948; Sörensen, 1954; Chapman and Haxo, 1966; Fott, 1968). It usually occurs in bogs, pools or lakes of pH 4.4-6.2 (Drouet and Cohen, 1935, 1937; Sörensen, 1954; Heywood, 1968). The habitats often contain *Sphagnum*, for example, the Cedar

Swamp near Woods Hole, Massachusetts, U.S.A. (Drouet and Cohen, 1935), Austwick, Moss and Helwith Moss in West Yorkshire, U.K. (Smith, 1933), and locations in Sweden (Sörensen, 1954) and Czechoslovakia (Fott and Nováková, 1965; Heywood, 1968). *Gonyostomum semen* was usually collected between the months of April and October and at times it was present in large numbers.

Gonyostomum latum was reported from small, shallow ponds in South Bohemia, Czechoslovakia (Fott, 1949). These pools had a pH of 6.7-7.0 and contained iron bacteria, floating aquatic vegetation, and two other species of chloromonads, *Merotricha bacillata* and *Vacuolaria viridis*. *Gonyostomum latum* (originally described as *Vacuolaria depressa*) was collected from ponds in Germany where it often occurred amongst water plants (Lauterborn, 1899).

Gonyostomum intermedium (Skuja, 1956) and *Gonyostomum ovatum* (Fott, 1952) are rare species which were described from plankton samples collected in Sweden and Czechoslovakia, respectively. *Gonyostomum ovatum* occurred in water containing abundant *Sphagnum* and having a pH of 3.2-5.0 (Fott, 1952).

Vacuolaria virescens has been collected in Europe and North America from ponds (Poisson and Hollande, 1943; Spencer, 1971), bogs (Tschermak-Woess, 1954; Graffius, 1966) and lakes (Skuja, 1948; Graffius, 1966). Its habitats are frequently rich in vegetation. This species appears to be tolerant of pH conditions since it has been recorded from both acid and alkaline water; for example, in Michigan, U.S.A., it occurred in a bog of pH 4.0-6.5 and a lake of pH 7.2-8.3 (Graffius, 1966). *Vacuolaria virescens* may be present in the plankton, or amongst the aquatic plants, or immediately above the bottom mud. This species has frequently been collected in the autumn or in the late winter and spring (Skuja, 1926; Poisson and Hollande, 1943; Spencer, 1971; Droop, personal communication). This may reflect both the higher nutrient levels at those times and the ability of *Vacuolaria virescens* to tolerate low temperatures. This latter property is well illustrated by its occurrence as macroscopic masses of palmelloid cells at the surface of a North Carolina pond in places where the thin layer of ice had been broken (Spencer, 1971). And in Cheshire, England, it became more frequent as autumn progressed attaining large numbers by late December (Droop, personal communication). *Vacuolaria virescens* variety *minuta* and variety *praegnans* occur in fish ponds in Swedish Lappland (Skuja, 1964).

Vacuolaria viridis was collected in March, 1889 from a swamp near Caen, France, where it occurred amongst plants of *Chara* (Dangeard, 1939). It was subsequently reported from two locations near Lnáře in South Bohemia, Czechoslovakia: one habitat was a deep woodland pool that was rich in bog plants

(Fott, 1935) and the other was the habitat of *Gonyostomum latum* as described earlier.

Vacuolaria penardii has been collected in Switzerland and Czechoslovakia but there are no details of the habitats (Penard, 1921; Fott, 1970). The organism described as *Trentonia flagellata* (Graffius, 1966) was probably *Vacuolaria penardii*; this species occurred in an acid bog (pH 4.0-6.5) in Michigan.

Members of the genus *Vacuolaria* have been reported from swamps, lakes, ponds and artificial bodies of water in and around São Paulo, Brazil (Skvortzov and Noda, 1968). This is a sub-tropical region at an elevation of 600-700 metres above sea level. The habitats were rich in organic material, and the organisms were most frequent during the coldest part of the year. Subsequently, a new species, *Vacuolaria tropicalis*, was described from an artificial pond (pH 5.0) in the São Paulo Botanical Garden, Brazil, where it occurred from December to February of the years 1968-1969 (Bicudo and Cardoso, 1973).

Merotricha bacillata has been reported from Russia where it occurred at the mouth of a river at a site which was rich in vegetation (Mereschowsky, 1879), Latvia, where it was abundant in the plankton above the mud (Skuja, 1934) and Czechoslovakia where it was present in ponds containing *Gonyostomum latum* and *Vacuolaria viridis* (Fott, 1949). In the Western Hemisphere this species occurred in a reservoir and artificual ponds in São Paulo, Brazil (Skvortzov, Bicudo and Bicudo, 1969), an acid bog (pH 4.0-6.5) in Michigan, U.S.A (Graffius, 1966) and a sphagnum bog in Massachusetts, U.S.A (Palmer, 1942).

CONCLUDING REMARKS

Investigations of the chloromonads have entered a challenging phase. Information has been obtained on the nutrition, biochemistry and ultrastructure of two species (*Gonyostomum semen* and *Vacuolaria virescens*) and the ultrastructure of a third species (*Chattonella subsalsa*). But even this information is incomplete; for example, the mechanism of mitosis has been studied in detail in only one species, and biochemical investigations have been confined to the chloroplast pigments and nuclear DNA. It is therefore apparent that future research must be directed towards a more comprehensive coverage of these three species as well as investigating other chloromonads that have been largely neglected thus far.

There is limited information on the ecology of the chloromonads and on some aspects of their environmental physiology such as phototaxis, diurnal migration, chloroplast movement and mitotic periodicity. But these findings are preliminary and much more detailed information is needed. Other aspects of the chloromonads remain obscure; for example, there do not appear to be any descriptions of their fossil record, life histories and meiosis. However, some of the gaps in our knowledge of the chloromonads may soon disappear. The publication within the last 15 years of approximately 20 papers dealing with the biochemistry and ultrastructure of the chloromonads is an indication that these fascinating organisms have aroused the interest of phycologists: there can be little doubt that they will receive even more attention in the future.

REFERENCES

Aaronson, S. and U. Behrens. 1974. Ultrastructure of an unusual contractile vacuole in several chrysomonad phytoflagellates. J. Cell Sci. 14: 1-9.

Atkinson, A. W., P. C. L. John and B. E. S. Gunning. 1974. The growth and division of the single mitochondrion and other organelles during the cell cycle of *Chlorella*, studied by quantitative stereology and three-dimensional reconstruction. Protoplasma 81: 77-109.

Bicudo, C. E. M. and M. B. Cardoso. 1973. *Vacuolaria tropicalis*, a new species of chloromonad from southern Brazil. Br. Phycol. J. 8: 339-342.

Biecheler, B. 1936a. Sur une chloromonadine nouvelle d'eau saumatre *Chattonella subsalsa* n. gen., n. sp. Archs. Zool. Exp. Gén. 78: 79-83.

Biecheler, B. 1936b. Sur la division de la Chloromonadine *Chattonella subsalsa* Biecheler. C. R. Soc. Biol. 123: 1126-1127.

Bouck, G. B. 1969. Extracellular microtubules. The origin, structure and attachment of flagellar hairs in *Fucus* and *Ascophyllum* antherozoids. J. Cell Biol. 40: 446-460.

Bouteille, M., M. Laval and A. M. Dupuy-Coin. 1974. Localization of nuclear functions as revealed by ultrastructural autoradiography and cytochemistry. In Busch, H. [Ed.] *The Cell Nucleus*. Academic Press, New York, Vol. 1. pp. 3-71.

Chadefaud, M. 1937. Sur l'organisation et les trichocystes de *Gonyostomum semen* (Ehr.) Diesing. C. R. Acad. Sci. 204: 1688-1690.

Chapman, D. J. and F. T. Haxo. 1966. Chloroplast pigments of Chloromonadophyceae. J. Phycol. 2: 89-91.

Christensen, T. 1964. The Gross Classification of Algae. In Jackson, D. F. [Ed.] *Algae and Man*. Plenum Press, New York. pp. 59-64.

Cienkowsky, L. 1870. Ueber Palmellaceen und einige Flagellaten. Arch. mikrosk. Anat. 7: 421-436.

Cowles, R. P. and C. E. Brambel. 1936. A study of the environmental conditions in a peat bog with special reference to the diurnal vertical distribution of *Gonyostomum semen*. Biol. Bull. Mar. Biol. Lab., Woods Hole 71: 286-298.

Dangeard, P. A. 1939. Nouvelles observations sur la *Vacuolaria virescens* Cienk. Botaniste 29: 311-334.

Dodge, J. D. 1973. The Fine Structure of Algal Cell. Academic Press, New York.

Drouet, F. and A. Cohen. 1935. The morphology of *Gonyostomum semen* from Woods Hole, Massachusetts. Biol. Bull. Mar. Biol. Lab., Woods Hole 68: 422-439.

Drouet, F. and A. Cohen. 1937. Further observations on *Gonyostomum semen*. Bot. Gaz. 98: 617-618.

Fott, B. 1935. Über den inneren Bau von *Vacuolaria viridis* (Dangeard) Senn. Arch. Protistenk. 84: 242-250.

Fott, B. 1949. *Corone*, a new genus of colonial Volvocales. Mém. Soc. r. Sci. Bohème 2: 1-9.
Fott, B. 1952. Mikroflora oravských rašelin. Preslia 24: 189-209.
Fott, B. 1968. VIII. Klasse:Chloromonadophyceae. In Huber-Pestalozzi, G. G. [Ed.] *Das Phytoplankton des Süsswassers*. 2nd ed. Vol 3. Stuttgart. pp. 79-93.
Fott, B. 1970. Taxonomische Übertragungen und Namensänderungen unter den Algen III. Chloromonadophyceae. Preslia 42: 16-20.
Fott, B. and M. Nováková. 1965. On the occurrence of a dystrophic alga *Phacomyxa sphagnophila* Skuja. Int. Revue Ges. Hydrobiol. 50: 85-89.
Fritsch, F. E. 1935. *The Structure and Reproduction of the Algae*. Cambridge Univ. Press, Cambridge, Vol. 1, pp. 721-723.
Gibbs, S. P., L. L. Chu and C. Magnussen. Evidence that *Olisthodiscus luteus* is a member of the Chrysophyceae. Phycologia (in press).
Graffius, J. H. 1966. Additions to our knowledge of Michigan Pyrrhophyta and Chloromonadophyta. Trans. Am. Microsc. Soc. 85: 260-270.
Grobe, B. and C. G. Arnold. 1975. Evidence of a large, ramified mitochondrion in *Chlamydomonas reinhardii*. Protoplasma 86: 291-294.
Guillard, R. R. L. and C. J. Lorenzen. 1972. Yellow-green algae with chlorophyllide c. J. Phycol. 8: 10-14.
Heath, I. B., A. D. Greenwood and H. B. Griffiths. 1970. The origin of flimmer in *Saprolegnia*, *Dictyuchus*, *Synura* and *Cryptomonas*. J. Cell Sci. 7: 445-461.
Heath, I. B. and W. M. Darley. 1972. Observations on the ultrastructure of the male gametes of *Biddulphia levis* Ehr. J. Phycol. 8: 51-59.
Heywood, P. 1968. Studies on the Chloromonads. Ph.D. Thesis, University of London.
Heywood, P. 1972. Structure and origin of flagellar hairs in *Vacuolaria virescens*. J. Ultrastruct. Res. 39: 608-623.
Heywood, P. 1973a. Intracisternal microtubules and flagellar hairs of *Gonyostomum semen* (Ehrenb.) Diesing. Br. Phycol. J. 8: 43-46.
Heywood, P. 1973b. Nutritional studies on the Chloromonadophyceae: *Vacuolaria virescens* and *Gonyostomum semen*. J. Phycol. 9: 156-159.
Heywood, P. 1973c. Observations on the nuclear cytology of *Vacuolaria virescens*. Br. Phycol. J. 8: 211A.
Heywood, P. 1973d. Ultrastructure of mitosis in *Vacuolaria virescens*. J. Phycol. 9: 15A.
Heywood, P. 1974a. Observations on interphase and mitotic nuclei of *Gonyostomum semen* (Ehrenb.) Diesing. Br. Phycol. J. 9: 220A.
Heywood, P. 1974b. Mitosis and cytokinesis in the chloromonadophycean alga *Gonyostomum semen*. J. Phycol. 10: 355-358.
Heywood, P. 1974c. Occurrence of microbodies in chloromonadophycean algae. Arch. Microbiol. 99: 265-269.
Heywood, P. 1976. Ultrastructural characterization of the interphase nucleus of *Gonyostomum semen*. Cytobios 17: 79-86.
Heywood, P. 1977a. Chloroplast structure in the chloromonadophycean alga *Vacuolaria virescens*. J. Phycol. 13: 68-72.
Heywood, P. 1977b. Evidence from serial sections that some cells contain large numbers of mitochondria. J. Cell Sci. 26: 1-8.
Heywood, P. 1978a. Ultrastructure of mitosis in the chloromonadophycean alga *Vacuolaria virescens*. J. Cell Sci. 31: 37-51.
Heywood, P. 1978b. Osmoregulation in the alga *Vacuolaria virescens*. Structure of the contractile vacuole and the nature of its association with the Golgi apparatus. J. Cell Sci. 31: 213-224.
Heywood, P. and M. B. E. Godward. 1972. Centromeric organization in the chloromonadophycean alga *Vacuolaria virescens*. Chromosoma 39: 333-339.
Heywood, P. and M. B. E. Godward. 1973. Chromosome number and morphology in *Vacuolaria virescens* (Chloromonadophyceae). Ann. Bot. 37: 423-425.

Heywood, P. and M. B. E. Godward. 1974. Mitosis in the alga *Vacuolaria virescens*. Am. J. Bot. 61: 331-338.

Heywood, P. and P. T. Magee. 1976. Meiosis in Protists. Some structural and physiological aspects of meiosis in algae, fungi, and protozoa. Bact. Rev. 40: 190-240.

Hollande, A. and M. Enjumet. 1956. Sur une invasion des eaux du port d'Alger par *Chattonella subsalsa* (=*Hornellia marina* Sub.) Biecheler. Remarques sur la toxicite de cette Chloromonadine. Bull. Stn. Agric. Pêche Castiglione (N.S.) 8: 271-280.

Hovasse, R. 1945. Contribution à l'étude des Chloromonadines: *Gonyostomum semen* Diesing. Archs. Zool. Exp. Gén. 84: 239-269.

Kirby, H. 1950. *Materials and Methods in the Study of Protozoa*. Cambridge Univ. Press, London.

Klein, R. M. and A. Cronquist. 1967. A consideration of the evolutionary and taxonomic significance of some biochemical, micromorphological and physiological characters in the Thallophytes. Q. Rev. Biol. 42: 105-296.

Koch, W. and E. Schnepf. 1967. Einige elektronenmikroskopische Beobachtungen an *Vacuolaria virescens* Cienk. Arch. Mikrobiol. 57: 196-198.

Lauterborn, R. 1899. Protozoen-Studien IV. Flagellaten aus dem Gebiete des Oberrheins. Z. Wiss. Zool. 65: 369-388.

Leedale, G. F., B. S. C. Leadbeater and A. Massalski. 1970. The intracellular origin of flagellar hairs in the Chrysophyceae and Xanthophyceae. J. Cell Sci. 6: 701-719.

Loeblich, A. R. and K. E. Fine. 1977. Marine chloromonads: more widely distributed in neritic environments than previously thought. Proc. Biol. Soc. Wash. 90: 388-399.

Margalef, R. 1968. Présence de *Chattonella subsalsa* Biechler dans le port de Barcelone. Rapp. Comm. Int. Mer. Médit. 19: 581-582.

McKanna, J. A. 1976. Fine structure of fluid segregation organelles of *Paramecium* contractile vacuoles. J. Ultrastruct. Res. 54: 1-10.

Mereschkowsky, C. 1879. Studien über Protozoen des nördlichen Russland. Arch. Mikrosk. Anat. 16: 153-248.

Mignot, J. P. 1967. Structure et ultrastructure de quelques Chloromonadines. Protistologica 3: 5-23.

Mignot, J. P. 1976. Compléments à l'étude des Chloromonadines. Ultrastructure de *Chattonella subsalsa* Biecheler flagellé d'eau saumatre. Protistologica 12: 279-293.

Mignot, J. P. and R. Hovasse. 1975. Sur la présence d'organites de type toxicystes chez le Protiste flagellé *Colponema loxodes* et leur comparaison avec les trichocystes de *Gonyostomum semen*. Ann. Stat. Biol. Besse-en-Chandesse 9: 201-211.

Palmer, C. M. 1942. Additional records for algae, including some of the less common forms. Butler Univ. Bot. Stud. 5: 224-234.

Penard, E. 1921. Studies on some flagellata. Proc. Acad. Nat. Sci. Philad. 73: 105-168.

Pickett-Heaps, J. D. 1971. The autonomy of the centriole: fact or fallacy? Cytobios 3: 205-214.

Poisson, R. and A. Hollande. 1943. Considérations sur la cytologie, la mitose et les affinités des Chloromonadines. Étude de *Vacuolaria virescens* Cienk. Annls. Sci. Nat. (Zool.) Sér. II 5: 147-160.

Provasoli, L. 1958. A catalogue of laboratory strains of free-living and parasitic protozoa. J. Protozool. 5: 1-38.

Rae, P. M. M. 1976. Hydroxymethyluracil in eukaryote DNA: a natural feature of the Pyrrophyta (Dinoflagellates). Science 194: 1062-1064.

Ragan, M. A. and D. J. Chapman. 1978. *A Biochemical Phylogeny of the Protists*. Academic Press, New York.

Schnepf, E. and W. Koch. 1966a. Golgi-Apparat und Wasserausscheidung bei *Glaucocystis*. Z. PflPhysiol. 55: 97-109.
Schnepf, E. and W. Koch. 1966b. Über die Entstehung der pulsierenden Vacuolen von *Vacuolaria virescens* (Chloromonadophyceae) aus dem Golgi-Apparat. Arch. Mikrobiol. 54: 229-236.
Skuja, H. 1926. Vorarbeiten zu einer Algenflora von Lettland. I. Acta Horti Bot. Univ. Latv. 1: 33-54.
Skuja, H. 1934. Beitrag zur Algenflora Lettlands. I. Acta Horti Bot. Univ. Latv. 7: 25-86.
Skuja, H. 1948. Taxonomie des Phytoplanktons einiger Seen in Uppland, Schweden. Symb. Bot. Upsal. 9: 1-399.
Skuja, H. 1956. Taxonomische und biologische Studien über das Phytoplankton Schwedischer Binnengewässer. Nova Acta R. Soc. Scient. Upsal. Ser. 4, 16: 1-404.
Skuja, H. 1964. Grundzüge der Algenflora und Algenvegetation der Fjeldgegenden um Abisko in Schwedisch-Lappland. Nova Acta R. Soc. Scient. Upsal. Ser. 4, 18: 1-465.
Skvortzov, B. V., C. E. M. Bicudo and R. M. T. Bicudo. 1969. First report of the occurrence of Chloromonadophyceae in Brazil. Rickia 4: 93-98.
Skvortzov, B. V. and M. Noda. 1968. On Brazilian and European species of genus *Vacuolaria*. J. Jap. Bot. 43: 69-76.
Smith, A. M. 1933. *Gonyostomum semen* Diesing. A flagellate now first recorded for the British Isles. Naturalist, Hull, 1933: 49-50.
Smith, G. M. 1950. *The Freshwater Algae of the United States*, 2nd ed. McGraw-Hill, New York.
Sörensen, I. 1954. *Gonyostomum semen* (Ehrenb.) Diesing - en vattenorganism av teoretisk och praktiskt intresse. Svensk Faun. Revy. 16: 47-52.
Spencer, L. B. 1971. A study of *Vacuolaria virescens* Cienkowski. J. Phycol. 7: 274-279.
Stokes, A. C. 1886. Notices of new freshwater Infusoria. Proc. Am. Phil. Soc. 23: 562-568.
Stokes, A. C. 1888. A preliminary contribution toward a history of freshwater Infusoria of the United States. J. Trenton Nat. Hist. Soc. 1: 71-344.
Subrahmanyan, R. 1954. On the life-history and ecology of *Hornellia marina* gen. et sp. nov., (Chloromonadineae), causing green discoloration of the sea and mortality among marine organisms off the Malabar Coast. Indian J. Fish. 1: 182-203.
Taylor, F. J. R. 1976. Flagellate phylogeny: a study in conflicts. J. Protozool. 23: 28-40.
Tschermak-Woess, E. 1954. Das sogenannte Alveolarplasma und die Schleimbildung bei *Vacuolaria virescens*. Öst. Bot. Z. 101: 328-333.

Copyright 1980 by Elsevier North Holland, Inc.
Cox, ed. Phytoflagellates

PHOTOSYNTHETIC CRYPTOPHYTES

ELISABETH GANTT
Radiation Biology Laboratory, Smithsonian Institution, 12441 Parklawn Drive, Rockville, Maryland 20852, U.S.A.

ABSTRACT

The colored cryptophytes are distinct from other algal groups by a unique periplast surrounding the cell, by the presence of chlorophyll *a* and c_2 and by the phycobiliproteins that are located in the intrathylakoidal lumens of the photosynthetic lamellae. Most species are obligate photoautotrophs and exist in marine to freshwater environments. Presence of a nucleomorph in the chloroplast-nuclear space may also prove to be a unique feature. Of the two flagella, one has two rows of mastigonemes and the other has one row. Large ejectosomes in the vestibular region, and small ones located in periplast grooves consist of two coiled ribbons. On ejection the main ribbon unwinds and becomes a rigid tube which is connected by a "hinge" to the small ribbon. During mitosis a barrel-shaped chromosome mass is surrounded by a sheath of endoplasmic reticulum, which is traversed by spindle microtubules. The spindle is short and apparently lacks polar structures. Cell division occurs by formation of cytoplasmic furrows at right angles to the spindle without involvement of microtubules. Sexual stages have not been identified. Symbiosis with other organisms is rare. A taxonomic reevaluation of members within this group is necessary, and should include comparative morphological and pigment studies made on clones grown under identical culture conditions.

INTRODUCTION

The cryptophyte algae are a relatively small group whose members, although recognized since the late 1800's (Pringsheim, 1944), have not been extensively studied. All species thus far described are small, single-celled biflagellates. Some of them possess chloroplasts and are photosynthetically active, while others lacking a chloroplast are colorless and exist as obligate heterotrophs. This chapter will deal primarily with the cryptophyte species which contain chloroplasts and exist as photoautotrophs or photoheterotrophs.

From a phylogenetic standpoint, it is interesting to consider how closely related the colorless forms are to the colored ones and which of them is more primitive. Both groups contain the same essential morphological features, but because the colorless forms lack plastids, apparently without even a remnant of what may be considered a vestigial plastid, they will not be covered here. Instead the reader is referred to selected references on their structural characteristics which are well delineated by Anderson (1962); Joyon (1963); Mignot (1965, 1967); Mignot *et al.* (1968, 1970); Schuster (1968).

Butcher (1967) prepared the most comprehensive treatise on the colored cryptophytes, in which the two largest genera were *Chroomonas* and *Cryptomonas*. He was the first to include ultrastructural characteristics in classification. Because of the small size of the unicells, it is apparent that more extensive electron microscopic investigations will be crucial in further studies and in re-evaluations of the species status within this group.

CHARACTERISTICS OF COLORED CRYPTOPHYTES.

General Characteristics.

1. Cells are generally asymmetric with a slipper or bean shape, often with a dorsi-ventral flattening. The cell sizes range from 2-20 μm long by 3-16 μm wide.
2. Two flagella of equal, or almost equal, length emerge from a vestibular region, usually near the anterior end. Flagellar propulsion occurs with the anterior end leading.
3. A periplast, containing thin plate thickenings arranged in regular rows, surrounds the entire cell up to the vestibular region.
4. Ejectosomes found in the cell periphery are small and those in the vestibular region are large.
5. A chloroplast-nuclear envelope encloses the chloroplast and nucleus.
6. Usually a single chloroplast is present with one or several pyrenoids. Its photosynthetic lamellae, occurring singly or in groups, contain electron dense material in the intrathylakoidal spaces.
7. Chlorophyll *a*, c_2 and phycobiliproteins are the usual photosynthetic pigments.
8. A nucleomorph, distinct from the nucleus, occurs between the chloroplast and the chloroplast-nuclear envelope.
9. Mitosis occurs with a partial breakdown of the nuclear envelope and has typical mitotic phases, but centrioles are lacking.
10. Sexual stages seem to be absent.

Most of the basic features of the cryptophytes are illustrated in Fig. 1. The cell consists of two major compartments. One compartment contains the nucleus, a chloroplast with its own limiting membrane, and a pyrenoid. This compartment is enclosed by the chloroplast nuclear envelope, also known as chloroplast-endoplasmic reticulum which occurs in many algal groups (Gibbs, 1962a, 1970). Its outer membrane has ribosomes attached and is continuous with the outer nuclear membrane. This compartment also contains a nucleomorph, starch, and endoplasmic reticulum. The second compartment contains an active Golgi apparatus, ejectosomes, the flagellar bases (not shown), rough endoplasmic reticulum, and a mitochondrial complex which extends throughout the cytoplasm (Santore and Greenwood, 1977). The depression at the flagellar end should be considered as a vestibular region and not as a gullet because evidence for any kind of direct ingestion or phagocytotic function is not available. The periplast with thin regular plates surrounds the cell up to the vestibular region. Mastigonemes are present in the chloroplast-nuclear envelope (perinuclear space).

Chloroplast Structure. A single chloroplast, which may be cup-shaped or divided into two or more lobes, is present in each cell where it often occupies a considerable portion of the cell volume. Only the endosymbiont of *Mesodinium rubrum* (Hibberd, 1977) has more than one chloroplast. The report of multiple chloroplasts in *Cyanomonas* (Pascher, 1913) was corrected by Rosenberg (1944) who pointed out that this alga was actually *Chroomonas nordstedtii*, which has one chloroplast.

The chloroplast structure is one of the most distinctive features of the cryptophytes and is essentially the same in *Hemiselmis*, *Chroomonas*, *Cryptomonas*, and *Rhodomonas* (Dodge, 1969; Gibbs, 1970; Lucas, 1970a, b; Wehrmeyer, 1970b; Gantt *et al.*, 1971; Hibberd, 1977). Four membranes surround each chloroplast, but only the inner two are the chloroplast limiting membranes. The thylakoids are unique because the intrathylakoidal space is filled by electron dense material (Fig. 2). Dodge (1969) and Wehrmeyer (1970b) initially suggested this as the site of the phycobiliproteins, for which Gantt *et al.* (1971) provided evidence by following pigment extraction and electron microscopy.

Thylakoid widths vary from 20-30 nm, but tend to be fairly uniform in the same cell. They are not fused into grana regions, but are sometimes very closely appressed (Glazer *et al.*, 1971). Pairing of the thylakoids is common; however, linear associations of thylakoids can vary from one to 10 or more. This is partly dependent on the species but mostly on the light conditions (Faust and Gantt, 1973; Vesk and Jeffrey, 1977). Branching of the thylakoid lamellae can occur, sometimes involving many neighboring lamella (Wehrmeyer, 1970b).

Eyespots are not universally present. Their occurrence seems to be mainly determined by culture conditions. When an eyespot is present it consists of one to several layers of osmiophilic bodies in the stroma region of the chloroplast (Fig. 2). Dodge (1969) in *Chroomonas mesostigmatica* noted about 35 osmiophilic bodies, with a diameter of 260 nm. He noted that these bodies were lined up against the chloroplast-limiting membrane facing toward the inside of the cell near a highly vesiculate region. A flattened vesicle often occurs appressed against the eyespot region. This has been found in *Cryptomonas calceiformis* (Lucas, 1970a), in *Chroomonas mesostigmatica* (Dodge, 1969) and *Chroomonas* sp. (Fig. 2). There does not appear to be a direct association of the eyespot with any other organelle.

The pyrenoid in the stroma region of the chloroplast is distinguishable by its homogeneous matrix and by the relative decrease of traversing thylakoids. Starch grains within the chloroplast-nuclear region often surround the pyrenoid, but they are outside of the chloroplast limiting membrane (Gibbs, 1962b) Thylakoid membranes traverse the pyrenoid region in most genera but not in *Mesodinium* (Hibberd, 1977) and *Cryptomonas* (Lucas, 1970a; Wehrmeyer, 1970b). In this latter genus, extensive cytoplasmic invaginations have been observed.

In *Mesodininium*, *Hemiselmis*, *Rhodomonas* and *Chroomonas* pyrenoids tend to be everted, i.e., to extend out of the chloroplast. Starch plates form around a pyrenoid thus pushing it out and resulitin in a stalked appearance. However, the stalked vs. non-stalked feature does not appear to be uniform enough within a population to serve as a species characteristic. Chemical analysis of starch produced by *Chroomonas salina* (Antia *et al.*, 1979), *Chilomonas* and other unicellular algae showed that it broadly resembles potato starch.

Photosynthetic pigments. Chlorophyll *a*, as in other algae, is the major photosynthetic pigment. Jeffrey (1976) established that chlorophyll c_2, but not chlorophyll c_1, occurred in 8 of 9 species of the genera *Chroomonas*, *Cryptomonas* and *Rhodomonas*. The exception was *Chroomonas mestostigmatica* which had both c_1 and c_2. In this regard, the cryptophytes are unlike the chrysophytes and haptophytes which have both forms.

Only the cryptophytes have phycobiliproteins as well as chlorophyll c_2 as photosynthetic accessory pigments. Action spectra determined by Haxo and Fork (1959) and Haxo (1960), and comparisons with the absorption of isolated pigments (ÓhEocha and Raftery, 1959) provided clear evidence of the phycobiliproteins as major contributors to the photosynthetic process of these algae. For maximum energy transfer, a prime requirement is a close proximity of the phycobili-

Fig. 1. *Chroomonas* sp. showing the basic structural characteristics of cryptophytes: chloroplast (CH), ejectosome (E), Golgi apparatus (G), nucleus (N), nucleomorph (NM), pyrenoid (P), starch (S), mastigonemes (arrows) in the chloroplast-nuclear envelope. (25,000 X). From Gantt *et al.*, 1971; permission of J. Cell Biol.

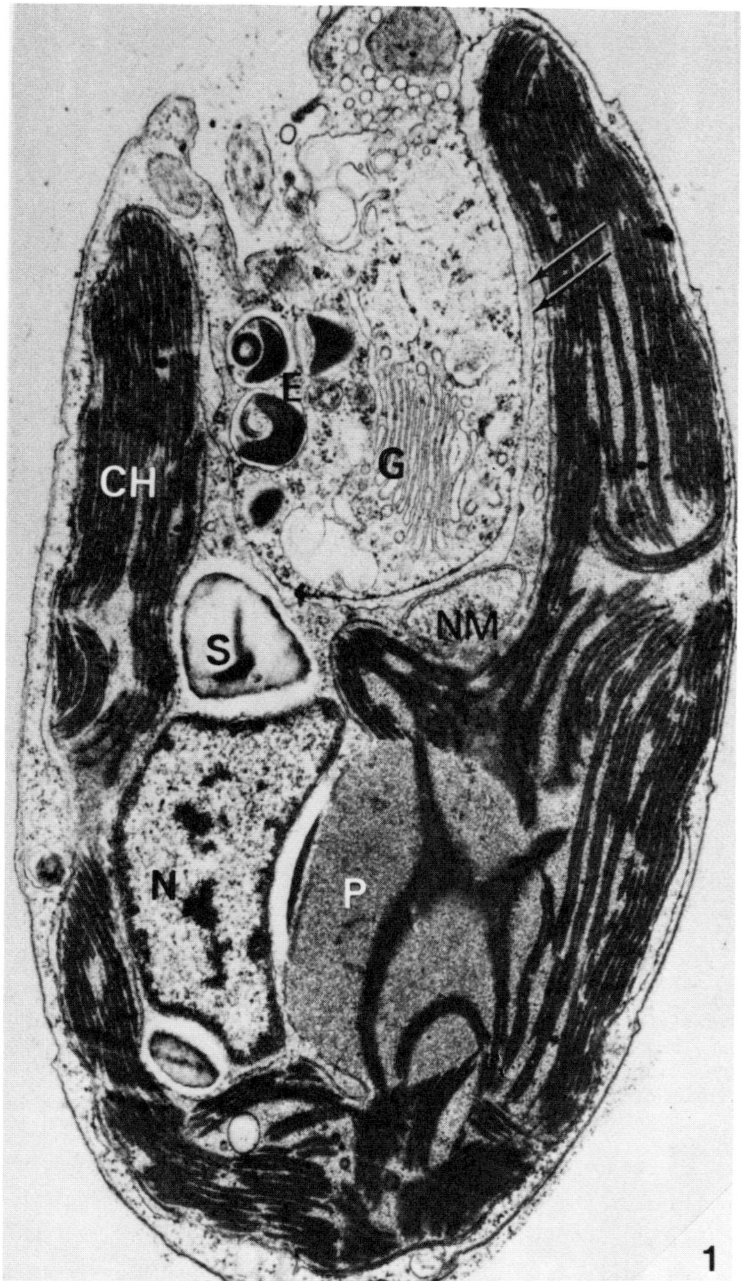

1

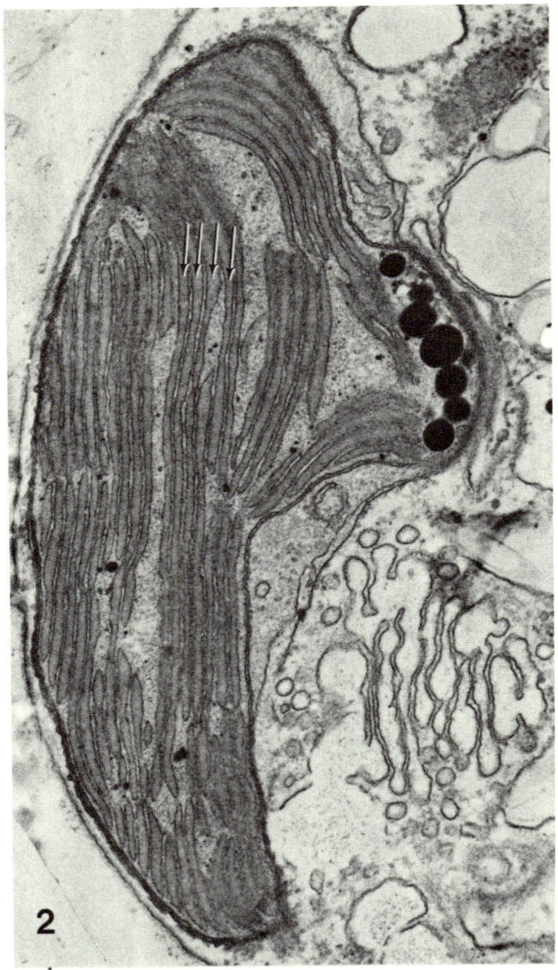

Fig. 2. Cross-section through a chloroplast of *Chroomonas* sp. with an eyespot in the stroma region. A cytoplasmic vesicle is appressed near the eyespot. Electron dense material in the intrathylakoidal space (arrows) is presumed to be the location of the phycobiliproteins.(39,000 X).

proteins and chlorophyll. This seems to have been satisfied by enclosure of the phycobiliproteins in the intrathylakoidal lumens. It has been possible to show indirectly that phycobiliproteins are located within the intrathylakoidal lumens in three genera (Gantt *et al.*, 1971). Unlike the red and blue-green algae this group does not contain phycobilisomes. Each species contains either phycoerythrin or phycocyanin, but not both. Allophycocyanin, which is the bridging pigment to chlorophyll *a* in red and blue-green algae, has not been found in cryptophytes. Chlorophyll c_2, according to energy transfer calculations made by MaColl and Berns (1978) does not seem to serve as an intermediary pigment between phycobilins and chlorophyll *a*, which suggests that it may contribute its energy directly to chlorophyll *a*. Location of the phycobiliproteins in the intrathylakoidal lumens raises the interesting implication that the thylakoids of the cryptophytes may be inverted when compared to the red and blue-green algae where the phycobilisomes are on the stroma side.

A summary of the photophysical and biochemical properties of cryptophytan phycobiliproteins (see references in Gantt, 1979) indicates that they possess the essential features of those from red and blue-green algae, except they do not tend to form aggregates. Extracted phycobiliproteins from numerous cryptophyte species has shown that there are at least two phycocyanin and three phycoerythrin forms. However, only one of the forms tends to occur per species. The phycocyanin forms have double peaks: PC I (585 and 645 nm) and PC II (588 and 615 nm). The phycoerythrin forms have single peaks: PE I (545 nm); PE II (545 nm); and PE III (566 nm). Whereas considerable future work is required for the full characterization of cryptophyte phycobiliproteins, even at this stage the information is sufficient to be used in separating certain strains into distinct species.

Nucleomorph. In the perinuclear space, outside the chloroplast-limiting membrane, a small organelle exists that has a nucleus-type of structure (Figs. 1, 6). A double membrane with pores surrounds a matrix in which are embedded bodies resembling small to gigantic ribosomes. In the endosymbiont *Mesodinium*, the nucleomorph is convincingly separate from the nucleus of the host and of the endosymbiont (Hibberd, 1977), leaving little doubt that it is independent. Its function is not known, but its location outside of the chloroplast near starch-forming regions, and its apparent well-timed division during mitosis, suggests that it may have a regulatory function. Nucleomorphs have not yet been reported in other algal groups.

Periplast. The periplast is another distinctive cryptophycean characteristic. Although it has some superficial similarities with the dinophycean plates and the euglenoid pellicle, its structure is fundamentally different (Dodge, 1973) from all algal divisions. By light microscopy surface striations with a spiral orientation are noticeable in the larger species, which by scanning and transmission microscopy can be seen to possess considerable additional details. The investigations carried out by Santore (1977) with scanning electron microscopy portend that examination of the periplast structure may come to be of critical taxonomic value. Of several species studied, two groups are discernibly different in the shape of the periplast plate areas, and whether or not the plate areas occur on the inside as well as outside of the plasma membrane.

In *Chroomonas* sp., *Hemiselmis rufescence* and probably *Chroomonas mesostigmatica*, representing one group, inner and outer thickenings are closely appressed to the plasma membrane. They cover the entire cell up to the vestibular region. These thickened regions are referred to as plates or plate areas (Dodge, 1969; Gantt, 1971; Santore, 1977). A plate area has a thickness of *ca.* 25 nm and a rectangular shape (Fig. 3). The plate areas are delineated by grooves running longitudinally and vertically to the cell axis. In the grooves the plasma mem-

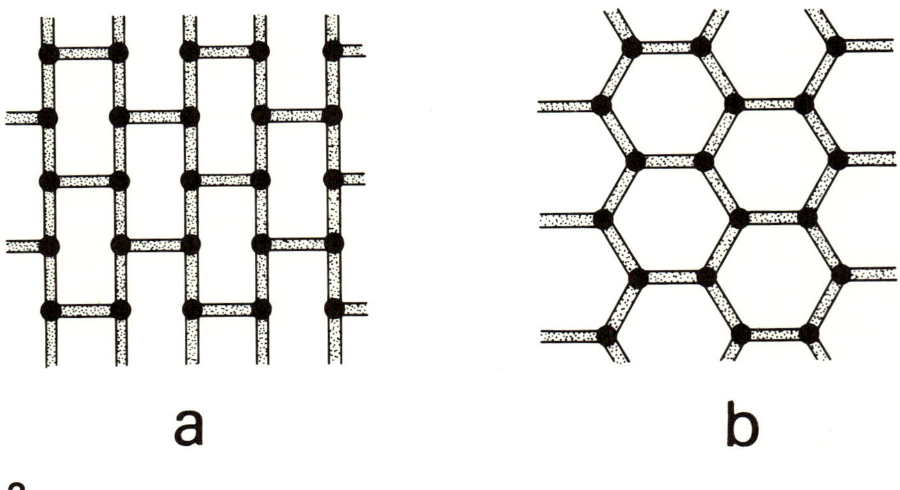

3

Fig. 3. Diagrammatic representation of periplast patterns formed in *Chroomonas* sp. (a) and in *Cryptomonas* (b). Ejectosome location is at corners of the plates. Anterior-posterior and vertical grooves (stipples) are without reinforced plate thickenings, thus allowing for changes in cell shape. (Adapted from Santore, 1977).

brane does not have the reinforced thickenings. This feature probably allows for considerable elasticity and alteration in cell shape, which can be observed in living cells. Adjacent plate areas are staggered by one third to one half their lengths, and the longitudinal rows of the periplastidial plate areas are joined to a raphe. In *Chroomonas* sp. about 14 rows attach to the raphe, but whether or not this number is significant for the strain must await future comparisons of other species. Near the raphe the plate areas are usually smaller in size. Peripheral ejectosomes tend to underlie the anterior corners of the plate areas and create a bulging appearance. The size of the plate areas vary considerably, but they tend to be larger (*ca.* 0.7 by 1.0 μm) than in the following group.

In several *Cryptomonas* strains and in *Chroomonas salina*, representing the second group, the periplast structure is distinguishably different (Lucas, 1970a; Hibberd *et al.*, 1971; Faust, 1974; Santore, 1977). A significant structural difference is seen in cross sections because distinct plates (15 nm thickness) occur inside of the plasma membrane. Furthermore, the plates are generally smaller (0.2-0.5 μm), more numerous and are spirally arranged. Whereas the plates of *Chroomonas salina* are rectangular, those in the *Cryptomonas* strains often appear to be hexagonal. The hexagonal appearance (Fig. 3) is enhanced by the location of numerous peripheral ejectosomes at the corners of the plates and by the apparently stretched membranes. According to Santore (1977), the plates are not intimately attached to the membrane along most of their length, but appear to be attached at their edges. It is interesting that plates detached by mild sonication appear as two sheets each with a polgonal shape and each composed of fibers in a parallel array (Faust, 1974).

The plates and plate areas probably serve a protective function by conferring a rigid armor. Its rigidty is removable, along with the plates, by treating cells with proteases. The removal of the plates by proteases suggests that they may be proteinaceous, or on the other hand, that they may be attached to the plasma membrane via a protein (Gantt, 1971; Faust, 1974). In addition, a fibrous layer commonly occurs on the outer periplast surface of some, if not all species. Fibers of various lengths (40-500 nm) have been recorded on *Cryptomonas* strains (Santore, 1977) and may prove to be characteristic of the individual.

Ejectosomes. Ejectosomes are specialized extrusive organelles which have also been referred to as trichocysts, or taeniobolocysts (see review by Hausmann, 1978). They occur in two size classes, the larger ones having a diameter of 0.5 μm and the smaller ones of *ca.* 0.1 μm. Large ejectosomes, being observable by light microscopy, have long been used as a distinguishing characteristic of the division, and their arrangement and number has been used in species designations (Pascher, 1913; Pringsheim, 1944, 1968; Butcher, 1967). They occur in the vestibular region, or along a furrow, in one or several rows. Observation of the smaller ejectosomes requires electron microscopy. They are located in the grooves of the periplast, usually near corners of periplast plates (Dodge, 1969, 1973; Santore, 1977).

On casual consideration they can be regarded as defensive organelles since they are ejected in what appear to be alarm responses, such as crowding and changes in pH. Observations of cells mounted in culture medium under a cover slip reveal that self destruction usually results in response to sticking to the cover slip or slide. It is presumed that in the natural environment they serve the cells for direct defense.

Except for size and location the ejectosomes are structurally identical. Anderson (1962) provided the first detailed electron micrographs for ejectosomes of the colorless *Chilomonas*, and the characteristics described appear common to all species (Lucas, 1970a, b; Mignot *et al.*, 1970; Schuster, 1970; Wehrmeyer, 1970a; Dodge, 1973). Each ejectosome is contained in a membrane vesicle with an opening to the exterior. This opening is normally closed by an overlapping flap. In open vesicles the top of the flap is often bulbous, suggesting that on its inflation withdrawal from the opening occurs (Gantt, 1971). The vesicle and the plasma membranes are continuous which means that the ejectosomes are actually on the cell's exterior. In the cryptophytes every ejectosome consists of two coils, each being a tightly coiled ribbon (Fig. 4). The basal coil is larger and when sectioned vertically exhibits a depression in which lies the smaller coil. On expulsion they rapidly uncoil and the edges of the ribbon round up to form a rigid tubular structure with a "hinge", which is the attachment region of the small ribbon. According to Mignot and colleagues (1970; 1973) the large coil unwinds first pushing the small spiral ahead, which unwinds later. The chemical nature of the ribbon material accord-

Fig. 4. Ejectosome structure of cryptophytes. a. A large ejectosome sectioned longitudinally through its center. The small coil on top is actually attached to the main coil. The parallel striations are sections of the coiled ribbon. b. In this cross-section, the ribbon is seen to be continuous. (Both 116,000 X). c. A diagrammatic view of an uncoiling ejectosome beginning with the unwinding of the main ribbon, its subsequent rounding creating a tubular structure and finally the unwinding of the small coil. Adapted from Hovasse and Mignot (1975). d. Uncoiled ejectosomes are rigid structures and hollow as indicated by penetration of the negative stain (phosphotungstic acid). The small unwound coil extends from the joint. (23,000 X).

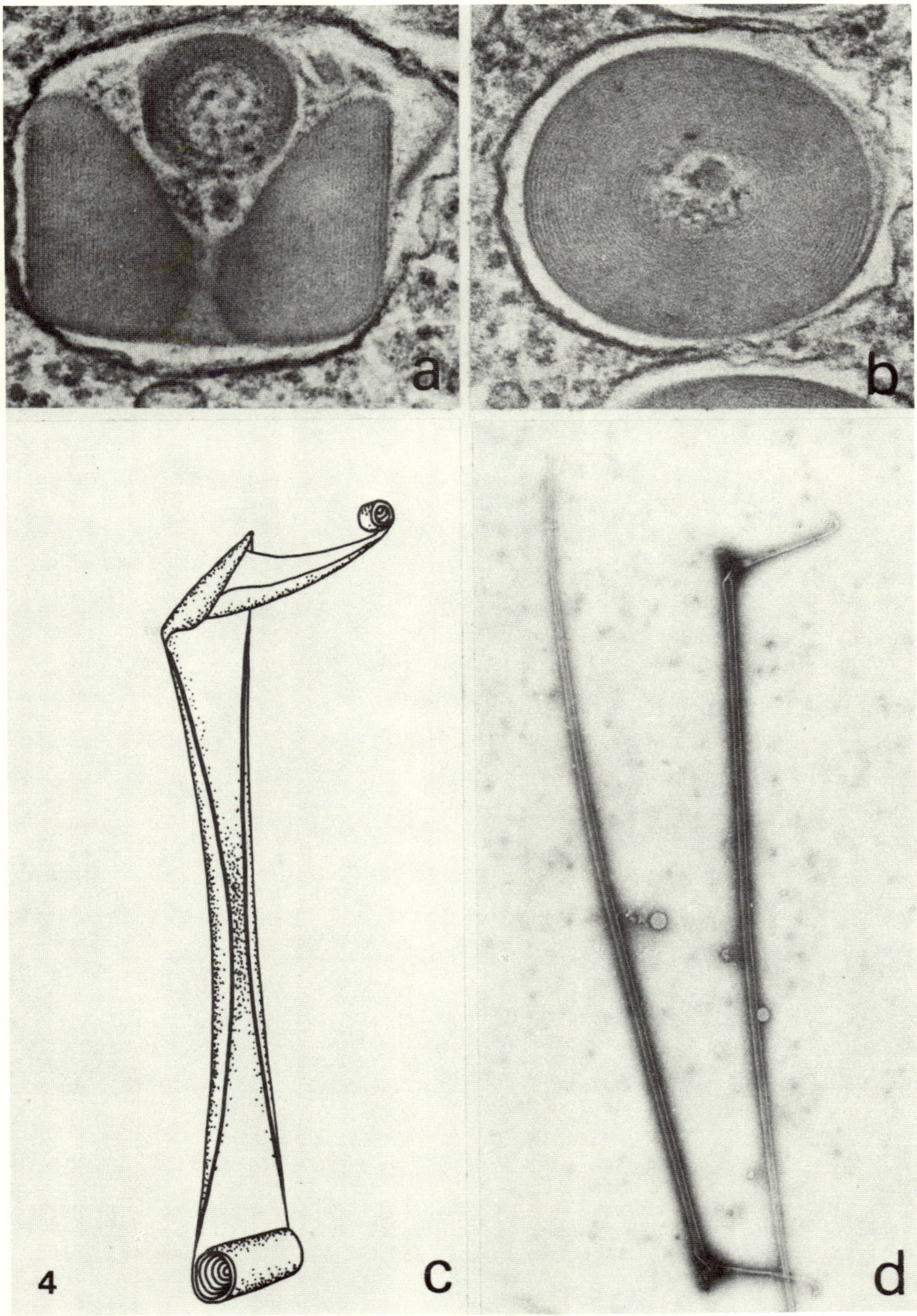
a
b
4
c
d

ing to their data is proteinaceous. Different hydration states of the proteins on the two sides of the ribbon may account for the unfolding process.

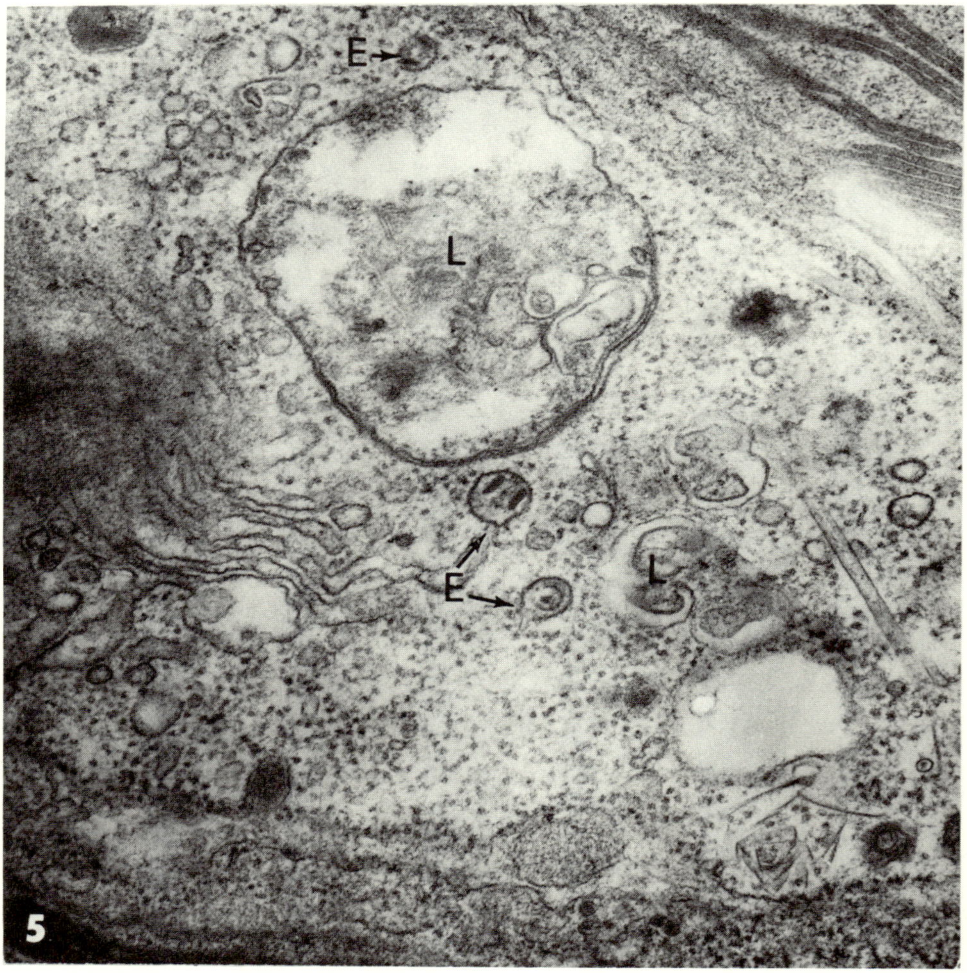

Fig. 5. Vesicles with juvenile ejectosomes (E) occur close to the Golgi apparatus, from which they are presumed to originate. Lysosomes (L) are filled with membranes and other cellular constituents. (49,000 X). (From Wehrmeyer, 1970a; permission of Protoplasma).

Origin of ejectosomes from the Golgi apparatus has been suggested because of the close proximity of vesicles containing incomplete small ejectosomes (Fig. 5) with the Golgi region (Wehrmeyer, 1970a). All free-living cryptophyte species studied by electron microscopy appear to have ejectosomes.

These ejectosomes are distinct from projectile organelles in any other algae or protozoan division with one qualified exception. Two species of prasinophytes (Manton, 1969; Norris and Pearson, 1975) have ejectosomes which, however, consist of only one main coil but lack the small coil. On this basis the two divisions appear to be related, but ejectosomes could also have evolved independently.

Mitosis and cytokinesis. The nucleus is small (*ca.* 3 μm) and in nondividing cells is generally located in the posterior region. Positive identification of individual chromosomes has not been made by either light or electron microscopy. By electron microscopy the nucleus, in addition to a nucleolar region, also contains other dense chromatin material randomly distributed in a less dense matrix.

Mitotic division has been elucidated entirely by Oakley and collaborators (Oakley and Dodge, 1973, 1976; Oakley and Bisalputra, 1977; Oakley, 1978; Oakley and Heath, 1978). From their detailed examination of *Cryptomonas* sp. (formerly classified as *Rhodomonas lens*) and *Chroomonas salina* they found that mitotic and cell division occur rapidly (10-15 minutes) and the processes in the two species are highly similar.

Nuclear division is first indicated by division of the flagellar bases followed by the migration of the nucleus from the posterior to the anterior region. Although the flagellar bases move apart laterally, they do not move far (< 180°) and never move to opposite sides. During prophase the chromatin aggregates into a mass, which is interspersed by chromatin-free channels. The nuclear membrane begins to break down, but prior to breakdown there is a 15 fold increase of the microtubules between the flagellar bases and the nucleus. Microtubules extending down from the flagellar bases innervate the nuclear area. However, they do not remain attached to the flagellar bases.

At metaphase (Fig. 6) a single barrel-shaped chromatin mass (with a 3.5 μm diameter and 0.7 μm thickness) occurs in the center of the spindle. The spindle is short (2.5 μm pole to pole) and rectangular, and the nuclear area is surrounded by endoplasmic reticulum which, on cursory examination, could be mistaken for a nuclear membrane. Some microtubules pass through chromatin-free channels and some are attached to the chromatin mass, but all extend from pole to pole. By this stage two nucleomorphs occur at opposite poles of the spindle;

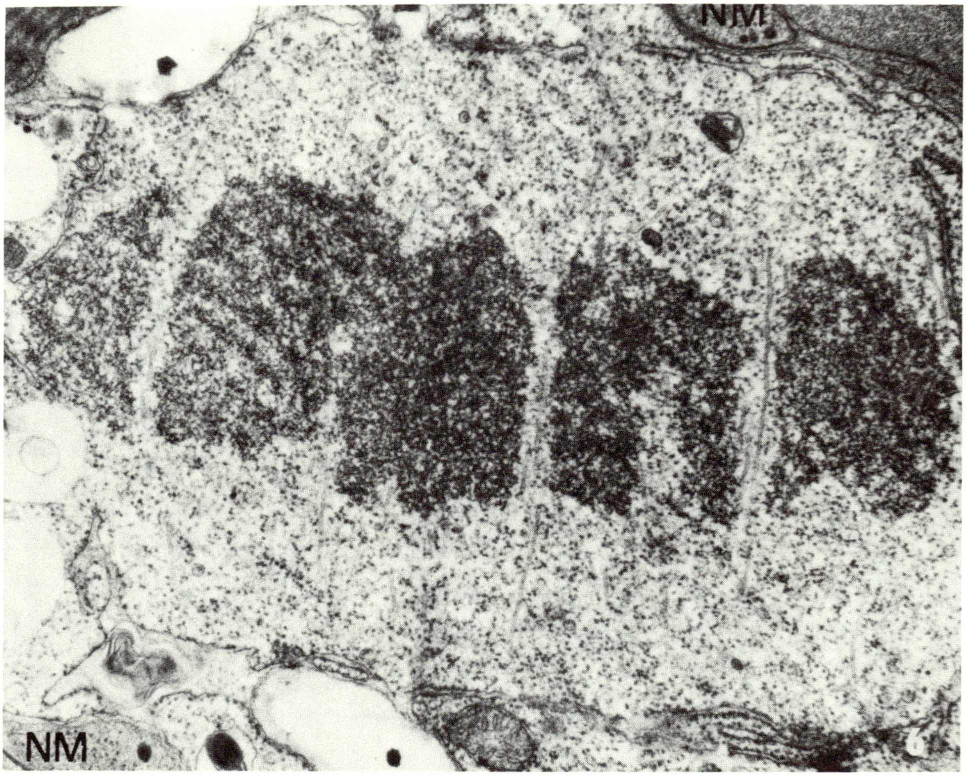

Fig. 6. A section through a metaphase stage of *Chroomonas salina* showing the chromatin mass in the center with microtubules of the spindle passing through chromatin-free channels. Rough endoplasmic reticulum delimits the nuclear area. Two nucleomorphs (NM) which had previously divided occur at top and bottom. (25,000 X). (From Oakley and Dodge, 1973; permission of Nature).

they presumably divided during prophase.

During anaphase the chromatin mass splits into two equal clumps that migrate toward the poles. Telophase quickly follows by the reformation of nuclear membrane from chloroplast-ER, which occurs as the nuclear masses push into the chloroplast. Centrioles are absent, as are any true organizing centers. Although the flagellar bases seem to function in the initial synthesis of the spindle, the function ceases by metaphase. A rhizoplast with typical striations has been observed in the polar region of the spindle, but its functional involvement is not clear.

Cell division occurs by furrowing, which usually begins in the peripheral cytoplasm during metaphase or anaphase, and is completed during telophase (Oakley and Dodge, 1976). In *Cryptomonas* sp. and *Chroomonas salina* the furrows form perpendicular to the spindle orientation from the anterior and posterior end. In the constriction region, just inside the plasma membrane, a thin amorphous mass is visible, which Oakley and Bisalputra (1977) postulate as possibly being actin. In *Cryptomonas ovata* var. *palustris* (Ward and Bowen, 1979), the furrowing begins at the posterior end and proceeds anteriorly toward the flagella. Cell division in this species is of longer duration (30 minutes to 3 hours) than in the species studied by Oakley and colleagues. A cell plate is not apparent during cytokinesis; nor is there any sign of microtubules in the furrow or in the final cytoplasmic constriction. Chloroplast division probably occurs during early nuclear division, because two hemispherically shaped chloroplasts are already present at the poles by the time the metaphase spindle is formed. Division of the Golgi apparatus is not conspicuous and presumably occurs also in the early mitotic stages. The flagella of the daughter cells remain active and their action may be responsible for the final separation of the daughter cells.

The cryptophytes share some similarities with the haptophytes and chrysophytes (Dodge, 1973) such as the shape of the spindle, the innervation of the microtubules on nuclear membrane breakdown, and its reformation from the chloroplast membrane. These similarities do not seem sufficient for including the groups in one phylogenetic division.

Flagella and other organelles. The structure of the flagella and those of several other organelles to be mentioned here is not peculiar to the Cryptophyceae, but is similar in several other algal groups.

All motile cryptophytes have two flagella, with the flagellar bases occurring in parallel. One flagellum is usually slightly shorter than the other, as in *Hemiselmis rufescence* where the flagellar lengths are 7 and 5.5 μm (Hibberd *et al.*, 1971) respectively. Mastigonemes (also known as flagellar hairs or flimmer hairs) often occur in two rows on the large flagellum and as one row on the small flagellum, with the flagellar tips remaining bare. During preparation for shadow casting, the mastigonemes sometimes shed (Hibberd *et al.*, 1971), but in sections they are better preserved (Dodge, 1969) and all species

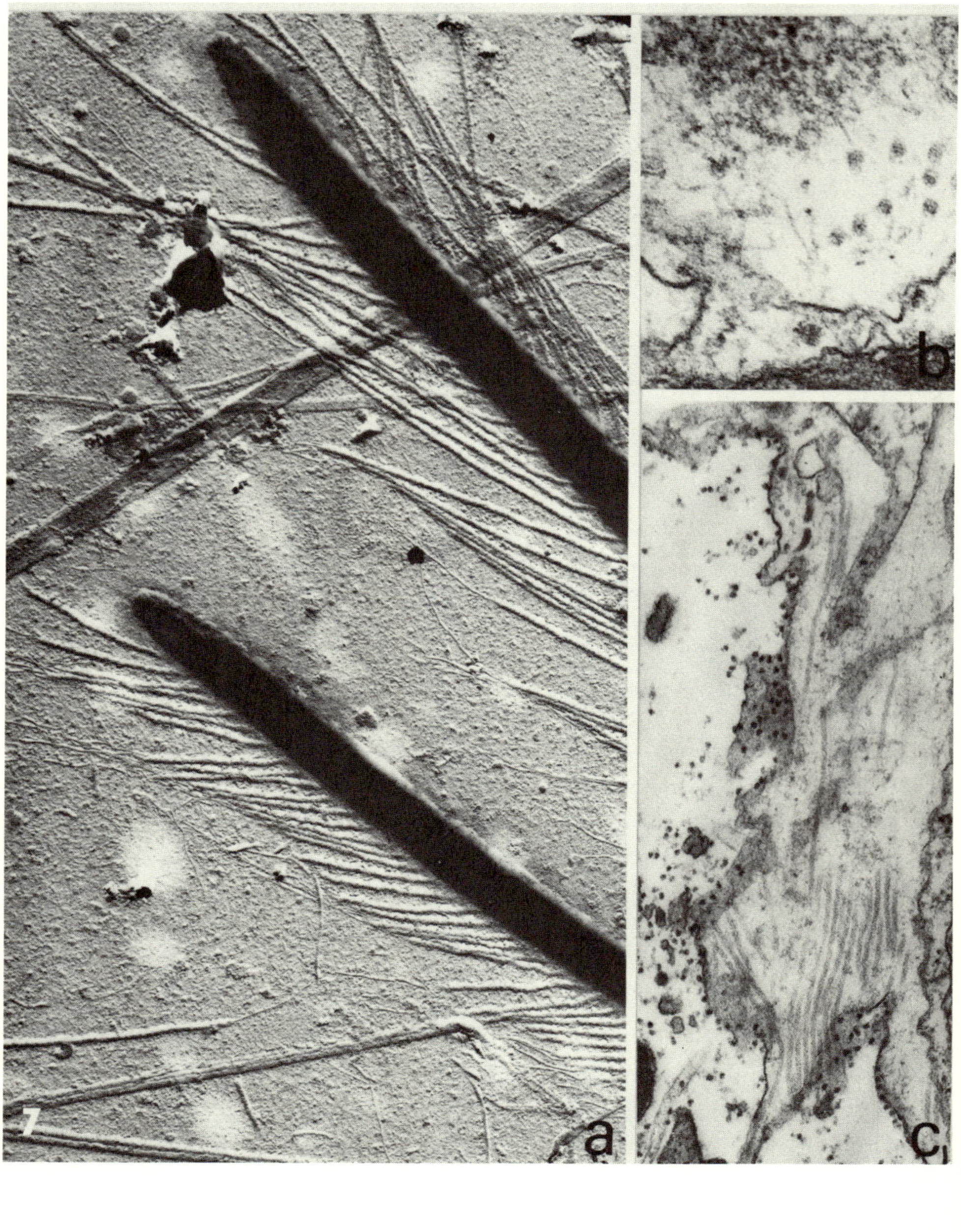
7
a
b
c

Fig. 7. Flagella and mastigonemes in *Cryptomonas*. a. A shadow whole mount of flagellar tips; one with two rows of mastigonemes and another with one row. Fine fibrils extend beyond the mastigoneme tubules.(19,000 X). b. Details of tubules and fibrils in the chloroplast-nuclear endoplasmic reticulum.(35,000 X). c. Long flimmer tubules in parallel alignment.(70,000 X). (From Heath *et al*, 1970; permission of Cell Sci.).

thus far examined have mastigonemes on both flagella. In *Cryptomonas ovata*, the larger flagellum has mastigonemes that are longer than those on the shorter flagellum with respective lengths of 2.5 and 1 μm (Hibberd *et al*., 1971). The mastigonemes typically consist of a rigid tubular portion and distal fibrils (Fig. 7). Production of mastigonemes occurs in the space of the chloroplast-nuclear envelope (Heath *et al*., 1970). In cross section the tubules appear hollow, and are somewhat smaller (16-29 nm) than microtubules. Their common presence suggests that they are continually produced and presumably added in their entirety to the flagellar region. Flagellar scales have not been found as in some members of the haptophytes, dinoflagellates and chrysophytes (Dodge, 1973). In section the flagella have the typical 9 + 2 axoneme structure and a regular flagellar base. A detailed comparison of the flagellar insertion and associated structures showed basically the same arrangement in *Rhodomonas*, *Cryptomonas*, and in the two colorless forms *Chilomonas* and *Cyathomonas* (Mignot, *et al*., 1968). The flagellar bases were shown to be joined by a straited fibrillar structure. A banded rhizoplast, with attachment points on both flagellar bases, extends toward the nucleus. The rhizoplast, as in the green flagellate *Platymonas* (Salisbury and Floyd, 1978), probably functions in flagellar activity possibly in altering the direction of cell motion. One microtubular strand runs parallel with the rhizoplast. Three others also emerge from the basal region and seem to be anchored in the vestibular region.

From individual sections one gets the mistaken impression that there are many small mitochondria, but it was recently shown by Santore and Greenwood (1977), that there is one mitochondrial complex per cell. In *Hemiselmis* Santore (1977) found at least 10 DNA areas per mitochondrium in serial sections. Similar careful studies on other algal species will probably show that this is not as unique an occurrence as this initial discovery indicates.

Lysosome-type structures (Fig. 5) described by Wehrmeyer (1970a) are probably equivalent to the "Corps-de-Maupas" (Lucas, 1970a, b), and have been found in all genera studied. Their contents and structure suggests that they have a phagocytic function. According to Lucas (1970a), there are about two per cell, and they are easily visualized by light microscopy after neutral red staining.

He postulated that they were related to multivesiculate bodies, because both have internal membraneous components, bounded by a single limiting membrane, and are closely associated.

In the anterior cell region, extensive vacuoles are present with a wide range of vesicles. These are not necessarily single vesicles, but are often a vesciculate network. This vesciculate region is often referred to as a contractile vacuole, which may be involved in osmoregulation. Unfortunately, definitive studies to support such an assumption are not available. Some of the smaller vesicles in the anterior region possess a fine fibrous coat and appear to originate from the Golgi apparatus (Dodge, 1973). Microbodies have not been identified, but it is quite probable that they exist in the cryptophytes as they do in most other algal groups.

Symbiosis. Symbiosis of a red colored cryptophyte and a ciliate host, collectively referred to as *Mesodinium rubrum*, has been clearly established from material collected from "red water" in marine environments. Laboratory culturing of the material has not been possible. The symbiont is photosynthetically active and thus allows the host to exist as a functional autotroph (Ryther, 1967). Examination of the pigment composition (Parsons and Blackbourne, 1968; Barber *et al.*, 1969) established the presence of phycoerythrin, chlorophyll *a* and *c*, (probably c_2; typical of most cryptophytes). Initial ultrastructural studies by Taylor *et al.* (1969) revealed a chloroplast morphology of associated thylakoid membranes, and a stalked pyrenoid surrounded by starch grains. Hibberd (1977) has convincingly shown that the symbionts are delineated from the host cytoplasm by being surrounded by a single membrane. Within the compartment is a large nucleus and numerous chloroplasts, some of which show membrane connections with the nucleus. The symbiont cytoplasm is distinctive by having a less dense rough endoplasmic reticulum than the host. Furthermore, the mitochondria within the symbiont cytoplasm do not exhibit the tubular-type cristae characteristic of the ciliate host. In its adaption the symbiont appears to have lost the periplast plates, its ejectosomes, probably its flagellar apparatus, and has reduced its Golgi apparatus.

The protist host of a blue-green algal type symbiont, *Cyanophora paradoxa* has been designated, among other things, as being of cryptophyte origin. There is little, if any, structural basis for this assumption. In fact, it lacks several characteristic features in that it does not have a periplast, instead of ejectosomes it has mucoid bodies (discobolocysts), and its flagellar apparatus and mitotic process is atypical of other crytophytes (Mignot *et al.*, 1969; Oakley and Dodge, 1976). For these reasons it cannot reasonably be considered as a cryptophyte.

Nutrition. Habitats of cryptophytes range from freshwater to marine. Marine collections (Throndsen and Heimdal, 1976) have shown that some species of *Cryptomonas* and *Hemiselmis* can be among the primary producers where they can attain cell concentrations of 3 x 10^4 cells/liter. Compared to some other phytoplankton species this is, however, a relatively small population density. In studies with *Cryptomonas ovata* var. *palustris*, Cloern (1977) found relatively low uptake rates of phosphate suggesting that in a low phosphate environment it would not be competitive with other algae. In laboratory cultures *Chroomonas* sp. attains cell densities of 5 x 10^9 cells/liter in a mineral medium (Faust and Gantt, 1973). When the mineral medium of *Chroomonas salina* was supplemented with glycerol, the cell concentration was increased to 7 x 10^9 cells/liter (Antia *et al.*, 1969).

Media for marine species vary greatly with the laboratory, but some species of *Chroomonas* and *Rhodomonas* have been successfully cultured in DV medium (Provasoli and MacLaughlin, 1963), Erdschreiber medium, f/2 medium (Guillard and Ryther, 1962), and the medium of Antia *et al.* (1969). Several species of *Cryptomonas*, *Hemiselmis* and *Chroomonas* can also be grown in 1/4 strength Erdschreiber medium (Lucas, 1970a, b). The freshwater species *Cryptomonas ovata* var. *palustris* grows successfully on M3 medium (Rayburn and Starr, 1974), and on the medium developed by Hoogenhout and Amesz (1965). Growth on agar thickened media is usually slow, and the cells tend to exist in a palmelloid state surrounded by thick mucilagenous material.

There has been long standing interest in the nutritional mode of the photosynthetic cryptophytes because of their obvious relatedness to the colorless forms which are obligate heterotrophs. Evidence for ingestion of organic particulate matter, or of whole cells, is not available. Wawrik (1970) in *Cryptomonas borealis* observed small cells within large cells, and postulated that the small cryptophyte had been ingested by the large cryptophyte. Since the ingestion process was not observed, it is equally likely that these cells may have been the products of a vegetative division process.

Heterotrophic growth thus far is known to occur in only one species. Antia and colleagues (Antia *et al.*, 1969; Cheng and Antia, 1970; Antia *et al.*, 1973) have shown that *Chroomonas salina* is capable of facultative heterotrophic growth when glycerol (0.25-1.0 M) is added to modified seawater. Growth on glycerol occurred both in light and darkness. Growth was greater in photoheterotrophic conditions than in photoautotrophic (without glycerol) or in chemoheterotrophic (in darkness with glycerol) conditions. On extended culturing in chemoheterotrophic conditions the cells become cream-colored suggesting a reduction of the photosynthetic pigments. Other species such as *Chroomonas* sp. are obligate phototrophs and adjust their photosynthetic pigment concentration with changes in light intensity (Faust and Gantt, 1973). In spite of the selective growth of *Chroomonas salina* on glycerol, it can be concluded that the colored cryptophytes should be considered as obligate phototrophs along with other plants. The colorless and colored forms probably diverged long ago in their development. It can be either assumed that the colorless forms resulted from a loss of the chloroplast and secondarily acquired heterotrophic capability, or that they existed as heterotrophs which acquired a chloroplast and lost their heterotrophic capability.

Evolutionary status. There seems to be little doubt that the colored cryptophytes are a distinct algal group and warrant a separate division. They have certain similarities in structure and pigment composition with other algal groups. Overemphasis of some of these similarities have led to suggestions of very close phylogenetic relationships, and even to taxonomic inclusion with dinoflagellates, haptophytes, chrysophytes, chloromonads, cyanophytes, rhodophytes, and prasinophytes. When viewed from a critical perspective, taking more than one or two characteristics into account, such inclusions are not justifiable.

Using primarily structural characteristics obtained by electron microscopy, Dodge (1973) diagrammed the relationships of the algae. According to his diagram, the cryptophytes show relationships with prasinophytes, haptophytes, chrysophytes, and chloromonads in that they have two or three characteristics in common. He was careful not to predict any evolutionary progression from his scheme because he pointed out that critical information is too limited.

In many phylogenetic schemes, based on biochemical characteristics (pigments, microtubular organelles, and symbiosis), the cryptophytes are generally considered as quite primitive, and are placed on a side branch off the main line. Various schemes are clearly summarized in Ragan and Chapman (1978). In regard to the cryptophytes these schemes are particularly weak because only limited biochemical or physiological data are available.

Lee (1972) assigned the cryptophytes a major status in his monophyletic scheme. He proposed an ancestral cryptophyte, which arose from an endosymbiotic association of a cyanophyte and a heterotrophic cryptophyte, and gave rise to all other groups. The presence of phycobiliprotein, major accessory pigments in red, blue-green, and cryptophyte algae, was used as a major characteristic in his evolutionary progression of endosymbiotic cyanelle → cryptophyte →

rhodophyte. Whereas, the phycobiliprotein of these groups have analogous functions and even show restricted immunological cross reactivity (MacColl *et al.*, 1976), other features are incompatible with such a proposed scheme. Cyanophytes and rhodophytes, unlike cryptophytes, have only chlorophyll *a*, have various phycobiliprotein types per alga, are capable of high aggregation states resulting in phycobilisomes that are located on the stroma side of the lamellae. Instead, in the cryptophytes the phycobiliproteins are located in the intrathylakoidal region and appear to be incapable of forming large aggregates. If evolution progressed from the more primitive cyanelle to the more advanced cryptophytes, then why did it again revert to a more primitive stage in the rhodophytes? It seems more probable that the cryptophyes diverged very early from one ancestral line, or that phycobiliprotein evolution occurred more than once. Elucidation of the phycobiliprotein biosynthetic pathway and its control in the three algal groups should bring a better understanding of their phylogenetic relationship.

Considerations for taxonomy. The most recent classification made by Butcher (1967), groups 84 species and varieties into 10 genera. In this scheme, *Rhodomonas* was totally dismissed as a genus. At least one-half of the species were newly described by him. Many of the samples were of material collected in the field. Because of this it is very likely that many of the species described will not remain, and a thorough reexamination of the cryptophyte genera and species is required.

For classification at the light microscope level 5 principal characteristics have been used: cell size; cell shape; chloroplast number and starch patterns; presence and direction of a furrow; and arrangement of large ejectosomes. By electron microscopy additional characteristics to be added are: the periplast plate patterns; flagellar length and mastigoneme ornamentation; cell surface hairs; and mitotic and cytokinetic patterns.

The taxonomy of the cryptophytes has always been problematical. The problem has been compounded by the small size of the cells, and by the nutritional variations existing in field collections and the absence of sexual stages that would allow crossing. Pringsheim (1968) stresses the importance of studying clones under identical conditions over extended periods. This emphasis should be well heeded by students interested in classification.

Furthermore, it is important to compare the growth rates of similar strains, under identical conditions, and to make structural and biochemical comparisons in actively growing and healthy cultures. The color of healthy cells is quite constant and is a highly reliable characteristic. Although significant changes in the phycobiliprotein/chlorophyll ratio can occur through adaptation to light (Faust and Gantt, 1973), the coloration of the cells is little affected. Extraction of phycobiliproteins (in 0.1 M phosphate buffer, pH 6.8) is extremely simple (Siegelman and Kycia, 1978) and should be routinely used in taxonomic characterizations of a species. Identification of chlorophyll forms, though more difficult, should be used when possible.

REFERENCES

Anderson, E. 1962. A cytological study of *Chilomonas paramecium* with particu- reference to the so-called trichocysts. J. Protozool. 9: 380-395.

Antia, N. J., J. Y. Cheng and F. J. R. Taylor. 1969. The heterotrophic growth of a marine photosynthetic cryptomonad (*Chroomonas salina*). Proc. 6th Intl. Seaweed Symp., Santiago de Compostela, Sept. 1968. pp. 17-29.

Antia, N. J., J. Y. Cheng, R. A. J. Fogle and E. Percival. 1979. Marine cryptomonad starch from autolysis of glycerol-grown *Chroomonas salina*. J. Phycol. 15: 57-62.

Antia, N. J., J. P. Kalley, J. McDonald and T. Bisalputra. 1973. Ultrastructure of the marine cryptomonad *Chroomonas salina* cultured under conditions of photoautotrophy and glycerol-heterotrophy. J. Protozool. 20: 377-385.

Barber, R. T., A. W. White and H. W. Siegelman. 1969. Evidence for a cryptomonad symbiont in the ciliate, *Cyclotrichium meunieri*. J. Phycol. 5: 86-88.

Butcher, R. W. 1967. An introductory account of the smaller algae of British Coastal Waters. IV. Cryptophyceae, Fisheries Investigations, London, Ser. IV: 1-54.

Cheng, J. Y. and N. J. Antia. 1970. Enhancement by glycerol of phototrophic growth of marine planktonic algae and its significance to the ecology of glycerol pollution. J. Fish. Res. Bd. Canada 27: 335-346.

Cloern, J. E. 1977. Effects of light intensity and temperature on *Cryptomonas ovata* (Cryptophyceae) growth and nutrient uptake rates. J. Phycol. 13: 389-395.

Dodge, J. D. 1969. The ultrastructure of *Chroomonas mesostigmatica* Butcher (Cryptophyceae). Arch. Mikrobiol. 69: 266-280.

Dodge, J. D. 1973. The Fine Structure of Algal Cells. Academic Press, New York. 261 pp.

Faust, M. A. 1974. Structure of the periplast of *Cryptomonas ovata* var. *palustris*. J. Phycol. 10: 121-124.

Faust, M. A. and E. Gantt. 1973. Effect of light intensity and glycerol on the growth, pigment composition and ultrastructure of *Chroomonas* sp. J. Phycol. 9: 489-495.

Gantt, E. 1971. Micromorphology of the periplast of *Chroomonas* sp. (Cryptophyceae). J. Phycol. 7: 177-184.

Gantt, E. 1979. Phycobiliproteins of cryptophyceae. In Hutner, H. S. and M. Levandowsky [Eds.] *Biochemistry and Physiology of Protozoa*, 2nd Ed. Academic Press, New York. pp. 121-137.

Gantt, E., M. R. Edwards and L. Provasoli. 1971. Chloroplast structure of the cryptophyceae: Evidence for phycobiliproteins within intrathylakoidal spaces. J. Cell Biol. 48: 280-290.

Gibbs, S. P. 1962a. Nuclear envelope chloroplast relationships in algae. J. Cell Biol. 14: 433-444.

Gibbs, S. P. 1962b. The ultrastructure of the pyrenoids of algae, exclusive of the green algae. J. Ultrastruct. Res. 7: 247-261.

Gibbs, S. P. 1970. The comparative ultrastructure of the algal chloroplast. Ann. N.Y. Acad. Sci. 175: 454-473.

Glazer, A. M., G. Cohen-Bazire and R. Y. Stanier. 1971. Characterization of phycoerythrin from a *Cryptomonas* sp. Arch. Mikrobiol. 80: 1-18.

Guillard, R. R. L. and J. H. Ryther. 1962. Studies on marine planktonic diatoms. I. *Cyclotella nana* Hustedt and *Detonula confervacea* (Cleve) Gran. Can. J. Microbiol. 8: 229-239.

Hausmann, K. 1978. Exclusive organelles in Protists. In Bourne, G. H. and J. F. Danielli [Eds.] *International Review of Cytology* 52. Academic Press, New York. pp. 197-276.

Haxo, F. T. 1960. The wavelength dependence of photosynthesis and the role of accessory pigments. In Allen, M. B. [Ed.] *Comparative biochemistry of Photoreactive Systems*. Academic Press, New York. pp. 339-360.

Haxo, F. T. and D. C. Fork. 1959. Photosynthetically active accessory pigments in Cryptomonads. Nature 184: 1051-1052.

Heath, I. B., A. D. Greenwood and H. B. Griffiths. 1970. The origin of flimmer in *Saprolegnia, Dictyuchus, Synura* and *Cryptomonas*. J. Cell Sci. 7: 445-461.

Hibberd, D. J. 1977. Observations on the ultrastructure of the cryptomonad endosymbiont of the red-water ciliate *Mesodinium rubrum*. J. Mar. Biol. Assn. U.K. 57: 45-61.

Hibberd, D. J., A. D. Greenwood and H. B. Griffiths. 1971. Observations on the ultrastructure of the flagella and periplast in the Cryptophyceae. Br. Phycol. J. 6: 61-72.

Hoogenhout, H. and J. Amesz. 1965. Growth rates of photosynthetic organisms in laboratory cultures. Arch. Mikrobiol. 50: 10-25.

Hovasse, R. and J.-P. Mignot. 1975. Trichocystes ét organites analogues chez les protists. Annee. Biol. 14: 397-422.

Jeffrey, S. W. 1976. The occurrence of chlorophyll c_1 and c_2 in algae. J. Phycol. 12: 349-354.

Joyon, L. 1963. Contribution à l'étude cytologique de quelques protozoaires flagellés. An. Fac. Sc. Clermont 22: 45-52.

Lee, R. E. 1972. Origin of plastids and the phylogeny of algae. Nature 237: 44-46.

Lucas, I. A. N. 1970a. Observations on the fine structure of the Cryptophyceae I. The genus *Cryptomonas*. J. Phycol. 6: 30-38.

Lucas, I. A. N. 1970b. Observations on the ultrastructure of representatives of the genera *Hemiselmis* and *Chroomonas* (Cryptophyceae). Br. Phycol. J. 5: 29-37.

MacColl, R. D. and D. S. Berns. 1978. Energy transfer studies on cryptomonad biliproteins. Photochem. Photobiol. 27: 343-349.

MacColl, R. D., D. S. Berns and O. Gibbons. 1976. Characterization of cryptomonad phycoerythrin and phycocyanin. Arch. Biochem. Biophys. 177: 265-275.

Manton, I. 1969. Tubular trichocysts in a species of *Pyramimonas* (*P. grossii* Parke). Österr. Bot. Z. 116: 378-392.

Mignot, J.-P. 1965. Étude ultrastructurale de (*Cyathomonas truncata*) From. (*Flagellé Cryptomonadine*). J. Micro. 4: 239-252.

Mignot, J.-P. 1967. Affininités des Euglénomonadines ét des Chloromonadines remarques sur la systématique des Euglénida. Protistologica III: 25-60.

Mignot, J.-P. and R. Hovasse. 1973. Novelle contribution a la connaissance de trichocystes les organites grillagés D *Entosiphon sulcatum* (Flagellata, Euglenida). Protestilogica 9: 373-391.

Mignot, J.-P., L. Joyon and F. G. Pringsheim. 1969. Quelques particularités structurales de *Cyanophora paradoxa* Korsch. protozoaire flagellé. J. Protozool. 16: 138-145.

Mignot, J.-P., R. Hovasse, and L. Joyon. 1970. Nouvelles données sur le fonctionnement des trichocystes (Taeniobolocystes) des Cryptomonadines. J. Micro. 9: 127-132.

Mignot, J.-P., L. Joyon and E. G. Pringsheim. 1968. Compléments a l'étude cytologique des Cryptomonadines. Protistologica IV: 493-506.

Norris, R. E. and B. R. Pearson. 1975. Fine structure of *Pyramimonas parkeae*, sp. nov. (Chlorophyta, Prasinophyceae). Arch. Protistenk. 117: 192-213.

Oakley, B. R. 1978. Mitotic spindle formation in *Cryptomonas* and *Chroomonas* (Cryptophyceae). Protoplasma 95: 333-346.

Oakley, B. R. and T. Bisalputra. 1977. Mitosis and cell division in *Cryptomonas* (Cryptophyceae). Can. J. Bot. 55: 2789-2800.

Oakley, B. R. and J. D. Dodge. 1973. Mitosis in the Cryptophyceae. Nature 244: 521-522.

Oakley, B. R. and J. Dodge. 1976. The ultrastructure of mitosis in *Chroomonas salina* (Cryptophyceae). Protoplasma 88: 241-254.

Oakley, B. R. and I. B. Heath. 1978. The arrangement of microtubules in serially sectioned spindles of the alga *Cryptomonas*. J. Cell Sci. 31: 53-70.

OhEocha, C. and M. Raftery. 1959. Phycoerythrins and phycocyanins of Cryptomonads. Nature 184: 1049-1051.

Pascher, A. 1913. Die Süsswasser Flora. Jena 2: G. Fischer, Verlag. pp. 99-107.

Parsons, T. R. and D. J. Blackbourn. 1968. Pigments of the ciliate *Mesodinium rubrum*. Neth. J. Res. 4: 27-31.

Pringsheim, E. G. 1944. Some aspects of taxonomy in the Cryptophyceae. New Phytol. 43: 143-150.

Pringsheim, E. G. 1968. Zur Kenntnis der Cryptomonaden des Süsswassers. Nova Hedwigia 16: 367-401.

Provasoli, L. and J. J. A. Maclaughlin. 1963. Limited heterotrophy of some photosynthetic dinoflagellates. In Oppenheimer, C. H. [Ed.] *Symposium in Marine Microbiology*. C. C. Thomas, Springfield, Illinois. pp. 15-113.

Rayburn, W. R. and R. C. Starr. 1974. Morphology and nutrition of *Pandorina unicocca* sp. nov. J. Phycol. 10: 42-49.

Ragan, N. A. and D. J. Chapman. 1978. A Biochemical Phylogeny of the Protists. Academic Press, New York. 317 pp.

Rosenberg, M. 1944. On a blue-green Cryptomonad, *Chroomonas* Nordstedtii Hansg. Ann. Bot. 8: 315-322.

Ryther, J. H. 1967. Occurrence of red water off Peru. Nature 214: 1318-1319.

Salisbury, J. L. and G. L. Floyd. 1978. Calcium induced contraction of the rhizoplast of a quadriflagellate green alga. Science 202: 975-977.

Santore, J. J. 1977. Scanning electron microscopy and comparative micromorphology of the periplast of *Hemiselmis rufescens*, *Chroomonas* sp., *Chroomonas salina* and members of the genus *Cryptomonas* (Cryptophyceae. Br. Phycol. J. 12: 255-270.

Santore, J. J. and A. D. Greenwood. 1977. The mitochondrial complex in Cryptophyceae. Arch. Mikrobiol. 112: 207-218.

Schuster, F. L. 1968. The gullet and trichocysts of *Cyathomonas truncata*. Exp. Cell Res. 49: 277-284.

Schuster, F. L. 1970. The trichocysts of*Chilomonas paramecium*. J. Protozool. 17: 521-526.

Siegelman, H. W. and J. H. Kycia. 1978. Algal Biliproteins. In Hellebust, J. A. and J. S. Craigie [Eds.] *Handbook of Phycological Methods: Physiological and Biochemical Methods*. Cambridge University Press, London. pp. 71-79.

Taylor, F. J. R., D. J. Blackbourn and J. Blackbourn. 1969. Ultrastructure of the chloroplasts and associated structures within the marine ciliate *Mesodinium rubrum* (Lohmann). Nature 224: 819-821.

Throndsen, J. and B. R. Heimdal. 1976. Primary production, phytoplankton and light in Straumsbukta near Tromsø. Astarte 9: 51-60.

Vesk, M. and S. W. Jeffrey. 1977. Effect of blue-green light on photosynthetic pigments and chloroplast structure in unicellular marine algae from six classes. J. Phycol. 13: 280-288.

Ward, B. and M. Bowen. 1979. Cytokinesis in *Cryptomonas ovata* (Cryptophyceae). Protoplasma 98: 275-277.

Wawrik, F. 1970. Mixotrophic bei *Cryptomonas borealis* Skuja. Archiv. Protistenk. 112: 312-313.

Wehrmeyer, W. 1970a. Struktur, Entwicklung und Abbau von Trichocysten in *Cryptomonas* and *Hemiselmis* (Cryptophyceae). Protoplasma 70: 295-315.

Wehrmeyer, W. 1970b. Zur Feinstruktur der Chloroplasten einiger photoautotropher Cryptophyceen. Arch. Mikrobiol. 71: 367-383.

Copyright 1980 by Elsevier North Holland, Inc.

FREE-LIVING DINOFLAGELLATES

KAREN A. STEIDINGER[1] and ELENOR R. COX[2]
[1]Florida Department of Natural Resources, Marine Research Laboratory, 100 Eighth Avenue S.E., St. Petersburg, Florida 33701 U.S.A., [2]Department of Biology, Texas A&M University, College Station, Texas 77843 U.S.A.

ABSTRACT

Dinoflagellates (Pyrrhophyta, Dinophyceae) are eukaryotes with at least one form in the life cycle being motile and usually having two dissimilar flagella. All dinoflagellates studied except one are haplonts with the diploid stage being a plano- or hypnozygote. The theca of motile stages consists of membrane layers and vesicles. The thecal vesicles may or may not contain a structural carbohydrate. A few species have body and flagellar scales. In addition to typical organelles and systems, these flagellates can contain trichocysts, pusules, nematocysts and eyespots. In photosynthetic species, chloroplasts usually contain lamellae of 2-3 thylakoids and can have interlamellar or projected pyrenoids. There is a wide variety of pyrenoid, pusule and eyespot types and not all species have these organelles. Motile stages contain a distinct nucleus with continually condensed chromosomes attached to the nuclear envelope. Nuclear division is a closed mitosis with extranuclear components. Environmental parameters, population age and isolate strain can affect ultrastructure characterization and interpretation. Intracellular chrysophyte symbionts have been observed for two species and an evolutionary hypothesis is projected.

INTRODUCTION

Dinoflagellates are eukaryotes of unique structure and diverse habit and habitat. Many are photosynthetic, free-living unicells ranging in size from <10 μm to <2 mm. At one point in the life cycle, there are motile cells that typically have two dissimilar flagella and a distinctive nucleus with condensed chromosomes.

Most dinoflagellates are of marine origin and range from autotrophic to heterotrophic nutrition. Almost all photosynthetic species have chlorophylls *a* and c_2, β carotene and xanthophylls. The dominant xanthophyll is peridinin; however, some photosynthetic species lack peridinin, e.g., *Peridinium balticum* (Levander) Lemmermann and *Ptychodiscus brevis* (Davis) Steidinger. Dinoflagellate food reserves are starch and oil. In one species the starch is a polyglucan similar to starches of higher plants, but more branched (Vogel and

Meeuse, 1968). Oil reserves are long-chained unsaturated fatty acids (C_{14}, C_{16}, C_{18}, C_{22}).

There are about 130 genera representing >1200 species in the Pyrrhophyta. The group is of interest to researchers investigating a variety of biological phenomena. For example, some species are bioluminescent, some are toxic, some are parasitic, some are intracellular symbionts of invertebrates, some are binucleate, some have unusual life cycles and many show circadian and diel rhythms. Dinoflagellates have asexual and sexual reproductive capacities and are well represented in the fossil record by resting stages called cysts, many of which probably represent hypnozygotes. Geologically, one order of this group dates back to the Silurian (≃ 440 million years ago) and the living cyst of *Gonyaulax polyedra* Stein appears identical to its fossil counterpart dating from > 50 million years ago.

THECA

Using light microscopy, two types of dinoflagellates usually can be distinguished morphologically, i.e., unarmored and armored. Unarmored forms have a cell covering of membranes, while armored forms have structural cellulose or other polysaccharides in vesicles between membranes. Armored species have walls divided into plates arranged in specific series which are used as taxonomic criteria (Fig. 1). Using electron microscopy the distinction is less clear because of gradations in basic organization. Loeblich (1970) reintroduced the term "amphiesma" for the cell covering of dinoflagellates to replace the widely used term "theca." The preferred term should be theca because all dinoflagellates are thecate and those in several major orders have epi- and hypothecae.

Dodge and Crawford (1970a) illustrated eight different types of thecae representing five categories: 1) no plates; 2) numerous thin plates (hundreds); 3) thick plates numbering 20 to 60; 4) thick plates with reticulations; and 5) only two large valves with up to eight small anterior plates. An exception to this representation includes *Pyrocystis* vegetative stages which have an outer layer, 24 layers of crossed parallel cellulose fibrils, a region of dispersed smaller fibrils and innermost, the plasmalemma (Swift and Remsen, 1970).

From our observations of dinoflagellate ultrastructure the most recurring pattern of thecal structure is an outer membrane, followed by thecal vesicles, another membrane which we interpret as the plasma membrane and then microtubules in unarmored forms. Thecal vesicles may or may not contain structural polysaccharides (plates). A 5th membrane probably representing a cytoplasmic

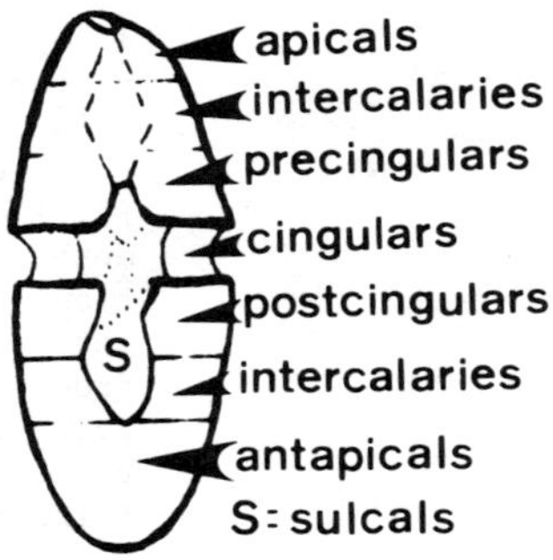

1

Fig. 1. Schematic diagram of armored form with major plate series. The apical', anterior intercalaries[a] and precingulars'' are components of the epitheca. The postcingulars''', posterior intercalaries[p] and antapicals'''' are components of the hypotheca.

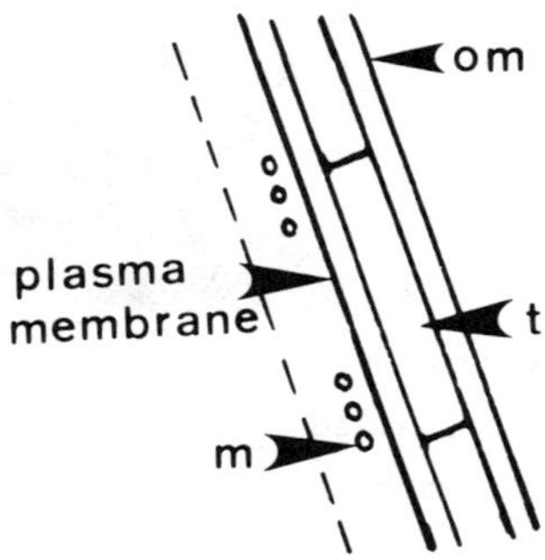

2

Fig. 2. Schematic diagram of 4 thecal membranes; (om = outer membrane; tv = thecal vesicles; m = microtubules; dotted membrane represents vesicular component, if present).

vesicular component can occur below the microtubules (Fig. 2). Armored species which shed both epitheca and hypotheca at cell division often have a pellicular layer (Fig. 3).

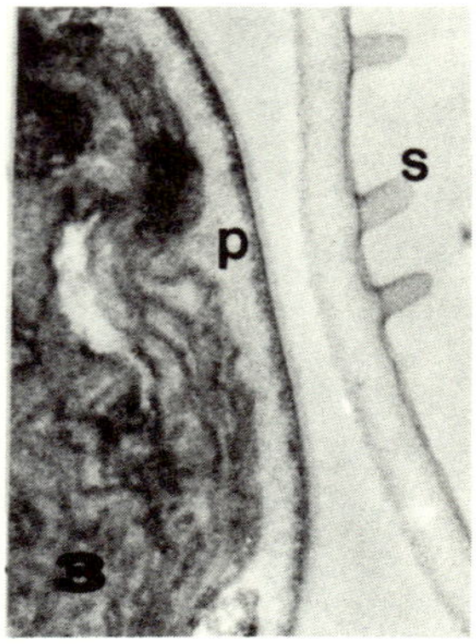

Fig. 3. UTex LB1595 *Peridinium loeblichii* (Cox and Arnott) Dale showing plate with spines (s) adjacent to protoplast of cell. A pellicle (p) separates the protoplast from the plate. The outer membrane and the thecal vesicles were removed in the fixation process. The plasma membrane is by our interpretation interior to the pellicular layer (x70,000). Reproduced from Cox and Arnott, 1971. Contributions in Phycology, [Eds.] Parker, B. C. and M. Brown. Allen Press.)

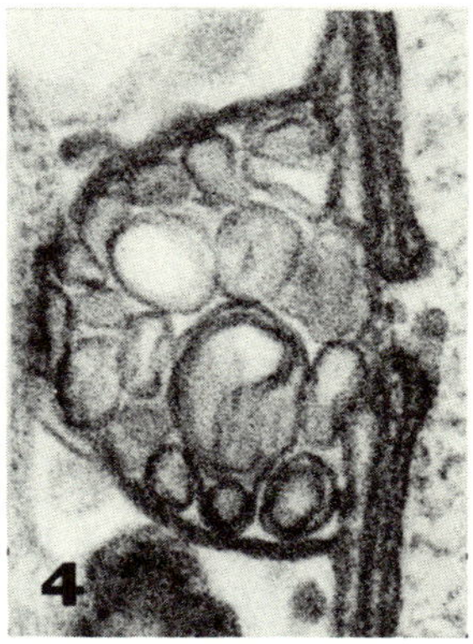

Fig. 4. Exocytotic vesicle with thecal pore (*Ptychodiscus brevis*) (x70,064).

Currently, researchers disagree on whether the outermost membrane or an inner membrane is the functional plasmalemma. Wetherbee (1975) presented convincing evidence for the plasmalemma being the outer membrane in his series of cytokinesis and wall formation in *Ceratium tripos* (O. F. Müller) Nitzsch. However, the mechanisms of ecdysis and plasmolysis in dinoflagellates can be used as an argument for an inner membrane being the plasma membrane (Steidinger *et al.*, 1973). The artificial production of viable *Gonyaulax polyedra* Stein spheroplasts by Adamich and Sweeney (1976) also suggests that the plasmalemma is not the outermost membrane. Additionally, numerous exocytotic or polyvesicular bodies (Fig. 4), some of which may represent lomasomes, occur at the innermost membrane site of several dinoflagellates. Lomasomes or plasmalemmasomes in algae have been discussed by Marchant and Robards (1968), Cole and Lin (1970) and Lewin *et al.* (1977). These vesicular bodies may be associated with membrane growth or enzymatic activity depending on their origins.

In those armored dinoflagellates which contain thick reticulated or spiny plates within the thecal vesicles (Category 4, Dodge and Crawford, 1970a), each plate has been found to possess a distinctive number of inner and outer faces (Figs. 5 and 6). An inner face is overlapped by an outer face of the adjacent plate. The morphological differences as seen in TEM replicas between the outer and inner faces of adjacent plates makes it possible to determine that growth of the thecal plates is localized in specific inner faces. At present the pattern of plate overlap has been studied only in species of the genera *Peridinium*, *Gonyaulax* and *Ornithocercus* (Dickensheets and Cox, 1971; Cox and Arnott, 1971; Taylor 1971, 1973; Gocht and Netzel, 1974; Dürr and Netzel, 1974; Postek and Cox, 1976) (Fig. 7).

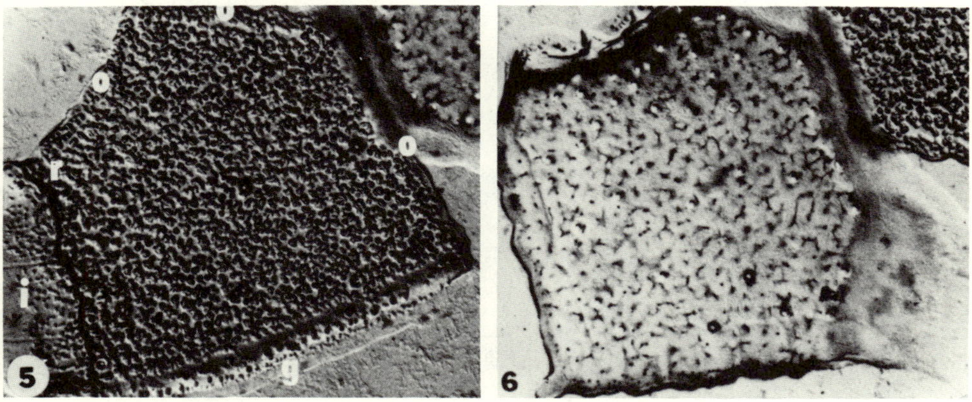

Figs. 5 and 6. Fig. 5. Replica of plate 3" of *Peridinium loeblichii* (Cox and Arnott) Dale which has three outer faces (o), one inner face (i) and one girdle face (g). Note ridge (r) and striations on inner face. Fig. 6. Replica of inside surface of plate 4"' showing that growth of inner face is unequal; here growth of the lower region being more than double the upper. (Reproduced from Cox and Arnott, 1971. Contributions in Phycology, [Eds.] Parker, B. C. and M. Brown. Allen Press.)

FLAGELLA

In most motile stages excluding the Prorocentrales, the transverse flagellum originates in a flagellar pore usually above the longitudinal flagellar pore and lies within the confines of the cingular groove, if present (Fig. 8). This flagellum has the typical axoneme structure; however, it also has a striated strand (hollow) similar to paraxial flagellar rods (solid) in other flagellates, contains packing material, and is covered by a sheath (Fig. 9). Taylor (1975) and Herman and Sweeney (1977) described a hemi-helical flagellum whereby the tension of the striated strand results in axoneme waves. There are fine hairs on the axoneme side which are attached to the theca preventing free helical movement. LeBlond and Taylor (1976) reported that transverse flagellar waves give the forward propulsion to dinoflagellate movement. Additionally, they

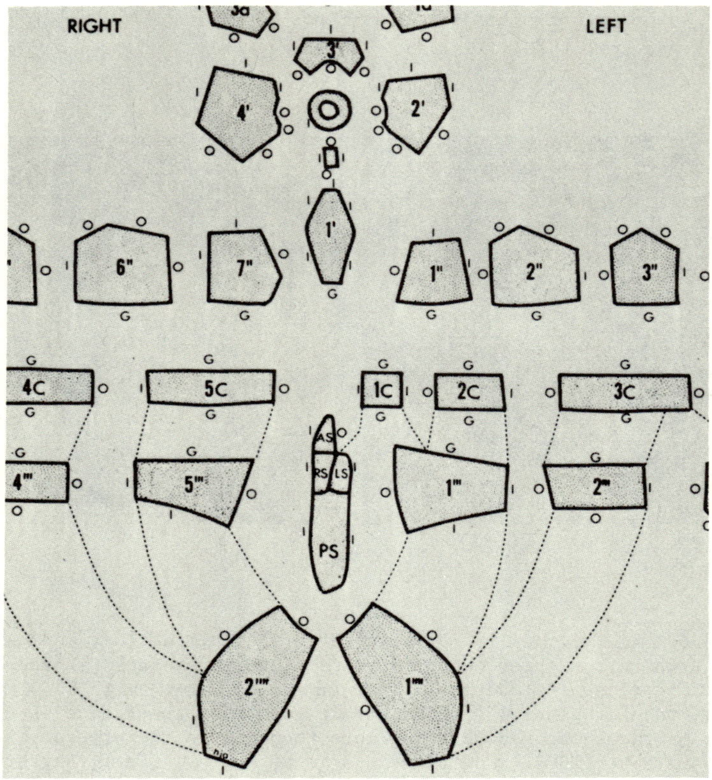

Fig. 7. Fig. 7 diagrammatically shows the relationship of the face of each thecal plate to the faces of the adjacent thecal plates in *Peridinium loeblichii* (Cox & Arnott) Dale. (Reproduced from Cox and Arnott, 1971. Contributions in Phycology, [Eds.] Parker, B. C. and M. Brown. Allen Press.)

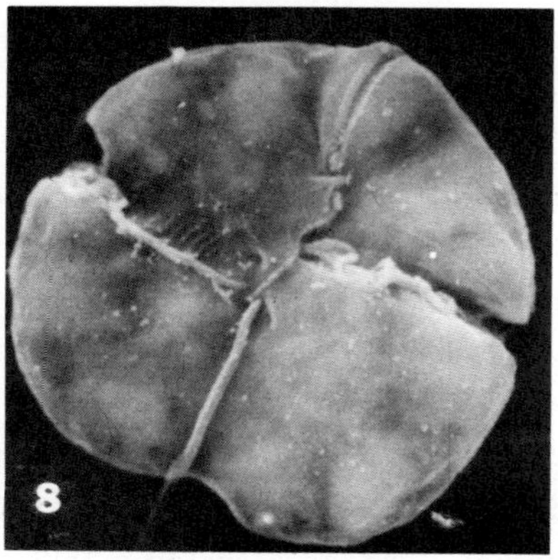

Fig. 8. Scanning electron micrograph of *Ptychodiscus brevis* showing placement of flagella and characteristic apical carina (x2,709).

Fig. 9. Transverse flagellum with straited strand (arrow) in *Ptychodiscus brevis* (x9,937).

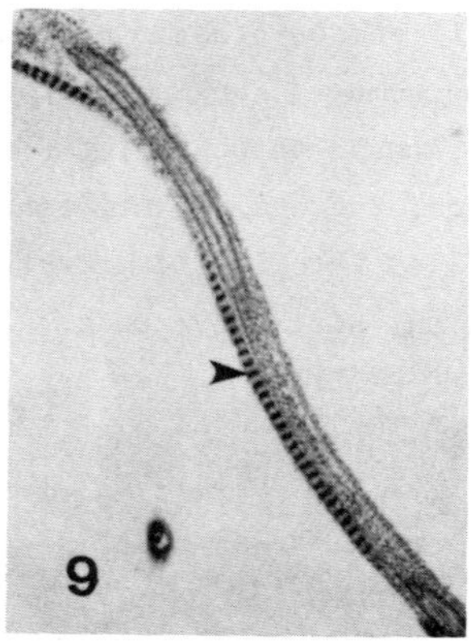

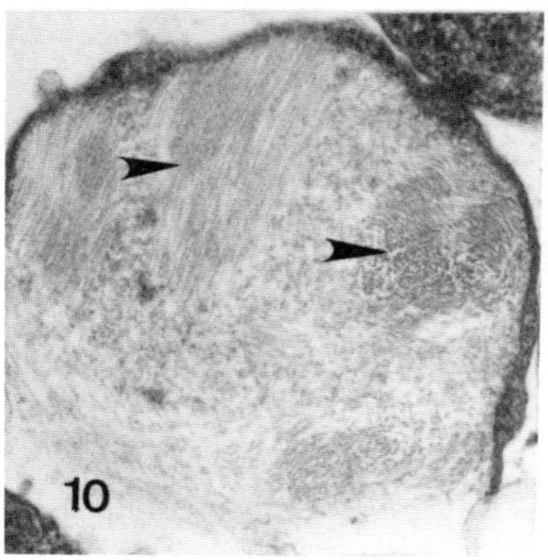

Fig. 10. Fibrous body with fibrils (arrows) in *Ptychodiscus brevis* (x29,285).

speculated that sulcal extensions may contribute to this thrust. Flagellar hairs may originate in fibrous bodies (Fig. 10) distributed throughout the cytoplasm (Dodge, 1973). Basal bodies of both the transverse and longitudinal flagella have the typical cartwheel structure. These are attached to a separate or joined striated root system. The root system does not attach to the nucleus, and it is not known whether the system is for anchorage, shock absorption or tactile reception.

Body and/or flagellar scales have been found external to the thecal complex in *Oxyrrhis marina* Dujardin, *Heterocapsa triquetra* (Ehrenberg) Stein and *Glenodinium* sp. (Clarice and Pennick, 1976; Pennick and Clarke, 1977). Some of these scales superficially resemble those of the prasinophytes. Taylor (1976) and Santore (1977) speculate that the presence of scales may not represent a homology, but rather parallel evolution in phytoflagellates.

NUCLEUS AND REPRODUCTION

Nucleus. All photosynthetic dinoflagellates studied to date are haplonts with chromosome numbers ranging up to 500 for the 1N stage. Chromosomes are continually condensed throughout the cell cycle and are uniform in size (Fig. 11). In division, the nuclear membrane and nucleolus are retained although there are parallel cytoplasmic microtubule-containing channels traversing the nucleus (Kubai and Ris, 1969; Dodge, 1973; Loeblich, 1976a; Loeblich and Hedberg, 1976). No centrioles have been found in free-living forms although cen-

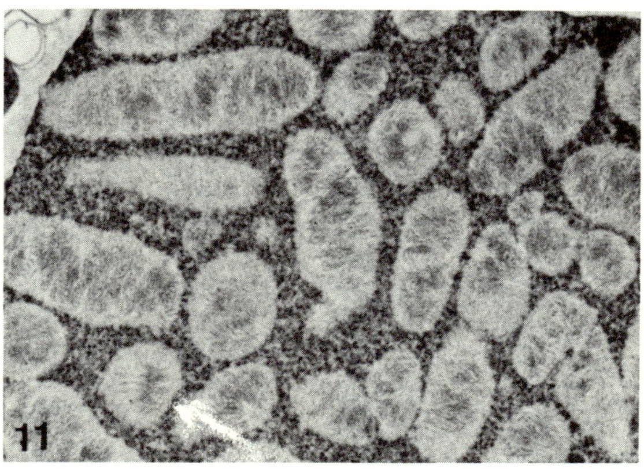

Fig. 11. *Ptychodiscus brevis* nucleus illustrating condensed chromosomes and DNA coiling (x9,045).

Fig. 12. Chromosome attachment to nuclear envelope in *Ptychodiscus brevis* (x59,947).

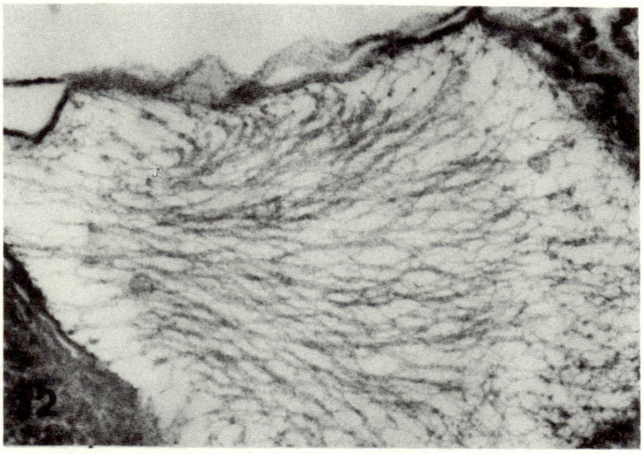

trioles have been reported for several parasitic species. In all dinoflagellates studied, the chromosomes are attached to the nuclear envelope (Fig. 12), and the extranuclear microtubules in the channels are thought to act in chromosome separation during division. Colchicine does not affect division in free-living species, although colchicine is known to disrupt microtubules. Recently, Silva (1978) has found endonuclear bacteria in two photosynthetic species.

There are several chromosome models available for describing the unique fibrillar coiling of the DNA molecules (Bouligand *et al.*, 1968; Soyer and Haapala, 1974; Loeblich, 1976; Livolant *et al.*, 1978). The most comprehensive model is the Livolant *et al.* modified, penny-stack model which illustrates the folding, twisting and overlap pattern of DNA strands. The question of polyteny and polyploidy is still being debated; however, recent models are based on 1 or 1-5 DNA molecules per chromosome. Renaturation studies reportedly do not support extensive polyploidy (Roberts *et al.*, 1974; Allen *et al.*, 1975). The similarity in appearance of the fibrillar DNA in prokaryotes and eukaryotic dinoflagellates may relate to the low levels of associated proteins. The low protein levels (Rizzo and Noodén, 1973) and physicochemical surroundings probably account for the coiled or folded profile appearances. Additionally, the associated proteins are different from typical eukaryotic histones. Ribonuclease has been used to uncoil the DNA configuration, EDTA produces DNA fibrillar clumping and uranyl salts stabilize the DNA pattern (Haller, 1961; Schmitter, 1971; Soyer and Haapala, 1974).

The varying chromosomal profiles seen in single sections can be affected by the amount of associated proteins; preparation schedules including *en bloc* staining, pH and dehydration, and the plane of section.

Reproduction and life histories. As more dinoflagellates are studied in culture, sexual life histories are being described. Species are either uni- or bisexual and most gametes are isogamous. Planozygotes and hypnozygotes have been reported with meiosis being zygotic or postzygotic (Steidinger, 1975; Pfiester, 1975, 1976, 1977; Loeblich, 1976; Turpin *et al.*, 1978; Walker and Steidinger, 1979). Meiosis in *Crypthecodinium cohnii*-like dinoflagellates was described by Beam *et al.* (1977) as a one division process. *Noctiluca* is the only known diplont and is unusual because the vegetative stage has a eukaryotic-type nucleus

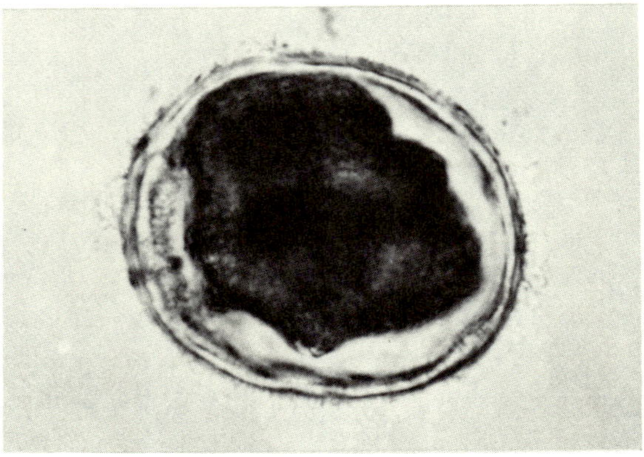

Fig. 13. *Gonyaulax monilata* hypnozygote with thick cell wall and dark protoplasm (x742).

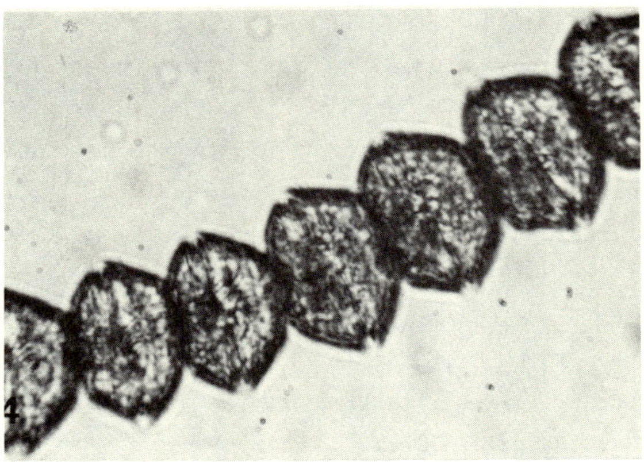

Fig. 14. *Gonyaulax monilata* chain of living daughter cells (x368).

called noctikaryotic without condensed chromosomes while the gametes have dinokaryotic nuclei (Zingmark, 1970).

Benthic resting stages (Fig. 14) often do not resemble their motile counterpart (Fig. 13); they lack flagella and motility, they have reduced cytoplasmic features with corresponding metabolic reduction and many are spherical, with or without projections. More than 70 extant species are known to produce resting cysts while the fossil record is represented by >1500 taxa. As the number of researchers interested in life cycle forms increases, more resting stages of extant species, particularly unarmored estuarine and neritic forms, will be described.

In one unarmored, freshwater dinoflagellate, hypnocyst formation was characterized by lipid accumulation, presence of an accumulation body, vacuoles with crystals, loss of pusule and Golgi, thicker cell wall and chromosomes which lost their characteristic fibrillar structure (Bibby and Dodge, 1972). Although some of these characters are typical of ageing cells and cells subjected to unfavorable environmental condition, e.g. lipid and waste accumulation, the processes in dinoflagellate encystment, which may represent zygote formation, are characteristic of surviving rather than dying populations.

Species complexes, such as those described by the team of Beam and Himes (1977, 1978), which involve sexually isolated strains, may someday be separated to species by mating behavior, electrophoresis, and immunoassay rather than by morphological and cytological characteristics. This systematic approach to sibling species was recently attempted in *Paramecium* and *Tetrahymena* by Sonneborn (1975) and Nanney and McCoy (1976). Other researchers recommend specifying strain type or isolate number.

CHLOROPLASTS, PYRENOIDS AND EYESPOTS

Chloroplasts, when present, typically have a triple membrane envelope without ER connections (Dodge, 1973). They can be lobed and irregular or discoid. Lamellae usually consist of 2-3 closely appressed thylakoids although the number can range from 1 to >8 (Fig. 15). Occasionally, grana-like structures have been reported. The chloroplast nucleoid areas can be scattered or localized in the stroma with a ribosomal groundplasm, DNA fibrils (Fig. 16) and a structure that appears similar to a nucleolus.

Not all dinoflagellates have distinct pyrenoids, but if they do, these structures are granular, proteinaceous and can vary considerably within a genus. Pyrenoids can be divided into five types: 1) simple, fusiform and interlamellar; 2) compound, interlamellar; 3) simple, stalked with starch cap; 4) complex, multiple stalked pyrenoid supported by chloroplast lobes, with or without a starch cap (Fig. 17); and 5) stalked, but matrix with invaginations of chloroplast envelope (Dodge and Crawford, 1971). Stalked pyrenoids are surrounded by the chloroplast envelope and some can contain a crystal lattice of proteins.

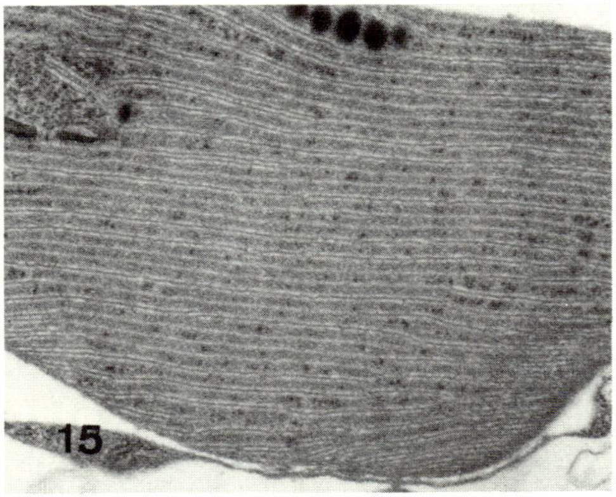

Fig. 15. *Ptychodiscus brevis*. Parallel lamellae consisting of 2-3 thylakoids per stack (x24,490).

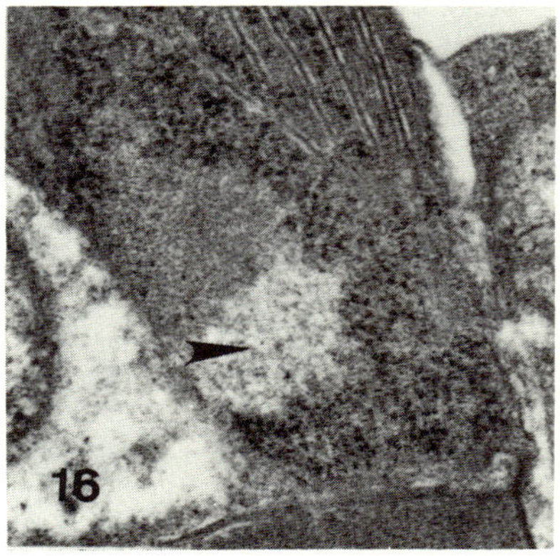

Fig. 16. *Ptychodiscus brevis*. Chloroplast nucleoid area (arrow) (x39,133).

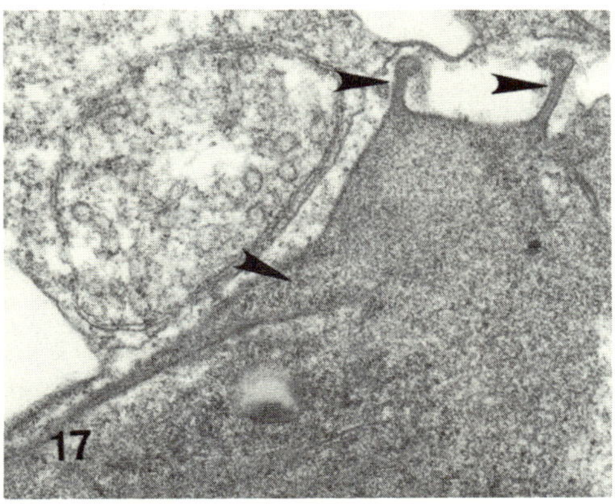

Fig. 17. *Ptychodiscus brevis*. Stalked pyrenoid with three discernable attachment points (arrows) (x37,177).

Pyrenoids have only been found associated with chloroplasts that do not contain grana and may be associated with the translocation of photosynthate in some species. Pyrenoid type has been used to differentiate between two species (or species complexes) of zooxanthellae in the encysted stage (Taylor, 1971).

Dodge (1975) distinguished five categories of chloroplasts: 1) more or less parallel lamellae; 2) same, but with peripheral lamellae; 3) girdle lamellae and internal pyrenoid; 4) fusion of thylakoid to envelope and internal pyrenoid; and 5) more or less radial arrangement of lamellae. According to Dodge, the presence of peridinin, parallel lamellae and stalked pyrenoids represents an advanced evolutionary state, and he believes most speciation took place while dinoflagellates were strictly heterotrophic. The unarmored species, *Ptychodiscus brevis* has either parallel or radial lamellae; multistalked pyrenoids and lacks peridinin as well as fucoxanthin (Jeffrey *et al.*, 1975; Steidinger *et al.*, 1978). This species does not contain endosymbiotic crysophytes and does not fit into the evolutionary scheme presented by Dodge.

Prezélin *et al.* (1976) and Prezelin (1976) demonstrated that in one dinoflagellate species photosynthetic rates can be maintained in lowered light by the accumulation of peridinin-chl-a-proteins (PCP) located on the surface of lamellae. These PCPs are thought to represent a photoadaptation mechanism which increases the potential for photon capture, similar to phycobiliproteins of the blue-green algae (cyanobacteria), red algae and cryptophytes.

Although dinoflagellates can osmotically and ionically regulate fluid contents and older cells accumulate lipids (both of which aid in bouyancy) they are motile and move at 1-2 m/hr. There are two distinct phases: 1) stop response for several seconds; and 2) directional movement in relation to stimuli. Light receptive pigments e.g., carotenoproteins and possibly a phytochrome (Forward, 1973, 1975), are involved. This photorecptor system and the organism's positive phototactic response account for rhythms whereby many species migrate toward the light at dawn. Vertical migration of dinoflagellates is usually confined to the uppermost 15 m of the water column.

Eyespots are uncommon in dinoflagellates, but among those that have them, there are four types: 1) mass of osmiophilic globules independent of chloroplast and located near the sulcus, not membrane bounded; 2) single layer of globules as part of the chloroplast periphery; 3) membrane bounded two-layered system of globules separated from the chloroplast and adjacent to a lamellar body; and 4) ocelli of a "lens", retinoid, and pigment granules (Dodge, 1973). It is assumed that these structures function in light reception. The ocellus of *Nematodinium* has a 0.6 μm resolution capability with a field of reception equal to 30° (Francis, 1967).

TRICHOCYSTS, NEMATOCYSTS AND MUCOCYSTS

Many dinoflagellates, both photosynthetic and heterotrophic, have ejectile organelles called trichocysts (Figs. 18a and 18b) which differ morphologically from those of cryptomonads, prasinophytes and ciliates. Mature trichocysts are found at the periphery of cells and number from only a few to hundreds.

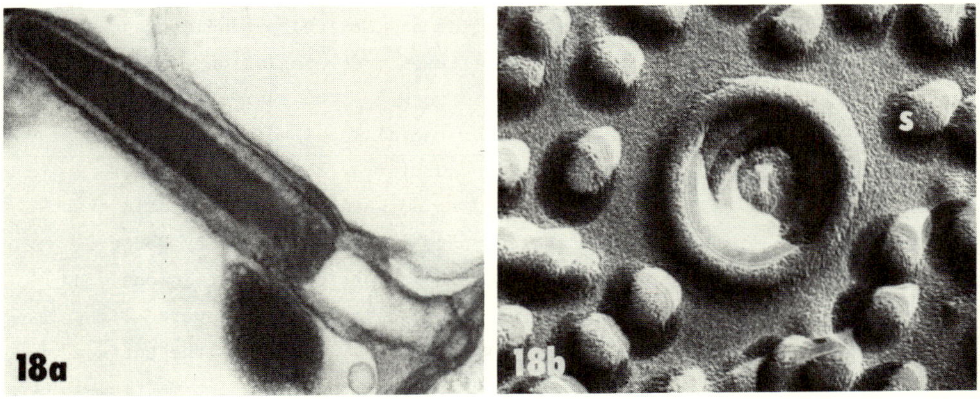

Figs. 18a and 18b. Fig. 18a. Mature trichocyst of *Ptychodiscus brevis* with thecal pore (x46,486). Fig. 18b. Replica of a plate in *Peridinium loeblichii*. T = trichocyst pore; s = spines which form ornamentation on outer surface of plate (x52,000). (Reproduced from Cox and Arnott, 1971. Contributions in Phycology, [Eds.] Parker, B. C. and M. Brown. Allen Press.)

There are basically two parts - the neck and body. The neck is attached to thecal membranes and contains twisted fibrils; the body can also contain twisted fibrils attached to the core. The main core is square in cross section and consists of proteinaceous granules. The mid-body can have tubules or fibrils between the core and the limiting membrane. The origin of fibrils in primed trichocysts could possibly be from fibrous bodies (Schmitter, 1971; Steidinger, 1979). Trichocysts develop in swollen vesicles near dictyosomes. Several theories have been proposed for the release mechanism. One suggests that a change in osmotic pressure allows water to solubilize the protein core which is then recrystallized and ejected. Speculations on function include reducing sinking rates, osmoregulation, attachment organelles, avoidance, secretion or stabilization. Mucocysts (Fig. 19) are also present in some species and perhaps contribute to the outer coat of mucopolysaccharide material above thecal membranes. Speculation on function of mucocysts could include those applied to other algae, e.g., protective, adhesive, increased frictional drag, etc. Another ejectile organelle is the nematocyst. This complex organelle has also been called a cnidocyst because of an overstated similarity to structures in cnidarians.

The arrangement of trichocyst pores is fairly consistent in certain species and is even used as a taxonomic character. The repeatability must be dictated by membrane phenomena. Satir (1975) demonstrated that fusion rosettes in plasma lemmas are specific sites of trichocyst and mucocyst attachment in many protozoa.

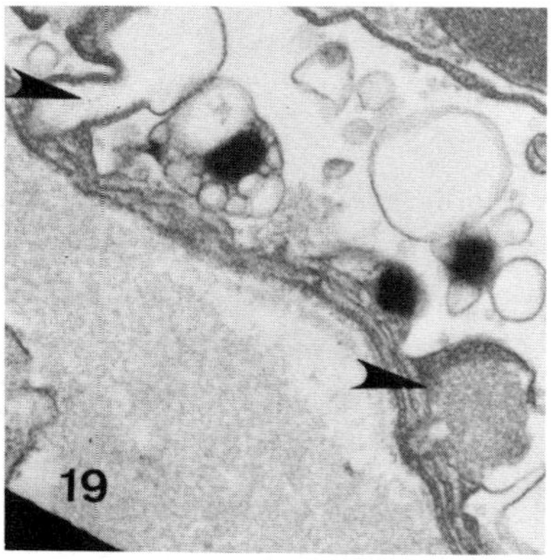

Fig. 19. *Ptychodiscus brevis* mucocysts (arrows) at theca, both with pores (x38,239).

PUSULES

Most, if not all freshwater dinoflagellates contain pusules as do some marine dinoflagellates and even parasitic forms. If present, they are usually connected to flagellar canals (Fig. 20) and consist of a two membrane system.

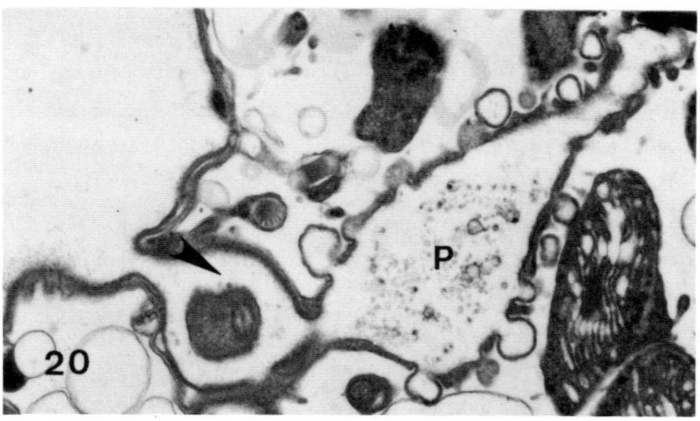

Fig. 20. *Ptychodiscus brevis* pusule (p) with connection to flagellar canal (arrow) (x17,232).

Dodge (1972) characterized seven types and divided them into two categories: 1) chamber with pusule vesicles, and 2) a system of tubules or sacs. The function of pusules in dinoflagellates is open to question. Cachon *et al.* (1970) suggested an excretory function and Dodge has suggested an osmoregulatory function to maintain ionic balance. Although the pusule is reportedly noncontractile, Biecheler (1952) noted distinct changes in volume, and the chambers of category (1) are usually filled with membrane profiles. Steidinger and Williams (1970) thought pusules were involved with phagocytosis; however, no bacteria or particulate matter has been observed within the system. In *Ceratium hirundinella* (O. F. Müller) Bergh there is no pusule connected to the flagellar canal, but tubules are found throughout the cytoplasm and food vacuoles have also been found (Dodge and Crawford, 1970b).

ULTRASTRUCTURAL VARIATION

Qualitative ultrastructural changes in micro- and macroalgae due to environmental or physiological parameters include: 1) grana formation with high CO_2 (Gergis, 1972); 2) chloroplast changes with light (Findley *et al.*, 1970; Kulandaivelu and Senger, 1976; Jeffrey and Vesk, 1977, 1978); 3) mitochondrial changes with heavy metals (Silverberg, 1976); 4) appearance of myelin figures with ageing (Aaronson, 1973; Gomez *et al.*, 1974); 5) increase in vacuolization with ageing (Colombo and Orsenigo, 1977; Oliveira and Bisalputra, 1977); 6) accumulation of lipid with ageing (McLean, 1968, 1969; Palisano and Walne, 1972); and 7) accumulation of plastiglobuli with ageing (Palisano and Walne, 1972). Changes with ageing are thought to occur due to internal nitrogen depletion, accumulation of waste products and loss of membrane bounded enzyme systems.

In old cultures of the symbiotic dinoflagellate, *Gymnodinium microadriaticum* (Fruedenthal) Taylor, the following qualitative variations were noted: 1) chloroplast lamellae were more appressed, stroma reduced; 2) increased frequency of waste storage in vacuoles; 3) loss of starch; and 4) breakdown of membrane structures and organelles (Taylor, 1968). The effects of blue-green light on dinoflagellate chloroplasts also indicates increased thylakoids and reduced chloroplast stroma (Vesk and Jeffrey, 1977). In the freshwater species *Peridinium cinctum* f. *westii* (Lemmermann) Lefevre, older specimens had fewer chloroplasts and thylakoids (Messer and Ben-Shaul, 1972).

Recently, Steidinger (1979) documented the following quantitative ultrastructural variation in *Ptychodiscus brevis* using morphometric techniques: 1) decrease in chloroplast area with increasing population age; 2) increased frequency of fibrous bodies with age; 3) significant differences in vesiculation with population age; 4) differences in thylakoids between culture and field specimens; and 5) chromosome area differences between strains (haploid vs autodiploid). In addition to analyzing subtle ultrastructure changes associated with specific parameters, e.g., culture strain, light intensity, salinity, population age, this technique could be applied to determining organelle function, e.g., is the pusule an osmoregulatory organelle? The technique is also useful in determining whether old culture strains have undergone ploidy (di-, poly- or aneuploidy) if chromosome counts are unattainable.

ORIGIN OF PHOTOSYNTHETIC DINOFLAGELLATES

Margulis (1968, 1970) and others have suggested that the eukaryotic cell evolved by a series of symbioses involving prokaryotic cells. Gibbs (1978) has proposed that the chloroplasts of *Euglena* have arisen from endosymbiotic green algal chloroplasts. Evidence is accumulating that the endosymbiotic theory may be useful in explaining the origin of the photosynthetic dinoflagellates.

In 1969, *Peridinium foliaceum* (Stein) Biecheler was shown to have fucoxanthin and not peridinin as the dominant xanthophyll. Dodge (1971) found that this species (as *Glenodinium foliaceum* Stein) possessed a typical dinoflagellate nucleus and a polymorphic body which was bounded by a double membrane with nuclear pores. He termed this body a "eukaryotic nucleus", although he did not speculate with any degree of certainty about its origin or function. Tomas *et al.* (1973) described a second dinoflagellate, *Peridinium balticum* (Levander) Lemmermann, which possessed a dinokaryotic and eukaryotic nucleus (Figs. 21, 22). Later, Tomas and Cox (1973) proposed that photosynthetic dinoflagellates may have evolved from phagotrophic ancestors which were invaded by an algal symbiont. The proposal of Tomas and Cox (1973) was based on chloroplast ultrastructure, pyrenoid and thylakoid characteristics and the presence of chloroplast endoplasmic reticulum. They described a single invaginating membrane which separated the chloroplasts, eukaryotic nucleus and cytoplasm from the dinoflagellate nucleus, eyespot, other organelles and cytoplasm. They stated, "The absorption of the symbiotic plasmalemma and the fusion of 2 of the 4 membranes surrounding the chloroplasts would result in a photosynthetic dinoflagellate with triple membrane bounded chloroplasts." Their preliminary pigment analysis showed fucoxanthin was present. This has been restated by Loeblich (1976), Dodge (1979) and Spero (1979). Gibbs' (1978) discussion of the origin of the photosynthetic dinoflagellates is particularly informative.

The eukaryotic nucleus of *Peridinium balticum* and *P. foliaceum* has several

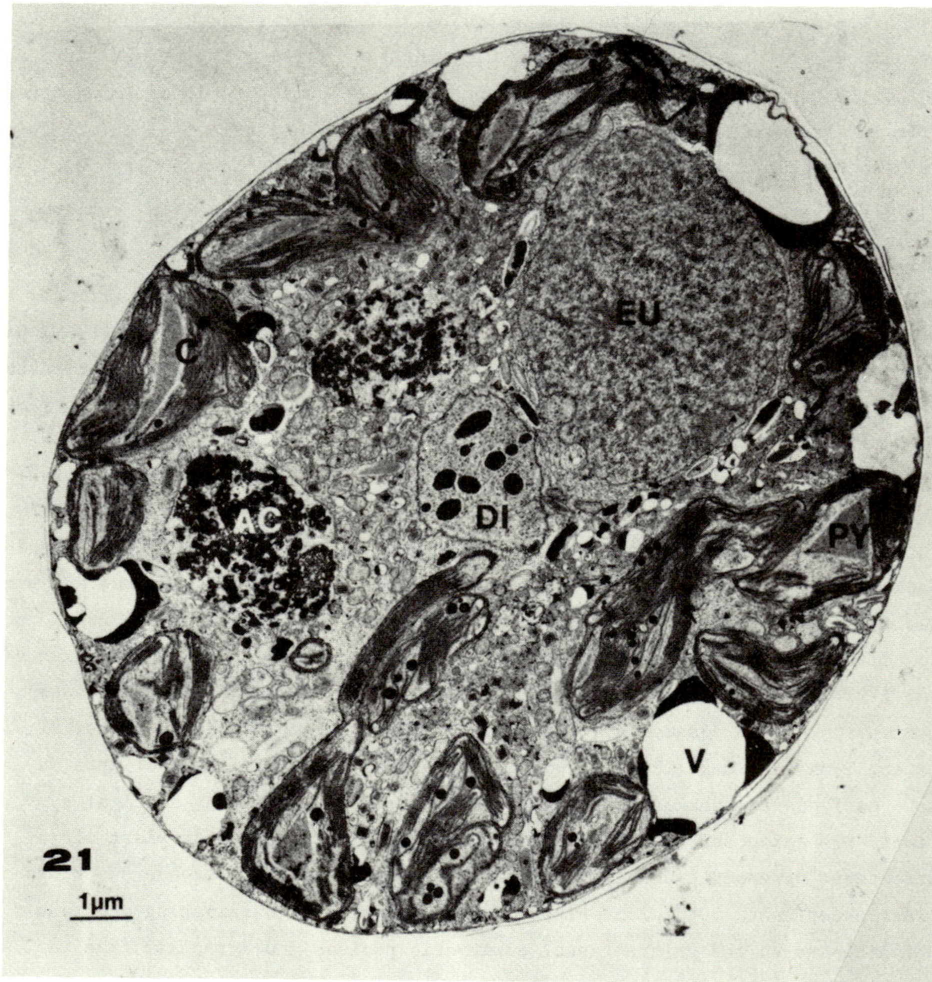

Fig. 21. A vegetative cell of *Peridinium balticum* illustrating the two distinctive nuclei, peripherally located chloroplasts, other organelles and structures. C, chloroplast; V, vacuole; PY, pyrenoid; DI, dinokaryotic nucleus; EU, eukaryotic nucleus; AC, accumulation body. Bar, 1 μm. (Reproduced from Tomas and Cox, 1973 with permission of Journal of Phycology.)

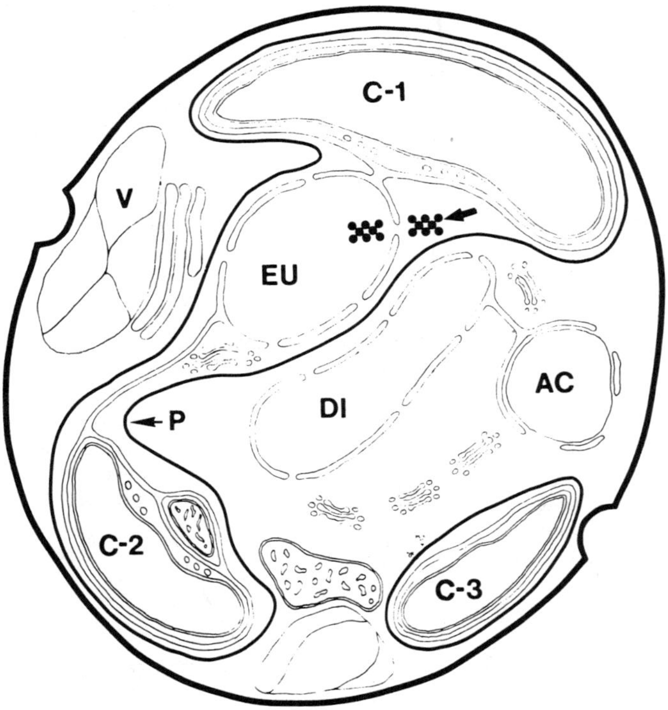

22

Fig. 22. A diagrammatic representation of the relationship between *Peridinium balticum* and its endosymbiont. The plasmalemma of the endosymbiont ramifies throughout the dinokaryotic cytoplasm and segregates the eukaryotic cytoplasm with its set of organelles from that of the dinoflagellate. The chloroplasts found within *P. balticum* belong to the endosymbiont as they are enclosed within its plasmalemma. V, vacuole; arrow at P, plasmalemma of chrysophyte symbiont; AC, accumulation body; DI, dinokaryotic necleus; EU, eukaryotic nucleus; C-1, C-2, C-3, chloroplasts; arrow, virus particles. (Reproduced from Tomas and Cox, 1973 with permission of Journal of Phycology.)

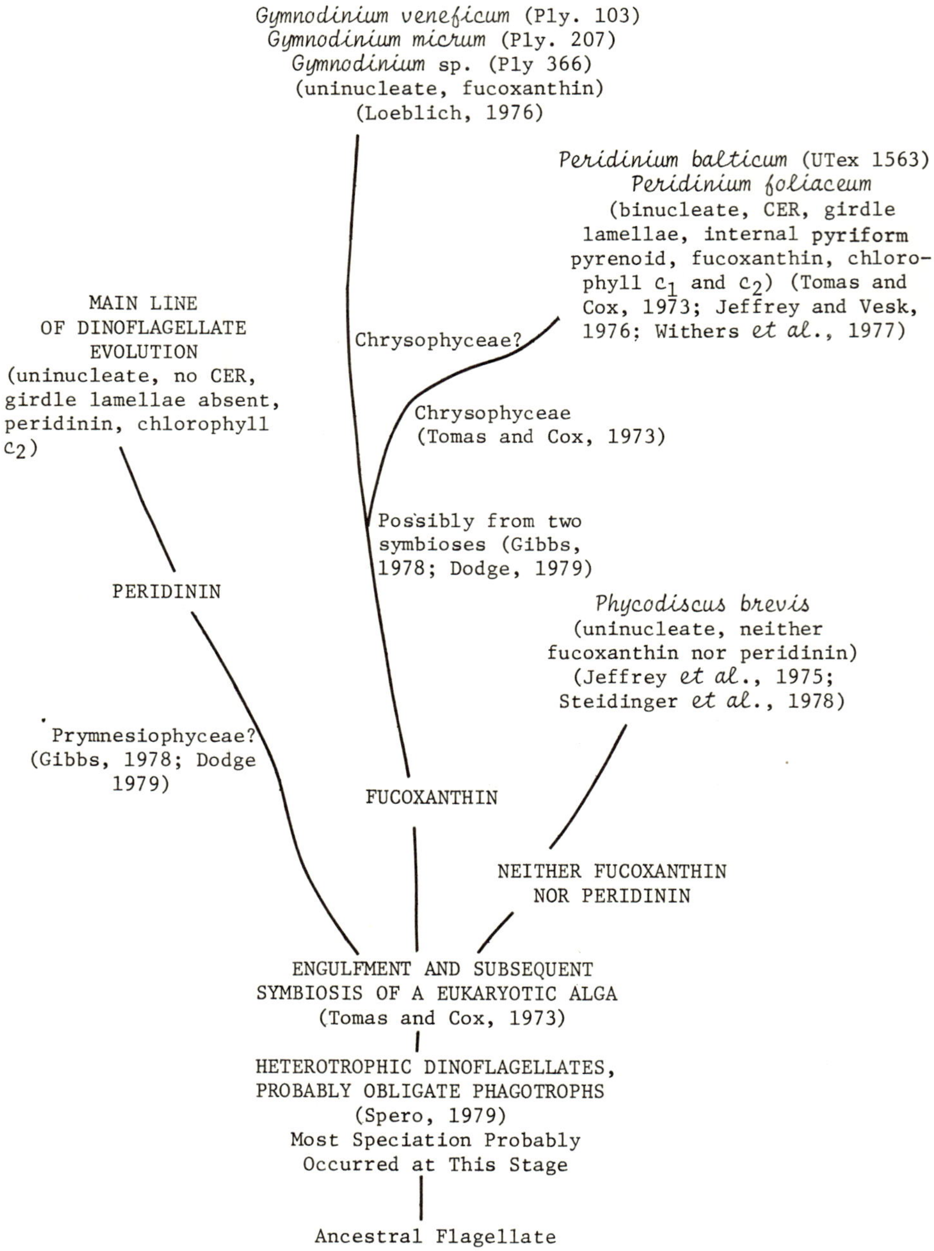

Fig. 23. A diagrammatic representation of possible phylogenetic relationships among the variously pigmented photosynthetic dinoflagellates.

anomalous characteristics which must be explained by more research. It is extremely polymorphic; its chromatin does not condense to form chromosomes. There is no spindle nor microtubules and its division is apparently by a fission-like process (Tippitt and Pickett-Heaps, 1976). Rizzo and Cox (1977) showed that the eukaryotic nucleus of *P. balticum* contained basic proteins with the same gel mobilities as four of the five histones of calf thymus.

Although plate tabulations differ, *Peridinium balticum* and *P. foliaceum* are closely related systematically and have been isolated together from natural populations. Distinctions between the two organisms are subtle (Withers *et al.*, 1977). *P. balticum* is more round, whereas *P. foliaceum* is extremely compressed dorso-ventrally, appearing concave.

Jeffrey and Vesk (1976) presented ultrastructural evidence for a membrane bounded endosymbiont in *Peridinium foliaceum* similar to the one found in *P. balticum* and stated that the chloroplast pigments of both species contain the full complement of chrysophyte chlorophylls (*a*, c_1 and c_2) and carotenoids (β-carotene, diatoxanthin, diadinoxanthin and fucoxanthin). They did not find the typical dinoflagellate carotenoids, peridinin and dinoxanthin. Mandelli (1968) and Withers and Haxo (1975) showed that *P. foliaceum* had fucoxanthin, and not peridinin, as the major carotenoid and that chlorophyll c_1 and c_2 were present. In a detailed study of the pigments of *P. balticum* Withers *et al.* (1977) reported that some differences in pigmentation occur between *P. balticum* and *P. foliaceum*, but that the pigments are essentially the same and differ from the characteristic dinoflagellate pigment complement which consists of peridinin and chlorophyll c_2, and others.

Steidinger *et al.* (1978) have studied the ultrastructure of *Pychodiscus brevis*. *P. brevis* lacks both peridinin and fucoxanthin (Jeffrey *et al.*, 1975) and is not binucleate. Ultrastructural details and pigment characteristics of such dinoflagellates as *P. brevis* present an interesting dilemma for workers who speculate on the origin of the photosynthetic dinoflagellates.

There is very good evidence that endosymbiosis has occurred in the dinoflagellates not just once, but several times and it is possible to propose that some of the varying carotenoid pigment patterns in the photosynthetic dinoflagellates are the result of the occurrence of symbiotic events at different times. The present line of evolution of the photosynthetic dinoflagellates does not appear to be from a single symbiosis, but from several which lead to the different carotenoid pigment patterns we find today (Loeblich, 1976; Dodge, 1979). Fig. 23 is a diagrammatic representation of possible phylogenetic relationships among the variously pigmented photosynthetic dinoflagellates.

REFERENCES

Aaronson, S. 1973. Crystalline structures and myelin-like features in *Ochromonas* mitochondria. Microbios 7: 231-233.

Adamich, M. and B. M. Sweeney. 1976. The preparation and characterization of *Gonyaulax* spheroplasts. Planta (Berl.) 130: 1-6.

Allen, J. R., T. M. Roberts, A. R. Loeblich III and L. C. Klotz. 1975. Characterization of the DNA from the dinoflagellate *Crypthecodinium cohnii* and implications of nuclear organization. Cell 6: 161-169.

Beam, C. A. and M. Himes. 1977. Sexual isolation and genetic diversification among some strains of *Crypthecodinium cohnii*-like dinoflagellates. Evidence of speciation. J. Protozool. 24: 532-539.

Beam, C. A., M. Himes, J. Himelfarb, C. Link and K. Shaw. 1977. Genetic evidence of unusual meiosis in the dinoflagellate *Crypthecodinium cohnii*. Genetics 87: 19-32.

Bibby, B. T. and J. D. Dodge. 1972. The encystment of a freshwater dinoflagellate: a light and electron-microscopial study. Br. Phycol. J. 7: 85-100.

Biecheler, B. 1952. Recherches sur les péridiniens. Bull. Biol. Fr. Belg. Suppl. 36: 1-149.

Bouligand, Y., M. Soyer and S. Puiseaux-Dao. 1968. La structure fibrillaire et l'onentation des chromosomes chez les dinoflagellés. Chromosoma (Berl.) 24: 251-287.

Cachon, J., M. Cachon and C. Greuet. 1970. Le systeme pusulaire de quelques péridiniens libres ou parasites. Protistologica 6: 467-476.

Clarice, K. J. and N. C. Pennick. 1976. The occurrence of body scales in *Oxyrrhis marina* Dujardin. Br. Phycol. J. 11: 345-348.

Cole, K. and S. Lin. 1970. Plasmalemmasomes in sporelings of the brown alga *Petalonia debilis*. Can. J. Bot. 48: 265-268.

Colombo, P. M. and M. Orsenigo. 1977. Sea depth effects on the alga photosynthetic apparatus. II. An electron microscopy study of the photosynthetic apparatus of *Halimeda tuna* (Chlorophyta, Siphonales) at -0.5m and -6.0m sea depths. Phycologia 16: 9-17.

Cox, Elenor R. and H. J. Arnott. 1971. The ultrastructure of the theca of the marine dinoflagellate, *Ensiculifera loeblichii* sp. nov. In Parker, B. C. and M. Brown [Eds.] *Contributions in Phycology*. Allen Press. pp. 121-136.

Dickensheets, R. E., Jr. and Elenor R. Cox. 1971. Thecal ultrastructure of *Scrippsiella faeroense*. Protoplasma 73: 139-143.

Dodge, J. D. 1971. A dinoflagellate with both a mesokaryotic and a eukaryotic nucleus. I. Fine structure of the nuclei. Protoplasma 73: 145-157.

Dodge, J. D. 1972. The ultrastructure of the dinoflagellate pusule: a unique osmoregulatory organelle. Protoplasma 75: 285-302.

Dodge, J. D. 1973. *The fine structure of algal cells*. Academic Press, London, 261 pp.

Dodge, J. D. 1975. A survey of chloroplast ultrastructrue in the Dinophyceae. Phycologia 14: 253-263.

Dodge, J. D. 1979. The Phytoflagellates: Fine Structure and Phylogeny. *Biochemistry and Physiology of the Protozoa*. Academic Press, London, pp. 7-57.

Dodge, J. D. and R. M. Crawford. 1970a. A survey of thecal fine structure in the Dinophyceae. Bot. J. Linn. Soc. 63: 53-67.

Dodge, J. D. and R. M. Crawford. 1970b. The morphology and fine structure of *Ceratium hirundinella* (Dinophyceae). J. Phycol. 6: 137-149.

Dodge, J. D. and R. M. Crawford. 1971. A fine-structural survey of dinoflagellate pyrenoids and food reserves. Bot. J. Linn. Soc. 64: 105-115.

Dürr, G. and H. Netzel. 1974. The fine structure of the cell surface in *Gonyaulax polyedra* (Dinoflagellata). Cell Tiss. Res. 150: 21-41.

Findley, D. L., P. L. Walne and R. W. Holton. 1970. The effects of light intensity on the ultrastructure of *Chlorogloea fritschii* Mitra grown at high temperature. J. Phycol. 6: 182-188.

Forward, R. B., Jr. 1973. Phototaxis in a dinoflagellate: action spectra as evidence for a two-pigment system. Planta (Berl.) 111: 167-178.

Forward, R. B., Jr. 1975. Dinoflagellate phototaxis: pigment system and circadian rhythm as related to diurnal migration. In Vernberg, F. J. [Ed.] *Physiological Ecology of Estuarine Organisms*. University of South Carolina Press, Columbia, South Carolina. pp. 367-381.

Francis, D. 1967. On the eyespot of the dinoflagellate *Nematodinium*. J. Exp. Biol. 47: 495-502.

Gergis, M. S. 1972. Influence of carbon dioxide supply on the chloroplast structure of *Chlorella pyrenoidosa*. Arch. Mikrobiol. 83: 321-327.

Gibbs, S. P. 1978. The chloroplasts of *Euglena* may have evolved from symbiotic green algae. Can. J. Bot. 56: 2883-2889.

Gocht, H. and H. Netzel. 1974. Rasterelektronenmikroskopische untersuchungen am panzer von *Peridinium* (Dinoflagellata). Arch. Protistenk. 116: 381-410.

Gomez, M. P., J. B. Harris and P. L. Walne. 1974. Studies of *Euglena gracilis* in ageing cultures. II. Ultrastructure. Br. Phycol. J. 9: 175-193.

Haller, G. de. 1961. Variations dans l'ultrastructure du noyau chez *Amphidinium elegans* (Gymnodinidae). Progress in Protozoology, Publ. House Czech. Acad. Sci. Prague. pp. 423-424.

Herman, E. M. and B. M. Sweeney. 1977. Scanning electron microscopic observations of the flagellar structure of *Gymnodinium splendens* (Pyrrophyta, Dinophyceae). Phycologia 16: 115-118.

Himes, M. and C. A. Beam. 1978. Further studies of breeding restrictions among *Crypthecodinium cohnii*-like dinoflagellates. Evidence of a new interbreeding group. J. Protozool. 25: 378-380.

Jeffrey, S. W., M. Sielicki and F. T. Haxo. 1975. Chloroplast pigment patterns in dinoflagellates. J. Phycol. 11: 374-384.

Jeffrey, S. W. and M. Vesk. 1976. Further evidence for a membrane-bound endosymbiont within the dinoflagellate *Peridinium foliaceum*. J. Phycol. 12: 450-455.

Jeffrey, S. W. and M. Vesk. 1977. Effect of blue-green light on photosynthetic pigments and chloroplast structure in the marine diatom *Stephanopyxis turris*. J. Phycol. 13: 271-279.

Jeffrey, S. W. and M. Vesk. 1978. Chloroplast structural changes induced by white light in the marine diatom *Stephanopyxis turris*. J. Phycol. 14: 238-240.

Kubai, D. A. and H. Ris. 1969. Division in the dinoflagellate *Gyrodinium cohnii* (Schiller). A new type of nuclear reproduction. J. Cell Biol. 40: 508-528.

Kulandaivelu, G. and H. Senger. 1976. Changes in the reactivity of the photosynthetic apparatus in heterotrophic ageing cultures of *Scenedesmus obliquus*. II. Changes in ultrastructure and pigment composition. Physiol. Plant. 36: 165-168.

LeBlond, P. H. and F. J. R. Taylor. 1976. The propulsive mechanism of the dinoflagellate transverse flagellum reconsidered. Biosystems 8: 33-39.

Lewin, J., R. E. Norris, S. W. Jeffrey and B. E. Pearson. 1977. An aberrant chrysophycean alga *Pelagococcus subviridis* gen. nov. et sp. nov. from the north Pacific Ocean. J. Phycol. 13: 259-266.

Livolant, F., M.-M. Giraud and Y. Bouligand. 1978. A goniometric effect observed in sections of twisted fibrous materials. Biol. Cellulaire 31: 159-168.

Loeblich, A. R., III. 1970. The amphiesma or dinoflagellate cell covering. Proc. N. Amer. Paleontol. Conv., Part G: 867-929.

Loeblich, A. R., III. 1976a. Dinoflagellate genetics and DNA characterization. Stadlv. Sympos. 8: 111-128. U. Missouri, Columbia.

Loeblich, A. R., III. 1976b. Dinoflagellate evolution: speculation and evidence. J. Protozool. 23: 13-28.

Loeblich, A. R., III and M. F. Hedberg. 1976. Desmokont mitosis: involvement of several microtubule containing channels. Bot. Mar. 19: 255-256.

Mandelli, E. F. 1968. Carotenoid pigments of the dinoflagellate *Glenodinium foliaceum* Stein. J. Phycol. 4: 347-348.

Marchant, R. and A. W. Robards. 1968. Membrane systems associated with the plasmalemma of plant cells. Ann. Bot. n.s. 32: 457-471.

Margulis, L. 1968. Evolutionary criteria in Thallophytes: a radical alternative. Science 161: 1020-1022.

Margulis, L. 1970. *Origins of Eukaryotic Cells*. Yale University Press. New Haven.

McLean, R. J. 1968. Ultrastructure of *Spongiochloris typica* during senescence. J. Phycol. 4: 277-283.

McLean, R. J. 1969. Rejuvenation of senescent cells of *Spongiochloris typica*. J. Phycol. 5: 32-37.

Messer, G. and Y. Ben-Shaul. 1972. Changes in chloroplast structure during culture growth of *Peridinium cinctum* f. *westii* (Dinophyceae). Phycologia 11: 291-299.

Nanney, D. L. and J. W. McCoy. 1976. Characterization of the species of the *Tetrahymena pyriformis* complex. Trans. Amer. Micros. Soc. 95: 664-682.

Oliveira, L. and T. Bisalputra. 1977. Ultrastructural and cytochemical studies on the nature and origin of the cytoplasmic inclusions of ageing cells of *Ectocarpus* (Phaeophyta, Ectocarpales). Phycologia 16: 235-244.

Palisano, J. R. and P. L. Walne. 1972. Acid phosphatase activity and ultrastructure of aged cells of *Euglena granulata*. J. Phycol. 8: 81-88.

Pennick, N. C. and K. J. Clarke. 1977. The occurrence of scales in the Peridinian dinoflagellate *Heterocapsa triquetra* (Ehrenb.) Stein. Br. Phycol. J. 12: 63-66.

Pfiester, L. A. 1975. Sexual reproduction of *Peridinium cinctum* f. *ovoplanum* (Dinophyceae). J. Phycol. 11: 259-265.

Pfiester, L. A. 1976. Sexual reproduction of *Peridinium willei* (Dinophyceae). J. Phycol. 12: 234-238.

Pfiester, L. A. 1977. Sexual reproduction of *Peridinium gatunense* (Dinophyceae). J. Phycol. 13: 92-95.

Postek, M. J. and Elenor R. Cox. 1976. Thecal ultrastructure of the toxic marine dinoflagellate *Gonyaulax catenella*. J. Phycol. 12: 88-93.

Prézelin, B. B. 1976. The role of peridinin-chlorophyll a-proteins in the photosynthetic light adaption of the marine dinoflagellate *Glenodinium* sp. Planta (Berl.) 130: 225-233.

Prézelin, B. B., A. C. Ley and F. T. Haxo. 1976. Effects of growth irradiance on the photosynthetic action spectra of the marine dinoflagellate *Glenodinium* sp. Planta (Berl.) 130: 251-256.

Rizzo, P. J. and Elenor R. Cox. 1977. Histone occurrence in chromatin from *Peridinium balticum*, a binucleate dinoflagellate. Science 198: 1258-1260.

Rizzo, P. J. and L. D. Noodén. 1973. Isolation and chemical composition of dinoflagellate nuclei. J. Protozool. 20: 666-672.

Roberts, T. M., L. C. Klotz, A. R. Loeblich,III and J. R. Allen. 1974. Measurements of molecular weight of dinoflagellate chromosomal DNA. J. Phycol. 10 (Suppl.): 15. (Abstract).

Santore, U. J. 1977. Scanning electron microscopy and comparative micromorphology of the periplast of *Hemiselmis rufescens*, *Chroomonas* sp., *Chroomonas salina* and members of the genus *Cryptomonas* (Cryptophyceae). Br. Phycol. J. 12: 255-270.

Satir, B. 1975. The final steps in secretion. Sci. Amer. 233: 28-37.

Schmitter, R. E. 1971. The fine structure of *Gonyaulax polyedra*, a bioluminescent marine dinoflagellate. J. Cell Sci. 9: 147-173.

Silva, E. S. 1978. Endonuclear bacteria in two species of dinoflagellates. Protistologica 14: 113-119.

Silverberg, B. A. 1976. Cadium-induced ultrastructural changes in mitochondria of freshwater green algae. Phycologia 15: 155-159.

Sonneborn, T. M. 1975. The *Paramecium aurelia* complex of fourteen sibling species. Trans. Amer. Micros. Soc. 94: 155-178.

Soyer, M. and O. K. Haapala. 1974. Structural changes of dinoflagellate chromosomes by pronase and ribonuclease. Chromosoma (Berl.) 47: 179-192.

Spero, H. J. 1979. A study of the life cycle, feeding and chemosensory behavior in the holozoic dinoflagellate *Gymnodinium fungiforme* Anissimova. M.S. Thesis. Texas A&M University. College Station, Texas. 95 pp.

Steidinger, K. A. 1975. Implications of dinoflagellate life cycles on initiation of *Gymnodinium breve* red tides. Environmental Letters 9: 129-139.

Steidinger, K. A. 1979. Quantitative ultrastructural variation between culture and field specimens of the dinoflagellate *Ptychodiscus brevis*. Ph.D. Dissertation. University of South Florida, Tampa, Florida. 70 pp.

Steidinger, K. A., E. W. Truby and C. J. Dawes. 1978. Ultrastructure of the red tide dinoflagellate *Gymnodinium breve*. I. General description. J. Phycol. 14: 72-79.

Steidinger, K. A. and J. Williams. 1970. *Memoirs of the Hourglass Cruises: dinoflagellates*. Fla. Dep. Nat. Resour. Mar. Res. Lab. Vol. II. 251 pp.

Swift, E. and C. C. Remsen. 1970. The cell wall of *Pyrocystis* spp. (Dinococcales). J. Phycol. 6: 79-86.

Taylor, D. L. 1968. *In situ* studies on the cytochemistry and ultrastructure of a symbiont marine dinoflagellate. J. Mar. Biol. Assn. U.K. 48: 349-366.

Taylor, D. L. 1971. Ultrastructure of the "zooxanthella" *Endodinium chattonii in situ*. J. Mar. Biol. Assn. U.K. 51: 227-234.

Taylor, F. J. R. 1971. Scanning electron microscopy of thecae of the dinoflagellate genus *Ornithocercus*. J. Phycol. 7: 249-258.

Taylor, F. J. R. 1973. Topography of cell division in the structurally complex dinoflagellate genus *Ornithocercus*. J. Phycol. 9: 1-10.

Taylor, F. J. R. 1975. Non-helical transverse flagella in dinoflagellates. Phycologia 14: 45-47.

Taylor, F. J. R. 1976. Flagellate phylogeny: a study in conflicts. J. Protozool. 23: 28-40.

Tippit, D. H. and J. D. Pickett-Heaps. 1976. Apparent amitosis in the binucleate dinoflagellate, *Peridinium balticum*. J. Cell Sci. 21: 273-289.

Tomas, R. N. and Elenor R. Cox. 1973. Observations on the symbiosis of *Peridinium balticum* and its intracellular alga. I. Ultrastructure. J. Phycol. 9: 304-323.

Tomas, R. N., Elenor R. Cox and K. A. Steidinger. 1973. *Peridinium balticum* (Levander) Lemmermann, an unusual dinoflagellate with a mesocaryotic and an eucaryotic nucleus. J. Phycol. 9: 91-98.

Turpin, D. H., P. E. K. Dobell and F. J. R. Taylor. 1978. Sexuality and cyst formation in Pacific strains of the toxic dinoflagellate *Gonyaulax tamarensis*. J. Phycol. 14: 235-238.

Vesk, M. and S. W. Jeffrey. 1977. Effect of blue-green light on photosynthetic pigments and chloroplast structure in unicellular marine algae from six classes. J. Phycol. 13: 280-288.

Vogel, K. and B. J. D. Meeuse. 1968. Characterization of the reserve granules from the dinoflagellate *Thecadinium inclinatum* Balech. J. Phycol. 4: 317-318.

Walker, L. W. and K. A. Steidinger. 1979. Sexual reproduction in the toxic dinoflagellate, *Gonyaulax monilata*. J. Phycol. 15: 312-315.

Wetherbee, R. 1975. The fine structure of *Ceratium tripos*, a marine armored dinoflagellate. I. The cell covering (theca). J. Ultrastruct. Res. 50: 58-64.

Withers, N. W., Elenor R. Cox, R. W. Tomas and F. T. Haxo. 1977. Pigments of the dinoflagellate *Peridinium balticum* and its photosynthetic endosymbiont. J. Phycol. 13: 354-358.

Withers, N. W. and F. T. Haxo. 1975. Chlorophyll c_1 and c_2 and extraplastidic carotenoids in the dinoflagellate *Peridinium foliaceum* Stein. Plant Sci. Lett. 5: 7-15.

Zingmark, R. G. 1970. Sexual reproduction in the dinoflagellate *Noctiluca miliarias* Suriray. J. Phycol. 6: 122-125.

Copyright 1980 by Elsevier North Holland, Inc.

PHYLOGENY OF PHYTOFLAGELLATES*

KENNETH D. STEWART AND KARL MATTOX
Department of Botany, Miami University, Oxford, Ohio 45056

I. INTRODUCTION

A consideration of the phylogeny of phytoflagellates must be concerned not only with the rich structural and biological diversity described in the preceding chapters, but also with the equally imposing lore of protozoa and fungi. It seems impossible for anybody to possess full command of the literature, not to mention firsthand knowledge of any but a small part of the field. It is nevertheless clear that the literature does not contain enough information to outline protistan evolution with any degree of confidence. It is often even difficult to imagine just what kind of additional knowledge would provide the necessary understanding. The protistologist-phylogenist, obliged to deal with both too much and too little, must temper his natural enthusiasm with the realization that his concepts are, if not actually false, almost certainly simplistic.

Despite the difficulties, the rapid accumulation of fine structural and biochemical information during the last two decades has inspired many reviews dealing with protistan evolution. The diversity of views expressed is exceeded only by that of the organisms themselves. Some might doubt the value of such speculations, but they are the result of the intellectual urge to explain individual phenomena and to organize the bewildering mass of information into a coherent and more manageable concept of historical development. It is often said that some of the more fundamental differences between phylogenetic ideas will forever remain matters of opinion. We take a more optimistic outlook that researchers will be around for a long time to come, and eventually there will be clear indications as to which ideas are closest to the truth. In the meantime, responsible attempts to reconstruct the phylogeny of microorganisms will continue to pinpoint a number of problems that can and should be investigated in the laboratory.

It seems neither feasible nor desirable to give equal consideration to all published views on phytoflagellate evolution. Readers interested in discussions of ideas different from the ones expressed below will readily find them in the cited literature. We have no more than the space necessary to summarize

*This work was supported by National Science Foundation Grant DEB78-03554.

some general ideas which, in our opinion, presently provide the best explanations of the known facts and to point out some areas where the lack of information allows little useful speculation. Discussion will be centered around two main problems: (1) What were the characteristics of early phytoflagellates? and (2) What are the relationships of phytoflagellate groups to each other?

II. WHAT WERE THE CHARACTERISTICS OF EARLY PHYTOFLAGELLATES?

This problem cannot be separated from that of which came first, phytoflagellates or zooflagellates. That question is of the most fundamental importance and cannot be ignored. If it is answered incorrectly, a consideration of other questions is seriously hampered at the outset and will likely lead to many conclusions just the opposite from the truth. If a person subscribes to the classical theory of the origin of chloroplasts and eucaryotic cells; that is, by direct filiation from photosynthetic procaryotes, he might expect the earliest eucaryotic cells and perhaps even the first phytoflagellates to be small cells, relatively simple insofar as intracellular organization is concerned (e.g., the possibly primordial organization occasionally mentioned for the tiny green flagellate *Micromonas pusilla*). Another, who favors the endosymbiontic theory of the origin of chloroplasts; that is, from the ingestion of photosynthetic procaryotes by non-pigmented forms, is likely to visualize the early eucaryotic plants as motile, perhaps even flagellated, photosynthetic predators, large and structurally complex enough to find, capture and digest other microorganisms. The first person might expect the early eucaryotic plant cells and possibly even the earliest phytoflagellates to be walled (e.g., the primitive position accorded *Chlamydomonas* by Klein and Cronquist 1967); the other would doubtlessly expect them to be naked or scaly, the better to ingest their prey. The classicist might expect the first phytoflagellates to have symmetrically inserted flagella, as in green algae, because that would seem to be the most efficient arrangement if the cell's shape had only to be concerned with motility and streamlining. A consideration of the endosymbiontic theory, however, might lead one to imagine asymmetrically inserted flagella in a cell whose exterior form accommodates an ingestion apparatus or region as well as the flagellar apparatus.

We think a comparison of form and function in phytoflagellates strongly supports their origin from zooflagellates and, by implication, the endosymbiontic origin of chloroplasts. We shall compare zooflagellates and phytoflagellates, rather than specifically discuss evidence of an endosymbiontic origin of chloroplasts, because the evidence and arguments there have been reviewed many times. See especially the thorough review by Taylor (1974) and the literature he cites. The direct filiation theory (Uzzell and Spolsky 1974; Raff and Mahler 1972) appears to be supported more by arguments than by evidence or a comparison of microorganisms.

It has been recognized since the mid-nineteenth century that pigmented and non-pigmented flagellates have close similarities which prevent their separation into "plants" and "animals." The modern recognition that the flagella of all those organisms are similar structures of such a complexity that they almost certainly had a single origin further indicates that zooflagellates gave rise to phytoflagellates, or vice versa, rather than their having independent origins. One reason for thinking that the main direction of evolution has been from zooflagellates to phytoflagellates is that all groups of phytoflagellates have characteristics more associable functionally with phagotrophic protozoa than with an autotrophic nutrition. This is true even in photosynthetic phytoflagellates not known to ingest solid food and, to some extent, even in the zoospores of "true plants," the multicellular higher algae. It is far easier to reconcile these facts with a zooflagellate ancestry than to suppose that all groups of phytoflagellates exhibit an equal and parallel trend toward becoming protozoa. "Animal-like" characteristics of phytoflagellates include phagocytosis, a nearly universal asymmetry of cell form, lack of a cell wall, flagellar pits, and some complex cytological features such as trichocysts and large striated fibers (rhizoplasts).

Most of the phytoflagellate groups; e.g., Prymnesiophyceae, Chrysophyceae, Cryptophyceae, Euglenophyceae, and Dinophyceae have phagocytic members. Even the flagellated stage of a coccoid chlorophytan, *Halosphaera minor*, has been reported to take in graphite and plant cells (Parke and den Hartog-Adams 1965).

All groups of phytoflagellates appear to be basically asymmetrical, and the same condition exists in the flagellated reproductive cells of multicellular algae. The two groups that most closely approach symmetry are the green algae and prymnesiophytes, but their cytoskeletons, the microtubular roots attached to the flagellar apparatus, are asymmetrical. The asymmetry of many chlorophytan flagellates and zoospores is slight but nevertheless distinct, and a few genera (e.g., *Heteromastix*, Manton et al. 1965; Mattox and Stewart 1977) are strongly asymmetrical. Only the flagellated reproductive cells of a very few advanced green algae (e.g., Oedogoniales) are completely radially symmetrical. Ringo's (1967) report of four equal roots in *Chlamydomonas* has been found to be mistaken; Moestrup (1978) has shown that two roots with two microtubules alternate with two roots with four microtubules, a typical situation for chlorophycean algae (see Melkonian 1978). The significance of this general asymmetry is not known, but the only explanation that occurs to us is that asymmetry was inherited from phagotrophic zooflagellates where the placement of an anterior ingestion apparatus or area resulted in asymmetrically inserted flagella and consequently unequal flagellar roots (e.g., *Cephalothamnium*, Hitchen 1974). A further indication that cellular asymmetry might be correlated with an anterior ingestion apparatus is that some of the most nearly externally symmetrical zooflagellates (e.g., *Gyromitus*, Swale and Belcher 1974, and *Trichonympha*) have an ingestion area at the posterior of the cell. A zooflagellate with two unequal microtubular flagellar roots like *Cephalothamnium* might have "doubled" to produce a more nearly symmetrical cell with four microtubular roots of alternating sizes around a flagellar pit, as has been described for the scaly green flagellates *Pyramimonas* (Moestrup and Thomsen 1974) and *Platymonas* (Melkonian 1979). See also the discussion of cell "doubling" and "halving" in green flagellates in Stewart and Mattox 1978.

Phytoflagellates of every group are naked or scaly, and that is true even in many which are not phagocytic, saprophytic or from saline habitats. Scales are also present in zooflagellates not apparently related to an algal group (e.g., *Cyathobodo*, Hibberd 1976a). The only phytoflagellates with a true wall are *Chlamydomonas* and closely related genera, and here it is clear that these are advanced flagellates that evolved from scaly ancestors. It appears, moreover, that the cell walls in chlamydomonad flagellates evolved from the fusion of body scales (Mattox and Stewart 1977). If phytoflagellates and their chloroplasts evolved directly from walled photosynthetic procaryotes, it is difficult to explain their fundamentally wall-less condition, especially in autotrophic genera of freshwater habitats. Furthermore, there is no evidence that phagocytic algal flagellates have lost walls. On the other hand, the wall-less condition of phytoflagellates is obviously consistent with a zooflagellate ancestry.

At least some of the genera in four phytoflagellate groups, green algae, cryptomonads, euglenoids and chloromonads, have their flagella inserted in pits. Even one prymnesiophyte, *Pavlova* (Green and Manton 1970), has a small pit. The function of the pits is unknown, but flagellar pits also occur in some phagotrophic zooflagellates (e.g., *Rhipidodendron*, Hibberd 1976b; bodonids, Hitchen 1974). The function of flagellar pits is not at all clear in these zooflagellates either, but in some cases it seems likely that the pit is a result of the fact that the flagellar attachment is situated less anteriorly than the exterior part or opening of an ingestion apparatus (e.g., *Cephalothamnium*, Hitchen 1974). Presently, there seems to be no better explanation for flagellar pits in phytoflagellates than that they are essentially remnants of a phagotrophic zooflagellate ancestry. Flagellar pits are present in some scaly chlorophytan flagellates (e.g., *Tetraselmis* = *Platymonas*) that can be considered primitive for other reasons, but the pits have apparently been lost in advanced green flagellates like *Chlamydomonas* (Stewart and Mattox 1978).

There are clear trends of evolution from phytoflagellates having the probable zooflagellate characteristics discussed above to organisms more "plant-like." To judge from the structure of flagellated reproductive cells, naked or scaly asymmetrical chrysophycean flagellates gave rise to diatoms and brown algae (see Hibberd 1976c). Scaly zoospores in some filamentous green algae demonstrate their origin from scaly, not walled, flagellates (Stewart and Mattox 1975b). There is also evidence that scaly green flagellates with flagellar pits gave rise to the more "plant-like," walled chlamydomonad flagellates (Mattox and Stewart 1977). In contrast, there is no evidence in algae or phytoflagellates for similarly complex trends of evolution in the reverse direction. Known cases of evolution from autotrophic to heterotrophic forms involve the simple loss of photosynthesis with no clear evidence that such forms gave rise to zooflagellate groups. For example, loss of photosynthetic pigments, <u>but not the loss of plastids</u>, has occurred in the colorless chrysophyte *Anthophysa* (Belcher and Swale 1972). Some chlorophytan flagellates (*Polytoma* and *Polytomella*) have become colorless saprophytes, <u>which still possess starch-containing plastids</u> (e.g., Lang 1963). There is no evidence that such flagellates gave rise to aplastidic heterotrophic groups; they merely demonstrate the same general capacity for heterotrophy shown by parasitic or saprophytic angiosperms. There is evidence that plastids have assumed physiological functions other than photosynthesis (e.g., the starch-storing by green algal plastids), and the complete loss of plastids by chlorophycean flagellates and others might be lethal (see discussion of chloroplast origin in Section II).

The classical concept that zooflagellates had their origin from photosynthetic flagellates has been weakened but is far from being abandoned. Many modern works, particularly those by protozoologists, speak of a phytoflagellate, or often more specifically a chrysophyte or "chromophyte" ancestry for zooflagellates (e.g., Hanson 1977; Sleigh 1973, p. 7; Taylor 1978). The view, developed before electron microscopy, that the simpler zooflagellates could be considered reduced or "aberrant" chrysophytes has persisted. It must be emphasized, therefore, that electron microscopy has failed to support an algal origin for a number of organisms for which that ancestry was previously assumed. Most notably, the works of Leadbeater and Manton (1974) and Hibberd (1975) have shown that choanoflagellates (botanically the Craspedophyceae and zoologically the

Choanoflagellida) have no affinity with any algal group and should be "removed from the plant kingdom." The series of drawings by Bold and Wynne (1978, p. 371), demonstrating a possible origin of choanoflagellates from chrysophytes, might seem reasonable from the gross external form of the cells but is untenable when fine structure is considered (see Hibberd 1975, 1976c). Differences between choanoflagellates and chrysophytes include features of the mitochondrial cristae, flagellar apparatus, flagellar hairs and general cellular organization. There are not many more structural ways for small flagellated unicells to differ. Similarly, fine structural studies have not given good reason for believing, as in the past, that such zooflagellates as *Rhipidodendron* (Hibberd 1976b), *Cyathobodo* (Hibberd 1976a), *Gyromitus* (Belcher and Swale 1974) and bodonids (e.g., Hitchen 1974) have their origin from any algal group. Electron microscopy of some other zooflagellates (e.g. *Cercomonas*, Mignot and Brugerolle 1975b; *Bicosoeca*, Moestrup and Thomsen 1976, Belcher 1975, Mignot 1974, Hibberd 1978a; and *Colponema*, Mignot and Brugerolle 1975a) has shown that they do indeed have some chrysophycean or "chromophytan" characteristics, but only tradition favors an interpretation that the direction of this relationship is from phytoflagellate to zooflagellate.

As mentioned earlier, it is not possible to discuss all the numerous ideas that have been expressed concerning protistan evolution, but comment on two recent articles (Taylor 1978; Cavalier-Smith 1978) seems necessary because of their views on some of the matters discussed above. Taylor assumes that chloroplasts evolved from endosymbionts, but surprisingly he also assumes that modern protozoa evolved from photosynthetic flagellates of either the "chromophyte" or "chlorophyte" lines. Taylor's views have the merit of accommodating both the endosymbiontic theory of the origin of chloroplasts and the classical view that zooflagellates evolved from phytoflagellates, but he gives little supporting evidence for the latter event. A number of points we made earlier, particularly the lack of evidence that phytoflagellates gave rise to aplastidic protozoan groups, argue against Taylor's idea as well as the purely classical viewpoint. It is possible, or even probable, that plastids have been lost completely in some relatively minor evolutionary lines (see discussion of chloroplast origin in Section II), but there is no reason to think that such events were of major importance in the origin of protozoa.

Cavalier-Smith (1978) has published a scheme whereby photosynthetic procaryotes gave rise, by direct filiation, to early eucaryotic cells which were both photosynthetic and phagocytic. Having conjured a cell that has almost everything to begin with, he has only to add flagella to get an organism from which he can derive most other groups by simple loss, thereby enabling himself in the later steps to follow his own rule that economic hypotheses should be favored. That might be convenient, but it is biologically unsound. Eucaryotic cells are characterized by an elaborate system of cytoplasmic motility (Stanier 1970; Pollard 1976) and by an endomembrane system that is an intracellular digestive system in phagotrophic organisms. The two systems are extremely complex structurally and functionally, and it seems likely that their early functions were capture, ingestion and utilization of food particles. Cavalier-Smith's scheme has the same weakness as all theories of direct filiation from photosynthetic procaryote to photosynthetic eucaryote; it must postulate selective pressures that could produce such complex systems in a photosynthetic organism whose survival was not strictly dependent on them. Cavalier-Smith's statement (1975) that such an organism would have the advantage of photosynthesizing during the day and feeding at night is a weak rationalization. To judge from present procaryotes, there were few if any "predators" before the origin of eucaryotic cells or their predecessors. That is reason to suspect that it was the exploitation of the "predator niche" that led to the origin of eucaryotic cells. It is not reasonable to expect that a cell that remained photosynthetic could compete with or evolve at the same rate as cells that had abandoned photosynthesis, or never had it, and were therefore subject to more strict selection in that direction. A similar kind of objection can be made to Taylor's (1978) and others' views that the characteristics of successful protozoan groups originated in photosynthetic organisms.

In conclusion, we think that the characteristics of phytoflagellates indicate an origin from phagotrophic zooflagellates with anterior ingestion apparatuses. If so, it would follow that the features of the earliest phytoflagellates, except for their newly acquired photosynthetic endosymbionts, were essentially the same as in such zooflagellates; that is, they were asymmetrical, phagotrophic, scaly or naked, and complexly organized internally. With some likely exceptions such as chrysophytes, they probably also possessed flagellar pits.

III. WHAT ARE THE RELATIONSHIPS OF THE PHYTOFLAGELLATE GROUPS TO EACH OTHER?

If the conclusions of Section II are correct, the relationship between any two phytoflagellate groups is that between their respective zooflagellate ancestors. Since it is likely that eucaryotic flagella had a single origin, the conclusions of Section II also imply that flagellated protozoa had a common origin, in contrast to the polyphyletic origin postulated by schemes which derive them from phytoflagellates. The problem of establishing the nature of relationships among flagellated algal classes would be, therefore, the same as trying to determine the characteristics of early flagellated protozoa and the pattern of their evolutionary diversification. The latter task is one of great scope and complexity, and the scarcity of relevant information does not warrant detailed speculations concerning the interrelationships of all of the many groups. Further discussion in this section will be confined to some general zoo- and phytoflagellate features which should be taken into account in phylogenetic considerations.

A comparison of protozoan flagellates with each other and then with algal flagellates is limited to those features which occur in all or most groups and which are known to show variation. These features should also be conservative as is best indicated by relatively small or infrequent variation within monophyletic groups. General considerations are, therefore, largely restricted to structural features of the mitotic spindle, mitochondria, cell covering or cell surface, and the flagellar apparatus with its internal and external appendages. Some less widespread features such as trichocysts might occasionally be useful in more specific considerations. The phylogeny of each of the mentioned structural features and some possibilities suggested by an origin of chloroplasts from endosymbionts will be discussed separately before a consideration of their implications with regard to the relationships between flagellate groups.

The mitotic spindle.

The structure, development and mechanisms of the mitotic spindle have been intensively studied in recent years in both protists and higher organisms. There has also been wide discussion of its phylogeny (see Kubai 1975, 1978 and the literature cited there). The mitotic spindle is a useful feature for comparing and characterizing groups of eucaryotic microorganisms, because it occurs in all that have been carefully studied and exhibits both variation and conservatism. Conservatism is indicated by the fact that variation in spindle development and structure is infrequent or minor within natural (monophyletic) groups. Although all mitotic spindles have some features in common, their variation between eucaryotic groups shows they, like other systems, have been subject to evolutionary modifications. In a comparison of flagellate groups, the value of the mitotic spindle is limited to the extent that we can determine the direction of evolutionary changes and whether particular changes occurred once only or independently in two or more lines.

The numerous variations in spindle development are sometimes striking and sometimes subtle (Kubai 1975, 1978; Fuller 1976; Heath 1978). Some variations are almost certainly of general phylogenetic importance while others are of minor importance, or important only with regard to certain groups. The most striking and probably also the most fundamental or ancient difference between mitotic spindles is between those where spindle microtubules lie outside the nuclear envelope during the entire course of development and those where spindle microtubules occur within the nucleoplasm for at least the later stages of mitosis. All mitotic spindles will be referred to hereafter by the designations given below:

Type I Spindle - Spindle microtubules remain outside an intact nuclear envelope for the entire course of mitosis. Examples: Dinophyceae and Hypermastigida.

Type II Spindle - Spindle microtubules occur in nucleoplasm during some or all stages of mitosis.

Subtype IIa Spindle - Spindle microtubules develop first in the cytoplasm but penetrate the nucleoplasm by the time the spindle is fully organized. The nuclear envelope can remain largely intact except for local interruptions or it may largely disperse. Examples: higher plants, higher animals, many green algae, diatoms, some protozoa.

Subtype IIb Spindle - Spindle microtubules develop within the nucleoplasm and remain there for the entire course of mitosis. The nuclear envelope typically remains intact. Examples: many fungi, Euglenophyceae, trypanosomes, Oomycetes, *Vaucheria*.

The distinction between IIa and IIb is somewhat of an oversimplification since in some cases spindle microtubules develop inwardly from bodies lying barely outside the poles of the nucleus, but in most cases the assignment of a spindle to one of the two subtypes is easily accomplished.

Although some have argued that the mitotic spindle originated inside the nucleus and that intranuclear, closed spindles (subtype IIb) are therefore more primitive (e.g., Pickett-Heaps 1974, 1975a; Cavalier-Smith 1978), the weight of developmental evidence and a comparison of microorganisms strongly favors a direction of evolution from a type I spindle to a type II. Firstly, organisms with a type I spindle possess two other unusual features that are very likely primitive. In both hypermastigids and dinoflagellates, the chromosomes are attached to the nuclear envelope, a feature reminiscent of the association of the procaryotic genophore with the plasmalemma. Secondly, dinoflagellate chromosomes often lack histones even in free-living forms, another procaryote-like feature. In our opinion, however, the most compelling consideration is that spindle development in higher plants, higher animals and many eucaryotic microorganisms (subtype IIa) may "recapitulate" phylogeny when it begins outside the nucleus but eventually penetrates the nucleoplasm (see Ris and Kubai 1974). Finally, there is the possibility that the mitotic spindle evolved from the flagellar apparatus (see discussion of the flagellar apparatus below), in which case the earliest spindles would have been extranuclear (type I).

A direction of evolution from a type I spindle to a type II would place the few groups with type I in a relatively primitive position, but it unfortunately has no implications which can be used to compare the many type II groups with each other. Not only is there no evidence that would indicate whether the type II spindle evolved once or a number of times, but there is also no reason to think that the distinction between subtypes IIa and IIb represents a single phylogenetic occurrence or is even very important. The developmental difference between subtypes IIa and IIb seems to be determined by whether or not the spindle organizing centers (sites of origin of spindle microtubules) lie outside the nucleus (subtype IIa) or either on or inside the nuclear envelope (subtype IIb). Matters discussed above would suggest that spindle subtype IIa is more primitive than IIb, because the spindle organizing centers are extranuclear in type I spindles. The phylogenetic movement of the spindle organizing center to or inside the nucleus was probably not, however, an unusual or complex change since it is accomplished ontogenetically in some IIa spindles (e.g., *Stigeoclonium*, Floyd et al. 1972a). Aldrich's (1969) work on the slime mold *Physarum* is a case in point. In the myxamoebal stage that possesses centrioles and can produce flagella, the mitotic spindle is subtype IIa, but in the plasmodial stage, where centrioles are absent, the spindle is subtype IIb. The most obvious explanation of Aldrich's results is that the spindle organizing center (MTOC) remains outside the nucleus in the myxamoebal stage, where it functions to produce centrioles and flagella, but moves to the inside of the nucleus in the centriole-less plasmodial stage and remains there for many nuclear divisions. The events in *Physarum* suggest that there is not much phylogenetic distance between IIa and IIb spindles. Since the flagellate organization is doubtless the ancestral condition of slime molds, it is of great interest that the events in *Physarum* also suggest that the centric, open subtype IIa spindle is the ancestral type here; that is, more primitive than the completely closed IIb spindle of the centriole-less plasmodial stage. This directional relationship between IIa and IIb spindles is also supported by the work of Peterson and Ris (1976) who have shown that the type IIb spindle of *Saccharomyces cerevisiae* has no unusual or primitive characteristics, and they conclude that completely closed, intranuclear spindles may be more advanced, or at least more specialized, than the open spindles of higher plants and animals.

The viewpoints expressed above lead to the conclusion that no type II spindle is likely to have many primitive structural characteristics. The lack of kinetochores or the absence of microtubules which terminate at and attach to chromosomes has sometimes been reported for type II spindles (e.g., *Mucor*, McCully

Robinow 1973; *Euglena*, Leedale 1968), and this has sometimes been considered an indication that the spindle is primitive or at least organized differently from the spindles of more advanced organisms. Better fixation and more thorough study have, however, shown chromosomal microtubules in organisms previously thought to lack them (e.g., *Fusarium*, Aist and Williams 1972; *Euglena*, Gillot and Treimer 1978; see also the discussion by Kubai 1975, p. 203). Moreover, a direction of evolution from a type I spindle to a type II would indicate that all type II spindles have kinetochores or at least microtubules that terminate at a point where they attach to chromosomes. This is because type I spindles themselves possess kinetochores, the sites in or on the nuclear envelope where chromosomes and extranuclear spindle microtubules are attached. These sites are sometimes highly differentiated (*Trichonympha*, Kubai 1973) or they may appear simply as dense regions (*Amphidinium*, Oakley and Dodge 1974). Kinetochores were not described in *Cryptothecodinium cohnii* (Kubai and Ris 1969), but the much-reproduced Fig. 24 of Kubai and Ris shows chromosomes attached to the nuclear envelope and microtubules terminating in dense material on the nuclear envelope, indicating that the spindle of *C. cohnii* is organized like other type I spindles in that respect. As mentioned earlier, careful work is beginning to indicate a general occurrence of chromosomal microtubules in type II spindles (see especially Peterson and Ris 1976) except in the special case of diatoms where the lack of chromosomal microtubules almost certainly represents a loss related to the highly specialized organization of the diatom spindle (see Pickett-Heaps and Tippit 1978).

One feature that might appear to show enough structural variation to be useful for comparison of groups is the nature of the organelle or substance that occupies spindle poles and/or from which spindle microtubules develop. It seems likely, however, that the material at all spindle poles is fundamentally the same (i.e., homologous) and that variations in appearance are superficial differences dependent on the degree of localization of spindle development and in some cases on whether or not cell division occurs in flagellated stages.

Spindle poles can consist of plaques or globular bodies (e.g., fungi, see Fuller 1976, Heath 1978; Sporozoa, Kelley and Hammond 1973; radiolarians, Hollande et al. 1969), tiny dense areas associated with centrioles (e.g., higher animals), ring-shaped structures (red algae, McDonald 1972), or simply granular material (e.g., *Chlamydomonas*, Johnson and Porter 1968; higher plants). When the spindle organizing structure of flagellates is large enough and structurally organized enough, it is recognizable as a rhizoplast or striated fiber (*Ochromonas*, Slankis and Gibbs 1972, Bouck and Brown 1973; *Platymonas*, Stewart et al. 1974; *Heteromastix*, Mattox and Stewart 1977). Rhizoplasts are attached to the flagellar basal bodies and usually extend to or along the sides of the nucleus. The rhizoplasts of the three genera mentioned are long, and the spindles produced by them have broad, flattened poles. More tapered spindles (as in metazoa) probably result when a region homologous to the rhizoplast is small, as in the tiny dense regions often seen at the poles and attached to centrioles.

It also seems probable that the granular mass at the poles of *Chlamydomonas* (Johnson and Porter 1968) and possibly higher plants consists of the same material as rhizoplasts but is less highly organized or regularly arranged. Although *Chlamydomonas* lacks recognizable rhizoplasts, the related *Carteria* does have them (Lembi 1975). In the interphase *Chlamydomonas* there is a ribosome-free granular region below the basal bodies in the position that would be occupied by rhizoplasts if they were present (see Fig. 13 in Ringo 1967). Apparently such regions were seen and called rhizoplasts by light microscopists who studied *Chlamydomonas* (see Kater 1929). The relationship between these granular or fibrillar regions and striated rhizoplasts is likely to be at least analogous to that between smooth muscle and striated myofibrils in metazoans (see discussion of rhizoplast contractility below). Since basal bodies lie near the spindle poles in *Chlamydomonas* (Coss 1974), it seems likely that the granular areas reported at the poles by Johnson and Porter (1968) are of the same nature as the region below the basal bodies at interphase.

The spindle pole bodies or "nucleus associated organelles" of some protozoans, higher fungi and diatoms are also probably homologous to rhizoplasts. This is almost certain in diatoms where the spindle pole body becomes attached to basal bodies when flagellated gametes are produced (Manton et al. 1970). Spindle pole bodies of fungi sometimes have a banded appearance similar to rhizoplasts, and some higher fungi are known to produce other structures with undeniable resemblances to rhizoplasts (e.g., Harvey 1974). Since rhizoplasts, when present in flagellates, are part of the flagellar apparatus, homologizing rhizoplasts and spindle pole bodies may be tantamount to postulating a flagellate ancestry for higher fungi. Such an ancestry is assumed for other reasons in the following discussion of the flagellar apparatus.

The relationship between rhizoplasts and mitotic spindles might have important implications with regard to the mechanism of anaphase spindle movements. In an important contribution, Salisbury and Floyd (1978) have recently shown the rhizoplast of *Platymonas* to be contractile. Addition of Ca^{++} and ATP to the medium causes a rhythmic contraction detectable with the light microsope and shorter wider rhizoplasts with different banding when the cells are prepared for electron microscopy. Some rhizoplasts appear to merely nucleate the growth of spindle microtubules (e.g., *Ochromonas*, Bouck and Brown 1973). *Heteromastix*, which is related to *Platymonas*, also has rhizoplasts at the poles (Mattox and Stewart 1977). In *Platymonas*, however, the rhizoplasts "dissolve" to form large granular or fibrillar regions from which spindle microtubules develop (Stewart et al. 1974). In short, it appears possible that the spindle of *Platymonas* is at least in part produced by a transformation of rhizoplasts. An elucidation of the molecular mechanism of rhizoplast contraction might, therefore, have a bearing on the mechanism of spindle movements. Another intriguing possibility is that the relationship between rhizoplasts and microtubules (see also Stewart and Mattox 1978) might indicate that the fibrillar components of rhizoplasts have some resemblance to microtubular protofilaments. Cachon et al. (1977) reported that an apparently contractile striated fiber of the protozoan *Sticholonche* contains no tubulin, but probably much remains to be understood about transformations of the proteins concerned with cytoplasmic rigidity and motility.

Although the comparison of mitotic spindles might be of little general value as an indicator of the relationship between groups with type II spindles, details of structure and development of spindles have proven conservative enough to provide some characterization of individual groups and valuable indications of phylogeny within at least one group. The characterization provided by spindles can in some instances allow identification of an organism; for example, an endosymbiont of a foraminiferan has been recognized by its mitotic spindle as a diatom although it lacks a typical frustule (Berthold 1978). The study and comparison of spindles has led to a very different view of phylogeny within the green algae and will eventually lead to a major revision of their classification (Pickett-Heaps 1975b; Stewart and Mattox 1975, 1978).

It seems that some of the evolutionary modification of type II spindles has been a response to a reduction of the space available for mitosis to occur. The collapsing telophase spindle and phycoplast of chlorophycean algae appear to have evolved along with a cell wall in chlamydomonad flagellates (Mattox and Stewart 1977). The cell wall prevents a rapid elongation of the cell and spindle during cell division, with the result that daughter nuclei do not become widely separated at telophase. Similarly, the very regularly organized spindle of diatoms might be the phylogenetic result of an increased rigidity necessary for the accurate growth of the cleavage furrow between daughter nuclei only slightly separated because of confinement by the frustule (see Fig. 4 in Pickett-Heaps et al. 1978). Although there are probably not many primitive features in type II spindles, the lack of a shortening of chromosome-to-spindle pole distance during anaphase in some forms (e.g. *Klebsormidium*, Floyd et al. 1972b) could be one example (see Ris and Kubai 1974). At least in some cases, a change to the more common chromosome-to-spindle pole shortening might also be a response to reduced space. In the scaly green flagellate *Heteromastix* (Mattox and Stewart 1977), there is much cell and spindle elongation during cell division but no shortening of chromosome-to-spindle pole distance for at least most of anaphase. In related, walled green flagellates (e.g., *Platymonas*, Stewart et al. 1974), with collapsing telophase spindles, phycoplasts and relatively little cell or spindle elongation, there is chromosome-to-pole shortening during anaphase.

The flagellar apparatus

The flagellar apparatus is one of the most useful features for characterizing groups of eucaryotic microorganisms. Most groups have flagella or are clearly related to groups that do. The evolutionary conservatism of the flagellar apparatus is indicated by relatively minor or at least infrequent variations within groups that can be considered monophyletic. The "9 + 2" organization of the flagellar axonemes and the nine triplets of basal bodies show only occasional minor variations and so are actually too conservative and too anciently established to be useful for comparison. More variable and more useful are the transition region between flagella and basal bodies (see Hibberd 1978b), the type, number and arrangement of rhizoplasts and microtubular roots, and other accessory structures such as paraflagellar rods and flagellar scales or hairs. As with the mitotic spindle, variations in the flagellar apparatus have so far been more useful in the characterization of groups (e.g., Hibberd 1978b) and in interpretation of trends of evolution within groups (e.g., Pickett-Heaps 1975; Stewart and Mattox 1978) than in interpretations of the relationships between groups.

One of the more fundamental and interesting problems with regard to the flagellar apparatus is its time of origin relative to that of other features which characterize eucaryotic cells. Two questions specifically come to mind: (1) Which came first, flagella or mitotic spindles? and (2) Do the flagella-less Rhodophyceae, Basidiomycetes and Ascomycetes primitively lack flagella or have they lost them? Since both the flagellar apparatus and the mitotic spindle possess microtubules, and since fairly typical spindles exist in the three groups mentioned, the questions are not really different. It has sometimes been proposed that microtubules and the mitotic apparatus arose within the nucleus (e.g., Pickett-Heaps 1974, 1975; Cavalier-Smith 1978), but there is no evidence for this nor, in our opinion, is it supported by a comparison of eucaryotic microorganisms. In some cases at least, such proposals appear to have been inspired by early notions that closed intranuclear spindles (subtype IIb) are the most primitive type. As discussed earlier, there is now no indication that

that is the case. Perhaps another reason was the idea that the nucleus is the most distinctive and significant feature of the eucaryotic cell. As discussed below, there are good reasons for a somewhat different view.

We think that a consideration of organization and development in flagellates strongly suggests that flagella, or at least a microtubule-containing precursor of flagella, evolved before the mitotic spindle and that the spindle itself evolved as an elaboration of the flagellar apparatus or its precursor. One indication that flagella evolved before mitotic spindles is that dinoflagellates and hypermastigids have fully evolved and complex flagellar apparatuses with the typical 9 + 2 axonemes even though their mitotic spindles (type I) are probably the most primitive in eucaryotes. In dinoflagellates, a primitive spindle and possibly primitive chromosomal organization, combined with otherwise typically eucaryotic features suggest, as Stanier (1970) has discussed, that cytoskeletal and motility systems are the essence of eucaryotic organization. The evolution of the spindle and chromosomal structure appears to have lagged behind those cytoplasmic systems. If a phagotrophic nutrition led to the origin of eucaryotic cells (see Section II), one would expect that the strongest selection operated on systems of cellular architecture, motility, phagocytosis and digestion. Changes in nuclear characteristics would not have been so important, and their evolution correspondingly less rapid. On the other hand, it is reasonable to suppose that a phagocytic cell surface might lead to the transfer of DNA from a surface membrane to an internal one. The internal membrane might also eventually become modified to compartmentalize the DNA if much of the remainder of the endomembrane system became a flowing system involved with phagocytosis and digestion.

There are some other reasons for thinking that the flagellar apparatus came somewhat earlier than the mitotic spindle and also that the spindle evolved from the flagellar apparatus. One of the most obvious is the fact that the spindle develops from the flagellar apparatus in the majority of flagellate groups. In most cases where the spindle develops at some distance from the basal bodies, it develops from a rhizoplast, which is also part of the flagellar apparatus. In some of the exceptions (*Cyanophora*, Pickett-Heaps 1972a), rhizoplast-like material could be present but not organized densely enough to be detectable in thin sections (see discussion of rhizoplasts above). The main other type of exception is exemplified by ciliates, where cellular organization has become so highly specialized that it is difficult to judge just how it might relate to more primitive organization.

The interpretation of the preceding paragraph could be reversed; that is, it could be claimed that the flagellar apparatus evolved from the spindle, because flagella develop from the spindle-organizing region. However, aside from the likelihood that the motility apparatus of eucaryotes evolved before modern nuclear characteristics, there are some developmental phenomena that better support the first interpretation. In *Trichonympha* (Kubai 1973), which has a type I spindle, the non-chromosomal complement of spindle microtubules (i.e., the "continuous spindle") develops before the nucleus migrates to its vicinity. This early development of the spindle occurs between the replicating units of the cytoskeleton and flagellar complex; Kubai (1973, Fig. 1) illustrates the spindle developing between daughter rostra. The events in *Trichonympha* suggest that the continuous spindle could have originally evolved as a rigid framework that facilitated cell elongation and the separation of replicating cytoskeletal and motility apparatuses. The late migration of the nucleus to the spindle in *Trichonympha* (Kubai 1973) may reflect phylogenetic events in that the utilization of the microtubular system for chromosomal separation could have been a later development. The events in hypermastigids are spectacular, but some of them are not unique. It is common for microtubules to occur between separating flagellar apparatuses as well as in the spindle (e.g. Johnson and Porter 1968; Deason and Darden 1971; Mattox and Stewart 1977; Floyd 1978). Microtubules also develop around separating centrioles in higher animal cells, and there is no indication that all of these are later utilized in the spindle.

Other organelles are attached to the flagellar apparatus and divide or are cleaved when the flagellar apparatus separates. The mitotic spindle could be viewed as one example of a more general phenomenon, one that became the most elaborate because of the number of chromosomes and the exact manner in which they separate. Microbodies are often attached to flagellar apparatuses or spindle poles and often apparently divide during mitosis and separation of flagellar apparatuses (Pickett-Heaps 1972b; Mattox and Stewart 1977; Bronchart and Demoulin 1977). The attachment of mitochondria to the flagella apparatus, to spindle poles or to some other part of the cytoskeleton is often detected (Powell 1975; Rogers et al., in press; LaFountain 1972). Even when a flagellar apparatus is not present and the cytoskeletal-motile apparatus consists mostly of a fundamental matrix of protein molecules, organelles appear to be anchored to it (Pollard 1976). The position and movement of the organelles are presumably controlled by the matrix. Since the chromosomes of *Trichonympha* appear to separate before the nucleus becomes associated with the spindle (Kubai 1973), a matrix probably provides the motile force. The association of the nucleus with the microtubules of a separating flagellar apparatus could well have been a later evolutionary development. A previous hypothesis that the mitotic spindle evolved from the flagellar apparatus (Marqulis 1970) was based on the idea that flagella evolved from endosymbionts, a concept entirely unsupported by evidence or developmental studies.

If the mitotic spindle did evolve from the flagellar apparatus, it follows that all organisms with a microtubular spindle had flagellated ancestors. This includes all groups of organisms where nuclear division has been observed with the electron microscope. Flagella must have had an extremely early origin. There is also some reason to think that flagella evolved before mitochondria (see discussion of mitochondria below). If any "preflagellar" eucaryotic organism is extant, it has yet to be found and carefully studied.

Although angiosperms, some gymnosperms, and the large green algal order Zygnematales demonstrate that all traces of flagella can be lost, it is still often considered possible or even likely that red algae and higher fungi did not have flagellated ancestors. This is largely because the latter two groups are not related in an obvious way to flagellated organisms. It is possible, however, that higher fungi are related to the zoosporic Chytridiomycetes (see Fuller 1976). It is also possible to explain the loss of flagella in higher fungi as it is explained in angiosperms; that is, as a response to a terrestrial habitat. Flagellated cells would not be effective agents of dispersal for the basically terrestrial higher fungi, and hyphal contact, provided by the characteristic matted growth, eliminates the necessity for flagellated gametes. Furthermore, if genetic exchange is accomplished by hyphal contact, greater nutritional advantage is obtained for the product. There is an apparent parallel in the filamentous Zygnematales, whose usual growth as floating tangled mats could have led to their conjugative sexual process.

The loss of flagella by the aquatic and largely sessile red algae is much more difficult to explain. It is possible that red algae had *Cyanophora*-like flagellated ancestors, as suggested by Schnepf and Brown (1971) but if so, the lack of flagella remains mysterious. The old idea that higher fungi had their origin from red algae has been newly discussed (Demoulin 1974; Kohlmeyer 1975), but some consideration should be given to the possibility of the reverse relationship. Since parasitic red algae are not known to lose plastids completely, an origin of red algae from higher fungi through the acquisition of blue-green algal endosymbionts is possible. It would also be more explanatory in that there would be reason for the lack of flagella in red algae. There are no present criteria to judge whether the biochemical differences between higher fungi and red algae are significant (see Ragan and Chapman 1978, p. 222-223), but certainly one would not expect the cell wall to remain unchanged if fungi did give rise to red seaweeds or vice versa. It is surprising that nobody has determined whether the characteristic AAA pathway of lysine synthesis in higher fungi also occurs in red algae.

Mitochondria

Perhaps the mitochondria of flagellate groups can be compared by more sophisticated criteria in the future, but the only feature presently known for a variety of organisms, and recognized to vary predictably, is the morphology of mitochondrial cristae. In simple terms, some groups have flat cristae and others tubular cristae. Although that might seem a trivial difference, some evidence is accumulating that it might correlate with a relatively early and fundamental evolutionary dichotomy of eucaryotes.

Evolutionary conservatism of cristal type is indicated by the fact that only one type or the other occurs in each group well enough studied to be considered natural (monophyletic). This has been recognized and used in interpretations since the earliest years of EM study of thin sections (e.g., Manton and Parke 1960) until the present (e.g., see Leadbeater and Manton 1974; Hibberd 1975; Moestrup and Thomsen 1976). Many of the groups with tubular cristae are listed on the left side of Fig. 1, and many with flat cristae on the right; the few groups in which all members lack mitochondria (e.g., hypermastigids) are omitted from Fig. 1.

Since no functional difference between the two kinds of mitochondrial cristae is known, the phylogenetic importance of the structural difference can be judged only by a consideration of the distribution of tubular and flat cristae with regard to other characteristics. The correlative occurrence of certain other characteristics suggests that groups with flattened cristae are more related to each other than to any group with tubular cristae. If so, the difference in cristal type represents a fundamental evolutionary divergence. Some reasons for thinking that groups with flattened cristae are related to each other are listed below. The support from any one line of evidence is fragmentary and not compelling, but the items taken together are more impressive than those that could be listed for alternative views.

1. Most flagellate groups with flattened cristae have a pellicular cell surface reinforced with superficial microtubules (euglenoids, *Cyanophora*, kinetoplastids and cryptomonads). Green algal flagellates are the only group to possess scales.
2. Two groups with flattened cristae, cryptomonads and green algae, possess similar rolled trichocysts of a type not known elsewhere (cf. Manton 1969).
3. A multilayered structure occurs in the flagellar apparatus of some green algal zoospores (Stewart and Mattox 1978), one euglenoid (Moestrup 1978), and

an organism related to *Cyanophora* (*Gloeochaete*, Kies 1976). Such structures are so far unknown in organisms with tubular cristae.

4. *Spongomonas*, a flagellate with flat cristae and of uncertain taxonomic position, cannot be assigned to the Kinetoplastida because of differences in cellular organization and the lack of kinetoplasts (Hibberd 1976b). An affinity with kinetoplastids is indicated, however, by the presence of an extracellular connection between the two flagella.

5. *Cyanophora*, another flagellate with flattened cristae and uncertain taxonomic position, has a mitotic spindle with some intriguing similarities to that of *Klebsormidium*, a filamentous green alga. In *Klebsormidium* there is virtually no shortening of chromosome-to-spindle pole distance during anaphase (Floyd et al. 1972b) and this appears to be true also of *Cyanophora* (see Fig. 11 and 15 of Pickett-Heaps 1972b). An unusual feature of *Klebsormidium* is that fragments of nuclear envelope coat the poleward side of chromosomes during anaphase (Pickett-Heaps 1972b; Floyd et al. 1972b). This is also the case in *Cyanophora* (Pickett-Heaps 1972a, Figs. 13, 14).

6. The DAP pathway of lysine synthesis occurs in some organisms with tubular cristae and in some with flat cristae, but the AAA pathway is known only in those with flat cristae (Chytridiomycetes, Ascomycetes, Basidiomycetes, Zygomycetes and *Euglena*, see Bartnicki-Garcia 1970).

7. Whenever chloroplasts are present in a group with flattened cristae, they contain either chlorophyll b or phycobilins. Whenever chloroplasts are present in a group with tubular cristae, they possess major brownish or yellowish pigments and always lack chlorophyll b and phycobilins.

The last item listed is one of great interest and will be discussed further below. As will be explained, the type of cristae and the kind of chloroplast pigments might be evolutionarily linked characteristics. The presence of two distinct types of mitochondria in eucaryotic microorganisms could be explained by one of two hypotheses. Firstly, one type of crista could have evolved from the other, or both from an ancestral type. Alternatively, the two types of mitochondria might be the result of two separate origins of mitochondria from two different kinds of endosymbionts. In the following discussion of chloroplasts, reasons are given for favoring the latter hypothesis.

Aspects of an endosymbiontic origin of chloroplasts

If it is assumed that chloroplasts evolved from endosymbionts, an explanation is required for the facts that chloroplasts of groups with tubular mitochondrial cristae never have chlorophyll b or phycobilins and that chloroplasts of groups with flat cristae never have pigment systems like those of groups with tubular cristae (see paragraph above). This can be explained easily only if chloroplasts arose once only or arose once only in a group with flat cristae and once only in a group with tubular cristae, and in either case, gave rise to all other chloroplastidic flagellates. Then it could be claimed that the different pigment systems are the result of evolutionary divergence, or that the chloroplasts in groups with flat and tubular cristae evolved from endosymbionts that had different pigment systems initially. Unfortunately for those who prefer the simplest possible explanations, there is mounting evidence that chloroplasts have evolved many times independently. If so, why should chloroplasts in organisms with flat cristae always have chlorophyll b or phycobilins, and chloroplasts in organisms with tubular cristae never have those pigments? Random or opportunistic acquisition of chloroplasts by protozoa would be expected to yield organisms with the combination of tubular cristae and phycobilins, or ones with flat cristae and brownish chloroplasts.

Before making an attempt to answer the questions raised in the foregoing paragraph, we shall recount some of the information supporting the view that there have been many instances of the origin of chloroplasts from endosymbionts. One of the earliest and most interesting lines of comparative evidence comes from studies on some dinoflagellates. Tomas and Cox (1973) demonstrated that *Peridinium balticum* has a eucaryotic endosymbiont, which possesses chloroplasts later shown to contain fucoxanthin. No other chloroplasts occur in *P. balticum.* The presence of three membranes around the chloroplasts of more typical photosynthetic dinoflagellates led Tomas and Cox to suggest that the third membrane is an evolutionary remnant that indicates that other photosynthetic dinoflagellates acquired their chloroplasts by the reduction of photosynthetic, eucaryotic endosymbionts. We believe that their suggestion is strongly supported by Tippit and Pickett-Heaps' (1976) demonstration that the nucleus of the endosymbiont in *P. balticum* divides "amitotically." The absence of a spindle demonstrates that the nucleus is degenerate. The lack of girdle lamellae and the presence of peridinin (rather than fucoxanthin) in most uninucleate, photosynthetic dinoflagellates suggests that the chloroplasts of dinoflagellates have resulted from multiple endosymbioses (see Gibbs 1978). The observations in dinoflagellates led to considerations that euglenoids, the only other group with three chloroplast membranes, have also obtained chloroplasts from reduced, eucaryotic endosymbionts, but this time from a green alga (Gibbs 1978; Stewart and Mattox 1975a). For somewhat different reasons, it now seems likely that the chloroplasts of cryptomonads originated from eucaryotic endosymbionts (Greenwood et al. 1977). Inside the chloroplast-chloroplast ER complex of cryptomonads, there is an unusual, small nucleus-like structure (nucleomorph), which appears to be a nucleus in an even more advanced stage of degeneracy than the one in *P. balticum*. In *Cyanophora*, sometimes called a cryptomonad, the chloroplast obviously had its origin from a procaryotic cell, and there is no reason to doubt that this is also the case in green algae. Except for dinoflagellates, it is more problematical whether the chloroplasts of phytoflagellate groups with tubular cristae evolved from procaryotes directly or from reduced eucaryotic cells.

To return to the original problem, why is it that organisms with flattened cristae never acquire and reduce a eucaryotic cell with brownish chloroplasts as probably occurred a number of times in dinoflagellates? Why is it that organisms with tubular cristae have apparently never reduced a eucaryotic green alga as may be the case in *Euglena*? Why is it that an organism with tubular cristae has never taken up with a blue-green alga to the extent that it is difficult to determine whether or not the photosynthetic organelle should be called a chloroplast or a blue-green algal endosymbiont, as in *Cyanophora* (see Schnepf and Brown 1971)? Part of a possible answer becomes apparent after a while. Flagellates are able to reduce endosymbiontic eucaryotic cells only if the endosymbionts are genetically close enough to the host cells. In other words, protozoa with tubular cristae can reduce only other eucaryotic cells with tubular cristae, and those with flat cristae only others with flat cristae. Whether or not the actual shape of the cristae is important, it is the same in related organisms. Whatever the primary origin of the correlation between pigment systems and the form of mitochondrial cristae, later endosymbiontic reductions of eucaryotic cells have maintained that correlation, because extensive reductions could occur easily only when there was a genetic "fit" between host and endosymbiont. Photosynthetic endosymbionts with flat cristae do occur in organisms with tubular cristae (e.g., cryptomonads or green algae in ciliates, Taylor et al. 1969; Hibberd 1977; Karakashian et al. 1968), but in no case are these endosymbionts known to undergo genetic and structural reduction comparable to those discussed earlier.

Even if the ideas above are a valid explanation of the pattern of reduction of eucaryotic endosymbionts, there remains the question of why chloroplast pigment systems and form of cristae became correlated in the first place. Even the evolutionary conversion of procaryotes to chloroplasts appears to hold to the general relationship. Blue-green algae occur as endosymbionts in organisms with tubular cristae (e.g., in the testate amoeba *Paulinella*, Kies 1974), but in no case are these known to become integrated to the degree that they could be considered true chloroplasts as in *Cyanophora*, which has flat cristae. Possible explanations of the primary origin of the correlation between chloroplast pigments and cristal form become much more speculative, but justified because they could prove testable if comparative biochemists should eventually extend their investigations from a few "standard" organisms to a carefully selected variety. The correlation obviously began when photosynthetic procaryotic endosymbionts become reduced to the status of chloroplasts. Since mitochondria almost certainly had an earlier origin than chloroplasts, one must consider the possibility that the type of mitochondrion had some influence on the kind of photosynthetic procaryote that could become thoroughly integrated. That line of reasoning and an extension of the ideas expressed with regard to the reduction of eucaryotic endosymbionts suggest that particular kinds of endosymbiontic, photosynthetic procaryotes can be more easily reduced to true chloroplasts because they are related to the endosymbionts that gave rise to mitochondria. Such a relationship could be explained if the endosymbionts that gave rise to mitochondria were originally photosynthetic. These early endosymbioses must have been based on the respiratory capabilities of the endosymbionts and led to a loss of their photosynthetic apparatus. They must also have eventually become genetically reduced and genetically integrated with the host cell in a manner not yet explainable. If so, later endosymbioses based on the endosymbiont's photosynthetic properties would more quickly or easily become genetically integrated if the endosymbionts were related to the original respiratory endosymbionts. If it seems far-fetched that the ancestors of mitochondria were photosynthetic, one need only consider that the TCA cycle and oxidative phosphorylation might well have originated in photosynthetic procaryotes. Most or many of the aerobic procaryotes available at the time mitochondria evolved could have been photosynthetic.

In conclusion, it is postulated that the endosymbionts that gave rise to mitochondria with flat cristae were photosynthetic and closely related to blue-green algae and/or procaryotes like *Prochloron* (see Lewin 1977). Later endosymbioses in descendant organisms led to chloroplasts with phycobilins or chlorophyll b. The tubular cristae of other groups are a reflection of the origin of mitochondria from a different sort of procaryote with a corresponding difference in the kind of photosynthetic organism later reduced to a chloroplast. From this viewpoint, some otherwise apparently trivial phenomena might take on added significance. We are reminded of the fact that the mitochondrial cristae and the thylakoids of *Cyanophora* have a similar appearance, both being arranged in concentric lamellae (Mignot et al. 1969). We doubt that the resemblance is coincidental. Finally, we must point out that the conclusion is clouded by two possibly important facts, for which no explanation is presently obvious. Firstly, although *Cyanophora*, green algae, and red algae appear to be clear examples of the origin of chloroplasts from procaryotic cells, there is no equally clear case of this in organisms with tubular cristae. All of the known chloroplastidic groups with tubular cristae have chloroplast ER or three chloroplast membranes. This could eventually prove to mean that all have obtained chloroplasts by reduction of eucaryotic cells (see Gibbs 1978). Secondly, a photosynthetic procaryote with pigmentation similar to that of eucaryotic algae with tubular cristae has not been discovered.

If chloroplasts have been relatively recently acquired, their loss would seem quite possible. This seems to be the case in both euglenoids and dinoflagellates. Aplastidic euglenoids with eyespots and similarities to *Euglena* are almost certainly examples of this (see Leedale 1978). There is, however, no clear case of complete plastid loss in groups that can be presumed to have acquired chloroplasts early (e.g., green algae; see also Gibbs 1978). This argues against a general origin of zooflagellates from photosynthetic flagellates (see Section I) even though zooflagellates might have obtained mitochondria by the reduction of photosynthetic procaryotes. In any case, the occurrence of flagella in groups with both cristal types indicates that flagella and therefore zooflagellates existed before the origin of the two cristal types; that is, before the origin of mitochondria.

Cell covering or cell surface

The present knowledge of cell covering in phytoflagellates provides little of use in determining relationships between groups. Phytoflagellates, as well as zooflagellates, are usually naked or scaly, and there are no reliable indications as to whether scales arose once or a number of times. In many cases, there is also no way to decide whether a scaleless group has always lacked scales or has lost them. Features of a scaly covering or the presence of a pellicular surface are often of some value, however, in the characterization of groups.

True cell walls are rare in phytoflagellates (except for cysts, zygotes or other non-motile stages), being known only in *Chlamydomonas* and closely related genera. In chlamydomonads the cell wall appears to have evolved from the fusion of scales. In other green algae and in algae of other classes, cell walls apparently evolved in non-flagellated stages only and did not result from scale fusion (Mattox and Stewart 1977; Domozych et al., in press). The frustule of diatoms is a special case; here it seems that the number of scales has been reduced to two, and these have become elaborately specialized.

It is of some interest that flagellate groups with flat mitochondrial cristae (see right side of Fig. 1) tend to have pellicular cell surfaces (Euglenophyceae, Cryptophyceae, Kinetoplastida, *Cyanophora*); the only scaly flagellates are found in the green algae. An implication is that the zooflagellate ancestor of green algae might be related to the Kinetoplastida but differs from them in having a scaly surface rather than a pellicle. To our knowledge, however, no scaly zooflagellate with flat cristae has yet been discovered.

Summary

The main ideas developed in the foregoing discussions are incorporated in Fig. 1 as the "trunk" and "main branch" of a phylogenetic tree. A flagellate is

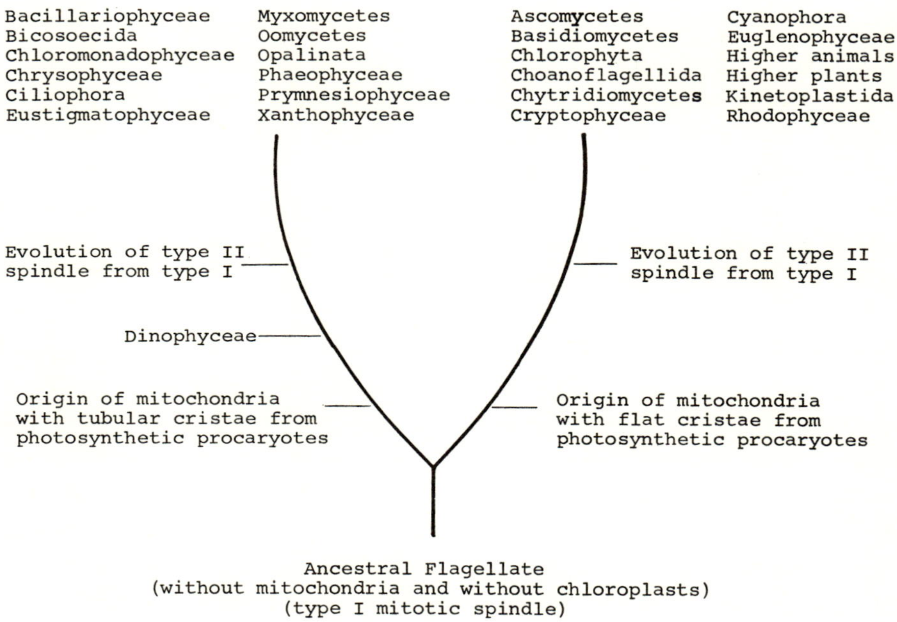

Fig. 1. The base of a phylogenetic tree. See text for explanation.

placed at the base of the tree because there is no evidence that any extant organisms are primitively without flagella (see discussion of the flagellar apparatus above). The difference between mitochondria with flat cristae and those with tubular cristae is presumed to be the result of their independent origin from different kinds of procaryotic endosymbionts (see earlier discussions of mitochondria and chloroplasts). If so, the presence of flagella both in groups with flat cristae and groups with tubular cristae indicates that flagella preceded the origin of mitochondria. Since dinoflagellates have a type I mitotic spindle and are related to the organisms on the left side of the tree on the basis of their mitochondrial cristae and pigmentation, it must be assumed that evolution from a type I spindle to a type II occurred independently at least once on each side of the tree. Present information allows few conclusions concerning detailed relationships between groups appearing on the same side of the tree, and they are therefore presented merely as lists. Some groups are omitted to keep the lists short. The hypermastigids and similar groups are omitted because there is yet no way to judge whether they have lost mitochondria or never had them.

Since tubular mastigonemes occur in many groups on the left side of the tree and in cryptomonads, *Cyanophora* and perhaps bodonids (see Hitchen 1974) on the right side, it is concluded that tubular mastigonemes had a very early origin. The tree does not explain the presence of a stellate structure in the flagellar transition region in both prymnesiophytes and green algae (see Manton 1964). The stellate structure could be very primitive and lost in most groups, or it could have been independently derived. In either case, the structure of the tree need not be changed.

There are several modern proposals for kingdom-level classification of eucaryotic organisms (see Leedale 1974 for a thorough discussion). All have serious defects owing largely to the fact that they are not based on a broad concept of basic trends of evolution in eucaryotic microorganisms. If evidence continues to support the dichotomy portrayed in Fig. 1, consideration should be given to classifying all eucaryotes into two kingdoms. We would propose that the groups on the left side of Fig. 1 be classified in the kingdom Dinobiota and the groups on the right in the kingdom Bodonobiota. In addition to the rationale presented earlier, there is a recent report on histones which supports Fig. 1. Glover and Gorovsky (1979) sequenced H4 histone in the ciliate *Tetrahymena*, and concluded that "a divergence of lines leading to ciliates is ancient and substantially antedates the divergence of lines leading to higher plants and animals." When histones and other conservative proteins are studied in enough organisms, combined structural and biochemical information should eventually lead to phylogenetic concepts that are something more than "matters of opinion."

REFERENCES

Aist, J. R. and P. H. Williams. 1972. Ultrastructure and time course of mitosis in the fungus *Fusarium oxysporum*. J. Cell Biol. 55: 368-389.

Aldrich, H. C. 1969. The ultrastructure of mitosis in myxamoebae and plasmodia of *Physarum flavicomum*. Am. J. Bot. 56: 290-299.

Bartnicki-Garcia, S. 1970. Cell wall composition and other biochemical markers in fungal phylogeny. In Phytochemical Phylogeny, ed. by J. B. Harborne. Academic Press, New York.

Belcher, J. H. 1975. The fine structure of the loricate colourless flagellate *Bicoeca planctonica* Kisselw. Arch. Protistenk. 117: 78-84.
Belcher, J. H. and E. M. F. Swale. 1972. The morphology and fine structure of the colorless colonial flagellate *Anthophysa vegetans* (O. F. Müller) Stein. Br. Phycol. J. 7: 53-59.
Berthold, W. U. 1978. Ultrastrukturanalyse der endoplasmatischen Algen von *Amphistegina lessonii* d' Orbigny, Foraminifera (Protozoa) und ihre systematische Stellung. Arch. Protistenk. 120: 16-62.
Bold, H. C. and M. J. Wynne. 1978. Introduction to the algae, structure and reproduction. Prentice Hall, Englewood Cliffs.
Bouck, G. B. and D. L. Brown. 1973. Microtubule biogenesis and cell shape in *Ochromonas*. I. The distribution of cytoplasmic and mitotic microtubules. J. Cell Biol. 56: 340-359.
Bronchart, R. and V. Demoulin. 1977. Unusual mitosis in the red alga *Porphyridium purpureum*. Nature 268: 80-81.
Cachon, J., M. Cachon, L. G. Tilney, and M. S. Tilney. 1977. Movement generated by interactions between the dense material and the ends of microtubules and non-actin-containing microfilaments in *Sticholonche zanclea*. J. Cell Biol. 72: 314-338.
Cavalier-Smith, T. 1975. The origin of nuclei and of eucaryotic cells. Nature 256: 463-468.
Cavalier-Smith, T. 1978. The evolutionary origin and phylogeny of microtubules, mitotic spindles and eucaryote flagella. BioSystems 10: 93-114.
Coss, R. A. 1974. Mitosis in *Chlamydomonas reinhardtii* basal bodies and the mitotic apparatus. J. Cell Biol. 63: 325-329.
Deason, T. D. and W. H. Darden, Jr. 1971. The male initial and mitosis in *Volvox*. In Contributions in Phycology, ed. by B. C. Parker and R. M. Brown, Jr. Allen Press, Lawrence, Kansas.
Demoulin, V. 1974. The origin of Ascomycetes and Basidiomycetes: the case for a red algal ancestry. Bot. Rev. 40: 315-345.
Domozych, D. S., K. D. Stewart, and K. R. Mattox. The comparative aspects of cell wall chemistry in the green algae. In press.
Floyd, G. L. 1978. Mitosis and cytokinesis in *Asteromonas gracilis*, a wall-less green monad. J. Phycol. 14: 440-445.
Floyd, G. L., K. D. Stewart, and K. R. Mattox. 1972a. Comparative cytology of *Ulothrix* and *Stigeoclonium*. J. Phycol. 8: 68-81.
Floyd, G. L., K. D. Stewart, and K. R. Mattox. 1972b. Cellular organization, mitosis, and cytokinesis in the ulotrichalean alga, *Klebsormidium*. J. Phycol. 8: 176-184.
Fuller, M. S. 1976. Mitosis in fungi. Int. Rev. Cytol. 45: 113-153.
Gibbs, S. P. 1978. The chloroplasts of *Euglena* may have evolved from symbiotic green algae. Can. J. Bot. 56: 2883-2889.
Gillot, M. A. and R. Triemer. 1978. The ultrastructure of cell division in *Euglena gracilis*. J. Cell Sci. 31: 25-35.
Glover, C. V. C. and M. A. Gorovsky. 1979. Amino-acid sequence of *Tetrahymena* histone H4 differs from that of higher eucaryotes. Proc. Nat. Acad. Sci. U.S.A. 76: 585-589.
Green, J. C. and I. Manton. 1970. Studies in the fine structure and taxonomy of flagellates in the genus *Pavlova*. I. A revision of *Pavlova gyrans*, the type species. J. Mar. Biol. Assn. U.K. 50: 1113-1130.
Greenwood, A. D., H. B. Griffiths, and U. J. Santore. 1977. Chloroplasts and cell compartments in Cryptophyceae. Br. Phycol. J. 12: 119-124.
Hanson, E. D. 1977. The origin and early evolution of animals. Wesleyan Univ. Press, Middletown, Conn.
Harvey, I. C. 1974. Striated fibres in the cytoplasm of the imperfect fungus *Pleiochaeta setosa* (Kirchn.) Hughes. Protoplasma 80: 371-380.
Heath, I. B. 1978. Experimental studies of mitosis in the fungi. In Nuclear Division in the Fungi, ed. by I. B. Heath. Academic Press, New York.

Hibberd, D. J. 1975. Observations on the ultrastructure of the choanoflagellate *Codosiga botrytis* (Ehr.) Saville-Kent, with special reference to the flagellar apparatus. J. Cell Sci. 17: 191-219.

Hibberd, D. J. 1976a. Observations on the ultrastructure of three new species of *Cyathobodo* Petersen et Hansen (*C. salpinx*, *C. intricatus* and *C. simplex*) and on the external morphology of *Pseudendromonas vlkii* Bourrely. Protistologica 12: 249-261.

Hibberd, D. J. 1976b. The fine structure of the colonial colorless flagellates *Rhipidodendron splendidum* Stein and *Spongomonas uvella* Stein with special reference to the flagellar apparatus. J. Protozool. 23: 374-385.

Hibberd, D. J. 1976c. The ultrastructure and taxonomy of the Chrysophyceae and Prymnesiophyceae (Haptophyceae): a survey with some new observations on the ultrastructure of the Chrysophyceae. Bot. J. Linn. Soc. 72: 55-80.

Hibberd, D. J. 1977. Observations on the ultrastructure of the cryptomonad endosymbiont of the red-water ciliate *Mesodinium rubrum*. J. Mar. Biol. Assn. U.K. 57: 45-61.

Hibberd, D. J. 1978a. *Bicosoeca accreta* Sp. Nov., a flagellate accumulating extraneous silica fragments. Br. Phycol. J. 13: 161-166.

Hibberd, D. J. 1978b. Possible phylogenetic value of the transitional helix in some chromophyte algal classes and some colourless protists. BioSystems 10: 115-116.

Hitchen, E. T. 1974. The fine structure of the colonial kinetoplastid flagellate *Cephalothamnium cyclopum* Stein. J. Protozool. 21: 221-231.

Hollande, A., J. Cachon, and M. Cachon. 1969. La dinomitose attractophorienne à fuseau endonucléaire chez les Radiolaires Thalanophysidae. Son homologie avec la mitose des Foraminiferes et avec celle des levures. C. R. Acad. Sci., Ser. D. 269: 179-182.

Johnson, U. G. and K. R. Porter. 1968. Fine structure of cell division in *Chlamydomonas reinhardi*. J. Cell Biol. 38: 403-423.

Karakashian, S. J., M. W. Karakashian, and M. A. Rudzinska. 1968. Electron microscopic observations on the symbiosis of *Paramecium bursaria* and its intracellular algae. J. Protozool. 15: 113-128.

Kater, J. McA. 1929. Morphology and division of *Chlamydomonas nasuta* with reference to the phylogeny of the flagellate neuromotor system. Univ. Calif. Publ. Zool. 33: 125-168.

Kelley, G. L. and D. M. Hammond. 1973. Fine structural aspects of nuclear division and merogony of *Eimeria ninakohlyakimovae* in cultured cells. J. Parasitol. 59: 1071-1079.

Kies, L. 1974. Elektronmikroskopische Untersuchungen an *Paulinella chromatophora* Lauterborn, einer Thekamobe mit blau-grünen Endosymbionten (Cyanellen). Protoplasma 80: 69-89.

Kies, L. 1976. Untersuchungen zur Feinstruktur und taxonomischen Einordnung von *Gloeochaete wittrockiana*, einer apoplastidalen capsalen Alge mit blaugrünen Endosymbionten (Cyanellen). Protoplasma 87: 419-446.

Klein, R. M. and A. Cronquist. 1967. A consideration of the evolutionary and taxonomic significance of some biochemical, micromorphological, and physiological characters in the thallophytes. Quart. Rev. Biol. 42: 105-296.

Kohlmeyer, J. 1975. New clues to the possible origin of Ascomycetes. Bioscience 25: 86-93.

Kubai, D. F. 1973. Unorthodox mitosis in *Trichonympha agilis*: Kinetochore differentiation and chromosome movement. J. Cell Sci. 13: 511-552.

Kubai, D. F. 1975. The evolution of the mitotic spindle. Int. Rev. Cytol. 43: 167-227.

Kubai, D. F. 1978. Mitosis and fungal phylogeny. In Nuclear Division in the Fungi, ed. by I. Heath. Academic Press, New York.

Kubai, D. F. and H. Ris. 1969. Division in the dinoflagellate *Gyrodinium cohnii*. A new type of nuclear reproduction. J. Cell Biol. 40: 508-528.

LaFountain, J. R. 1972. An association between microtubules and aligned mitochondria in *Nephrotoma* spermatocytes. Exptl. Cell Res. 71: 325-328.

Lang, N. J. 1963. Electron microscopic demonstration of plastids in *Polytoma*. J. Protozool. 10: 333-339.

Leadbeater, B. S. C. and I. Manton. 1974. Preliminary observations on the chemistry and biology of the lorica in a collared flagellate (*Stephanoeca diplocostata* Ellis). J. Mar. Biol. Assn. U.K. 54: 269-276.

Leedale, G. F. 1968. The nucleus in *Euglena*. In The Biology of Euglena, ed. by D. E. Buetow. Academic Press, New York. pp. 185-272.

Leedale, G. F. 1974. How many are the kingdoms of organisms? Taxon 23: 261-270.

Leedale, G. F. 1978. Phylogenetic criteria in euglenoid flagellates. BioSystems 10: 183-187.

Lembi, C. A. 1975. A rhizoplast in *Carteria radiosa* (Chlorophyceae). J. Phycol. 11: 219-221.

Lewin, R. A. 1977. *Prochloron*, type genus of the Prochlorophyta. Phycologia 16: 217-219.

Manton, I. 1964. The possible significance of some details of flagellar bases in plants. J. Roy. Microsc. Soc. 82: 279-285.

Manton, I. 1969. Tubular trichocysts in a species of *Pyramimonas* (*P. grossii* Parke). Osterr. Bot. Z. 116: 378-392.

Manton, I. and M. Parke. 1960. Further observations on small green flagellates with special reference to possible relatives of *Chromulina pusilla* Butcher. J. Mar. Biol. Assn. 39: 275-298.

Manton, I., D. G. Rayns, H. Ettl, and M. Parke. 1965. Further observations on green flagellates with scaly flagella: the genus *Heteromastix* Korschikoff. J. Mar. Biol. Assn. U.K. 45: 241-255.

Manton, I., K. Kowallik, and H. A. von Stosch. 1970. Observations on the fine structure and development of the spindle at mitosis and meiosis in a marine centric diatom (*Lithodesmium undulatum*). IV. The second meiotic division and conclusion. J. Cell Sci. 7: 407-444.

Margulis, L. 1970. Origin of eucaryotic cells. Yale University Press, New Haven and London.

Mattox, K. R. and K. D. Stewart. 1977. Cell division in the scaly green flagellate *Heteromastix angulata* and its bearing on the origin of the Chlorophyceae. Am. J. Bot. 64: 931-945.

McCully, E. K. and C. F. Robinow. 1973. Mitosis in *Mucor hiemalis*. Arch. Mikrobiol. 94: 133-148.

McDonald, K. 1972. The ultrastructure of mitosis in the marine red alga *Membranoptera platyphylla*. J. Phycol. 8: 156-165.

Melkonian, M. 1978. Structure and significance of cruciate flagellar root systems in green algae: Comparative investigations in species of *Chlorosarcinopsis* (Chlorosarcinales). Pl. Syst. Evol. 130: 265-292.

Melkonian, M. 1979. An ultrastructural study of the flagellate *Tetraselmis cordiformis* Stein (Chlorophyceae) with emphasis on the flagellar apparatus. Protoplasma 98: 139-151.

Mignot, J.-P. 1974. Étude ultrastructurale des *Bicoeca*, protistes flagellés. Protistologica 10: 543-565.

Mignot, J.-P. and G. Brugerolle. 1975a. Étude ultrastructurale du Flagelle phagotrophe *Colponema loxodes* Stein. Protistologica 11: 429-444.

Mignot, J.-P. and G. Brugerolle. 1975b. Étude ultrastructurale de *Cercomonas* DuJardin (= *Cercobodo* Krassilstchick) protiste flagellé. Protistologica 11: 547-554.

Mignot, J. P., L. Joyon, and E. G. Pringsheim. 1969. Quelques particularitiés structurales de *Cyanophora paradoxa* Korsch., protozoaire flagellé. J. Protozool. 16: 138-145.

Moestrup, Ø. 1978. On the phylogenetic validity of the flagellar apparatus in green algae and other chlorophyll a and b containing plants. BioSystems 10: 117-144.

Moestrup, Ø. and H. A. Thomsen. 1974. An ultrastructural study of the flagellate *Pyramimonas orientalis* with particular emphasis on golgi apparatus activity and the flagellar apparatus. Protoplasma 81: 247-269.

Moestrup, Ø. and H. A. Thomsen. 1976. Fine structural studies on the flagellate genus *Bicoeca*. I. *Bicoeca maris* with particular emphasis on the flagellar apparatus. Protistologica 12: 101-120.

Oakley, B. R. and J. D. Dodge. 1974. Kinetochores associated with the nuclear envelope in the mitosis of a dinoflagellate. J. Cell Biol. 63: 322-325.

Parke, M. and I. den Hartog-Adams. 1965. Three species of *Halosphaera*. J. Mar. Biol. Assn. U.K.: 525-536.

Peterson, J. B. and H. Ris. 1976. Electron microscopic study of the spindle and chromosome movement in the yeast *Saccharomyces cerevisiae*. J. Cell Sci. 22: 219-242.

Pickett-Heaps, J. D. 1972a. Cell division in *Cyanophora paradoxa*. New Phytol. 71: 561-567.

Pickett-Heaps, J. D. 1972b. Cell division in *Klebsormidium subtillissimum* (formerly *Ulothrix subtillissima*) and its possible phylogenetic significance. Cytobios 6: 167-183.

Pickett-Heaps, J. D. 1974. Evolution of mitosis and the eucaryote condition. BioSystems 6: 37-45.

Pickett-Heaps, J. D. 1975a. Aspects of spindle evolution. Ann. N. Y. Acad. Sci. 253: 352-361.

Pickett-Heaps, J. D. 1975b. Green Algae: Structure, reproduction and evolution in selected genera. Sinauer Associates, Inc., Sunderland, Mass.

Pickett-Heaps, J. D. and D. H. Tippit. 1978. The diatom spindle in perspective. Cell 14: 455-467.

Pickett-Heaps, J. D., D. H. Tippit, and J. A. Andreozzi. 1978. Cell division in the pennate diatom *Pinnularia*. II. Later stages in mitosis. Biol. Cell. 33: 79-84.

Pollard, T. D. 1976. Cytoskeletal functions of cytoplasmic contractile proteins. J. Supramol. Struc. 5: 317-334.

Powell, M. J. 1975. Ultrastructural changes in nuclear membranes and organelle associations during mitosis of the aquatic fungus *Entophlyctis* sp. Can. J. Bot. 55: 1668-1685.

Raff, R. A. and H. R. Mahler. 1972. The non symbiotic origin of mitochondria. Science 177: 575-582.

Ragan, M. A. and D. J. Chapman. 1978. A biochemical phylogeny of the protists. Academic Press, New York.

Ringo, D. L. 1967. Flagellar motion and fine structure of the flagellar apparatus in *Chlamydomonas*. J. Cell Biol. 33: 543-571.

Ris, H. and D. F. Kubai. 1974. An unusual mitotic mecahnism in the parasitic protozoan *Syndinium* sp. J. Cell Biol. 60: 702-720.

Rogers, C. E., K. R. Mattox, and K. D. Stewart. The zoospores of *Chlorokybus atmophyticus*, a charophyte with sarcinoid growth habit. Am. J. Bot. In press.

Salisbury, J. L. and G. L. Floyd. 1978. Calcium-induced contraction of the rhizoplast of a quadriflagellate green alga. Science 202: 975-977.

Schnepf, E. and R. M. Brown, Jr. 1971. On relationships between endosymbiosis and the origin of plastids and mitochondria. In Origin and Continuity of Cell Organelles, ed. by J. Reinert and H. Ursprung. Springer Verlag, New York.

Slankis, T. and S. P. Gibbs. 1972. The fine structure of mitosis and cell division in the chrysophycean alga *Ochromonas danica*. J. Phycol. 8: 243-256.

Sleigh, M. A. 1973. The biology of protozoa. American Elsevier Publ. Co., New York.

Stanier, R. Y. 1970. Some aspects of the biology of cells and their possible evolutionary significance. Symp. Soc. Gen. Microbiol. 20: 1-38.

Stewart, K. D. and K. R. Mattox. 1975a. Comparative cytology, evolution and classification of the green algae with some consideration of the origin of other organisms with chlorophylls a and b. Bot. Rev. 41: 104-135.

Stewart, K. D. and K. R. Mattox. 1975b. Some aspects of mitosis in primitive green algae: phylogeny and function. BioSystems 7: 310-315.

Stewart, K. D. and K. R. Mattox. 1978. Structural evolution in the flagellated cells of green algae and land plants. BioSystems 10: 145-152.

Stewart, K. D., K. R. Mattox, and C. D. Chandler. 1974. Mitosis and cytokinesis in *Platymonas subcordiformis*, scaly green monad. J. Phycol. 10: 65-79.

Swale, E. M. F. and J. H. Belcher. 1974. *Gyromitus disomatus* Skuja - a free-living colourless flagellate. Arch. Protistenk. 116: 211-220.

Taylor, F. J. R. 1974. Implications and extensions of the serial endosymbiosis theory of the origin of eucaryotes. Taxon 23: 229-258.

Taylor, F. J. R. 1978. Problems in the development of an explicit hypothetical phylogeny of the lower eucaryotes. BioSystems 10: 67-89.

Taylor, F. J. R., D. J. Blackbourn, and J. Blackbourn. 1969. Ultrastructure of the chloroplasts and associated structures within the marine ciliate *Mesodinium rubrum* (Lohmann). Nature 224: 819-821.

Tippit, D. H. and J. D. Pickett-Heaps. 1976. Apparent amitosis in the binucleate dinoflagellate *Peridinium balticum*. J. Cell Sci. 21: 273-289.

Tomas, R. N. and E. R. Cox. 1973. Observations on the symbiosis of *Peridinium balticum* and its intracellular alga. I. Ultrastructure. J. Phycol. 9: 304-323.

Uzzell, T. and C. Spolsky. 1974. Mitochondria and plastids as endosymbionts: A revival of special creation? American Scientist 62: 334-344.

GENUS INDEX

Acrosiphonia centralis 134
Amphidinium 443
  A. *carterae* 44
Amphipleura pellucida 346
Anisomonas longifilis 92
Ankyra 116
Anthophysa 217, 225, 437
Apedinella 224, 226
  A. *spinifera* 227
Apistonema 308
Archaemonadaceae 237
Archaias angulatus 41
Ascoglena 185
Ascomycetes 446, 451, 456
Astasia longa 167, 173, 176
  A. *ocellata* 178, 181
Asteromonas 34, 86, 107, 110, 118, 126, 128, 130, 136
  A. *gracilis* 89, 95, 118, 119, 125
  A. *propulsa* 135
Astrephomenaceae 156
Astrephomene 156, 157, 160
  A. *gubernaculifera* 156, 157, 160
Bacillariophyceae 228, 248, 356, 456
Basidiomycetes 446, 451, 456
Bicosoeca 438
Bicosoecida 456
Bipedinomonas 86, 87, 92, 93, 94, 98, 114, 115, 117
  B. *angulata* 94, 105, 108, 109, 114, 117, 119, 125, 127, 130, 131
  B. *rotunda* 94, 103, 109, 116, 119
Bodonobiota 457
Botrydiopsis 246
  B. *alpina* 246, 251
  B. *arhiza* 246
  B. *intercedens* 266
Botrydium 261, 267
Brachiomonas 18
Bumilleria sicula 253, 257
Bumilleriopsis filiformis 255, 266
Carteria 5, 6, 9, 14, 16, 18, 22, 23, 32, 42, 45, 61, 63, 67, 89, 102, 444
  C. *acidicola* 41
  C. *crucifera* 23
  C. *eugametos* 23
  C. *nivale* 77
  C. *olivieri* 13, 23
  C. *radiosa* 13, 22, 23, 24, 104
Cephalothamnium 436
Ceratium hirundinella 422
Ceratium tripos 410
Cercomonas 438
Chaetosphaeridium 87, 111
Chara 111, 374
Charophyceae 87, 90, 93, 111, 136
Charophyta 87, 166
Chattonella 351, 352, 356, 368
  C. *subsalsa* 352, 354, 356, 357, 360, 363, 366, 367
Chilomonas 384, 390, 397
Chlainomonas 61, 67
  C. *kolii* 62, 63, 67, 71
  C. *rubra* 62, 64, 66, 67, 70, 71, 80
Chlamydobotrys 159
Chlamydomonadaceae 6, 9, 46
Chlamydomonas 5, 6, 7, 8, 9, 11, 14, 15, 16, 18, 19, 20, 21, 22, 23, 24, 27, 29, 30, 31, 32, 34, 38, 39, 40, 41, 42, 43, 44, 45, 61, 63, 64, 65, 66, 67, 70, 71, 73, 78, 79, 80, 81, 89, 90, 102, 103, 104, 110, 116, 118, 126, 128, 155, 434, 435, 436, 437, 443, 444, 455
  C. *acidophila* 41, 42
  C. *eugametos* 11, 14, 31, 34
  C. *globosa* 41
  C. *hedleyi* 41
  C. *moewusii* 14, 31, 32, 33, 34, 36
  C. *nivalis* 61, 62, 65, 66, 69, 70, 71, 72, 78, 79, 80
  C. *reinhardtii* 5, 7, 9, 11, 14, 15, 16, 17, 18, 20, 21, 23, 24, 26, 27, 28, 29, 30, 31, 32, 33, 34, 36, 38, 40, 42, 44, 45, 46
  C. *sphagnophila* 41
Chlorella 5, 9
Chlorobrachis 159
Chlorobotrys 320
  C. *regularis* 323, 324
Chlorococcales 42, 67, 68, 69
Chlorococcum 43, 134
Chlorodendron 85, 86
Chlorogonium 6, 7, 8, 18, 30, 40, 46, 103
  C. *elongatum* 42
  C. *euchlorum* 42
Chloromonadales 351

Chloromonadineae 351
Chloromonadophyceae 351, 456
Chloromonadophyta 351
Chloromonas 61, 63, 65, 66, 67, 71, 78, 81
  C. brevispina 61, 63, 64, 65, 67, 68, 69, 70, 80
  C. nivalis 63, 64, 66, 67, 69, 70, 77, 78
  C. pichinchae 66, 67, 69, 70, 71, 73, 74, 75, 76, 77, 78, 79, 80
Chloronephris 86
Chlorophyceae 87, 93, 134, 135, 136, 244, 248, 320
Chlorophyta 61, 85, 86, 87, 100, 112, 113, 166, 456
Choanoflagellida 438, 456
Chromophyta 335
Chromulina 80, 215, 218, 222, 223
  C. chionophilia 80
  C. placentula 225
  C. psammobia 225
Chromulinales 214
Chroomonas 382, 383, 384, 386, 388, 389, 399, 400
  C. mesostigmatica 383, 384, 388
  C. nordstedtii 383
  C. salina 384, 389, 393, 394, 395, 399, 400
Chrysamoeba radians 219, 225
Chrysapiales 214
Chrysochromulina 280, 282, 291, 298, 300, 301, 304, 308
  C. acantha 293
  C. camella 300, 302
  C. chiton 276, 284, 292, 295, 298, 302
  C. cymbium 300
  C. ephippium 306
  C. ericina 300
  C. kappa 275, 276
  C. parkeae 300
  C. pringsheimii 300, 304
  C. strobilus 291, 296, 300
Chrysococcales 214
Chrysococcus 219
  C. cordiformis 225
Chrysomonadina 213
Chrysonema 214
Chrysophaerales 214
Chrysophaerella 218
Chrysophyceae 159, 213, 214, 215, 216, 217, 218, 219, 223, 224, 225, 226, 227, 228, 234, 236, 237, 248, 250, 251, 253, 256, 260, 262, 274, 293, 310, 326, 338, 356, 435, 456
Chrysophycophyta 335
Chrysophyta 80, 104, 335
Chrysosphaerella 217
Chrysotila stipitata 276
Chytridiomycetes 449, 451, 456
Ciliophora 456
Cladophora 134
Coccolithus 309
  C. huxleyi 44
  C. pelagicus 309
Coccosphaerales 274, 309
Codium fragile 135
Colacium 166, 185, 186, 202
Coleochaete 19, 111
Colpomena 352, 438
Coronaceae 153
Corone 152, 153
Corymbellus aureus 276, 301
Craspedomonadales 213
Craspedophyceae 437
Cryocystis 67
  C. brevispina 68
  C. granulosa 63, 77
  C. japonica 68
Cryodactylon glaciale 67, 68
Crypthecodinium cohnii 416
Cryptoglema pigra 169
Cryptomonas 382, 383, 384, 388, 389, 393, 395, 397, 399
  C. borealis 399
  C. calceiformis 384
  C. ovata 397
  C. ovata var. *palustris* 395, 399
Cryptophyceae 245, 251, 326, 395, 435, 455, 456
Cryptothecodinium cohnii 443
Crystallolithus 309
  C. hyalinus 309
Cyanomonas 383
Cyanophora 447, 449, 450, 451, 453, 454, 455, 456, 457
  C. paradoxa 398
Cyathobodo 436, 438
Cyathomonas 397
Cyrtophora 237
Cysta 237
Dangeardinella 86
Diacronema vlkianum 282, 286, 289, 293, 294, 312
Dicrateria inornata 289
Dictyocha 336, 337, 338, 340, 349
  D. fibula 335, 336, 337, 338, 340, 342, 346, 349
  D. octonaria 336
  D. speculum 336

Dictyochales 213, 214
Dicylindricum 230
Dinobiota 457
Dinobryon 41, 217, 225, 228, 230, 231, 232, 234, 237
D. cylindricum 228, 229, 231, 232, 234, 235
D. divergens 228
D. sertularia 220
Dinophyceae 245, 407, 435, 441
Distephanus 336, 340
D. speculum 336
Dolichomastix 91, 93, 95, 96
Dunaliella 6, 8, 9, 11, 20, 24, 32, 34, 41, 44, 89, 107, 110, 126, 128
D. primolecta 14, 104, 119
Dysmorphococcus 6, 11, 30, 46
Ebridians 335
Ecballocystis 132
Ellipsoidion acuminatum 323, 325, 327, 328
Emiliania huxleyi 278, 289, 293, 301, 309
Enteromorpha intestinalis 134
Entosiphon 167, 198, 200, 202
Epipyxis 217
Euchlorophyceae 86
Eudorina 20, 150, 152, 156
E. elegans 150, 155
Euglena 5, 165, 166, 167, 168, 169, 170, 171, 172, 173, 174, 175, 176, 177, 178, 182, 183, 184, 186, 198, 202, 203, 280, 443, 451, 453, 455
E. acus 176
E. gracilis 167, 170, 173, 174, 175, 176, 178, 181, 182, 183, 184, 185, 202
E. gracilis var. *bacillaris* 172, 173, 177, 178, 179, 184
E. granulata 167
E. spirogyra 185, 202
Euglenophyceae 118, 175, 248, 435, 441, 455, 456
Euglenophyta 87, 166
Eustigmatophyceae 134, 175, 266, 268, 319, 320, 332, 333, 456
Fibrocapsa 352
Fritschiella 19
Fucus 253
Fusarium 443
Glaucocystis 364
Glenodinium 414
G. foliaceum 423
Gonium 147, 148, 156
G. multicoccum 148
G. octonarium 148
G. pectorale 148
G. sociale 148
Gonium-Volvox 147, 148, 152, 158
Gonyaulax 411
G. monilata 416
G. polyedra 408, 410
Gonyostomum 351, 352, 366, 368
G. intermedium 352, 374
G. latum 374, 375, 352
G. ovatum 352, 360, 374
G. semen 352, 354, 356, 357, 358, 360, 364, 366, 367, 371, 372, 373, 374, 375
Gymnodinium microadriaticum 422
Gyromitus 86, 436, 438
Haematococcaceae 6, 46, 158
Haematococcus 6, 7, 8, 9, 11, 18, 32, 39, 43, 44, 46, 158
Halosphaera 86, 91, 93, 98, 100, 116, 117, 121, 123, 132
H. minor 116, 435
H. parkeae 100
H. russellii 97, 116, 121
H. viridis 88, 116
Haptophyceae 274, 356
Hemiselmis 383, 384, 397, 399
H. rufescence 388, 397
Heterocapsa triquetra 414
Heterokontae 244
Heteromastix 92, 94, 130, 435, 443, 444, 446
H. angulata 94
H. longifilis 92
"*Heteromastix sp.* (P198)" 135, 136
Heterosigma 352
Hornellia marina 352, 373
Hyalocardium 86
Hyalophacus 175
Hyaloselene 352
Hymenomonas 282, 288, 301, 308
H. carterae 276, 281, 288, 289, 300, 302, 308, 312
H. pringsheimii 298, 300, 301
H. roseola 288, 312
Hypermastigida 441
Imantonia rotunda 289
Isochrysidales 213, 274
Isochrysis galbana 288
Kinetoplastida 451, 455, 456
Klebsormidium 19, 111, 445, 451
Kuzininia 86
Lagynion 237

Loxophyceae 86
Lundiella 152, 153
Mallomonas 217, 218, 236
  M. caudata 219, 228, 231, 233
  M. papillosa 225
  M. striata 222
Mallomonopsis 218, 223
  M. ouradion 222
  M. paxillata 222
Mantoniella 86, 91, 93, 94, 95, 97, 107, 109, 115, 125, 126, 127, 128, 129, 130
  M. squamata 91, 95, 96, 104, 109, 113, 117, 119, 125, 126, 127, 130, 131, 132, 133, 134, 135
Marthasterites 302
Mastigosphaera 153
Menoidium 185
Merotricha 351, 352, 366, 372
  M. bacillata 352, 356, 366, 374, 375
Mesodinium 384, 387
  M. rubrum 383, 398
Mesostigma 86, 91, 98, 103, 114, 115, 119, 123
  M. viride 90, 91, 98, 103, 104, 108, 117, 119, 123
Meyeriella 86
Micromonas 86, 87, 93, 95, 107, 115
  M. pusilla 89, 90, 94, 103, 109, 111, 114, 117, 118, 125, 133, 134, 135, 136, 138, 434
Mischococcus sphaerocephalus 246, 251, 257
Monochrysis lutheri 311
Monodus subterraneus 266, 323, 332, 333
Monomastix 86, 93, 107, 124, 130, 352
  M. aff. *minuta* 90, 96, 105, 108, 109, 111, 116, 117, 123
Mucor 442
Myochloris 86
Myriophyllum spicatum 40
Myxomycetes 456
Najas flexilis 40
Nannochloris 134
Nematodinium 419
Nephroselmis 86, 91, 93, 95, 96, 108
  N. gilva 91, 95, 96, 117, 119
Nitella 111
Nitzschia 41
Noctiluca 416
Ochromonadales 214, 218
Ochromonas 214, 215, 216, 217, 220, 224, 226, 228, 236, 443, 444
  O. danica 217, 218, 220, 221, 234, 235
  O. minima 227
  O. tuberculata 225, 228, 236
Ochrosphaera 288
  O. neopolitana 308
  O. verrucosa 308
Oedogoniaceae 46
Oedogoniales 19, 116
Oedogonium 5
Olisthodiscus 352
  O. luteus 224, 227, 234, 235, 236
Oocystis 67
  O. lacustris 68
Oomycetes 441, 456
Opalinata 456
Ophiocytium majus 248, 251, 253, 255, 257, 259, 263, 265, 266, 268
Ornithocercus 411
Oxyrrhis marina 414
Pachysphaera 86, 91, 93, 98, 100, 132
  P. marshalliae 100
Pandorina 147, 148, 153, 154, 155 156, 160
  P. charkowiensis 148
  P. morum 148, 149, 155
Paramecium 417
Paranema 167, 174, 198, 200
Paraphysomonas 218, 219
Pascheriella 159
Pascherina 159
Paulinella 453
Pavlova 218, 223, 274, 280, 436
  P. gyrans 283, 286, 294, 307
  P. lutheri 283, 288, 311, 312
  P. salina 286
Pavlovales 273, 274, 278, 282, 286, 288, 293, 302, 304, 307
Pedinella 220, 224, 226
  P. hexacostata 219
Pedinellaceae 219
Pedinomonas 86, 93, 104, 107, 114, 115, 118, 119, 129, 130
  P. major 90, 96, 103, 117, 123
  P. minor 89, 90, 94, 104, 105, 109, 112, 117, 118, 119, 123, 125, 129, 130
  P. noctilucae 94, 103, 117
  P. tuberculata 89, 95, 116
Peranema trichophorum 198, 200
Peridinium 411

Peridinium balticum 407, 423, 427, 453
P. brevis 413
P. cinctum f. westii 422
P. foliaceum 423, 427
P. loeblichii 410, 411, 412, 420
Phacotaceae 6, 9, 46
Phacus 175, 185
P. longicaudus 173
Phaeaster pascheri 218
Phaeocystis 293
P. pouchetii 311
Phaeophyceae 253, 293, 356, 456
Phaeoplaca 214
Phaeoplacales 214
Phaeothamniales 214
Physarum 442
Platychrysis pigra 304
Platydorina 152, 156
P. caudata 152
Platymonas 23, 85, 86, 92, 93, 99, 100, 103, 104, 105, 106, 107, 111, 112, 113, 115, 116, 117, 120, 123, 124, 125, 126, 128, 129, 132, 133, 134, 135, 397, 436, 437, 443, 444, 445, 446
P. convolutae 99, 117
P. subcordiformis 100, 106, 113, 125, 126, 127, 128, 129, 130, 131
P. tetrathele 98, 99, 100, 125, 136
Pleodorina 150, 155
P. californica 151
P. illinoisensis 151
Pleurochloris 322
P. commutata 322, 325
Pocillomonas 86
Polyblepharidaceae 6
Polyblepharides 7, 86
Polyedriella helvetica 322, 327, 328, 330
Polysiphonia lanosa 134
Polytoma 5, 6, 7, 8, 11, 15, 18, 19, 20, 29, 32, 38, 40, 45, 89, 103, 437
P. uvella 7, 8
Polytomella 6, 7, 8, 9, 11, 19, 21, 22, 23, 26, 27, 28, 29, 30, 32, 39, 40, 45, 437
P. agilis 13, 22, 23, 29, 108
Prasinates 85
Prasinocarteria 86
Prasinochlamydomonas 86
Prasinocladus 85, 86, 92, 98, 100, 111, 115, 116, 117, 125, 129, 132, 133, 134, 135, 137
P. ascus 118
P. marinus 99, 116, 117
Prasinochloris 86, 132
P. sessilis 90
Prasinophyceae 5, 85, 86, 87, 88, 94, 95, 98, 99, 101, 103, 104, 105, 106, 109, 112, 113, 114, 118, 124, 125, 129, 130, 134, 135, 136, 137, 223
Prasinophyta 87
Prasinovolvocales 86
Prochloron 454
Prorocentrales 411
Protoaceromonas 86
Protochroomonas 86
Prymnesiales 274
Prymnesiophyceae 135, 213, 217, 218, 274, 282, 295, 296, 300, 302, 310, 311, 435, 456
Prymnesium 274, 276, 278, 281, 282, 284, 292, 295, 298, 310
P. parvum 276, 281, 280, 295, 302, 310, 311, 312
Pseudobumilleriopsis pyrenoidosa 255
Pseudocardiomonas 86
Pseudokephyrion 217, 219
P. pseudospirale 219
Pseudopedinella ambisua 219
P. hexacosta 226
P. pyriformis 219, 227
Pseudoscourfieldia 86, 87, 93, 98, 99, 106, 107, 115
P. longifilis 92, 96, 99, 105, 119, 127, 130, 131, 135
P. marina 92, 96, 99, 105, 114, 117, 118, 119
Pterosperma 86, 90, 91, 98, 100, 101, 132
Ptychodiscus brevis 407, 410, 414, 418, 419, 420, 421, 422, 427
Pyramichlamys 97, 117
Pyramidomonas 85
Pyramimonas 85, 86, 88, 91, 92, 93, 94, 97, 98, 99, 103, 104, 105, 106, 107, 108, 111, 113, 114, 115, 116, 117, 118, 119, 120, 121, 122, 123, 124, 125, 126, 127, 128, 130, 131, 132, 133, 134, 135, 136, 137, 138, 223, 436
P. amylifera 97, 105, 114, 118, 121, 122, 130, 133, 135, 136

Pyramimonas aff. disomata 97, 117
P. grossii 97, 121, 124, 130, 132
P. inconstans 136
P. montana 89, 97, 102, 103, 113, 117, 118, 119, 123
P. obovata 97, 104, 114, 121, 135
P. orientalis 97, 104, 121
P. parkeae 97, 102, 104, 113, 114, 120, 121, 122, 123, 124, 125, 127, 130, 131
P. aff. pluriocculata 97
P. reticulata 103, 114, 117, 119, 123, 130
P. reticulata var. *conica* 130
P. tetrarhynchus 94, 97, 120, 121, 123, 130
P. virginica 97, 114, 121, 124
Pyrobotrys 159
Pyrocystis 408
Pyrrhophyta 407, 408
Raphidonema nivale 70, 73, 75
Raphidophyceae 351
Reckertia 352
Rhabdomonas 185
Rhipidodendron 436, 438
Rhizochromulina 215, 224
R. marina 219, 220, 225
Rhizochrysidales 214
Rhodomonas 383, 384, 397, 399
R. lens 393
Rhodophyceae 446, 456
Ruttnera spectabilis 276, 288
Saccharomyces cerevisiae 442
Sarcinochrysis 276
Scenedesmus 41
Scotiella 67
S. polyptera 77
Scourfieldia 86
S. caeca 90, 94, 105, 138
S. complanata 89
Smithsonimonas 61

Spermatozopsis 87, 90, 130, 138
S. exsultans 90
Sphaerellopsis rubra 80
Sphaeropsis 86
Sphagnum 373, 374
Sphalaeromantis 216, 217, 218, 219, 223
S. salina 222
S. tetragona 222, 224
Spondylomoraceae 153, 158, 159
Spondylomorum 159
Spongomonas 451
Stephanoon 153
Stephanoptera 86, 136
S. gracilis 136
Stephanosphaera 158
S. pluvialis 158
Stichococcus 134
Stichogloeales 214
Sticholonche 445
Stigeoclonium 19, 442
Streptomonas 86
Strombomonas 185
Swarchewskiella 86
Syncrypta 217
S. glomerifera 218
Synura 214, 217, 218, 222, 223
S. echinulata 222
S. petersenii 219, 222
S. sphagnicola 222
S. spinosa 222
S. uvella 222
Tetrahymena 417, 457
Tetralithus 302
Tetraselmis 85, 86, 437
Tetraspora 19, 20, 39
Tetratoma 61
Thallochrysidales 214
Thaumatomastix 352
Trachelomonas 175, 185, 186, 187, 191, 192, 202
T. bulla 185, 191, 192, 194
T. grandis 186
T. hispida 186
T. hispida var. *coronata* 185, 186, 188, 191, 192, 194, 196, 198
T. lefevrei 185, 192, 194, 196
T. oblonga var. *punctata* 187
T. volvocina 185
T. zorensis 185, 188, 192, 194
Trentepohlia 111
Trentonia 352
T. flagellata 352, 375
Tribonema 248, 261, 262, 267
T. aequale 356
T. minus 266
T. viride 256
Trichloris 86, 88
Trichonympha 436, 443, 448
Trochiscia 67
T. cryophila 68
T. nivalis 68
T. rubra 68
Ulotrichaceae 46
Ulvaphyceae 87, 90, 93, 126, 127, 136
Uroglena 231
U. americana 228
U. volvox 228
Uva 159

Vacuolaria 351, 352, 368, 375
V. depressa 374
V. penardii 352, 375
V. tropicalis 352, 375
V. virescens 352, 353, 354, 356, 357, 358, 360, 363, 364, 366, 367, 368, 371, 372, 374, 375
V. viridis 352, 363, 371, 372, 374, 375
Vaucheria 244, 246, 250, 251, 253, 256, 257, 260, 261, 262, 264, 267, 268, 441
V. woroniniana 256
Vischeria 320
V. punctata 322, 325, 328, 330
V. stellata 322, 324, 325, 330
Volvocaceae 147, 150, 153, 154, 155, 156, 157, 158
Volvocales 42, 104, 147
Volvox 19, 20, 147, 150, 153, 154, 155, 156, 158
V. aureus 156
V. capensis 156
V. carteri 152, 154
V. carteri f. nagariensis 156
V. carteri f. weismannia 156
V. dissipatrix 156
V. gigas 156
V. obversus 156
V. rousseletii 156
Volvulina 152, 156, 157, 160
V. steinii 152, 153, 157, 160
Xanthophyceae 135, 243, 244, 245, 248, 251, 253, 256, 257, 260, 261, 262, 267, 268, 293, 320, 321, 324, 326, 329, 333, 356, 456
Zygnematales 449
Zygomycetes 451

SUBJECT INDEX

Biochemistry
  Chlorophytes (unicellular) 40
  Prymnesiophytes (=Haptophytes) 310
  Xanthophytes 266
Cell Covering
  Chloromonads 351
  Chlorophytes (colonial) 153
  Chlorophytes (snow algae) 63
  Chlorophytes (unicellular) 9, 15
  Chrysophytes 218
  Cryptophytes (photosynthetic)
    periplast 387
  Dinoflagellates (free-living)
    theca 408
  Euglenoid flagellates
    lorica 185
    pellicle 183
  Eustigmatophytes 321
  Phytoflagellates 455
  Prasinophytes 94
    aflagellate stages 99
    scales in one layer 95
    scales in more than one layer 96
  Prymnesiophytes (=Haptophytes)
    scale formation and composition 296
    scales 300
  Silicoflagellates
    protoplast 337
    skeleton 337
  Xanthophytes 245
Characteristics - General
  Chloromonads 351
  Chlorophytes (colonial) 147
  Chlorophytes (snow algae) 61
  Chlorophytes (unicellular) 5, 7, 15
  Chrysophyceae 213, 214, 225
  Cryptophytes (photosynthetic) 381, 382
  Dinoflagellates (free-living) 407
  Euglenoid flagellates 165, 166
  Eustigmatophytes 319, 320, 324
  Phytoflagellates 434
  Prasinophytes
    general 85
    internal characteristics 101
    living cells 87
  Prymnesiophytes (-Haptophytes) 273, 274, 282, 295
  Silicoflagellates 335
  Xanthophytes
    general 243
    motile cell 250
Chloroplasts
  Chloromonads 354
  Chlorophytes (unicellular) 14
  Chrysophytes 224
  Cryptophytes (photosynthetic) 383
  Dinoflagellates (free-living) 417
  Endosymbiontic origin of 451
  Euglenoid flagellates 168
  Eustigmatophytes 320, 329
  Prasinophytes 112
  Prymnesiophytes (-Haptophytes) 292
  Xanthophytes 244, 253
Colonial Chlorophytes
  Astrephomenaceae 156
  Development (colonial) 154
  Haematococcaceae 158
  Spondylomoraceae 158
  Volvocaceae 147
Culture Media
  Chloromonads 352
  Chlorophytes (snow algae) 80
  Chlorophytes (unicellular) 44
  Chrysophytes 236
  Cryptophytes 399
  Eustigmatophytes 333
  Prymnesiophytes (=Haptophytes) 312
  Xanthophytes 267
Cultures
  Chloromonads 352
  Chlorophytes (colonial) 160
Cytokinesis
  Cryptophytes (photosynthetic) 393
  Prasinophytes 125
Ejectosomes
  Cryptophytes (photosynthetic) 389
Endoplasmic reticulum
  Chloromonads 357
  Chlorophytes (unicellular) 8
  Cryptophytes (photosynthetic) 381
  Euglenoid flagellates 168
  Prasinophytes 118
  Prymnesiophytes (=Haptophytes) 302
  Xanthophytes 255
Evolutionary relationships
  Cryptophytes (photosynthetic) 400
  Dinoflagellates (photosynthetic) 423
  Phytoflagellates 433
Eyespots
  Chlorophytes (unicellular) 11
  Dinoflagellates (free-living) 417
  Euglenoid flagellates 175
  Prymnesiophytes (=Haptophytes) 292
  Xanthophytes 253
Flagella
  Chloromonads 357
  Chlorophytes (unicellular) 18, 24

Chrysophytes 220
Cryptophytes (photosynthetic) 395
Dinoflagellates (free-living) 411
Euglenoid flagellates 173
paraflagellar body 175
Eustigmatophytes 326
Phytoflagellates 446
Prasinophytes 39
basal bodies and associated flagellar roots 101
flagellar appendages 90
microtubular roots 108
transition zone 103
Prymnesiophytes (=Haptophytes) 280
haptonemata 282
reduced haptonemata 286
Xanthophytes 251
Flavins
Euglenoid flagellates 176
Fossil Record
Chlorophytes (snow algae) 81
Chlorophytes (unicellular) 46
Chrysophytes 237
Gametogenesis
Xanthophytes 264
Golgi Bodies
Chrysophytes 225
Eustigmatophytes 331
Prasinophytes 118
Prymnesiophytes (=Haptophytes) 295
Xanthophytes 256
Habitat
Chloromonads 373
Chlorophytes (snow algae) 70
Chlorophytes (unicellular) 41
Chrysophytes 236
Cryptophytes (photosynthetic) 399
Eustigmatophytes 332
Prasinophytes 85
Prymnesiophytes (=Haptophytes) 311
Silicoflagellates 335
Xanthophytes 267
Isolation and Chemical Characterization
Euglenoid flagellates 176
Life Histories
Chlorophytes (snow algae) 67
Chlorophytes (unicellular) 29
Chrysophyceae 227
Dinoflagellates (free-living) 416
Eustigmatophytes 332
Prasinophytes 129
Prymnesiophytes (=Haptophytes) 307
Xanthophytes 260
Lipids
Euglenoid flagellates 177
Meiosis
Chlorophytes (snow algae) 70
Chlorophytes (unicellular) 32, 38
Chrysophytes 234
Microbodies
Chloromonads 357
Mitochondria
Chloromonads 357
Eustigmatophytes 331
Phytoflagellates 450
Prasinophytes 117
Prymnesiophytes (=Haptophytes) 295
Xanthophytes 256
Mitosis
Chloromonads 386
Chlorophytes (snow algae) 70
Chlorophytes (unicellular) 32
Chrysophytes 234
Cryptophytes (photosynthetic) 393
Euglenoid flagellates 173
Phytoflagellates 440
Prasinophytes 125
Prymnesiophytes (=Haptophytes) 310
Xanthophytes 261
Movement Patterns
Chlorophytes (colonial) 154
Chlorophytes (snow algae) 66
Chlorophytes (unicellular) 27
Chrysophytes 226
Eustigmatophytes 332
Prymnesiophytes (=Haptophytes) 290, 304
Xanthophytes 260
Nucleus
Chloromonads 367
Chrysophytes 225
Dinoflagellates (free-living) 414
Euglenoid flagellates 173
Eustigmatophytes 331
Prasinophytes 116
Prymnesiophytes (=Haptophytes) 295
Xanthophytes 256
Nucleomorph
Cryptophytes (photosynthetic) 387
Nutrition
Cryptophytes (photosynthetic) 399
Palmelloid Formation
Chlorophytes (unicellular) 32
Physiology
Chlorophytes (unicellular) 40
Prymnesiophytes (=Haptophytes) 310
Xanthophytes 266
Pigments
Chloromonads 356

Cryptophytes (photosynthetic) 384
Euglenoid flagellates 177
Eustigmatophytes 320
Prasinophytes 133
Xanthophyceae 244
Pusules
Dinoflagellates (free-living) 421
Pyrenoids
Chlorophytes (unicellular) 14
Dinoflagellates (free-living) 417
Eustigmatophytes 329
Prymnesiophytes (=Haptophytes) 292
Xanthophytes 253
Research Needed
Chloromonads 375
Chlorophytes (snow algae) 80
Chlorophytes (unicellular) 45
Chrysophytes 237
Euglenoid flagellates 202
Eustigmatophytes 333
Prymnesiophytes (=Haptophytes) 213
Xanthophytes 268
Resting Spores
Chlorophytes (unicellular) 32
Rhizoplasts
Prasinophytes 104
Scale Reservoir
Prasinophytes 118
Sexual Reproduction
Chlorophytes 155
Chlorophytes (unicellular) 30, 34
Eustigmatophytes 332
Xanthophytes 260
Skeleton Formation
Silicoflagellates 338
Storage Metabolites
Eustagmatophytes 324
Xanthophytes 248
Symbiosis
Cryptophytes (photosynthetic) 398
Dinoflagellates (free-living) 423
Trichocysts
Chloromonads 366
Euglenoid flagellates 198
Prasinophyceae 123
Ultrastructure
Chlorophytes (snow algae) 65
Chlorophytes (unicellular) 9, 39
Chrysophytes 217, 235
Dinoflagellates (free-living) 422
Euglenoid flagellates 175
Eustigmatophytes 326
Prymnesiophytes (=Haptophyte) 280
Silicoflagellates 337
Xanthophytes 251
Vesicles
Chloromonads 363
Eustigmatophytes 331
Prasinophytes 123
Prymnesiophytes 304
Xanthophytes 259
Zoospores
Xanthophytes 262, 264

JOHN RYLANDS
UNIVERSITY
LIBRARY OF
MANCHESTER